Quadrature Theory

The Theory of Numerical Integration on a Compact Interval

Mathematical
Surveys
and
Monographs

Volume 178

Quadrature Theory

The Theory of Numerical Integration on a Compact Interval

Helmut Brass
Knut Petras

American Mathematical Society
Providence, Rhode Island

2010 *Mathematics Subject Classification.* Primary 41A55; Secondary 65D30, 65D32.

For additional information and updates on this book, visit
www.ams.org/bookpages/surv-178

Library of Congress Cataloging-in-Publication Data

Brass, Helmut.
Quadrature theory : the theory of numerical integration on a compact interval / Helmut Brass, Knut Petras.
p. cm. – (Mathematical surveys and monographs ; v. 178)
Includes bibliographical references and index.
ISBN 978-0-8218-5361-0 (alk. paper)
1. Gaussian quadrature formulas. 2. Numerical integration. I. Petras, Knut, 1962– II. Title.

QA299.4.G3B73 2011
515′.43–dc23 2011030201

∞ The paper used in this book is acid-free and falls within the guidelines
established to ensure permanence and durability.
Visit the AMS home page at `http://www.ams.org/`

10 9 8 7 6 5 4 3 2 1 16 15 14 13 12 11

Contents

Preface

Methods for the approximate calculation of definite integrals are covered in every book on numerical analysis. Our intention here is to provide a complementary treatment of this topic by presenting a coherent theory that encompasses many deep and elegant results as well as a large number of interesting (solved and open) problems.

The inclusion of the word "theory" in the title highlights our emphasis on concepts rather than numerical recipes. Thus, no computer programs and only a few numerical examples are given in the book. The focus on theory does not, however, mean that we pass over concrete practical problems, merely that we choose to restrict our attention to problems for which a guaranteed result can be obtained in a systematic manner. Systematic analyses of this kind rely on certain properties of the integrand, over and beyond the knowledge of finitely many function values. Such additional information about the integrand (called "co-observations") forms the central organizing principle for our theory, and distinguishes our book from other texts on quadrature. A wide variety of co-observations are examined in this monograph, as we believe such information will be very useful for solving problems in practical contexts.

While quadrature theory is often viewed as a branch of numerical analysis, its influence extends much further: it has been the starting point of many far-reaching generalizations in various directions, as well as a testing ground for new ideas and concepts; in fact, in many instances the extensions seem more "natural" than the original motivating problem. We shall discuss such generalizations, although the classical problem will remain our guiding star throughout the book.

Working on quadrature has given us great pleasure over the years, and we hope we can convey our enthusiasm for the subject to the readers of this book.

The mathematical prerequisites for engaging with this text are knowledge (at the level taught in most undergraduate courses) of linear algebra, advanced calculus and real analysis.

We thank our wives for their patience and assistance in many ways.

We are grateful to Alice Yew for her help in editing our manuscript.

Helmut Brass and Knut Petras
Technische Universität Braunschweig

CHAPTER 1

Introduction

In this chapter we give a concise introduction to quadrature theory by discussing some examples, concepts and results.

The aim of quadrature theory is to compute an approximate value for the definite integral

$$I[f] := \int_a^b f(x)\,dx.$$

A *quadrature rule* is a map Q, determined by numbers a_ν (the *coefficients*) and $x_\nu \in [a,b]$ (the *nodes*) for $\nu = 1, 2, \dots, n$, which assigns to a function f the value

$$Q[f] := \sum_{\nu=1}^{n} a_\nu f(x_\nu).$$

We shall always require that $a \le x_1 < x_2 < \cdots < x_n \le b$. If the a_ν and x_ν are chosen suitably, then $Q[f]$ is an approximation of $I[f]$; this is an immediate consequence of the definition of the integral as a limit of Riemann sums.

The calculation of $Q[f]$ for any fixed Q is easy, especially if done on a computer. However, such a computation would be mathematically meaningful only if one has a good understanding of—and preferably a bound on—the *error*

$$R[f] := I[f] - Q[f].$$

Any bound on the error must be based on properties of f. These properties, as well as the set of functions having them, will be referred to as the *co-observation* of the problem at hand, denoted by $\mathcal{C}$. Defining

$$\rho(Q, \mathcal{C}) := \sup\left\{\left|R[f]\right| : f \in \mathcal{C}\right\},$$

we have the guaranteed inclusion

$$I[f] \in \left[Q[f] - \rho(Q,\mathcal{C}),\ Q[f] + \rho(Q,\mathcal{C})\right].$$

As a concrete example, let us consider the "midpoint rule" of order n, which is given by

$$a_\nu := h, \qquad x_\nu := a + \left(\nu - \tfrac{1}{2}\right)h, \qquad \text{where } h := \frac{b-a}{n};$$

that is,

$$Q_n^{\mathrm{Mi}}[f] = h \sum_{\nu=1}^{n} f\left(a + (\nu - \tfrac{1}{2})h\right).$$

Because it represents one of the simplest Riemann sums, Q_n^{Mi} is quite a natural choice of quadrature rule. To use Q_n^{Mi} as a tool for computing $I[f]$, accompanied by error bounds, we need to select a co-observation. In quadrature theory, the following co-observations are very popular:

$$\mathcal{C}_M^{(r)} := \{f : \sup_{a\le x\le b} |f^{(r)}(x)| \le M\}.$$

It can be shown quite easily (see Section 4.3) that

$$\rho(Q_n^{\mathrm{Mi}}, \mathcal{C}_M^{(1)}) = \frac{(b-a)^2}{4n} M, \tag{1.1}$$

$$\rho(Q_n^{\mathrm{Mi}}, \mathcal{C}_M^{(2)}) = \frac{(b-a)^3}{24n^2} M, \tag{1.2}$$

$$\rho(Q_n^{\mathrm{Mi}}, \mathcal{C}_M^{(r)}) = \infty \quad \text{for } r = 3, 4, \ldots$$

Let us use these results to compute

$$I[f_0] := \int_0^{2\pi} e^{\sin x}\, dx \tag{1.3}$$

with an error not exceeding $0.5 \cdot 10^{-8}$. The integrand f_0 is an element of $\mathcal{C}_{1.46}^{(1)}$ and of $\mathcal{C}_{2.72}^{(2)}$ (we could take any bounds for $\sup|f_0'(x)|$ and $\sup|f_0''(x)|$ as the "M" in each case; 1.46 and 2.72 are the best values to two decimal places). Thus, upon taking $b - a = 2\pi$ in (1.1) and (1.2), we obtain the bounds

$$|R_n^{\mathrm{Mi}}[f]| \le \frac{14.41}{n} \qquad \text{for } f \in \mathcal{C}_{1.46}^{(1)},$$

$$|R_n^{\mathrm{Mi}}[f]| \le \frac{28.12}{n^2} \qquad \text{for } f \in \mathcal{C}_{2.72}^{(2)}.$$

If $n = 1$, the first bound is better; otherwise, the second one is better. Now, $28.12n^{-2} < 0.5 \cdot 10^{-8}$ if $n > 74993$, which tells us we should compute $Q_{74994}^{\mathrm{Mi}}[f_0]$. It turns out, however, that this would be a huge waste of time: $Q_5^{\mathrm{Mi}}[f_0]$ would do just as well! But how could we know this before embarking on a long computation?

It should not be surprising that a (best) upper bound for all of $\mathcal{C}_{2.72}^{(2)}$ would be very rough for a specific $f_0 \in \mathcal{C}_{2.72}^{(2)}$. The following theorem allows, at least asymptotically, the possibility of identifying such cases of *overestimation*:
For $f \in \mathcal{C}_M^{(2)}$,

$$\lim_{n\to\infty} n^2 R_n^{\mathrm{Mi}}[f] = \frac{(b-a)^2}{24}[f'(b) - f'(a)]. \tag{1.4}$$

This result shows that the error bound (1.2) is realistic when $f'(b) \neq f'(a)$ holds, because in that case we have

$$\lim_{n\to\infty} \frac{|R_n^{\mathrm{Mi}}[f]|}{\rho(Q_n^{\mathrm{Mi}}, \mathcal{C}_M^{(2)})} > 0.$$

On the other hand, if $f'(b) = f'(a)$, as in example (1.3), then $\rho(Q_n^{\mathrm{Mi}}, \mathcal{C}_M^{(2)})$ is asymptotically much larger than $|R_n^{\mathrm{Mi}}[f]|$. This suggests a way to improve the situation for (1.3): include the condition $f'(b) = f'(a)$ in the definition of $\mathcal{C}$. To see an example of what can be gained by modifying the definition of $\mathcal{C}$ in this manner, let $\mathcal{C} := \mathcal{C}_M^{(4)} \cap \{f : f'(b) = f'(a)\}$; then

$$\rho(Q_n^{\mathrm{Mi}}, \mathcal{C}) = \frac{7(b-a)^5}{5760n^4} M.$$

By using this new co-observation, more realistic bounds on the error may be obtained, and we can continue with this sort of improvement. We note, however, that there is another reason to consider different kinds of co-observation. In the examples

$$\int_0^1 \sqrt{1-x^4}\,dx \tag{1.5}$$

and

$$\int_0^{e^{-2}} \frac{dx}{\ln x}, \tag{1.6}$$

the integrands do not belong to any of the classes $\mathcal{C}_M^{(r)}$, so we have to look for other types of co-observation. The set defined by

$$\mathcal{C}_M^{\mathrm{Var}} := \{f \in C[a,b] : \operatorname{Var} f \le M\},$$

where $\operatorname{Var} f$ denotes the total variation of f on $[a,b]$, will prove to be useful. In Section 4.4 we shall show that

$$\rho(Q_n^{\mathrm{Mi}}, \mathcal{C}_M^{\mathrm{Var}}) = \frac{(b-a)}{2n} M. \tag{1.7}$$

This result is of rather universal applicability, because in practical problems one hardly expects to encounter bounded functions with $\operatorname{Var} f = \infty$, and in most cases (for instance when we know that f consists of a finite number of monotonic pieces and we have a bound for $\sup\{|f(x)| : a \le x \le b\}$) some bound on $\operatorname{Var} f$ can be obtained easily. To understand the problem of overestimation in our examples (1.5)–(1.6), we shall again use an "inverse" theorem. An inverse theorem, such as (1.4), is one that deduces some property of the function from the sequence of errors. Here we use the result "If f is concave and $R_n^{\mathrm{Mi}}[f] = O(n^{-1-\beta})$ with $\beta \in \,]0,1]$, then $f \in \operatorname{Lip}\beta$". From this we conclude that $R_n^{\mathrm{Mi}}[f] = O(n^{-1-\beta})$ cannot hold for $\beta > \frac{1}{2}$ in (1.5) and cannot hold for $\beta > 0$ in (1.6); therefore the bound (1.7) is satisfactory.

Our choice of using $Q_n^{\mathrm{Mi}}[f]$ to approximate $I[f]$ was somewhat arbitrary; perhaps we can do better by applying another rule. For a comparison to make sense, the rules being compared have to use the same information, which means the same set of function values (the "observation") and the same co-observation. Denote by $\mathbf{Q}(\{x_1,\dots,x_n\})$ the set of rules having (exactly) the nodes $x_1,\dots,x_n$. We define

$$\rho^{\mathrm{opt}}(\mathcal{C},\{x_1,\dots,x_n\}) := \inf\left\{\rho(Q,\mathcal{C}) : Q \in \mathbf{Q}(\{x_1,\dots,x_n\})\right\}$$

and say that $Q^{\mathrm{opt}} \in \mathbf{Q}(\{x_1\dots,x_n\})$ is *optimal* in $\mathcal{C}$ if

$$\rho(Q^{\mathrm{opt}},\mathcal{C}) = \rho^{\mathrm{opt}}(\mathcal{C},\{x_1,\dots,x_n\}).$$

It can be seen from the following result that the investigation of further rules is appropriate: Let $x_{\nu,n} = a + (2\nu-1)(b-a)/(2n)$; then

$$\rho^{\mathrm{opt}}\left(\mathcal{C}_M^{(r)},\{x_{1,n},\dots,x_{n,n}\}\right) = O(n^{-r}) \quad \text{as } n \to \infty.$$

If $f \in \mathcal{C}_M^{(r)}$ for $r \ge 3$ and $f'(b) \ne f'(a)$, then (1.4) shows that it is possible to save a great deal of work, at least asymptotically, provided we use a sequence of optimal rules instead of Q_n^{Mi}.

Can the precision be improved further? If our aim is to compute $I[f]$ based on knowledge of $f(x_i)$ for $i = 1, \ldots, n$ and $f \in \mathcal{C}$, why restrict our algorithms to quadrature rules? Maybe there is a function Q of n variables such that

$$\sup\left\{\left|I[f] - Q[f(x_1), \ldots, f(x_n)]\right| : f \in \mathcal{C}\right\} < \rho^{\text{opt}}\left(\mathcal{C}, \{x_1, \ldots, x_n\}\right)\ ?$$

It is an important result that no such function exists if $\mathcal{C}$ is a convex and symmetric set in the space of continuous functions ($\mathcal{C}$ is symmetric if and only if $f \in \mathcal{C}$ implies $-f \in \mathcal{C}$). Almost all co-observations of interest (in particular $\mathcal{C}_M^{(r)}$) are of this kind, so there is no need to generalize the notion of quadrature rule.

Up to now we have considered the nodes to be fixed. This is reasonable if the values $f(x_1), \ldots, f(x_n)$ were known from the beginning, for example as results of a previous computation or as data that were experimentally determined. In this case we should select some Q from $\mathbf{Q}(\{x_1, \ldots, x_n\})$, preferably an optimal one. However, if we also have freedom in the choice of nodes, there is a possibility of achieving further improvement in our results.

Let us denote by $\mathbf{Q}_n$ the set of all quadrature rules with n nodes. We say that $\widetilde{Q} \in \mathbf{Q}_n$ is a *best rule* in $\mathcal{C}$ if

$$\rho(\widetilde{Q}, \mathcal{C}) = \rho_n^{\text{best}}(\mathcal{C}) := \inf\{\rho(Q, \mathcal{C}) : Q \in \mathbf{Q}_n\}.$$

This definition is quite natural if we consider all $Q \in \mathbf{Q}_n$ as being equivalent in terms of computational cost.

We illustrate the definition with two examples. In $\mathcal{C}_M^{(1)}$, the midpoint rule is the best. In $\mathcal{C}_M^{(2)}$, the uniquely determined best rule is given by

$$Q[f] = h\Big[(\tau + \tfrac{1}{2})f(a + \tau h) + \sum_{\nu=1}^{n-2} f\big(a + (\nu + \tau)h\big) + (\tau + \tfrac{1}{2})f(b - \tau h)\Big]$$

where

$$h := \frac{b-a}{n-1+2\tau} \qquad \text{and} \qquad \tau := \frac{\sqrt{3}}{4},$$

and we have

$$\rho_n^{\text{best}}(\mathcal{C}_M^{(2)}) = \frac{(b-a)^3}{32(n-1+2\tau)^2}M.$$

A comparison with (1.2) reveals no great advantage of the optimal rule over the midpoint rule in $\mathcal{C}_M^{(2)}$. For $r \geq 3$, the best rules in $\mathcal{C}_M^{(r)}$ (whose existence is not self-evident but can be proved) are not known explicitly; however, they can be determined by a complicated iterative process for each particular n.

A "best" rule is the best possible with respect to some important measure of quality, but there are other criteria that should be taken into account as well; so the adjective "best" does not necessarily indicate an unconditional recommendation. The main role of a best rule is to serve as a benchmark against which to compare rules $Q \in \mathbf{Q}_n$ selected by other principles: the quotient

$$\frac{\rho(Q, \mathcal{C})}{\rho_n^{\text{best}}(\mathcal{C})} \tag{1.8}$$

should be small, although actually achieving its minimality is not of great practical importance.

Among other criteria to be taken into consideration when choosing a quadrature rule Q, the *availability* of Q ranks first. If a time-consuming algorithm to determine

Q has to precede the computation of $Q[f]$, then the advantages of quadrature rules (e.g. simplicity and universality) over other methods for evaluating $I[f]$ are lost. Hence the set of rules applied in computing practice is limited to a small number of well-documented ones. The selection of these rules, using as a guideline the smallness of (1.8) for some (possibly large) set of co-observations, is a major task in quadrature theory.

The central notions of quadrature theory can be generalized widely. Within the scope of the classical theory, however, there is one main generalization of fundamental importance, and that is the introduction of weight functions. Let w be a fixed integrable function and define

$$I_w[f] := \int_a^b f(x)w(x)\,dx.$$

Evidently, $I_w[f] = I[fw]$, but note that this reduction is not appropriate if f is a smooth function while w is not. Take the example of Fourier coefficients:

$$I_{cp}[f] := \frac{1}{\pi}\int_0^{2\pi} f(x)\cos(px)\,dx, \quad p \in \mathbb{N}.$$

If p is large, the integrand oscillates at high frequency and approximations obtained from the midpoint rule will be poor, as judged by (1.1), (1.2) or (1.7). To take advantage of the smoothness properties of f, it would be preferable to construct quadrature rules that are specially adapted to I_{cp}. To prepare the ground for dealing with many similar situations of practical interest, we shall develop the theory starting with a general I_w rather than just I.

CHAPTER 2

The Abstract Framework

Our study of quadrature theory begins with a very general framework, formulated mainly in terms of linear algebra, which will help us to discern the fundamental notions and principles. In this chapter we derive results that shed light on the place occupied by quadrature rules within the collection of all algorithms for computing a definite integral based on a common set of information. These results will serve to justify our focusing attention on quadrature rules in the rest of the book.

2.1. Standard estimation framework

DEFINITION 2.1.1. A *standard estimation framework* (SEF) comprises:

(1) a real linear space V;
(2) a linear functional I defined on V;
(3) a linear map $O : V \to \mathbb{R}^n$ (observation); and
(4) a non-empty convex symmetric[1] set $\mathcal{C} \subset V$ (co-observation).

In what follows, we shall always assume an SEF. In this context, we wish to approximate $I[f]$ based solely on knowledge of $O[f]$ and the assumption that $f \in \mathcal{C}$. As outlined in Chapter 1, in quadrature theory we use a special SEF with:

(1) $V = C[a, b]$;
(2) $I[f] = \int_a^b f(x)w(x)\,dx$, where w is a fixed integrable function;
(3) $O[f] = \big(f(x_1), \ldots, f(x_n)\big)$; and
(4) various choices of $\mathcal{C}$.

The main body of quadrature theory is based on such SEFs. Sometimes we generalize the observation by including values of derivatives as additional information, which is a reasonable thing to do if these values are obtainable at low computational cost, for example as known Taylor coefficients. Another type of observation that is occasionally of practical interest consists of mean values of f over small intervals; see Omladič *et al.* (1976), Pittnauer and Reimer (1979), Omladič (1992), Motornyi (1998) and Bojanov and Petrov (2001, 2005).

By varying V, I, O and $\mathcal{C}$, a broad spectrum of problems in numerical mathematics can be subsumed under the notion of an SEF, for example interpolation ($I[f] := f(u)$ with u fixed), numerical differentiation ($I[f] = f'(u)$ with u fixed), and the computation of Cauchy principal value integrals, line integrals, volume integrals etc.

[1] A symmetric set $\mathcal{C}$ is such that $-\mathcal{C} = \mathcal{C}$, i.e. $-f \in \mathcal{C} \iff f \in \mathcal{C}$.

Considering problems within an SEF enables us to compare algorithms and make precise the meaning of a "best" algorithm. The central idea is that there exists some information or the possibility of obtaining some kind of information, and we should make the best of it. The theory of information-based complexity (see, e.g., Woźniakowski 1985, Traub *et al.* 1988 or Novak 1988) discusses problems of this sort from a more general standpoint.

Next, let us define our available information.

DEFINITION 2.1.2. For $f \in \mathcal{C}$,

$$\text{Info}(f) := \{g : O[g] = O[f]\} \cap \mathcal{C}.$$

The aim now is to calculate the set

$$I[\text{Info}(f)]$$

of all possible values of I that are compatible with the given information Info(f). Since Info(f) is the intersection of two convex sets and hence also convex, and since I is linear, the set $I[\text{Info}(f)]$ is an interval. We characterize this interval by its endpoints.

DEFINITION 2.1.3. For $z \in O(\mathcal{C})$,

$$\begin{aligned} \hat{Q}[z] &:= \sup\{I[f] \,:\, O[f] = z,\, f \in \mathcal{C}\}, \\ \underset{\vee}{Q}[z] &:= \inf\{I[f] \,:\, O[f] = z,\, f \in \mathcal{C}\}. \end{aligned}$$

Often we shall write simply $\underset{\vee}{Q}[f]$ in place of $\underset{\vee}{Q}(O[f])$, but it is important to remember that $\underset{\vee}{Q}$ is defined on $O(\mathcal{C}) \subset \mathbb{R}^n$, and similarly for $\hat{Q}$.

We are interested in situations where the error of at least one algorithm can be bounded by a finite number. Therefore, the interval $I[\text{Info}(f)]$ of all possible values has to be bounded.

General Assumption. *Unless explicitly stated otherwise, we shall always assume that we are in an SEF where for each* $f \in \mathcal{C}$, $\hat{Q}[f] < \infty$ *and* $\underset{\vee}{Q}[f] > -\infty$.

The best we can do in this situation is to choose the midpoint of $I[\text{Info}(f)]$ as an estimate for $I[f]$. This estimate can deviate from the true value by at most half of the width of $I[\text{Info}(f)]$, and no smaller deviation can be guaranteed for any estimate.

DEFINITION 2.1.4. The *strongly optimal estimate* is

$$Q^{\text{so}}[f] := \frac{1}{2}\left(\hat{Q}[f] + \underset{\vee}{Q}[f]\right).$$

The *intrinsic error* is

$$\rho^{\text{intr}}[f] := \frac{1}{2}\left(\hat{Q}[f] - \underset{\vee}{Q}[f]\right),$$

and the *error of the SEF* is defined by

$$\rho^{\text{opt}} := \sup\{\rho^{\text{intr}}[f] \,:\, f \in \mathcal{C}\}.$$

We can view ρ^{opt} as the *a priori* error of the estimation problem, i.e. the smallest uncertainty that can be guaranteed by an appropriate algorithm before the evaluation of $O[f]$. This quantity can be used for, among other things, evaluating the observation functional O. On the other hand, $\rho^{\text{intr}}[f]$ is the *a posteriori* error bound.

Before we try to understand the notions introduced so far via an example, let us point out the equality

$$\check{Q}[f] = -\hat{Q}[-f], \tag{2.1}$$

which follows readily from the linearity of I and the symmetry of $\mathcal{C}$.

EXAMPLE 2.1.1 (Secrest 1964). We study the SEF given by

(1) $V = C[a,b]$;
(2) $I[f] = \int_a^b f(x)\,dx$;
(3) $O[f] = \big(f(x_1), \ldots, f(x_n)\big)$ where $a \le x_1 < x_2 < \cdots < x_n \le b$;
(4) $\mathcal{C} = \text{Lip}_M 1 := \big\{f : |f(x) - f(y)| \le M|x-y| \text{ for } x, y \in [a,b]\big\}$.

For $x \in [x_\nu, x_{\nu+1}]$, $f \in \mathcal{C}$ yields

$$f(x) - f(x_\nu) \le M(x - x_\nu), \tag{2.2}$$
$$f(x) - f(x_{\nu+1}) \le M(x_{\nu+1} - x), \tag{2.3}$$

which means that

$$f(x) \le \min\big\{f(x_\nu) + M(x - x_\nu),\ f(x_{\nu+1}) + M(x_{\nu+1} - x)\big\}.$$

Let x^* be such that $f(x_\nu) + M(x^* - x_\nu) = f(x_{\nu+1}) + M(x_{\nu+1} - x^*)$. On the interval $[x_\nu, x^*]$ the first bound (2.2) is smaller, while on $[x^*, x_{\nu+1}]$ the second bound (2.3) is smaller. We choose the better bound in each case and obtain

$$\begin{aligned}\int_{x_\nu}^{x_{\nu+1}} f(x)\,dx \le\ & \frac{x_{\nu+1} - x_\nu}{2}[f(x_{\nu+1}) + f(x_\nu)] \\ & + \frac{M}{4}(x_{\nu+1} - x_\nu)^2 - \frac{[f(x_{\nu+1}) - f(x_\nu)]^2}{4M}.\end{aligned} \tag{2.4}$$

Using analogous estimates on the intervals $[a, x_1]$ and $[x_n, b]$, we get

$$\begin{aligned}\hat{Q}[f] \le\ & f(x_1)(x_1 - a) + \frac{M}{2}(x_1 - a)^2 + \sum_{\nu=1}^{n-1} \frac{(x_{\nu+1} - x_\nu)}{2}[f(x_{\nu+1}) + f(x_\nu)] \\ & + \frac{M}{4}\sum_{\nu=1}^{n-1}(x_{\nu+1} - x_\nu)^2 - \frac{1}{4M}\sum_{\nu=1}^{n-1}[f(x_{\nu+1}) - f(x_\nu)]^2 \\ & + f(x_n)(b - x_n) + \frac{M}{2}(b - x_n)^2.\end{aligned}$$

In fact, equality holds. This can be seen by constructing a broken line function $f_o \in \text{Info}(f)$ such that equality holds in (2.2) and (2.3) on the respective intervals.

A similar expression for $\underset{\vee}{Q}[f]$ follows from (2.1). Finally, we obtain

$$
\begin{aligned}
&Q^{\mathrm{so}}[f]\\
&= f(x_1)(x_1-a)+\sum_{\nu=1}^{n-1}\frac{x_{\nu+1}-x_\nu}{2}[f(x_{\nu+1}+f(x_\nu)]+f(x_n)(b-x_n)\\
&= \left(\frac{x_1+x_2}{2}-a\right)f(x_1)+\sum_{\nu=2}^{n-1}\frac{x_{\nu+1}-x_{\nu-1}}{2}f(x_\nu)+\left(b-\frac{x_{n-1}+x_n}{2}\right)f(x_n).
\end{aligned}
$$

We call this algorithm the "generalized trapezoidal rule"; it will come up frequently. For the special case where $x_\nu = a+(\nu-\frac{1}{2})(b-a)/n$, we obtain the "midpoint rule"

$$
Q_n^{\mathrm{Mi}}[f] := \frac{b-a}{n}\sum_{\nu=1}^{n} f(x_\nu), \tag{2.5}
$$

whereas $x_\nu = a+(\nu-1)(b-a)/(n-1)$ gives the usual "trapezoidal rule"

$$
Q_n^{\mathrm{Tr}}[f] := \frac{b-a}{n-1}\left[\frac{1}{2}f(a)+\sum_{\nu=2}^{n-1}f(x_\nu)+\frac{1}{2}f(b)\right].
$$

In the general case, note that

$$
\begin{aligned}
\rho^{\mathrm{intr}}[f] = \frac{M}{2}(x_1-a)^2 &+ \frac{M}{4}\sum_{\nu=1}^{n-1}(x_{\nu+1}-x_\nu)^2\\
&- \frac{1}{4M}\sum_{\nu=1}^{n-1}[f(x_{\nu+1})-f(x_\nu)]^2+\frac{M}{2}(b-x_n)^2
\end{aligned}
$$

and therefore

$$
\rho^{\mathrm{opt}} = \frac{M}{4}\Big[2(x_1-a)^2+\sum_{\nu=1}^{n-1}(x_{\nu+1}-x_\nu)^2+2(b-x_n)^2\Big].
$$

Here is a useful property of the quantities introduced above.

LEMMA 2.1.1. $\underset{\vee}{Q}$ is a convex function while $\hat{Q}$ and ρ^{intr} are concave functions on $\mathcal{C}$.

PROOF. For given $f_1, f_2 \in \mathcal{C}$ and $\varepsilon > 0$, there exist $g_i \in \mathrm{Info}(f_i)$ satisfying

$$
\underset{\vee}{Q}[f_i]+\varepsilon \geq I[g_i] \quad \text{for } i = 1, 2,
$$

and for arbitrary $\alpha \in [0, 1]$ we have

$$
\begin{aligned}
\underset{\vee}{Q}[\alpha f_1+(1-\alpha)f_2] &\leq I[\alpha g_1+(1-\alpha)g_2] = \alpha I[g_1]+(1-\alpha)I[g_2]\\
&\leq \alpha\,\underset{\vee}{Q}[f_1]+(1-\alpha)\,\underset{\vee}{Q}[f_2]+\varepsilon. \qquad (2.6)
\end{aligned}
$$

Convexity of $\underset{\vee}{Q}$ follows, since the inequalities in (2.6) hold for each $\varepsilon > 0$. The concavity of $\hat{Q}$ follows in a similar fashion, and ρ^{intr} is concave because it is the difference of a concave function and a convex one. □

An important consequence is the following theorem, which says that the zero vector is the worst possible observation.

THEOREM 2.1.1. Denote by 0 the zero element of V. Then

$$\rho^{\text{opt}} = \rho^{\text{intr}}[0] = \sup\{I[f] \,:\, f \in \mathcal{C} \cap \operatorname{Ker} O\}. \tag{2.7}$$

PROOF. Let us first note that there exists an $f \in \mathcal{C}$ and that, by convexity and symmetry, we also have $0 = [f + (-f)]/2 \in \mathcal{C}$. From equation (2.1), we obtain that for each $f \in \mathcal{C}$,

$$\rho^{\text{intr}}[f] = \rho^{\text{intr}}[-f].$$

Thus, the concavity of ρ^{intr} gives

$$\rho^{\text{intr}}[0] = \rho^{\text{intr}}[(f + (-f))/2] \geq \frac{1}{2}\Big(\rho^{\text{intr}}[f] + \rho^{\text{intr}}[-f]\Big) = \rho^{\text{intr}}[f],$$

which proves the first equality in (2.7). The second equality follows immediately from (2.1). □

The next result shows that, in order to determine $I[\text{Info}(f)]$ for a given $f \in \mathcal{C}$, it is sufficient to investigate linear algorithms. By a linear algorithm we mean an algorithm that maps $O[f]$ linearly into $\mathbb{R}$. Since $O[f] \in \mathbb{R}^n$, such an algorithm can be represented by an element of $A \in \mathbb{R}^n$. For convenience, we use the notation

$$AO[f] := A \cdot (O[f])^{\mathrm{T}}.$$

In the following, we shall use some results from the geometry of convex sets; see, for instance, Webster (1994). We denote by $\operatorname{ri}(K)$ the relative interior of the convex set K, i.e. the set of interior points if we view K as a subset of its affine hull.

THEOREM 2.1.2 (Heindl 1982). Let $f \in \mathcal{C}$ and $O[f] \in \operatorname{ri}(O[\mathcal{C}])$. Then

$$\hat{Q}[f] = \inf_{A\in\mathbb{R}^n} \Big\{ AO[f] + \sup_{g\in\mathcal{C}}\big(I[g] - AO[g]\big) \Big\}, \tag{2.8}$$

$$\underset{\vee}{Q}[f] = \sup_{A\in\mathbb{R}^n} \Big\{ AO[f] + \inf_{g\in\mathcal{C}}\big(I[g] - AO[g]\big) \Big\}. \tag{2.9}$$

Moreover, the infimum and supremum over $A \in \mathbb{R}^n$ are attained.

PROOF. By a standard result on convex functions, there exists a hyperplane supporting $\underset{\vee}{Q}$ at $O[f]$, i.e. a linear mapping $A : \mathbb{R}^n \to \mathbb{R}$ that satisfies

$$\underset{\vee}{Q}[h] \geq AO[h - f] + \underset{\vee}{Q}[f] \tag{2.10}$$

for all $h \in \mathcal{C}$. Hence

$$\begin{aligned} \underset{\vee}{Q}[f] &\leq AO[f] + \underset{\vee}{Q}[h] - AO[h] \\ &= AO[f] + \inf\{I[g] - AO[g] \,:\, g \in \text{Info}(h)\}, \end{aligned}$$

and this holds for all $h \in \mathcal{C}$ such that

$$\underset{\vee}{Q}[f] \leq \sup_{A\in\mathbb{R}^n} \Big\{ AO[f] + \inf_{g\in\mathcal{C}}\big(I[g] - AO[g]\big) \Big\}.$$

On the other hand, for each $A \in \mathbb{R}^n$ we have

$$\begin{aligned}\underset{\vee}{Q}[f] &= AO[f] + \inf\{I[g] - AO[g] \,:\, g \in \operatorname{Info}(f)\} \\ &\geq AO[f] + \inf\{I[g] - AO[g] \,:\, g \in \mathcal{C}\}\end{aligned}$$

and, in particular,

$$\underset{\vee}{Q}[f] \geq \sup_{A \in \mathbb{R}^n} \{AO[f] + \inf_{g \in \mathcal{C}} (I[g] - AO[g])\}.$$

Equation (2.9) is therefore proved, with equality holding for the $A \in \mathbb{R}^n$ given in (2.10). The proof for $\hat{Q}[f]$ is similar. □

The applicability of this theorem is demonstrated by the following example.

EXAMPLE 2.1.2 (von Mises 1933). Let

(1) $V = C[-1, 1]$;
(2) $I[f] = \int_{-1}^{1} f(x)\, dx$;
(3) $O[f] = \big(f(-1), f(0), f(1)\big)$; and
(4) $\mathcal{C} = \{f : \sup_{-1 \leq x \leq 1} |f''(x)| \leq M\}$.

The SEF so defined is rather special, but the method that follows can easily be generalized (though the calculations may become much more involved).

First let us note that the General Assumption is satisfied; that is, elements of $\operatorname{Info}(f)$ cannot have arbitrarily large integrals. Take $\xi \in [-1, 1]$; then for $f \in \mathcal{C}$ we have

$$\left| \frac{f(\xi) - f(-1)}{\xi + 1} - \frac{f(1) - f(\xi)}{1 - \xi} \right| \leq 2M.$$

Multiplying through by $|(1 - \xi)(1 + \xi)| \leq 1$ and using the triangle inequality yields

$$|f(\xi)| \leq M + \tfrac{1}{2}|f(-1)(1 - \xi) + f(1)(1 + \xi)| \leq M + \max\{|f(-1)|, |f(1)|\}.$$

Now let $O[f] \in \operatorname{ri}(O[\mathcal{C}])$. We have to determine

$$\hat{Q}[f] = \inf_{A \in \mathbb{R}^3} \Big\{ AO[f] + \sup_{g \in \mathcal{C}} \big(I[g] - AO[g]\big) \Big\},$$

and then we can apply equation (2.1).

For $p \in \mathbb{P}_1$ and arbitrary $\alpha \in \mathbb{R}$, we have that $\alpha p \in \mathbb{P}_1$ also. Hence, if $I[p] - AO[p] \neq 0$ for a polynomial $p \in \mathbb{P}_1$, we obtain

$$\sup_{f \in \mathcal{C}} \big(I[f] - AO[f]\big) = \infty.$$

Therefore such A are not of interest. An easy calculation shows that any other $A \in \mathbb{R}^3$ can be written as

$$AO[f] = 2f(0) + \eta D[f] \quad \text{where } D[f] := f(-1) - 2f(0) + f(1).$$

The functional AO is a method for estimating the functional I, so $I - AO$ represents its associated error. A systematic error theory, the so-called Peano kernel theory (see Section 4.2), shows that

$$\sup_{f \in \mathcal{C}} \big(I[f] - AO[f]\big) = \begin{cases} (1 - 3\eta)M/3 & \text{if } \eta \leq 0, \\ (8\eta^3 - 3\eta + 1)M/3 & \text{if } 0 \leq \eta \leq 1/2, \\ (3\eta - 1)M/3 & \text{if } \eta \geq 1/2. \end{cases}$$

Equation (2.8) then gives

$$\hat{Q}[f] = 2f(0) + \frac{M}{3} - \frac{2M}{3}\left(\frac{M - D[f]}{2M}\right)^{3/2}$$

and, finally,

$$Q^{\text{so}}[f] = 2f(0) + \frac{M}{3}\left[\left(\frac{M + D[f]}{2M}\right)^{3/2} - \left(\frac{M - D[f]}{2M}\right)^{3/2}\right], \tag{2.11}$$

$$\rho^{\text{intr}}[f] = \frac{M}{3} - \frac{M}{3}\left[\left(\frac{M + D[f]}{2M}\right)^{3/2} + \left(\frac{M - D[f]}{2M}\right)^{3/2}\right], \tag{2.12}$$

$$\rho^{\text{opt}} = \frac{2 - \sqrt{2}}{6} M. \tag{2.13}$$

The condition $O[f] \in \text{ri}(O[\mathcal{C}])$ is equivalent to $|D[f]| < M$. It is obvious that the result can be extended to all $f \in \mathcal{C}$ satisfying $|D[f]| = M$. Therefore, we have full generality.

Up to now, we have endeavoured to make full use of our information Info(f). However, there are other criteria that can be used to evaluate algorithms. It will be necessary to consider algorithms other than Q^{so}, since our simple example already shows that calculation of Q^{so} may be difficult. In fact, there are very few situations in which Q^{so} is known explicitly.

Definition 2.1.5. Let A be a real function defined on $O(\mathcal{C})$. Then

$$Q := A \circ O$$

is called an *estimation rule.* The set of all estimation rules will be denoted by $\mathbf{Q}^{\text{gen}}$. The (worst-case) *error* of an estimation rule $Q \in \mathbf{Q}^{\text{gen}}$ is defined to be

$$\rho(Q) := \sup_{f \in \mathcal{C}} \big| I[f] - Q[f] \big|.$$

Note that errors of linear estimation rules have already, in some sense, played a role in Theorem 2.1.2. Another example is

$$\rho(Q^{\text{so}}) = \rho^{\text{opt}}. \tag{2.14}$$

Estimation rules with the above property earn their own name:

Definition 2.1.6. An estimation rule $Q^{\text{opt}} \in \mathbf{Q}^{\text{gen}}$ is said to be *optimal* if

$$\rho(Q^{\text{opt}}) = \inf\{\rho(Q) : Q \in \mathbf{Q}^{\text{gen}}\}.$$

Theorem 2.1.3.

$$\rho^{\text{opt}} = \inf\{\rho(Q) : Q \in \mathbf{Q}^{\text{gen}}\}. \tag{2.15}$$

Proof. Let $f \in \mathcal{C}$. From the definition of Q^{so} we obtain that for arbitrary $Q \in \mathbf{Q}^{\text{gen}}$,

$$\big[Q^{\text{so}}[f] - \rho^{\text{intr}}[f],\, Q^{\text{so}}[f] + \rho^{\text{intr}}[f]\big] \subseteq \big[Q[f] - \rho(Q),\, Q[f] + \rho[Q]\big].$$

Thus $\rho^{\text{intr}}[f] \le \rho(Q)$ and hence $\rho^{\text{opt}} \le \rho(Q)$. Since Q was arbitrary, the left-hand side of (2.15) does not exceed the right-hand side. Equality follows from (2.14). □

The estimation rule Q^{so}, while always optimal, is often complicated. A key result is that among all the optimal rules there is always a simple one.

DEFINITION 2.1.7. The estimation rule $Q = A \circ O$ is said to be *linear* if A is a linear map defined on $\mathbb{R}^n$. We denote the set of all linear estimation rules by $\mathbf{Q}$.

THEOREM 2.1.4 (Smolyak 1965). For each SEF, there exists a linear estimation rule that is optimal.

PROOF. Note that $(0,\dots,0) \in \mathrm{ri}(O[\mathcal{C}])$. Taking $f \equiv 0$ in Theorem 2.1.2, we obtain the existence of an $A \in \mathbb{R}^n$ satisfying

$$\hat{Q}[0] = \sup_{g \in \mathcal{C}} \big(I[g] - AO[g]\big).$$

Theorem 2.1.1 then yields

$$\rho^{\mathrm{opt}} = \rho^{\mathrm{intr}}(0) = \tfrac{1}{2}\big[\hat{Q}[0] - \underset{\vee}{Q}(0)\big] = \hat{Q}(0).$$

The above two equations combined with the symmetry of $\mathcal{C}$ give the theorem. $\square$

We now know that for any estimation rule, we cannot guarantee a smaller *a priori* error than for the best linear rule—linear rules seem to be the simplest reasonable estimation rules. Therefore, in what follows, it makes sense to consider (almost) exclusively linear estimation rules.

The problems of uniqueness and construction of optimal linear rules are often difficult. Sometimes the following result may be helpful.

THEOREM 2.1.5. If

$$\lim_{\varepsilon \to 0} \frac{\rho^{\mathrm{intr}}[\varepsilon f] - \rho^{\mathrm{intr}}[0]}{\varepsilon} = 0$$

for all $f \in \mathcal{C}$, then there exists exactly one linear optimal rule Q^{opt}, and for all $f \in \mathcal{C}$ we have

$$\lim_{\varepsilon \to 0} \frac{Q^{\mathrm{so}}[\varepsilon f]}{\varepsilon} = Q^{\mathrm{opt}}[f].$$

PROOF. Let Q be one of the linear optimal rules in Theorem 2.1.4, and take $\varepsilon \in [-1,1]$. Then

$$\hat{Q}[\varepsilon f] - Q[\varepsilon f] \le \rho^{\mathrm{opt}} = \hat{Q}[0];$$

that is,

$$\hat{Q}[\varepsilon f] - \hat{Q}[0] \le \varepsilon Q[f]$$

and therefore

$$\lim_{\varepsilon \downarrow 0} \frac{\hat{Q}[\varepsilon f] - \hat{Q}[0]}{\varepsilon} \le Q[f], \qquad \lim_{\varepsilon \uparrow 0} \frac{\hat{Q}[\varepsilon f] - \hat{Q}[0]}{\varepsilon} \ge Q[f].$$

These limits exist because concave functions have one-sided derivatives; if they coincide for each $f \in \mathcal{C}$, then they must both be equal to $Q[f]$ and the uniqueness of the optimal rule is proved.

Note that

$$\frac{\hat{Q}[\varepsilon f] - \hat{Q}[0]}{\varepsilon} = \frac{Q^{\mathrm{so}}[\varepsilon f]}{\varepsilon} + \frac{\rho^{\mathrm{intr}}[\varepsilon f] - \rho^{\mathrm{intr}}[0]}{\varepsilon}.$$

Thus, the theorem follows if we can prove the existence of

$$\lim_{\varepsilon\to 0} \frac{Q^{\mathrm{so}}[\varepsilon f]}{\varepsilon}. \tag{2.16}$$

The existence of this limit is a consequence of

$$\begin{aligned} \frac{Q^{\mathrm{so}}[\varepsilon f]}{\varepsilon} &= \frac{1}{2}\left(\frac{\hat{Q}[\varepsilon f]-\hat{Q}[0]}{\varepsilon}+\frac{\underset{\vee}{Q}[\varepsilon f]+\hat{Q}[0]}{\varepsilon}\right) \\ &= \frac{1}{2}\left(\frac{\hat{Q}[\varepsilon f]-\hat{Q}[0]}{\varepsilon}+\frac{\hat{Q}[-\varepsilon f]-\hat{Q}[0]}{-\varepsilon}\right). \end{aligned}$$

As above, we see that the one-sided limit as $\varepsilon \downarrow 0$ of the right-hand side exists. The other one-sided limit, as $\varepsilon \uparrow 0$, is obtained by exchanging the two summands in brackets and using equation (2.1). This proves the existence of (2.16). □

EXAMPLE 2.1.2 (continued). By using the explicit formulas for Q^{so} and ρ^{intr} given in (2.11) and (2.12), the limits in Theorem 2.1.5 can easily be calculated, giving us the unique linear optimal rule

$$Q^{\mathrm{opt}}[f] = \frac{\sqrt{2}}{4}f(-1) + \left(2-\frac{\sqrt{2}}{2}\right)f(0) + \frac{\sqrt{2}}{4}f(1), \tag{2.17}$$

which is simpler than Q^{so}. However, the optimal formula can be much worse than strongly optimal in the sense that

$$\sup_{f\in\mathcal{C}} \frac{I[f]-Q^{\mathrm{opt}}[f]}{\rho^{\mathrm{intr}}[f]} = \infty. \tag{2.18}$$

This can be seen, for instance, by considering $f(x) = Mx^2/2$.

Let us mention a further aspect relating to linear rules. For fixed $f \in \mathcal{C}$ and any $Q \in \mathbf{Q}^{\mathrm{gen}}$, the "uncertainty interval"

$$\big[Q[f]-\rho(Q),\, Q[f]+\rho[Q]\big]$$

contains the exact value $I[f]$. We can reduce the uncertainty by intersecting intervals associated with different estimation rules. Theorem 2.1.2 implies that the smallest possible uncertainty interval, $I[\mathrm{Info}(f)] = \big[\underset{\vee}{Q}[f],\, \hat{Q}[f]\big]$, is the intersection of all the uncertainty intervals associated with *linear* rules.

EXAMPLE 2.1.3. Suppose that we are given the function values $f(-1) = 0$, $f(0) = 0$ and $f(1) = 3/2$ together with the information $\sup_{-1\le x\le 1} |f''(x)| \le 2$. These conditions are satisfied by $f(x) = 3(x^2+x)/4$, for example. What can we say about $\int_{-1}^{1} f(x)\,dx$?. Example 2.1.2 gives a complete answer: the value of the integral is contained in the interval

$$\left[\frac{7\sqrt{14}-16}{24},\, \frac{16-\sqrt{2}}{24}\right] = [0.424\ldots,\, 0.607\ldots], \tag{2.19}$$

and no better result is possible. If we do not know $Q^{\rm so}$ (as is the case in almost all real-world problems), then we have to use another estimation rule, preferably an optimal one. Using (2.13) and (2.17), we obtain the uncertainty interval

$$\left[\frac{17\sqrt{2}-16}{24}, \frac{16+\sqrt{2}}{24}\right] = [0.335\ldots, 0.725\ldots].$$

If we were to use the trapezoidal rule (see Example 2.1.1), we would obtain $Q_3^{\rm Tr}[f] = 3/4$. This is remarkable, for (2.19) shows that there is no function from Info$[f]$ with this integral. The interval of uncertainty is now

$$\left[\frac{5}{12}, \frac{13}{12}\right] = [0.416\ldots, 1.083\ldots]$$

(here we have used $\rho(Q_{n+1}^{\rm Tr}) = (b-a)^3M/(12n^2)$ in the SEF with $\mathcal{C} = \mathcal{C}_M^{(2)}$). Another frequently used rule is Simpson's rule

$$Q_3^{\rm Si}[f] := \frac{1}{3}[f(-1) + 4f(0) + f(1)]$$

(see Examples 2.2.1 and 4.3.1), which gives the uncertainty interval

$$\left[\frac{49}{162}, \frac{113}{162}\right] = [0.302\ldots, 0.697\ldots].$$

Observe that intersection of the last two intervals, which were obtained from very simple and popular rules, leads to a better result than application of the sophisticated rule (2.17). Examples like this show that one should be wary of putting too much faith in optimal rules.

2.2. Linear rules that are exact on a subspace

As a consequence of Smolyak's theorem (Theorem 2.1.4), from now on we will consider linear estimation rules only. We shall continue to take a fixed SEF as our basis.

In nearly all practical contexts, large analytical problems arise in the course of calculating optimal rules; see, for example, Köhler (1988a), in which a simple and natural SEF is considered. Ever since the early stages of the development of quadrature theory, another principle has been used, which is based on the expectation that an estimation rule that produces the correct value for elements from a particular subspace is likely to give a good approximation for all elements of V. Before giving a formal statement of the principle, we need a definition.

DEFINITION 2.2.1. The linear functional

$$R := I - Q$$

is called the *remainder* of the estimation rule $Q \in \mathbf{Q}$.

Basic Principle. *Choose a subspace $U \subset V$ and find $Q \in \mathbf{Q}$ such that*

$$R[U] = \{0\}. \tag{2.20}$$

Equation (2.20) means that Q is "exact on U". In quadrature theory, U is typically chosen to be a space of polynomials (Chapter 5), trigonometric polynomials

(Chapter 8), natural spline functions (Section 4.7), other types of splines (Theorem 3.3.4, Example 3.3.1 and Theorem 7.3.2), other piecewise polynomials (see (7.31) and Section 7.4), rational functions (Example 2.3.2) or general Chebyshev systems (Section 3.4).

We may not prescribe arbitrary subspaces U; in a permissible subspace, the functional I that we want to calculate has to be linearly dependent on the observation functionals. An alternative formulation of the principle is given by the following theorem.

THEOREM 2.2.1. There is a $Q \in \mathbf{Q}$ satisfying equation (2.20) if and only if

$$I[U \cap \operatorname{Ker} O] = \{0\}. \tag{2.21}$$

PROOF. (i) If Q satisfies equation (2.20), then

$$I[U \cap \operatorname{Ker} O] = Q[U \cap \operatorname{Ker} O] = \{0\}.$$

(ii) Conversely, suppose that (2.21) holds. Let $\{O[u_i] : i = 1, \ldots, m\}$ be a basis of $O[U]$ and let $U_0 := \operatorname{span}\{u_1, \ldots, u_m\}$. Define a linear map A on $O[U]$ by

$$AO[u_i] = I[u_i] \quad \text{for } i = 1, \ldots, m$$

and, if $m < n$, extend it to $\mathbb{R}^n$. Then the estimation rule $Q := AO$ satisfies $R[U_0] = \{0\}$. Take an arbitrary $u \in U$; then there exists $u_0 \in U_0$ such that $O[u] = O[u_0]$. Therefore

$$R[u] = R[u - u_0] = I[u - u_0] - Q[u - u_0] = 0 - 0 = 0$$

and (2.20) is established. □

It seems natural to consider spaces U in (2.20) that are of dimension $\dim U = n$, since we have n degrees of freedom in the choice of Q, i.e. in the linear map $A : \mathbb{R}^n \to \mathbb{R}$ defining Q ($= AO$). The following well-known result from linear algebra is useful.

LEMMA 2.2.1. Let U be a subspace of V with $\dim U = n$. Then the following statements are equivalent:

(2.22) $U \cap \operatorname{Ker} O = \{0\}$.

(2.23) $\dim O[U] = n$.

(2.24) For each $x \in \mathbb{R}^n$, there exists $u \in U$ satisfying $O[u] = x$.

Using this lemma, we obtain the following theorem.

THEOREM 2.2.2. Let U be an n-dimensional subspace of V that satisfies (one of) the equivalent conditions of Lemma 2.2.1. Then there exists exactly one $Q \in \mathbf{Q}$ that has property (2.20).

PROOF. Since U is n-dimensional, any $n+1$ linear functionals must be linearly dependent in U. The preceding lemma tells us that the components of O are linearly independent. Therefore I has to be a linear combination of the components of O; that is, $I = AO$ on U for a certain unique A. □

DEFINITION 2.2.2. The estimation rule Q in Theorem 2.2.2 is called a *projection rule corresponding to U*.

This designation is explained by the following result.

THEOREM 2.2.3. Suppose that the assumptions of Theorem 2.2.2 are satisfied. Then the resulting estimation rule has the representation

$$Q = I \circ P$$

where P is a projection (i.e. a linear map defined on V that satisfies $P \circ P = P$) onto U with

$$O \circ P = O. \tag{2.25}$$

This projection is defined uniquely by U and (2.25).

PROOF. According to Lemma 2.2.1, the restriction of O to U is regular, so that

$$P := \left(O\big|_U\right)^{-1} \circ O$$

is defined. One can readily verify that P is a projection and that (2.25) holds. From the definition of P, it follows that $I \circ P \in \mathbf{Q}$. The projection property means that

$$Q[u] = (I \circ P)[u] = I[u] \quad \text{for all } u \in U.$$

Now let P_1 be another projection that satisfies (2.25). Then

$$O\big(P[f] - P_1[f]\big) = 0 \quad \text{for all } f \in V$$

and the lemma yields $P = P_1$. □

Taking $V = C[a,b]$ and $O[f] = \big(f(x_1), \ldots, f(x_n)\big)$, equation (2.25) becomes

$$P[f](x_i) = f(x_i) \quad \text{for } i = 1, \ldots, n.$$

This is an interpolation property, and Theorem 2.2.3 says, for instance, that the exactness for $\mathbb{P}_{n-1}$ can be obtained by applying I to the interpolation polynomial.

EXAMPLE 2.2.1. Let

$$V = C[a,b], \qquad I[f] = \int_a^b f(x)\,dx, \qquad O[f] = \Big(f(a), f\big((a+b)/2\big), f(b)\Big).$$

Let $\mathcal{C}$ be arbitrary and take $U = \mathbb{P}_2$. Each $Q \in \mathbf{Q}$ has the form

$$Q[f] = a_1 f(a) + a_2 f\left(\frac{a+b}{2}\right) + a_3 f(b).$$

Upon choosing basis elements $u_i \in \mathbb{P}_2$ given by $u_i(x) = x^i$, equation (2.20) reads

$$\begin{array}{ccccccc} a_1 & + & a_2 & + & a_3 & = & b-a\,, \\ a_1 a & + & a_2\left(\dfrac{a+b}{2}\right) & + & a_3 b & = & \dfrac{b^2-a^2}{2}\,, \\ a_1 a^2 & + & a_2\left(\dfrac{a+b}{2}\right)^2 & + & a_3 b^2 & = & \dfrac{b^3-a^3}{3}\,. \end{array}$$

The solution yields Simpson's rule

$$Q_3^{\mathrm{Si}}[f] = \frac{b-a}{6}\left[f(a) + 4\,f\left(\frac{a+b}{2}\right) + f(b)\right]. \tag{2.26}$$

On the other hand, the projection P from Theorem 2.2.3 is given by

$$\begin{aligned} P[f](x) = \frac{2}{(b-a)^2}\Big[& f(a)\left(x - \frac{a+b}{2}\right)(x-b) \\ & - 2f\left(\frac{a+b}{2}\right)(x-a)(x-b) + f(b)(x-a)\left(x - \frac{a+b}{2}\right)\Big]. \end{aligned} \tag{2.27}$$

Application of the functional I, i.e. integration, again yields the rule (2.26).

Theorem 2.2.1 shows that, moreover, Simpson's rule satisfies $R[\mathbb{P}_3] = \{0\}$, since

$$\int_a^b (x-a)\left(x - \frac{a+b}{2}\right)(x-b)\,dx = 0.$$

A projection P is said to be *based on* O if there exists a factorization $P = L \circ O$ for some map L. Not every $Q \in \mathbf{Q}$ that is exact on U (where $\dim U = n$) can be represented in the form $I \circ P$ with P being a projection onto U based on O. If $U \cap \operatorname{Ker} O \neq \{0\}$, then there is no such projection.

EXAMPLE 2.2.2. Let

$$V = C[0, 2\pi], \qquad I[f] = \int_0^{2\pi} f(x)\,dx,$$

$$O[f] = \big(f(x_1), f(x_2), \ldots, f(x_{2m})\big), \qquad x_\nu = \left(\nu - \tfrac{1}{2}\right)\frac{\pi}{m}$$

and

$$U = \operatorname{span}\{1,\ \cos x,\ \cos 2x, \ldots,\ \cos mx,\ \sin x,\ \sin 2x, \ldots,\ \sin{(m-1)x}\}.$$

Equation (2.22) is not satisfied, as $u_0(x) = \cos mx$ shows; hence there is no projection rule corresponding to U. Nevertheless, Q_{2m}^{Mi} is exact for U. This follows from the fact that

$$\sum_{\nu=1}^{2m} e^{ikx_\nu} = e^{ik\pi/(2m)} \sum_{\nu=0}^{2m-1} e^{i\nu k\pi/m} = e^{ik\pi/(2m)}\,\frac{e^{i2k\pi} - 1}{e^{ik\pi/m} - 1} = 0$$

for $k = 1, 2, \ldots, 2m-1$, upon separating the real and imaginary parts.

It is possible to prove that $U \cap \operatorname{Ker} O = \operatorname{span}\{u_0\}$, so that the existence of a rule which is exact for U would be a consequence of Theorem 2.2.1, but the proof given above leads to a more complete result.

The advantage of constructions obtained via application of (2.20) is that we need only solve a system of linear equations for the coefficients of A in $Q = AO$. We are free to use any U in (2.21), but evidently we would do better by taking the co-observation into account. It seems natural to quantify the disadvantage of applying an arbitrary rule Q by comparing its error with that of the optimal rule, so we define

$$\operatorname{loss}(Q) := \frac{\rho(Q)}{\rho^{\mathrm{opt}}}.$$

Ideally, we would like to combine easy constructibility with small loss. The realization of this ideal is embodied in the concept of reconstruction.

DEFINITION 2.2.3. A projection P based on O from V onto $U \subset V$ is called a *reconstruction* if

$$P[f] \in \text{Info}(O[f]) \quad \text{for all } f \in \mathcal{C}. \tag{2.28}$$

THEOREM 2.2.4. *For a reconstruction P,*

$$\left|I[f] - I \circ P[f]\right| \leq 2\rho^{\text{intr}}[f]\,. \tag{2.29}$$

PROOF. $I \circ P$ is an estimation rule. Furthermore, we have

$$\underset{\vee}{Q}[f] - \hat{Q}[f] \leq I[f] - (I \circ P)[f] \leq \hat{Q}[f] - \underset{\vee}{Q}[f]$$

and so the assertion is proved. □

EXAMPLE 2.2.3. We use the SEF of Example 2.2.1 specified by

$$\mathcal{C} = \{f \,:\, f'' \text{ is continuous and } \sup_{x\in[a,b]} |f''(x)| \leq M\}.$$

The projection (2.27) is then a reconstruction, since $O \circ P = O$ is obvious and

$$\left(P[f]\right)''(x) = \frac{4}{(b-a)^2}\left[f(a) - 2f\left(\frac{a+b}{2}\right) + f(b)\right] = f''(\xi) \quad \text{for } \xi \in [a,b].$$

The resulting inequality

$$\sup_{f\in\mathcal{C}}\left|\frac{I[f] - Q^{\text{Si}}[f]}{\rho^{\text{intr}}[f]}\right| \leq 2 \tag{2.30}$$

demonstrates a certain superiority of Simpson's rule over the optimal rule, since the corresponding left-hand side of (2.30) for the optimal rule is unbounded; see (2.18). This unboundedness holds for all linear estimation rules apart from Q^{Si}.

Unfortunately, there are only few cases where an SEF allows reconstruction. Hence, the elegant method of Theorem 2.2.4 is rarely applicable. A less restrictive notion is that of a δ-reconstruction.

DEFINITION 2.2.4. A projection P based on O onto $U \subset V$ is called a δ-*reconstruction* if for all $f \in \mathcal{C}$ we have:

(i) $P[f] \in \delta\mathcal{C}$ $(:= \{\delta f \,:\, f \in \mathcal{C}\})$;
(ii) $OP[f] = O[f]$.

THEOREM 2.2.5. *A δ-reconstruction P satisfies*

$$\left|I[f] - I \circ P[f]\right| \leq (1+\delta)\rho^{\text{opt}}[f]. \tag{2.31}$$

PROOF. We have

$$\frac{f - P[f]}{\delta + 1} \in \mathcal{C},$$

and an application of Theorem 2.1.1 gives

$$\begin{aligned} \left|I[f] - (I \circ P)[f]\right| &= \left|I[f - P[f]]\right| \\ &\leq (\delta+1)\sup\{I[g] \,:\, g \in \mathcal{C} \cap \operatorname{Ker} O\} \\ &= (\delta+1)\rho^{\text{opt}} \end{aligned}$$

as claimed. □

Note that (δ-)reconstructions are independent of the functional I under consideration. If we have a reconstruction, then we can use any linear functional and always obtain a good method in the sense of (2.29) or (2.31). In particular, we have $\operatorname{loss}(I \circ P) \le 2$ for reconstructions and $\operatorname{loss}(I \circ P) \le 1 + \delta$ for δ-reconstructions.

Examples of δ-reconstructions can be found in the following works: Brass (1995b) for quadratic and cubic splines, Brass (1996) for periodic splines, and Fischer (1997) for trigonometric polynomials.

2.3. Strong optimality: inner product spaces

We return to the construction of strongly optimal rules Q^{so}; the construction will be done according to the principle based on (2.20). An appropriate subspace U has to be chosen such that the resulting estimation rule is strongly optimal. This is not possible for an arbitrary SEF since, in general, Q^{so} is nonlinear (see, e.g., Brass 1998a). Golomb and Weinberger (1959) showed by means of an elegant theory that such a choice is possible if $\mathcal{C}$ is a ball in a Hilbert space.

We need something slightly weaker than a Hilbert space.

DEFINITION 2.3.1. An SEF will be called an *inner product SEF* if:

- a semi-definite inner product $\varphi(\cdot\,,\cdot)$ is defined on V;
- the co-observation is given by

$$\mathcal{C} = \{f \in V : \varphi(f,f) \le M^2\} \quad \text{for some } M > 0;$$

- the observation O satisfies

$$\{f \in V : \varphi(f,f) = 0\} \cap \operatorname{Ker} O = \{0\}. \tag{2.32}$$

The key step is the choice of an appropriate subspace into which we want to project.

DEFINITION 2.3.2. A space

$$S := (\operatorname{Ker} O)^{\perp} \ \big(:= \{s \in V : \varphi(g,s) = 0 \text{ for all } g \in \operatorname{Ker} O\}\big) \tag{2.33}$$

of dimension n is called a *spline space* for V and O.

Note that $s \in S \cap \operatorname{Ker} O = (\operatorname{Ker} O)^{\perp} \cap \operatorname{Ker} O$ implies $\varphi(s,s) = 0$, and (2.32) then shows that $s = 0$. Hence, by Theorems 2.2.2 and 2.2.3, we may project into spline spaces.

We furthermore introduce a semi-norm $\|\cdot\|$, defined by

$$\|f\| := \varphi(f,f)^{1/2}.$$

The next theorem says that in an inner product SEF, the projection rule corresponding to the spline space is always strongly optimal. This result is independent of the linear functional under consideration.

THEOREM 2.3.1 (Golomb and Weinberger 1959). *In an inner product SEF with spline space S, the strongly optimal rule Q^{so} is the unique linear estimation rule such that*

$$R[S] = \{0\}.$$

(where R is the remainder). It can be written as

$$Q^{\text{so}} = I \circ P,$$

where P is the projection onto S satisfying

$$O \circ P = O.$$

Moreover, P is an orthogonal projection, i.e.

$$\varphi(f - P[f], s) = 0 \quad \text{for all } f \in V \text{ and } s \in S, \tag{2.34}$$

and furthermore

$$\rho^{\text{intr}}[f] = \left(1 - \frac{\|P[f]\|^2}{M^2}\right)^{1/2} \rho^{\text{opt}}. \tag{2.35}$$

PROOF. For the projection P onto S (which exists), equation (2.34) holds since $f - P[f] \in \operatorname{Ker} O$ and $S \perp \operatorname{Ker} O$.

We have to prove strong optimality of the projection rule. Define the set W by

$$\text{Info}[f] = P[f] + W.$$

Then

$$\begin{aligned} W &= \{w \in \operatorname{Ker} O : \|w + P[f]\|^2 \ (= \|w\|^2 + \|P[f]\|^2) \ \le M^2\} \\ &= \{w \in \operatorname{Ker} O : \|w\|^2 \le M^2 - \|P[f]\|^2\}, \end{aligned}$$

and so

$$\begin{aligned} \hat{Q}[f] &= (I \circ P)[f] + \sup\{I[w] : w \in W\} \\ &= IP[f] + \left(\frac{M^2 - \|P[f]\|^2}{M^2}\right)^{1/2} \sup\{I[v] : v \in \mathcal{C} \cap \operatorname{Ker} O\}. \end{aligned}$$

Equation (2.1) yields $\underset{\vee}{Q}[f]$ and hence the strong optimality of $I \circ P$. Now, equation (2.35) follows from Theorem 2.1.1. □

EXAMPLE 2.3.1. Let V be the space of all continuous and piecewise continuously differentiable functions on $[a, b]$, and define an inner product on V by

$$\varphi(f, g) := \int_a^b f'(x) g'(x)\, dx.$$

Furthermore, let $I[f] = \int_a^b f(x)\, dx$, $O[f] = \big(f(x_1), \ldots, f(x_n)\big)$ for $a \le x_1 < x_2 < \cdots < x_n \le b$, and

$$\mathcal{C} = \Big\{f : \int_a^b [f'(x)]^2\, dx \le M^2\Big\}.$$

We have

$$|f(x) - f(x_1)| = \left|\int_{x_1}^x f'(u)\, du\right| \le \left(|x - x_1| \int_{x_1}^x \left[f'(u)\right]^2 du\right)^{1/2} \le M|b - a|^{1/2}.$$

Therefore $|f - f(x_1)|$ is uniformly bounded for all $f \in \mathcal{C}$, so that $\hat{Q} - \underset{\vee}{Q}$ is uniformly bounded. If $\varphi(f, f) = 0$, this bound is zero and $O[f] = 0$ implies, in particular, that $f(x_1) = 0$. Hence, equation (2.32) holds and we have an inner product SEF.

The corresponding spline space is the space of "natural spline functions of order one", i.e. the space of polygonal lines with vertices at $x_1, \ldots, x_n$ which are constant on $[a, x_1]$ and on $[x_n, b]$. Integration by parts shows that for each such natural spline s, $\varphi(f, s) = \int_a^b f'(x) s'(x)\, dx$ is a linear combination of $f(x_1), \ldots, f(x_n)$. We thus obtain $s \in (\operatorname{Ker} O)^\perp$. Obviously, for each $(y_1, \ldots, y_n) \in \mathbb{R}^n$ we can find an

$s \in S$ satisfying $s(x_i) = y_i$. Hence we have $\dim S = n$, and Theorem 2.3.1 is applicable. The projected element $P[f]$ is the natural spline function of order one that interpolates f at $x_1, \ldots, x_n$. Using simple geometric arguments, one can show that Q^{so} is the generalized trapezoidal rule in Example 2.1.1. Formula (2.35) reads as follows:

$$\rho^{\text{intr}}[f] = \left(1 - \sum_{\nu=1}^{n-1} \frac{[f(x_{\nu+1}) - f(x_\nu)]^2}{M^2(x_{\nu+1} - x_\nu)}\right)^{1/2} \rho^{\text{opt}},$$

where ρ^{opt} can be determined via Peano kernel methods (see Section 4.2). Here we just note the result:

$$\rho^{\text{opt}} = M\left(\frac{(x_1 - a)^3}{3} + \frac{1}{12}\sum_{\nu=1}^{n-1}(x_{\nu+1} - x_\nu)^3 + \frac{(b - x_n)^3}{3}\right)^{1/2}.$$

In many situations in numerical analysis, a small norm indicates good quality. A nice property that projections into spline spaces have is that they map onto the interpolating elements of minimal semi-norm ("interpolating" means having the same observation).

THEOREM 2.3.2. Under the assumptions of Theorem 2.3.1, for any $f, g \in V$ such that $O[g] = O[f]$ we have

$$\|g\| \geq \|P[f]\|,$$

with equality holding if and only if $g = P[f]$.

PROOF. Let $O[g] = O[f]$, i.e. $g - P[f] \in \operatorname{Ker} O$, and let $P[f] \in S$. It follows that

$$\begin{aligned}
\|g\|^2 &= \|P[f] + g - P[f]\|^2 \\
&= \|P[f]\|^2 + 2\underbrace{\varphi(P[f],\, g - P[f])}_{=0} + \|g - P[f]\|^2 \geq \|P[f]\|^2.
\end{aligned}$$

Equality holds if and only if $\|g - P[f]\| = 0$, i.e. if $g = P[f]$ (see the assumption (2.32)). □

Thus, the interpolating spline function from Example 2.3.1 is the "flattest" interpolating element in the sense that among all the $g \in V$ which interpolate f at $x_1, \ldots, x_n$, the minimal value of

$$\|g'\|_2^2 = \int_a^b \left[g'(x)\right]^2 dx$$

is attained by $g = P[f]$.

We will now strengthen our assumptions on V by requiring V to be a Hilbert space. A new tool, namely the Riesz representation theorem (see any text on functional analysis), now becomes available to us.

Riesz Representation Theorem. *Suppose l is a continuous linear functional on a Hilbert space V; then there exists one and only one $\widetilde{l} \in V$ (called the representer of l) such that*

$$l(x) = \varphi(x, \widetilde{l}) \quad \textit{for all } x \in V.$$

Furthermore,

$$\|l\| = \|\widetilde{l}\|.$$

An immediate consequence of this theorem is the following result.

THEOREM 2.3.3. Let V be a Hilbert space and let $o_1, \ldots, o_n$ be continuous and linearly independent functionals. Then a spline space for V and $O = (o_1, \ldots, o_n)$ is given by $\operatorname{span}(\widetilde{o}_1, \ldots, \widetilde{o}_n)$.

The spaces V of greatest relevance to us consist of real-valued functions f defined on $X \subset \mathbb{R}^d$. In such spaces we can define functionals l_y by

$$l_y[f] := f(y) \quad \text{for } y \in X.$$

If these functionals l_y are continuous, the Riesz representation theorem implies the existence of a function $K(\cdot, \cdot)$ on $X \times X$ such that

$$f(y) = \varphi(f, K(\cdot, y)) \quad \text{for } f \in V,\ y \in X.$$

This function K is called the *reproducing kernel.* Determination of the reproducing kernel is made simpler if an orthonormal basis of H is available, i.e. a sequence $u_i \in H$, $i = 1, 2, \ldots$, with

$$f = \sum_{i=1}^{\infty} \varphi(u_i, f) u_i \quad \text{for all } f \in H.$$

We then have

$$K(x, y) = \sum_{i=1}^{\infty} u_i(x) u_i(y) \quad \text{for } x, y \in X,$$

and from this we easily deduce that

$$\widetilde{l}(y) = l[K(\cdot, y)]$$

for all continuous linear functionals on spaces of the aforementioned type. For more information on the theory of reproducing kernels, see Meschkowski (1962). We now illustrate the theory by an example.

EXAMPLE 2.3.2. Fix $\rho > 0$, and let $p_\nu(x) = x^\nu$,

$$\mathcal{H}_\rho = \Big\{ f : f = \sum_{\nu=0}^{\infty} \alpha_\nu p_\nu \text{ with } \alpha_\nu \in \mathbb{R} \text{ and } \sum_{\nu=0}^{\infty} \alpha_\nu^2 \rho^{2\nu} < \infty \Big\}$$

and

$$\varphi\Big(\sum_{\nu=0}^{\infty} \alpha_\nu p_\nu \,,\ \sum_{\nu=0}^{\infty} \beta_\nu p_\nu \Big) = \sum_{\nu=0}^{\infty} \alpha_\nu \beta_\nu \rho^{2\nu}.$$

Evidently, $\mathcal{H}_\rho$ is a Hilbert space of holomorphic functions, and

$$u_i(x) = x^i \rho^{-i} \quad \text{for } i = 0, 1, \ldots$$

is an orthonormal basis for $\mathcal{H}_\rho$. Hence

$$K(x, y) = \sum_{i=0}^{\infty} x^i y^i \rho^{-2i} = \frac{\rho^2}{\rho^2 - xy}$$

is the reproducing kernel, and the spline space for $O[f] = (f(x_1), \ldots, f(x_n))$ is $S = \operatorname{span}\{K(x_i, y) : i = 1, \ldots, n\}$. Projecting onto S means interpolating, and we can construct $P[f]$. By analogy with the Lagrange interpolation formula, the interpolant can be written in the form

$$P[f](x) = \sum_{\lambda=1}^{n} f(x_\lambda) \frac{z(x)}{(x - x_\lambda) z'(x_\lambda)} \cdot \frac{v(x_\lambda)}{v(x)}$$

$$\text{where} \qquad z(x) = \prod_{\nu=1}^{n} (x - x_\nu), \qquad v(x) = \prod_{\nu=1}^{n} (\rho^2 - x x_\nu).$$

Now, $I \circ P$ is the strongly optimal rule. In the special case where $I[f] = \int_{-1}^{1} f(u)du$ and $\rho > 1$, we can obtain (by expanding in partial fractions) a fairly explicit expression

$$Q^{\text{so}}[f] = \sum_{\nu=1}^{n} a_\nu f(x_\nu)$$

$$\text{where} \qquad a_\lambda = \rho^{-2n+2} \sum_{\kappa=1}^{n} \frac{v(x_\kappa) v(x_\lambda)}{z'(x_\kappa) z'(x_\lambda)} \cdot \frac{1}{\rho^2 - x_\kappa x_\lambda} \cdot \frac{1}{x_\kappa} \ln \frac{\rho^2 + x_\kappa}{\rho^2 - x_\kappa};$$

see Wilf (1967) and Valentin (1968). For the determination of best nodes, see Engels (1977).

Further examples of how the theory presented in this section can be applied are given in Barnhill (1967, 1968), Rabinowitz and Richter (1970a, 1970b) and Chawla and Raina (1972).

As mentioned before, the availability of an orthonormal basis $u_1, u_2, \ldots$ for the underlying Hilbert space often simplifies things considerably. The following theorem is further evidence of this.

THEOREM 2.3.4. Let Q be any estimation rule in a given Hilbert space SEF. Then

$$\rho(Q) = M \left[\sum_{i=1}^{\infty} \left(R[u_i] \right)^2 \right]^{1/2}.$$

PROOF. We apply Parseval's equation

$$\varphi(f, f) = \sum_{i=1}^{\infty} \varphi^2(f, u_i)$$

(which holds for any f from the Hilbert space and is an easy consequence of the definitions) to $f = \widetilde{R}$. This gives

$$\rho(Q) = \sup\{R[f] : \varphi(f, f) \le M^2\} = M \|R\| = M \|\widetilde{R}\| = M \left[\varphi(\widetilde{R}, \widetilde{R}) \right]^{1/2}$$

$$= M \left[\sum_{i=1}^{\infty} \varphi^2(\widetilde{R}, u_i) \right]^{1/2} = M \left[\sum_{i=1}^{\infty} \left(R[u_i] \right)^2 \right]^{1/2}$$

as claimed. □

The formula

$$\rho(Q) = M \left(\sum_{i=1}^{\infty} \left(I[u_i] - Q[u_i] \right)^2 \right)^{1/2}$$

leads to an apparently new principle for the construction of estimation rules, namely: for a given Hilbert space SEF, choose $Q \in \mathbf{Q}$ such that

$$\sum_{i=1}^{\infty} \left[I[u_i] - Q[u_i] \right]^2$$

is as small as possible; see Wilf (1964). The minimum, however, is ρ^{opt}, and it is attained by a rule that can be constructed using our basic principle of requiring exactness on a suitable subspace.

2.4. Varying the observation

We wish to generalize our model so that we are not restricted to assuming a fixed observation. Instead, our approach will be to take a (large) set of observations and try to choose one that is appropriate for the particular estimation problem at hand.

DEFINITION 2.4.1. A *generalized SEF* is a quadruple $(V, I, \Omega, \mathcal{C})$ where V, I and $\mathcal{C}$ are as in the definition of a SEF (Definition 2.1.1) and Ω is a set of linear functionals on V.

In this section we shall always assume a generalized SEF.

For any $O \in \Omega^n$, the quadruple $(V, I, O, \mathcal{C})$ is an SEF. All notions from the previous sections can be used here, if the additional dependence on O is included; for example, we now write $\rho^{\mathrm{opt}}(O)$ or $\mathbf{Q}(O)$.

DEFINITION 2.4.2. Define $\rho_n^{\mathrm{best}} := \inf\{\rho^{\mathrm{opt}}(O) \,:\, O \in \Omega^n\}$.
Then $Q_n^{\mathrm{best}} := Q \in \bigcup_{O \in \Omega^n} \mathbf{Q}(O)$ is called a *best estimation rule* if $\rho(Q) = \rho_n^{\mathrm{best}}$.

EXAMPLE 2.4.1. Consider a generalized SEF given by $V = C[a,b]$, $I[f] = \int_a^b f(x)\,dx$, $\Omega = \{l_x : x \in [a,b] \text{ such that } l_x[f] = f(x)\}$ and $\mathcal{C} = \mathrm{Lip}_M 1$. In Example 2.1.1, we determined $\rho^{\mathrm{opt}}(O)$ for arbitrary $O \in \Omega^n$. We thus obtain

$$\begin{aligned}
\rho_n^{\mathrm{best}} &= \frac{M}{4} \left\{ \inf_{a \le x_1 \le x_2 \le \ldots \le x_n \le b} \left(2(x_1 - a)^2 + \sum_{\nu=1}^{n-1} (x_{\nu+1} - x_\nu)^2 + 2(b - x_n)^2 \right) \right\} \\
&= \frac{M(b-a)^2}{4n}.
\end{aligned}$$

Here, the infimum can be determined with standard methods, and we see that it is attained for $x_\nu = a + (2\nu - 1)(b-a)/(2n)$. Hence, the midpoint rule is the best choice in the given generalized SEF.

We know that a precision of $\rho_n^{\mathrm{best}} + \varepsilon$, where $\varepsilon > 0$, can be achieved with a linear rule that is independent of the particular $f \in \mathcal{C}$. Let us now try to improve our results by making a choice of $O \in \Omega^n$ that does take the properties of f into account. In our SEF model, information about the particular $f \in \mathcal{C}$ can be obtained only through $o \in \Omega$; therefore a choice of $O = (o_1, \ldots, o_n) \in \Omega^n$ has to be sequential. This means that, while O is not known *a priori*, the o_i can be chosen based on knowledge of $o_1[f], \ldots, o_{i-1}[f]$.

DEFINITION 2.4.3. An *n-stage sequential* (or *adaptive*) *algorithm* Alg is an algorithm given by:

(i) choice functions $g_i : \Omega^i \times \mathbb{R}^i \to \Omega$ for $i = 1, \ldots, n-1$;
(ii) termination functions $h_i : \Omega^i \times \mathbb{R}^i \to \{0, 1\}$ for $i = 1, \ldots, n-1$;
(iii) estimation functions $q_i : \Omega^i \times \mathbb{R}^i \to \mathbb{R}$ for $i = 1, \ldots, n$;
(iv) a functional $o_1 \in \Omega$.

The application of an n-stage sequential algorithm consists of a number of steps; this number (not more than n) is controlled by the algorithm itself. The final output is an estimate $Q[f]$. We set $g_0 := o_1$ and $h_0 := 0$. The kth step consists of calculating $o_k := g_{k-1}(o_1, \ldots, o_{k-1}, o_1[f], \ldots, o_{k-1}[f])$, followed by $o_k[f]$ and $h_k(o_1, \ldots, o_k, o_1[f], \ldots, o_k[f])$. If the latter value is zero, we proceed to the $(k+1)$st step. Otherwise, the process is terminated with the estimate

$$Q[f] = q_k(o_1, \ldots, o_k, o_1[f], \ldots, o_k[f]).$$

EXAMPLE 2.4.1 (continued). Define $\widetilde{\Omega} := \{\widetilde{o}_1, \ldots, \widetilde{o}_n\}$ where $\widetilde{o}_\nu[f] = f\big(a + (2\nu - 1)(b-a)/(2n)\big)$, and let O_{k-1} be the set of observations that has already been chosen after the $(k-1)$st step. The kth step proceeds as follows. The choice function g_k selects an $o_k \in \widetilde{\Omega} \setminus O_{k-1}$, which is applied to f. Then, $\rho^{\text{intr}}[O_k, f]$ is calculated as in Example 2.1.1. If this number is at most $M(b-a)^2/(4n)$, the termination function h_k is set to 1 and the algorithm terminates with the estimate

$$Q[f] := Q^{\text{so}}[O_k, f].$$

Otherwise, we set h_k to 0 and proceed to the $(k+1)$st step.

Obviously, this algorithm guarantees that

$$\big|I[f] - Q[f]\big| \le \rho_n^{\text{best}} = \frac{M(b-a)^2}{4n}$$

with at most n but often fewer function evaluations. The optimal algorithm from Example 2.4.1 (the earlier part) always requires n function values.

Different selections of the choice functions g_i yield many variants of the algorithm. One can seek to choose g_k so that O_k is always "as equally distributed as possible", or try to "maximize the reduction of the uncertainty" of $I[f]$ in each step. Sukharev (1979) devised an algorithm that is optimal with respect to the necessary number of function evaluations; however, one has to pay for this minimality with choice functions that are quite involved, so it is doubtful that a gain in the overall efficiency of computation can be achieved with Sukharev's algorithm.

To evaluate an algorithm, the quantity

$$\rho(\text{Alg}) := \sup\{\big|I[f] - Q[f]\big| \; : \; f \in \mathcal{C}\}$$

is crucial. The following remarkable result says, essentially, that "adaptiveness does not help in the worst case".

THEOREM 2.4.1 (Bakhvalov 1971). For each n-stage algorithm Alg, we have

$$\rho(\text{Alg}) \ge \rho_n^{\text{best}}.$$

PROOF. Let $f = 0$ $(\in \mathcal{C})$.

The algorithm chooses the observation $O = (o_1, \ldots, o_k)$, where $k \le n$. Application of Theorem 2.1.1 then gives

$$\rho(\text{Alg}) \ge \rho^{\text{intr}}(O, f) = \rho^{\text{intr}}(O, 0) = \rho^{\text{opt}}(O) \ge \inf_{O \in \Omega^k} \rho^{\text{opt}}(O) \ge \rho_n^{\text{best}}$$

as asserted. □

The first result comparing adaptive and non-adaptive algorithms was obtained by Kiefer (1957). For more results in this area, we refer the reader to Gal and Micchelli (1980), Sukharev (1985), Huerta (1986), Sukharev (1987), Novak (1988), Chuyan and Sukharev (1990), and Novak (1992).

Theorem 2.4.1 supplies a further argument for concentrating on linear estimation rules.

CHAPTER 3

Norm and Kernel of the Remainder Functional

In this chapter, we define the norm of an estimation rule and show why it is important for the norm to be small. Further, we prove the existence of rules with small norm that are exact on a prescribed subspace.

3.1. Norm of an estimation rule

Throughout this chapter, we assume an SEF as defined in Chapter 2. We take the arbitrary linear space V to be $C[X]$, the space of continuous real functions on a compact subset X of $\mathbb{R}^d$. We choose an observation of the form

$$O[f] = \big(f(x_1), \ldots, f(x_n)\big),$$

where the $x_i \in X$ are pairwise distinct. When discussing generalized SEFs, we shall take $\Omega = \{l_x : x \in X\}$ where $l_x[f] := f(x)$.

As seen in Chapter 2, it is reasonable to focus on linear estimation rules, i.e. rules Q of the form

$$Q[f] = Q_n[f] = \sum_{\nu=1}^{n} a_\nu f(x_\nu).$$

In the case where $X = [a, b]$, these estimation rules are just the quadrature rules defined at the beginning of Chapter 1; they will be the main objects of our study. Choosing a more general X in this chapter, however, does not cause greater technical difficulties and in fact leads to very interesting results.

The space $C[X]$ is equipped with the norm

$$\|f\| := \sup\{|f(x)| \,:\, x \in X\}.$$

The norm of a continuous linear functional on $C[X]$ is defined by

$$\|L\| := \sup\big\{\big|L[f]\big| \,:\, f \in C[X] \text{ such that } \|f\| \le 1\big\}.$$

For obvious reasons, we shall assume $\|I\| < \infty$. Evidently, $\|Q\| \le \sum_{\nu=1}^{n} |a_\nu|$ and, moreover, there always exists an $f \in C[X]$ with $f(x_\nu) = \operatorname{sgn} a_\nu$ and $\|f\| = 1$ such that equality holds, i.e.

$$\|Q\| = \sum_{\nu=1}^{n} |a_\nu|. \tag{3.1}$$

For us, the most interesting case is

$$I[f] = \int_X f(x) w(x)\, dx, \tag{3.2}$$

for which we can readily verify that

$$\|I\| = \int_X |w(x)|\, dx.$$

For the remainder functional $R = I - Q$, we have

$$\big|\|I\| - \|Q\|\big| \le \|R\| \le \|I\| + \|Q\|.$$

If I is of the form (3.2), we even have

$$\|R\| = \|I\| + \|Q\|.$$

To prove this equality, for arbitrary $\varepsilon > 0$ we modify an f satisfying $I[f] > \|I\| - \varepsilon$ and $\|f\| \le 1$ in an arbitrarily small neighbourhood of each x_i so that $f(x_i) = -\operatorname{sgn} a_i$.

The size of $\|Q\|$ relative to $\|I\|$ is a measure of quality for Q. Note that there exists an $f \in C[X]$ satisfying $\|f\| \le 1$ and $Q[f] = \|Q\|$. If $\|Q\|$ is much larger than $\|I\|$, the approximation $Q[f]$ for $I[f]$ is poor. A striking illustration is given by the next example.

EXAMPLE 3.1.1. Let $X = [-1, 1]$ and $I[f] = \int_{-1}^{1} f(x)\, dx$, and take the observation to be $O[f] = \big(f(x_1), \ldots, f(x_{100})\big)$ where $x_\nu = -1 + 2(\nu - 1)/99$. This choice of the x_ν seems to give a reasonable distribution of information over the interval of integration. Consider the estimation rule Q based on O that is defined uniquely by $R[\mathbb{P}_{99}] = 0$. The coefficients a_ν can be obtained by solving a linear system of equations, and we obtain from (3.1) that

$$1.2 \cdot 10^{24} < \|Q\| < 1.3 \cdot 10^{24}.$$

If $\|f\| \le 1$, there is a constellation for which the estimate gives a number of around 10^{24}, while it is obvious that $|I[f]| \le 2$.

Of course, whether a "bad" function has to be considered at all depends also on the co-observation. However, there is another argument against using estimation rules of large norm, namely *error propagation.*

When calculating $Q[f]$, the values of $f(x_\nu)$ are usually not known exactly but, rather, we know slightly perturbed values $f(x_\nu) + \varepsilon_\nu$. Often, it is possible to find a bound ε on the sizes $|\varepsilon_\nu|$ of the perturbations. Hence, we can determine $Q[f]$ only up to an uncertainty $\|Q\|\varepsilon$; in other words, the input error bound is multiplied by $\|Q\|$. Thus, in Example 3.1.1, even if $\rho(Q)$ is small and everything is calculated to machine precision, the output will be of no value.

The concept of the norm also lies behind the principle, stated in Section 2.2, of seeking $R[U] = \{0\}$ for some subspace U. This idea essentially consists of hoping that $f \approx u$ ($\in U$) will imply $R[f] \approx R[u]$ ($= 0$), which would be true only if $\|R\|$ is small. Thus, it does not make much sense to require $R[U] = \{0\}$ without also requiring that $\|R\|$ be small. A lower bound for estimation rules that satisfy $R[U] = \{0\}$ is obviously

$$\|Q\| \ge \sup\big\{\big|I[u]\big| \;:\; u \in U \text{ such that } \|u\| \le 1\big\}. \tag{3.3}$$

We will show in the next section that this lower bound is attained for appropriate observations. Nevertheless, let us emphasize that, in principle, nothing can be said against estimation rules whose norms differ only slightly from the lower bound (3.3).

For our purposes, the most useful applications of the norm arise from the following simple inequality.

THEOREM 3.1.1 (Lebesgue's inequality). Let $R[U] = \{0\}$. Then

$$\left|R[f]\right| \le \|R\| \operatorname{dist}(f, U),$$

where $\operatorname{dist}(f, U) := \inf_{u \in U} \|f - u\|$.

PROOF. In the inequality

$$\left|R[f]\right| = \left|R[f - u]\right| \le \|R\| \, \|f - u\|,$$

take the infimum over all $u \in U$. □

This theorem is important and useful because, in approximation theory, the quantity

$$\sup\{\operatorname{dist}(f, U) \,:\, f \in \mathcal{C}\}$$

has already been investigated for many different classes $\mathcal{C}$ and subspaces U. We refer the reader to the monographs by Timan (1963), Cheney (1966), Meinardus (1967), Korneichuk (1991), and DeVore and Lorentz (1993) for details.

Later, we shall consider several applications of Lebesgue's inequality; a survey is given in Brass (1992a). Here is a first example.

EXAMPLE 3.1.2 (Locher and Zeller 1968). Let $X = [a, b]$ and $I[f] = \int_a^b f(x)\,dx$. Furthermore, let Q have positive coefficients and satisfy $R[\mathbb{P}_{r-1}] = \{0\}$ for $r > 0$. This assumption implies that

$$\|Q\| = \sum_{\nu=1}^{n} a_\nu = Q[1] = I[1] = b - a.$$

We choose the classical co-observation

$$\mathcal{C}_M^{(r)} := \{f \in C[a, b] \,:\, \|f^{(r)}\| \le M\}.$$

A standard result from approximation theory, which can be obtained readily via elementary methods, is

$$\sup\{\operatorname{dist}(f, \mathbb{P}_{r-1}) \,:\, f \in \mathcal{C}_M^{(r)}\} = \frac{(b - a)^r}{r!\, 2^{2r-1}} M. \tag{3.4}$$

Therefore, Theorem 3.1.1 yields

$$\rho(Q) \le \frac{(b - a)^{r+1}}{r!\, 2^{2r-2}} M.$$

Although this result can be improved slightly, its simple proof and its potential for generalization are remarkable.

DEFINITION 3.1.1. An estimation method is a sequence of estimation rules with fixed V, fixed I and increasing n (number of observation points). A quadrature method is an estimation method with $V = C[a, b]$.

For notational simplicity, we shall take the sequence of numbers of observation points to be $1, 2, 3, \ldots$. Using a more general increasing sequence of integers will not engender additional difficulties.

DEFINITION 3.1.2. An estimation method is said to be *convergent* if

$$\lim_{n\to\infty} Q_n[f] = I[f]$$

for all $f \in V$.

A simple example of a convergent quadrature method is the midpoint method. A principal result of convergence theory is the following theorem of Polya (1933) (see also Hahn 1918).

THEOREM 3.1.2. The estimation method $(Q_n)_{n\in\mathbb{N}}$ is convergent if and only if the following two conditions are satisfied:

(i) $\sup\limits_n \|Q_n\| < \infty$.

(ii) There is a set G which is dense in V and such that

$$\lim_{n\to\infty} Q_n[g] = I[g] \quad \text{for all } g \in G.$$

PROOF. Let us begin with the sufficiency of the two conditions. Let $f \in V$. For $\varepsilon > 0$, there is a $g \in G$ satisfying $\|f - g\| < \varepsilon$. We then have

$$\begin{aligned} |R_n[f]| &= |R_n[f-g] + R_n[g]| \\ &\le \|R_n\|\,\|f-g\| + |R_n[g]| \\ &\le (\|I\| + \|Q_n\|)\,\varepsilon + |R_n[g]|. \end{aligned}$$

Condition (i) shows that the first summand is bounded by a constant times ε, and condition (ii) tells us that the second summand is bounded by ε for sufficiently large n.

The necessity is an immediate consequence of the following fundamental theorem from functional analysis.

Principle of Uniform Boundedness. *Let $(L_n)_{n\in\mathbb{N}}$ be a sequence of linear functionals defined on a Banach space V. If $(L_n[f])_{n\in\mathbb{N}}$ is convergent for each $f \in V$, then $\sup_\nu \|L_\nu\| < \infty$.* □

For practical purposes, it is important to have not merely convergence but "fast convergence". If we assume only that $f \in C[X]$, then the next theorem says that convergence can be arbitrarily slow.

THEOREM 3.1.3. Let

$$I[f] = \int_X f(x)w(x)\,dx,$$

where the integrable function w is not vanishing almost everywhere. Let $(Q_n)_{n\in\mathbb{N}}$ be an estimation method for I and $(\varepsilon_n)_{n\in\mathbb{N}}$ a positive null sequence. Then there is an $f \in C[X]$ for which

$$\lim_{n\to\infty} \frac{R_n[f]}{\varepsilon_n} = 0 \tag{3.5}$$

does not hold.

PROOF. Apply the uniform boundedness principle to $L_n := \varepsilon_n^{-1} R_n$. If (3.5) is true for all $f \in C[X]$, then $\|R_n\| \le \varepsilon_n \cdot \text{const}$; but this would contradict

$$\|R_n\| = \|I\| + \|Q_n\| \ge \|I\|,$$

so there must be an f for which (3.5) does not hold. □

For more precise results pertaining to fast convergence, see Lipow and Stenger (1972).

3.2. The interpolation theorem

As mentioned in the previous section, our basic principle of constructing quadrature rules that satisfy

$$R[U] = \{0\} \tag{3.6}$$

on a given subspace $U \subset C[X]$ is well-founded only if $\|R\|$, and thus $\|Q\|$, is small. The purpose of this section is to show that for an arbitrary finite-dimensional subspace U, we can find a $Q \in \mathbf{Q}$ that satisfies (3.6) with the smallest possible norm $\|I\|$. We will actually prove a more general result which has further applications.

THEOREM 3.2.1 (Pták 1958). *Let U be an m-dimensional subspace of $C[X]$, let I be a linear functional defined on U, and let*

$$M = \sup \{ |I[u]| \ : \ u \in U \text{ and } \|u\| \leq 1 \}$$

(i.e. M is the norm of I on U). Then there exist $x_1, \ldots, x_m \in X$ and $a_1, \ldots, a_m \in \mathbb{R}$ such that

$$I[u] = \sum_{\nu=1}^{m} a_\nu u(x_\nu) \quad \text{for all } u \in U \tag{3.7}$$

and

$$\sum_{\nu=1}^{m} |a_\nu| = M.$$

To prove this theorem, we need three preliminary lemmas.

LEMMA 3.2.1. *Let W be an m-dimensional linear space, and let $f = \sum_{\nu=1}^{s} \alpha_\nu w_\nu$ with $w_\nu \in W$. Then there is a representation*

$$f = \sum_{\nu=1}^{m} \beta_\nu v_\nu \tag{3.8}$$

with $\sum_{\nu=1}^{m} |\beta_\nu| \leq \sum_{\nu=1}^{s} |\alpha_\nu|$ and $\{v_1 \ldots, v_m\} \subset \{w_1, \ldots, w_s\}$. If all the a_ν are positive, then the β_ν may be chosen so that they are also positive.

PROOF. Let $s > m$; otherwise the statement is trivial. The $w_1, \ldots, w_{m+1}$ are linearly dependent. Hence there are γ_ν satisfying

$$\sum_{\nu=1}^{m+1} \gamma_\nu w_\nu = 0 \quad \text{and} \quad \sum_{\nu=1}^{m+1} |\gamma_\nu| > 0,$$

and for arbitrary τ we have that

$$f = \sum_{\nu=1}^{m+1} (\alpha_\nu - \tau\gamma_\nu) w_\nu + \sum_{\nu=m+2}^{s} \alpha_\nu w_\nu \,. \tag{3.9}$$

The function h given by

$$h(\tau) := \sum_{\nu=1}^{m+1} |\alpha_\nu - \tau\gamma_\nu| \quad \text{for } \tau \in \mathbb{R}$$

represents a broken line with vertices at $\alpha_\nu\gamma_\nu^{-1}$ if $\gamma_\nu \neq 0$. The minimum is attained at one of these vertices, say τ_0. Upon choosing $\tau = \tau_0$, one summand in (3.9) vanishes and we have $h(0) \geq h(\tau_0)$. Repeated application of this procedure gives (3.8).

If all the α_ν are positive, we easily see that τ can always be chosen so that one coefficient in (3.9) vanishes while all the other coefficients remain non-negative. □

The *convex hull* of a set $M \subset \mathbb{R}^m$, denoted by $\mathrm{conv}(M)$, is the union of all convex combinations of points in M. In other words,

$$\mathrm{conv}(M) := \left\{ \sum_{i=1}^{p} \alpha_i x_i \;:\; x_i \in M,\ \alpha_i \geq 0 \text{ with } \sum_{i=1}^{p} \alpha_i = 1,\ p \in \mathbb{N} \right\}. \tag{3.10}$$

LEMMA 3.2.2. If M is compact, then $\mathrm{conv}(M)$ is compact.

PROOF. Choose any $x_0 \in M$. Then

$$\mathrm{conv}(M) = \left\{ \sum_{i=1}^{p} \alpha_i x_i \;:\; x_i \in M,\ \alpha_i \geq 0 \text{ with } \sum_{i=1}^{p} \alpha_i = 1,\ p \in \mathbb{N} \right\}$$

$$= \left\{ x_0 + \sum_{i=1}^{p} \alpha_i (x_i - x_0) \;:\; x_i \in M,\ \alpha_i \geq 0 \text{ with } \sum_{i=1}^{p} \alpha_i = 1,\ p \in \mathbb{N} \right\}. \tag{3.11}$$

According to the previous lemma, we can find β_i such that

$$\sum_{i=1}^{p} \alpha_i (x_i - x_0) = \sum_{i \in A} \beta_i (x_i - x_0),$$

where $\beta_i \geq 0$ with $\sum_{i=1}^{m} \beta_i \leq 1$ and $A \subset \{1, 2, \ldots, p\}$ is a set of indices containing at most m elements. Therefore,

$$\sum_{i=1}^{p} \alpha_i x_i = x_0 + \sum_{i \in A} \beta_i (x_i - x_0) = \Big(1 - \sum_{i \in A} \beta_i\Big) x_0 + \sum_{i \in A} \beta_i x_i.$$

Thus, in (3.11), we can restrict to $p \leq m + 1$. Now, suppose that a sequence

$$y_\nu = \sum_{i=1}^{m+1} \alpha_i^{(\nu)} x_i^{(\nu)} \quad \text{such that} \quad \alpha_i^{(\nu)} \geq 0 \text{ with } \sum_{i=1}^{m+1} \alpha_i^{(\nu)} = 1 \quad \text{and} \quad x_i^{(\nu)} \in M$$

is given. By the usual method, we can select a convergent subsequence: take a subsequence $\nu_1, \nu_2, \ldots$ such that $\alpha_1^{(\nu_i)}$ converges, then a convergent subsequence of $x_1^{(\nu_i)}$, and so on. □

LEMMA 3.2.3 (Separation theorem). Let A be a convex compact subset of $\mathbb{R}^m$, and let $y \in \mathbb{R}^m \setminus A$. Then there exists a strictly separating hyperplane; that is, there exist $\beta, \alpha_1, \ldots, \alpha_m \in \mathbb{R}$ such that

$$\sum_{i=1}^{m} \alpha_i a_i < \beta \qquad \text{for all } (a_1, \ldots, a_m) \in A$$

and

$$\sum_{i=1}^{m} \alpha_i y_i > \beta \qquad \text{for } y = (y_1, \ldots, y_m).$$

This lemma is a standard result from convex geometry (see, e.g., Webster 1994: p. 69), so we omit its proof.

PROOF OF THEOREM 3.2.1. Let $\{u_1, \ldots, u_m\}$ be a basis for U, and let

$$D = \{(Mu_1(x), Mu_2(x), \ldots, Mu_m(x)) \,:\, x \in X\} \\ \cup \{(-Mu_1(x), -Mu_2(x), \ldots, -Mu_m(x)) \,:\, x \in X\}.$$

Being a union of two compact sets, D is compact. According to Lemma 3.2.2, $A := \operatorname{conv}(D)$ is also compact. Suppose

$$y := (I[u_1], \ldots, I[u_m]) \notin A.$$

Applying the separation theorem (Lemma 3.2.3) gives $\beta, \alpha_1, \ldots, \alpha_m$ such that

$$\sum_{i=1}^{m} \alpha_i I[u_i] > \beta$$

and

$$\sum_{i=1}^{m} \alpha_i Mu_i(x) < \beta, \qquad -\sum_{i=1}^{m} \alpha_i Mu_i(x) < \beta \qquad \text{for all } x \in X.$$

This means that for $u := \sum_{i=1}^{m} \alpha_i u_i$ we have

$$I[u] > \beta \quad \text{and} \quad M\|u\| < \beta,$$

which yields the contradiction

$$\beta < I[u] \leq M\|u\| < \beta.$$

Hence $y \in A$, and an application of Lemma 3.2.1 gives

$$y = \sum_{i=1}^{r} \beta_i M(u_1(x_i), \ldots u_m(x_i)) - \sum_{i=r+1}^{m} \beta_i M(u_1(x_i), \ldots u_m(x_i))$$

with non-negative β_i satisfying $\sum_{i=1}^{m} \beta_i \leq 1$. Thus, we have shown the existence of a_i (where $a_i = M\beta_i$ for $i \leq r$ and $a_i = -M\beta_i$ for $i > r$) that satisfy (3.7) and

$$\sum_{i=1}^{m} |a_i| = M \sum_{i=1}^{m} \beta_i \leq M.$$

For a sequence $v_1, v_2, \ldots$ with $\|v_\nu\| \leq 1$ and $I[v_\nu] \to M$, we have

$$\sum_{i=1}^{m} |a_i| \geq \sum_{i=1}^{m} a_i v_\nu(x_i) = I[v_\nu] \to M.$$

So we have proved $\sum_{i=1}^{m} |a_i| = M$ and therefore Theorem 3.2.1. □

Next, we aim to show that if I is a positive linear functional, then the a_ν in (3.7) can be chosen to be non-negative. Before doing so, we need to introduce a few more definitions. For $f \in C[X]$, we will take $f > 0$ to mean that $\inf\{f(x) : x \in X\} \geq 0$ and $f(x) \not\equiv 0$. We will say that f is *strictly positive* if the infimum is strictly positive.

DEFINITION 3.2.1. The linear functional I is said to be *positive* (respectively, *strictly positive*) on $U \subset C[X]$ if

$$u > 0 \implies I[u] \geq 0 \quad \text{(respectively, } I[u] > 0).$$

We say simply that I is positive if it is positive on the whole of $C[X]$.

THEOREM 3.2.2 (Tchakaloff 1957). Let U be an m-dimensional subspace of $C[X]$ that contains a strictly positive function u^*. Let I be a positive linear functional defined on U. Then there exist $x_1, \ldots, x_m \in X$ and non-negative numbers $a_1, \ldots, a_m$ such that

$$I[u] = \sum_{i=1}^{m} a_i u(x_i) \qquad \text{for all } u \in U.$$

PROOF. (i) Suppose that the constant functions are in U. We then apply Theorem 3.2.1 and obtain from

$$\sum_{i=1}^{m} a_i = I[1] = \sup\{|I[u]| : \|u\| \leq 1\} = \sum_{i=1}^{m} |a_i|$$

that the a_i are non-negative.

(ii) Under the weaker assumption in the statement of the theorem, we apply (i) to the functional $I^* := I[u^* \cdot\,]$ and the subspace $U^* := \operatorname{span}\{1, u_2/u^*, \ldots, u_m/u^*\}$. Here, $u_2, \ldots, u_m$ are chosen so that $U = \operatorname{span}\{u^*, u_2, \ldots, u_m\}$. This yields

$$I[u] = I\left[u^* \frac{u}{u^*}\right] = I^*\left[\frac{u}{u^*}\right] = \sum_{i=1}^{m} \frac{a_i}{u^*(x_i)} u(x_i) \quad \text{for all } u \in U,$$

which is the desired representation. □

In fact, even the assumption of a strictly positive function can be omitted if it is possible to extend the range of I to the space spanned by U and $f(x) \equiv 1$ without destroying the positivity. (In particular, this is true for the quadrature case; see, e.g., Filippi and Esser 1970). This may be seen as follows: Apply Theorem 3.2.2 to the extended functional I. Restricting again to U, we obtain an $(m+1)$-term representation of the m-dimensional space U. By Lemma 3.2.1, this can then be reduced to an m-term representation.

Another variant of Theorem 3.2.2 was established by Nürnberger (1985): If X has at most m components, then the assumption $\dim U = m$ can be replaced by by $\dim U = m + 1$. Easy examples show that this alternative is not valid if there exists no strictly positive function in U.

The first result in this area was obtained, as early as 1896, by Kowalewski. He proved Theorem 3.2.2 (and somewhat more) in the special case where $I[f] = \int_a^b f(x)\,dx$ and $u^*(x) = 1$.

3.3. Quadrature formulas and one-sided approximation

The explicit determination of those quadrature rules whose existence is given by Tchakaloff's theorem is usually not a straightforward task. One way of approaching this is connected to the problem of one-sided best approximation.

In this section we assume the following setting.

General Assumption. *Let U be a finite-dimensional subspace of $C[X]$ that contains a strictly positive function u^*. Let I be a strictly positive linear functional on U. Finally, let f be a real bounded function on X.*

The problem of one-sided best approximation consists of determining the *best approximation from below*, $\underline{u} \in U$, which is defined by

$$\sup\{I[u] \,:\, u \leq f,\; u \in U\} = I[\underline{u}],$$

or the *best approximation from above*, $\overline{u} \in U$, defined by

$$\inf\{I[u] \,:\, u \geq f,\ u \in U\} = I[\overline{u}].$$

THEOREM 3.3.1. Best approximations from below and from above always exist.

PROOF. It suffices to prove the existence of $\underline{u}$. The set

$$M := \{u \in U \,:\, u \leq f\}$$

is non-empty, since $-\alpha u^*$ is a member for sufficiently large α. Furthermore, we have $\sup\{I[u] \,:\, u \in M\} < \infty$, since there is a β such that $u \leq f \leq \beta u^*$ for all $u \in M$ and therefore $I[u] \leq \beta I[u^*]$. Hence, there is a sequence $\left(u^{(\nu)}\right)_{\nu \in \mathbb{N}}$ satisfying

$$\lim_{\nu \to \infty} I[u^{(\nu)}] = \sup\{I[u] \,:\, u \in M\}.$$

Now, take a basis $u_1, \ldots, u_m$ of U and let

$$u^{(\nu)} = \sum_{i=1}^{m} \alpha_i^{(\nu)} u_i.$$

If the coefficients $\alpha_i^{(\nu)}$ remain bounded, we may choose a convergent subsequence, whose limit will be the desired approximation. If the sequence of coefficients were unbounded, we would be able to find a $\kappa \in \{1, \ldots, m\}$ such that

$$\max_{1 \leq i \leq m} |\alpha_i^{(\nu_\mu)}| = |\alpha_\kappa^{(\nu_\mu)}| \longrightarrow \infty \quad \text{as } \mu \to \infty$$

for an appropriate sequence $\nu_1, \nu_2, \ldots$ of integers. In this case, let us define $v_\mu := |\alpha_\kappa^{(\nu_\mu)}|^{-1} u^{(\nu_\mu)}$, select a convergent subsequence of the v_μ, and denote the limit by v. We then have $I[v] = 0$ and, since $u^{(\nu)} \leq f$, we deduce that $v \leq 0$. If v were negative, we would have also had $I[v] < 0$, owing to the strict positivity of I. Hence, it must be that $v = 0$. But this cannot be true, since v or $-v$ has the representation $u_\kappa + \sum_{i \neq \kappa} \gamma_i u_i$ for some γ_i. □

We denote the set of zeros of a function g by $\mathrm{Zero}(g)$, and we will always assume finiteness of $\mathrm{Zero}(f - \underline{u})$.

THEOREM 3.3.2. Let f be continuous, or let f be non-positive and zero outside some finite set. If $\underline{u}$ is a best approximation of f from below, then there exist $a_\nu \geq 0$ and $x_\nu \in \mathrm{Zero}(f - \underline{u})$ (where $\nu = 1, 2, \ldots, n \leq \dim U$) such that

$$I[u] = \sum_{\nu=1}^{n} a_\nu u(x_\nu) \quad \text{for all } u \in U.$$

An analogous assertion holds for the best approximation of f from above.

The proof requires the following two lemmas.

LEMMA 3.3.1. Let $\underline{u}$ be a best approximation of f from below, where f satisfies the assumptions of Theorem 3.3.2. If $u \in U$ is such that

$$u(x) \geq 0 \quad \text{for all } x \in Z := \mathrm{Zero}(f - \underline{u}),$$

then $I[u] \geq 0$.

PROOF. Suppose that, on the contrary, there is an $u_0 \in U$ such that $u_0(x) \geq 0$ for all $x \in Z$ but $I[u_0] < 0$. Indeed, we may even assume that $u_0(x) > 0$ for all $x \in Z$, since we can always add on a positive multiple of u^* without violating $I[u_0] < 0$. Next, let Z_0 be an open set containing Z such that $u_0(x) \geq c > 0$ for all $x \in Z_0$. Since $f - \underline{u}$ is bounded away from zero on $X \setminus Z_0$ by a positive constant, we can find a $d > 0$ such that $g := f - (\underline{u} - du_0)$ is still positive on $X \setminus Z_0$. Of course, g is also positive on Z_0. So $\underline{u} - du_0$ is a better approximation from below than $\underline{u}$, since $I[\underline{u} - du_0] = I[\underline{u}] - dI[u_0] > I[\underline{u}]$. This is a contradiction. □

LEMMA 3.3.2 (Farkas's lemma; see, e.g., Webster 1994: p. 156).
Let $e, e_1, \dots, e_m \in \mathbb{R}^n$ be such that for any $x \in \mathbb{R}^n$, $e \cdot x \geq 0$ and $e_i \cdot x \geq 0$ for all i. Then there exist $\lambda_1, \dots, \lambda_m$ such that

$$e = \sum_{i=1}^{m} \lambda_i e_i.$$

PROOF OF THEOREM 3.3.2. Suppose that $u_1, \dots, u_n$ form a basis of U, and let $\operatorname{Zero}(f - \underline{u}) = \{t_1, \dots, t_m\}$. Setting

$$e = \big(I[u_1], \dots, I[u_n]\big), \quad e_i = \big(u_1(t_i), \dots, u_n(t_i)\big) \quad \text{and} \quad x = (\xi_1, \dots, \xi_n)$$

in Lemma 3.3.2, we have

$$e \cdot x = I[u] \quad \text{and} \quad e_i \cdot x = u(t_i) \quad \text{for } u = \sum_{\nu=1}^{n} \xi_\nu u_\nu.$$

Lemma 3.3.1 tells us that Farkas's lemma (Lemma 3.3.2) is applicable here, giving

$$I[u_\sigma] = \sum_{\nu=1}^{m} \lambda_\nu u_\sigma(t_\nu).$$

The theorem then follows from Lemma 3.2.1. □

As an application of Theorem 3.3.2, we now prove the existence of a $Q \in \mathbf{Q}_n$ satisfying $R[U] = \{0\}$, where U is some specified subspace of dimension $2n$. This reduction in the number of function evaluations by a factor of two is of obvious practical interest. Our first example, with $U = \mathbb{P}_{2n-1}$ (which yields the classical Gaussian quadrature formulas), will be considered in much greater detail in Chapter 6.

THEOREM 3.3.3. *Let I be strictly positive on $\mathbb{P}_{2n-1} \subset C[a, b]$. Then there exists a non-negative $Q \in \mathbf{Q}_n$ that is exact on $\mathbb{P}_{2n-1}$.*

PROOF. Choose $X = [a, b]$, $U = \mathbb{P}_{2n-1}$ and $f(x) = x^{2n}$ in Theorem 3.3.2. The assertion will follow if we are able to show that $Z := \operatorname{Zero}(f - \underline{u})$ has at most n elements. Counting multiplicities, $f - \underline{u}$ can have at most $2n$ zeros, where zeros in $]a, b[$ are counted twice. Therefore, it might be possible for Z to have more than n elements only if $f - \underline{u}$ has $n - 1$ zeros in $]a, b[$ and an additional zero at each of a and b. In order to remain within the upper bound $2n$, the latter two zeros (at a and at b) would have to be sign changes. Since there cannot be any other zeros, the inequality $f(x) - \underline{u}(x) \leq 0$ has to hold for all $x \geq b$, which is not possible for a polynomial $f - \underline{u}$ with leading coefficient 1. □

Another application of Theorem 3.3.2 concerns spline spaces. We begin with the definition of such a space.

DEFINITION 3.3.1. Let $r \in \{1, 2, \ldots\}$ and $a = \xi_0 < \xi_1 < \xi_2 < \cdots < \xi_m < \xi_{m+1} = b$ be given. A *spline function of degree r with knots $\xi_1 < \cdots < \xi_m$* is a function $f \in C^{(r-1)}[a, b]$ such that[1]

$$f\big|_{[\xi_\nu, \xi_{\nu+1}]} \in \mathbb{P}_r \quad \text{for } \nu = 0, 1, \ldots, m.$$

It can easily be proved that the set $\mathcal{S}_r(\xi_1, \ldots, \xi_m)$ of all spline functions of degree r with knots $\xi_1, \ldots, \xi_m$ is a linear space of dimension $m + r + 1$.

THEOREM 3.3.4 (Karlin and Studden 1966). Let I be strictly positive on $U = \mathcal{S}_r(\xi_1, \ldots, \xi_{2n-r-1}) \subset C[a, b]$, where $a \leq \xi_i \leq b$. Then there exists a non-negative $Q \in \mathbf{Q}_n$ that is exact on U.

This theorem was stated without proof in Schoenberg (1958). For its proof, we need the following two simple lemmas, whose proofs are left to the reader.

LEMMA 3.3.3. Suppose that $s \in C[a, b]$ alternates between being strictly convex and strictly concave on the intervals $[\eta_\nu, \eta_{\nu+1}]$, where $a = \eta_0 < \eta_1 < \cdots < \eta_p = b$. Then s has at most $p + 1$ zeros. If s is strictly convex on $[\eta_0, \eta_1]$ and $s(a) \leq 0$, then s has at most p zeros on $]a, b[$.

LEMMA 3.3.4. Suppose that $s \in C^1[a, b]$ has exactly λ zeros in $]a, b[$. If $s(a) \geq 0$ and $s'(a) > 0$, then s' has at least λ zeros in $]a, b[$.

PROOF OF THEOREM 3.3.4. In Theorem 3.3.2, take $X = [a, b]$, U to be our spline space and $f \in C^r[a, b]$ such that

$$f^{(r+1)}(x) = (-1)^{\nu+r+1} \quad \text{for } x \in]\xi_\nu, \xi_{\nu+1}[, \text{ where } \xi_0 = a \text{ and } \xi_{2n-r} = b.$$

Set $d := f - \underline{u}$; then $d^{(r-1)}$ is continuous and alternates between being strictly convex and strictly concave on the intervals $[\xi_\nu, \xi_{\nu+1}]$. According to Lemma 3.3.3, $d^{(r-1)}$ has at most $2n - r + 1$ zeros and, by Rolle's theorem,

$$d^{(\kappa)} \text{ has at most } 2n - \kappa \text{ zeros.} \tag{3.12}$$

We have to show that d has at most n zeros. Let us suppose that $r > 1$; for the $r = 1$ case, only minor modifications are necessary. First, note that $d^{(\kappa)}$ has only isolated zeros.

Case 1: d has more than n zeros on $[a, b[$. Then, for sufficiently small $\varepsilon > 0$, $d - \varepsilon$ has at least $2n + 1$ zeros, so that d' must have at least $2n$ zeros, which contradicts (3.12).

Case 2: d has $n - 1$ zeros in $]a, b[$ and one each at a and at b. Then, for sufficiently small $\varepsilon > 0$, $d - \varepsilon$ has $2n$ zeros, so that $d^{(\kappa)}$ has at least $2n - \kappa$ zeros in $]a, b[$. From (3.12), it follows that $d^{(\kappa)}$ has exactly $2n - \kappa$ zeros in $]a, b[$. Now, we prove by induction that

$$(-1)^{\kappa+1} d^{(\kappa)}(a) \geq 0 \quad \text{for } \kappa = 1, 2, \ldots, r - 1.$$

This holds at least for $\kappa = 1$, because $d(x) \geq 0$ and $d(a) = 0$. The induction step consists of applying Lemma 3.3.4 to $s := (-1)^{\kappa+1} d^{(\kappa)}$. We have therefore proved that $(-1)^r d^{(r-1)}(a) \geq 0$. Application of Lemma 3.3.3 to $s := (-1)^{r+1} d^{(r-1)}$ then yields $(-1)^{r+1} d^{(r-1)}(a) > 0$ and again a contradiction.

Since both cases lead to a contradiction, d cannot have more than n zeros. □

---

[1]The notation $f\big|_{[u,v]}$ means the restriction of the function f to the interval $[u, v]$.

EXAMPLE 3.3.1. Nikolov (1993) determined the quadrature rules in Theorem 3.3.4 explicitly for $r = 1$ and $I[f] = \int_0^1 f(x)\,dx$. For equidistant knots

$$\xi_\nu = \frac{\nu}{2n-1}, \qquad \nu = 1, \dots, 2n-2,$$

his result reads

$$Q[f] = \frac{2}{2n-1}\left[\frac{3}{4} f\left(\frac{2}{6n-3}\right) + \sum_{i=2}^{n-1} f\left(\frac{4i-3}{4n-2}\right) + \frac{3}{4} f\left(\frac{6n-5}{6n-3}\right)\right].$$

One can directly verify that $R[\mathcal{S}_1(\xi_1, \dots, \xi_{2n-2})] = \{0\}$.

Our next application of Theorem 3.3.2 does not concern a specific subspace U but, rather, a structural property of the coefficients in a quadrature formula.

THEOREM 3.3.5. Let $\mathbf{Q}^+(I, U)$ be the set of non-negative quadrature formulas that coincide with I on $U \subset C[a, b]$. Denote by $A(\eta, Q)$ the coefficient of $f(\eta)$ in $Q \in \mathbf{Q}^+(I, U)$ (which may be zero). Then

$$\sup\{A(\eta, Q) \,:\, Q \in \mathbf{Q}^+(I, U)\} = \inf\{I[u] \,:\, u \in U,\ u > 0,\ u(\eta) = 1\},$$

and the supremum and infimum are attained.

PROOF. Apply Theorem 3.3.2 with $X = [a, b]$,

$$f(x) = \begin{cases} 1 & \text{for } x = \eta, \\ 0 & \text{for } x \neq \eta, \end{cases}$$

and approximation from above. We obtain a $\overline{u} \in U$ satisfying $\overline{u}(\eta) = 1$ and $u > 0$, so there exists $\overline{Q} \in \mathbf{Q}^+(I, U)$ having as nodes the zeros of $\overline{u} - f$ (i.e. η and the zeros of $\overline{u}$).

For each $Q \in \mathbf{Q}^+(I, U)$, we have

$$A(\eta, Q) \leq Q[\overline{u}] = I[\overline{u}] = \inf\{I[u] \,:\, u \in U,\ u > 0,\ u(\eta) = 1\}.$$

In particular,

$$A(\eta, \overline{Q}) = \overline{Q}[\overline{u}] = I[\overline{u}]$$

and so the theorem is proved. □

The next section discusses explicit examples of this theorem.

3.4. Krein's theory

The main purpose of this section is to prove the existence of rules of "double precision", as was done in Section 3.3 for the special subspaces $U = \mathbb{P}_{2n-1}$ and $U = \mathcal{S}_r(\xi_1, \dots, \xi_{2n-r-1})$ (see Theorems 3.3.3 and 3.3.4). Following Krein (1951), we work with Haar spaces. Besides "double precision rules", we shall also investigate some closely related rules whose extremal properties make them interesting.

DEFINITION 3.4.1. An m-dimensional subspace U of $C[a, b]$ is called a *Haar space* if none of its non-vanishing elements has more than $m - 1$ zeros.

We can restrict this definition to $C[a, b]$ rather than considering a general $C[X]$, since Haar spaces exist essentially only on the real line. A prototypical Haar space is $\mathbb{P}_{m-1}$; other examples can be found in Karlin and Studden (1966) and Zielke (1979).

First, let us note some properties of Haar spaces. In the following, U will always denote an m-dimensional Haar space. A fundamental property is that interpolation is always possible.

LEMMA 3.4.1. Suppose $a \le x_1 < x_2 < \cdots < x_m \le b$. For each choice of $\{y_1, \ldots, y_m\}$, there exists exactly one $u \in U$ such that $u(x_i) = y_i$ for $i = 1, \ldots, m$.

PROOF. Let $\{u_1, \ldots, u_m\}$ be a basis of U. Then, writing $u = \sum_{i=1}^n \alpha_i u_i$, we obtain the linear system $\sum_{i=1}^n \alpha_i u_i(x_j) = y_j$ for the unknown coefficients α_i. The lemma follows since, by the definition of a Haar space, the corresponding homogeneous system has only the trivial solution. □

Let us now count double zeros also. We say that $u \in U$ has a double zero $x_0 \in (a, b)$ if $u(x_0) = 0$ and $u(x_0 + \varepsilon)u(x_0 - \varepsilon) > 0$ for all sufficiently small $\varepsilon > 0$. All other zeros are called simple zeros.

LEMMA 3.4.2. Suppose that $u \in U$ has δ double and σ simple zeros. Then either $2\delta + \sigma \le m - 1$ or $u = 0$.

PROOF. Suppose that $2\delta + \sigma > m - 1$ and $u \ne 0$. We can further assume $\delta + \sigma < m$, because otherwise $u = 0$ would follow from the definition. According to Lemma 3.4.1, we can find $u_0 \in U$ such that

$$u_0(x) = \begin{cases} 0 & \text{if } x \text{ is a simple zero of } u, \\ 1 & \text{if } x \text{ is a double zero of } u \text{ and} \\ & \quad u \text{ is non-positive in a neighbourhood of } x, \\ -1 & \text{if } x \text{ is a double zero of } u \text{ and} \\ & \quad u \text{ is non-negative in a neighbourhood of } x. \end{cases}$$

For all sufficiently small ε, the function $u + \varepsilon u_0$ has $2\delta + \sigma \ge m$ zeros, but this is impossible. □

LEMMA 3.4.3. Let L be a functional on $C[a, b]$ of the form

$$L[f] = \sum_{i=1}^p \alpha_i f(x_i), \quad a \le x_1 < \ldots < x_p \le b,$$

which vanishes on U.

(i) If $p \le m$, then we have $\alpha_1 = \cdots = \alpha_p = 0$.
(ii) If $p > m$ and $L \ne 0$, then the sequence $\alpha_1, \ldots, \alpha_p$ has at least m sign changes.

PROOF. (i) Suppose that $p \le m$ but $\alpha_1 \ne 0$. According to Lemma 3.4.1, we can find $u \in U$ with $u(x_1) = 1$ and $u(x_2) = \cdots = u(x_p) = 0$. Obviously, $L[u] \ne 0$, which is a contradiction.

(ii) Let $p > m$ and suppose that the sign changes among $\alpha_1, \ldots, \alpha_p$ occur at α_{ν_k} for $k = 1, \ldots, r$, where $r < m$. (In other words, we assume $a_{\nu_k - 1} a_{\nu_k} < 0$ for $k = 1, \ldots, r$; note that we may assume that $\alpha_i \ne 0$ for all i, since otherwise the corresponding summand could be omitted from the definition of L). We construct $u \in U$ with $u(x_{\nu_1}) = \cdots = u(x_{\nu_r}) = 0$, and prescribe another $m - 1 - r$ zeros as follows: if $m-1-r$ is even, choose the additional zeros from $]x_1,x_2[$; otherwise, choose $m-2-r$ zeros from $]x_1,x_2[$ and the remaining one at a. Finally, prescribe the value 1 at an additional point, so that u does not vanish identically. Thus, all non-vanishing summands in $\sum_{i=1}^{p}\alpha_i u(x_i)$ are of the same sign, and there is at least one such non-vanishing summand. This gives $L[u]\neq 0$, which is a contradiction. □

The notion of index of a set will be very convenient for what follows. We denote by $|A|$ the number of elements of a set A.

DEFINITION 3.4.2. Let $M\subset[a,b]$ be finite. Then

$$\operatorname{ind}(M):=2\,\big|M\cap\,]a,b[\,\big|+\big|M\cap\{a,b\}\big|$$

is called the *index* of M. The index $\operatorname{ind}(Q)$ of a quadrature formula Q is defined to be the index of the set of nodes of Q.

LEMMA 3.4.4. If $\operatorname{ind}(M)<m$, then there exists $u\in U$ such that $u>0$ and $u(x)=0$ for $x\in M$.

PROOF. Let us assume that m is odd; the case of even m is treated very similarly. Let $\{u_1,\dots,u_m\}$ be a basis of U.

(i) Assume that M consists of $k\leq(m-1)/2$ interior points $x_1,\dots x_k$. Choose another $(m-1)/2-k$ points $x_{k+1},\dots,x_{(m-1)/2}$. For each $\varepsilon>0$, construct $u^{(\varepsilon)}=\sum_{i=1}^{m}\alpha_i^{(\varepsilon)}u_i\in U$ such that

$$u^{(\varepsilon)}(a)>0\quad\text{ and }\quad u^{(\varepsilon)}(x_i)=0,\quad u^{(\varepsilon)}(x_i+\varepsilon)=0\quad\text{for } i=1,2,\dots,\frac{m-1}{2},$$

and normalize it by requiring $\sum_{i=1}^{m}|\alpha_i^{(\varepsilon)}|=1$. Now let ε tend to zero and choose from the $u^{(\varepsilon)}$ a convergent subsequence; denote its limit by $u^{(0)}$. It can then be seen that $u^{(0)}$ is non-negative throughout $[a,b]$, that $u^{(0)}(x_i)=0$, and that $u^{(0)}\neq 0$ because of the normalization.

(ii) Now assume that M consists of $k\leq(m-3)/2$ interior points $x_1,\dots,x_k$ and one or two boundary points, Again, we add another $(m-3)/2-k$ interior points and construct $u^{(\varepsilon)}$, with the same normalization as in (i), such that

$$u^{(\varepsilon)}(a)=u^{(\varepsilon)}(b)=u^{(\varepsilon)}(x_i)=u^{(\varepsilon)}(x_i+\varepsilon)=0\quad\text{for } i=1,2,\dots$$

and $u^{(\varepsilon)}(z)<0$ where $z\in(x_1,\min\{x_2,x_1+\varepsilon\})$. The rest of the proof is as in part (i). □

LEMMA 3.4.5. There is a strictly positive $u\in U$.

PROOF. Assume that m is even; the argument for odd m is similar. Choose disjoint sets M_1 and M_2, each consisting of $(m-2)/2$ interior points. According to Lemma 3.4.4, construct a non-negative function u_1 satisfying $u_1(M_1)=\{0\}$ and $u_1(a)=0$, and construct a non-negative function u_2 satisfying $u_2(M_2)=\{0\}$ and $u_2(b)=0$. Lemma 3.4.2 tells us that both functions have only the prescribed zeros, so $u:=u_1+u_2$ must be strictly positive. □

Now that these preliminary results have been established, we return to quadrature theory. Again, we denote by $\mathbf{Q}^+(I,U)$ the set of all non-negative quadrature rules that coincide with the strictly positive functional I on U, and we take U to be an m-dimensional Haar space.

DEFINITION 3.4.3. $Q\in\mathbf{Q}^+(I,U)$ is said to be of *Krein type* if $\operatorname{ind}Q\leq m+1$.

By Lemma 3.4.4 and the strict positivity of I, we see that there is no Krein-type rule with $\operatorname{ind} Q < m$. Obviously, a Krein-type rule has no more than $\lfloor (m+3)/2 \rfloor$ nodes and no more than $\lfloor (m+1)/2 \rfloor$ interior nodes. If a and b are nodes, the rule has $\lfloor (m-1)/2 \rfloor$ interior nodes.

Rules of Krein type yield a lot of information about the set $\mathbf{Q}^+(I, U)$. Before proving the existence of such rules, let us give an example of the information they can provide.

THEOREM 3.4.1. Let Q^{Kr} be of Krein type and let $Q \in \mathbf{Q}^+(I, U)$ be a rule different from Q^{Kr}. Then Q has a node between any two interior nodes of Q^{Kr}. Denote the first and last nodes of Q^{Kr} by y and z, respectively. If $\operatorname{ind} Q = m$ and $y > a$ (respectively, $z < b$), then there is another node of Q in $]a, y]$ (respectively, in $]z, b]$).

PROOF. Apply Lemma 3.4.3 to $L := Q^{\mathrm{Kr}} - Q$. Positive coefficients of L are generated by Q^{Kr} and negative ones by Q. If the assertions of the theorem were not true, we would have too few sign changes in the sequence of coefficients of L. □

COROLLARY 3.4.1. Interior nodes of different Krein-type rules strictly interlace.

Note that we have not proved the existence of Krein-type rules yet. The following fundamental existence theorem says that the set of Krein-type rules is a one-parameter family.

THEOREM 3.4.2 (Krein 1951). For each $\eta \in \,]a, b[$, there is exactly one Krein-type rule Q^η having η as a node. For $\eta \in \{a, b\}$, there is exactly one Krein-type rule Q^η of index m having η as a node. These rules can be characterized uniquely by

$$A(\eta, Q^\eta) = \sup\{A(\eta, Q) \,:\, Q \in \mathbf{Q}^+(I, U)\}.$$

PROOF. According to Theorem 3.3.5, there exists a rule $\widetilde{Q}^\eta$ satisfying

$$A(\eta, \widetilde{Q}^\eta) = I[\widetilde{u}],$$

where $\widetilde{u} > 0$ and $\widetilde{u}(\eta) = 1$. Let M be the set of nodes of $\widetilde{Q}^\eta$. Obviously, $\widetilde{u}$ has to satisfy $\widetilde{u}(M \setminus \{\eta\}) = \{0\}$. From Lemma 3.4.2, we obtain that $\operatorname{ind}(M \setminus \{\eta\}) < m$, i.e. $\operatorname{ind}(M) \le m$ if $\eta \in \{a, b\}$ and $\operatorname{ind}(M) \le m+1$ otherwise. Hence, the existence of a Krein-type rule Q^η is proved: there is at least one example, namely $\widetilde{Q}^\eta$. By the corollary above, the Krein-type rule is unique if $\eta \in \,]a, b[$. If there exist two different Krein-type rules of index m which both have a as one of their nodes, their difference would contradict Lemma 3.4.3. □

We illustrate the derivation of Krein-type rules with the following example.

EXAMPLE 3.4.1. Choose $[a, b] = [-1, 1]$, $I[f] = \int_{-1}^{1} f(x)\, dx$ and $U = \mathbb{P}_2$. Then:

(i) $\eta \in [-1, -\frac{1}{3}\,] \cup [\,\frac{1}{3}, 1]$ gives

$$Q^\eta[f] = \frac{2}{1+3\eta^2}\, f(\eta) + \frac{6\eta^2}{1+3\eta^2}\, f\Big(-\frac{1}{3\eta}\Big).$$

(ii) $\eta \in [-\frac{1}{3}, \frac{1}{3}\,]$ gives

$$Q^\eta[f] = \frac{1+3\eta}{3+3\eta}\, f(-1) + \frac{4}{3-3\eta^2}\, f(\eta) + \frac{1-3\eta}{3-3\eta}\, f(1).$$

For $\eta = \pm\frac{1}{3}$ and $\eta = \pm 1$, these rules are of index 3 ($= m$); otherwise they are of index 4.

First, we express $Q^\eta[f]$ as

$$Q^\eta[f] = \alpha_1 f(-1) + \alpha_2 f(\eta) + \alpha_3 f(1)$$

and determine the α_i via $R[\mathbb{P}_2] = \{0\}$. This yields (ii). If $\eta \notin [-\frac{1}{3}, \frac{1}{3}]$, the resulting rule is not positive and therefore cannot be of Krein type. Hence, in this case we set

$$Q^\eta[f] = \alpha_1 f(\eta) + \alpha_2 f(\xi)$$

and determine α_1, α_2 and ξ appropriately. This yields (i).

Among the Krein-type rules, the rules with $\operatorname{ind} Q = m$ are of special interest. They have the following properties.

THEOREM 3.4.3. There are exactly two Krein-type rules of index m. For even m, one of these rules has no boundary node, while the other has a and b among its nodes. For odd m, one of the rules has a (but not b) as a node, and the other has b (but not a) as a node.

DEFINITION 3.4.4. Let m be even. The Krein-type rule of index m having $m/2$ interior nodes is called a *Gaussian quadrature rule*, denoted by Q^{G}. The Krein-type rule of index m that has $(m-2)/2$ interior nodes and two boundary nodes is called a *Lobatto quadrature rule*, denoted by Q^{Lo}.

DEFINITION 3.4.5. Let m be odd. The Krein-type rules of index m having $(m-1)/2$ interior nodes are called *Radau quadrature rules*. The rule with a as a node is denoted by $Q^{\mathrm{Ra},a}$, and the rule with b as a node is denoted by $Q^{\mathrm{Ra},b}$.

PROOF OF THEOREM 3.4.3. According to Theorem 3.4.2, there is exactly one Krein-type rule of index m having the node a (respectively, b). From the definition of the index, it follows immediately that for even m, b (respectively, a) must also be a node. Similarly, for odd m, b (respectively, a) cannot be a node. Thus, we have already proved the assertions of the theorem for the Lobatto and Radau rules. Furthermore, we see that there can be rules of index m without boundary nodes only when m is even. Lemma 3.4.3 tells us that there can be at most one such rule. Therefore, we only have to show the existence of a Gaussian rule for $m = 2p$, which we shall do in three steps.

(i) Denote by $\overline{\mathbf{Q}}^a$ the set of Krein-type rules without a node at a. First, we show that

$$S := \{y \,:\, Q^\eta \in \overline{\mathbf{Q}}^a \text{ for all } \eta \in \,]a, y[\,\} \neq \emptyset.$$

If this were false, we would be able to find $\varepsilon_1, \varepsilon_2, \dots$ with $\varepsilon_\nu \downarrow 0$ such that $Q^{a+\varepsilon_\nu} \notin \overline{\mathbf{Q}}^a$. Since the existence of a strictly positive $u \in U$ implies that the coefficients of Krein-type rules are uniformly bounded, we may select a convergent subsequence of the $Q^{a+\varepsilon_\nu}$. In the limit, the smallest interior node vanishes, which means that we have created a rule of index no more than $m-1$, and this is a contradiction.

(ii) Now, define $\eta_0 := \sup S$. As in step (i), we may select a sequence of $Q^{\eta_0-\varepsilon_\nu} \in \overline{\mathbf{Q}}^a$ with $\varepsilon_\nu \downarrow 0$ that converges to Q^{η_0}. This shows that $Q^{\eta_0} \in \overline{\mathbf{Q}}^a$.

(iii) Suppose that there exists a sequence of $Q^{\eta_0+\varepsilon_\nu}$, with $\varepsilon_\nu \downarrow 0$, having nodes at a and b. The index of these rules has to be m. In the limit, we obtain Q^{η_0}.

Since we already proved in step (ii) that $Q^{\eta_0} \in \overline{\mathbf{Q}}^a$, i.e. that the node a vanishes, the index of this formula should be $m-1$, which is not possible. Hence, there can be no such sequence of rules with a node at b, and thus Q^{η_0} also has no node at b. Therefore, Q^{η_0} is the Gaussian rule. □

The rules described in Theorem 3.4.3 are "extremal" in many respects. An important result along these lines is the following.

THEOREM 3.4.4. Let Q_1 and Q_2 be the two Krein-type rules of index m for the m-dimensional Haar space U. Furthermore, suppose that U and $f \in C[a,b]$ span an $(m+1)$-dimensional Haar space. Then, for each $Q \in \mathbf{Q}^+(I,U)$ distinct from Q_1 and Q_2, we have that

$$\min\{Q_1[f], Q_2[f]\} < Q[f] < \max\{Q_1[f], Q_2[f]\}. \tag{3.13}$$

PROOF. Let $\underline{u}$ be best approximation of f from below. By Lemma 3.4.2, the zero set of $f - \underline{u}$ has index not exceeding m. By Theorem 3.3.2, there exists a quadrature rule having this zero set as its set of nodes. It is therefore a Krein-type rule of index m; let us call it Q_1. By the positivity of Q, we obtain that

$$0 < Q[f-\underline{u}] = Q[f] - I[\underline{u}] = Q[f] - Q_1[\underline{u}] = Q[f] - Q_1[f].$$

This is the first inequality in (3.13). The second inequality is proved by using the best approximation from above. □

Note that the above technique immediately yields the existence of a Gaussian rule if the Haar space can be extended to a larger subspace by adding an appropriate function f. We used this in the proof of Theorem 3.3.3, with $U = \mathbb{P}_{2n-1}$. In the general case, one has to prove the existence of such an f. This is possible but difficult; see Zielke (1979).

Jetter (1982) uses, instead, a smoothing procedure that yields a new Haar space, for which the extendibility is easily seen.

CHAPTER 4

Co-observations

In this chapter, we present for several types of co-observation the associated error theory and some optimality results. Our main tool will be Peano kernel theory, which is reviewed in Section 4.2.

4.1. Survey

From now on, we shall take the underlying space to be $C[a,b]$, and I will denote the linear functional

$$I[f] := \int_a^b f(x)w(x)\,dx.$$

A set $\mathcal{C} \subset C[a,b]$ will be useful as a co-observation only if enough theory can be derived from it to calculate, or at least estimate, $\rho(Q)$. Of course, whether or not we can be satisfied with a particular choice of Q depends on (approximate) knowledge of $\rho^{\text{opt}}(O)$ and/or ρ_n^{best}. The above conditions are satisfied by the "classical" co-observations

$$\mathcal{C}_M^{(r)} := \{f \in C[a,b] \,:\, \|f^{(r)}\| \le M\}, \tag{4.1}$$

leading to

$$|R[f]| \le \rho(Q, \mathcal{C}_1^{(r)})\, \|f^{(r)}\|. \tag{4.2}$$

By writing $f^{(r)}$, we are implicitly assuming that $f^{(r-1)}$ is absolutely continuous (see Section 4.3). The co-observation (4.1) will be investigated in Section 4.3 and, in greater detail, in Chapter 5. In order to apply (4.2), we need to know bounds for (higher) derivatives. For combinations of functions whose derivatives are known (such as the elementary functions and many special functions), such bounds can often be generated automatically—see, e.g., Moore (1966) or Corliss and Rall (1991); for the applied methods, see Appendix C. When there are difficulties in applying these procedures, the derivative-free error bounds introduced by Davis (1953), based on assuming holomorphy of f in a region $\mathcal{G}$ (enclosing $[a,b]$) of the complex plane, can be used. A typical example is

$$\mathcal{C}_M(\mathcal{G}) := \{f \,:\, f \text{ is holomorphic in } \mathcal{G} \text{ and } \sup_{z\in\mathcal{G}} |f(z)| \le M\}. \tag{4.3}$$

In this case we have to estimate f in $\mathcal{G}$, which can sometimes be easier than calculating bounds for derivatives—consider, for example,

$$f(x) = e^{-x^2}\sqrt{x^2+5}, \qquad \mathcal{G} = \{z \in \mathbb{C} \,:\, |z| < 2\}.$$

Aside from obtaining technically simple error bounds, a study of the co-observation (4.3) is of independent interest owing to the fact that many practical problems naturally lead to holomorphic integrands.

In any case, for problems such as the numerical calculation of

$$\int_0^1 \sqrt{1-x^4}\,dx$$

or

$$\int_0^{e^{-2}} \frac{dx}{|\ln x|}, \tag{4.4}$$

we need further types of co-observation. A useful class is

$$\mathcal{C}_M^{\mathrm{Var}} := \{f \in C[a,b] \,:\, \operatorname{Var} f \le M\}, \tag{4.5}$$

where $\operatorname{Var} f$ means the total variation of f on $[a,b]$. This co-observation is of particular interest since it enables us to deduce certain bounds even when only rough information about the quantitative behaviour of f is available. Suppose, for example, that f consists of k monotone pieces and that $\|f\|_\infty \le B$; then $\operatorname{Var} f \le 2kB$ and we can estimate errors.

There are, of course, continuous functions that do not belong to any of the aforementioned classes. Although such functions may not be of great practical relevance, it is still natural to ask whether there might be universally applicable error bounds. A quantity that is defined for any continuous function is the "modulus of continuity".

DEFINITION 4.1.1. Let $f \in C[a,b]$. The function $\omega(f\,;\cdot)$ given by

$$\omega(f;\delta) := \sup\{|f(x)-f(y)| \,:\, |x-y| \le \delta\}$$

is called the *modulus of continuity* of f.

We now define classes of functions whose moduli of continuity have a common majorant.

DEFINITION 4.1.2. Let $\omega \in C[0,b-a]$ be an increasing sub-additive function (that is, a function satisfying $\omega(\delta_1+\delta_2) \le \omega(\delta_1)+\omega(\delta_2)$) with $\omega(0)=0$. Then, we define

$$\mathcal{C}_\omega^{\mathrm{mc}} := \{f \,:\, \omega(f;\delta) \le \omega(\delta)\}. \tag{4.6}$$

For the special case of $w=1$, in Section 4.5 we will determine for all classes $\mathcal{C}_\omega^{\mathrm{mc}}$ the values ρ_n^{best} and ρ^{opt}. Since each function $f \in C[a,b]$ is in one of these classes, we can always obtain error bounds. However, such bounds are often unrealistically rough because the classes $\mathcal{C}_\omega^{\mathrm{mc}}$ are very large. We illustrate this remark by considering (4.4). In this case we have $\omega(f;\delta) = 1/|\ln\delta|$, and the smallest class of type (4.6) is the one with $\omega(\delta) = 1/|\ln\delta|$. Using the best quadrature rule $Q_n = Q_n^{\mathrm{best}}$ with n nodes for this class, we obtain

$$\rho(Q_n^{\mathrm{best}}) \ge \frac{e^{-2}}{4+2\ln 2n}.$$

In order to guarantee an error of at most $0.5\cdot 10^{-3}$, we have to choose $n > 10^{57}$, which in practical terms is impossible. On the other hand, from the co-observation (4.5) we obtain

$$\rho(Q_n) \le \frac{1}{4e^2 n} \le \frac{0.034}{n}$$

with the same quadrature rule. This shows that accurate computation of the integral (4.4) to three or four digits is no problem with the aid of a computer.

Most error bounds in numerical analysis come from using $\mathcal{C}_M^{(r)}$ as the co-observation. Therefore, we will be placing special emphasis on this class. In quadrature theory, however, it is often advantageous to replace the boundedness of $f^{(r)}$ by a more restrictive condition, namely boundedness of the total variation of $f^{(r)}$. On the one hand, we lose only a little in terms of generality, because bounded functions $f^{(r)}$ with $\operatorname{Var} f^{(r)} = \infty$ rarely occur in practice. On the other hand, the corresponding ρ_n^{best} will be smaller by a factor of order n^{-1}; see (4.7) and (4.8) below. So, we define

$$\mathcal{C}_M^{\text{Var},r} := \{f \in C[a,b] \,:\, \operatorname{Var} f^{(r)} \leq M\},$$

and from this class we obtain estimates of the type

$$\big|R[f]\big| \leq \rho(\mathcal{C}_1^{\text{Var},r}, Q) \operatorname{Var} f^{(r)}.$$

Besides the aforementioned properties, convexity of a function is also of importance. An arbitrary non-trivial class of convex functions is not necessarily symmetric, so the theory of Chapter 2 may not be applicable. Nonetheless, notions such as "optimal" or "best" rules still make sense; see Section 4.10. Note that in (4.4) we have a concave integrand f_0, but considering the integrand $-f$ would allow us to apply the results on convex functions. For the famous Gaussian rule Q_n^{G}, this gives an error estimate

$$\big|R_n[f_0]\big| \leq \frac{2-\ln 2}{2+\ln 2} \cdot \frac{1}{4e^2n^2}$$

(see Corollary 6.5.1).

Finally, let us mention the co-observation

$$\widetilde{\mathcal{C}}_M^{(r)} := \mathcal{C}_M^{(r)} \cap \{f \,:\, f \text{ is } (b-a)\text{-periodic}\}.$$

Periodicity often helps to simplify calculations, and its practical importance, e.g. in the calculation of Fourier coefficients, justifies a thorough investigation.

For a given quadrature problem, it is usually possible to consider the integrand as a member of several different co-observations. For example, if f is holomorphic, then all but (perhaps) the last two co-observations mentioned earlier are possible.

We now give a brief overview of the magnitudes of best error bounds. For simplicity, we restrict our attention to the case where $w(x) = 1$, but similar results hold for much more general weight functions. All the bounds presented here have the best possible order:

$$\rho_n^{\text{best}}(\mathcal{C}_M^{(r)}) \leq \frac{c_r(b-a)^{r+1}}{n^r}M, \quad \text{where } c_r \text{ depends on } r \text{ only;} \tag{4.7}$$

$$\rho_n^{\text{best}}(\mathcal{C}_M(\mathcal{G})) \leq cMq^n, \quad \text{where } q \in \,]0,1[\text{ and } c \text{ depends on } \mathcal{G} \text{ only;}$$

$$\rho_n^{\text{best}}(\mathcal{C}_M^{\text{Var}}) = \frac{(b-a)}{2n}M\,;$$

$$\rho_n^{\text{best}}(\mathcal{C}_\omega^{\text{mc}}) \leq (b-a)\,\omega\Big(\frac{b-a}{2n}\Big)\,;$$

$$\rho_n^{\text{best}}(\mathcal{C}_M^{\text{Var},r}) \leq \frac{c_r(b-a)^{r+1}}{n^{r+1}}M, \quad \text{where } c_r \text{ depends on } r \text{ only.} \tag{4.8}$$

Asymptotically, the larger r is the better is $\mathcal{C}_M^{(r)}$, and $C_M(\mathcal{G})$ is the best class because of the exponential convergence. Although this is important to know for the choice of $\mathcal{C}$, it should be mentioned that it is only an asymptotic statement. In particular, if the sequence $M_r := \|f^{(r)}\|_\infty$ ($r = 1, 2, \ldots$) increases fast, an estimate with small r may be advantageous, as it can guarantee a given precision with less effort (which means a smaller n). We will see such behaviour—not for the best but for the Gaussian method—in Example 6.6.3.

If there is no co-observation specified, then other concepts have to be introduced to enable the selection of a reasonable quadrature rule, because otherwise all rules are considered to be of equal value. Such concepts include the "most plausible interpolant" introduced by Holladay (1957). Among all possible curves through the points $(x_i, f(x_i))$, we usually prefer a smooth one; hence it would seem reasonable to determine the "smoothest" function p satisfying $p(x_i) = f(x_i)$ and use the value $I[p]$ as an estimate. To make this idea precise, we need to introduce a way of measuring (non-)smoothness. Intuitively, non-smoothness or "waviness" is correlated to the rapidity of changes in the first derivative. Therefore, the size of the second derivative is relevant. For practical reasons, it is preferable to define

$$\text{waviness of } f := \int_a^b \left[f''(x)\right]^2 dx,$$

because this leads to simple algorithms. We will examine this issue in more detail in Section 4.7.

Another concept that can be used to develop a co-observation-free theory is the "universality" of a quadrature formula. By a universal rule we mean a rule that is not designed for a particular co-observation but rather for a large set $\mathbf{C}$ of co-observations. For example, we may require that

$$\sup_{\mathcal{C} \in \mathbf{C}} \frac{\rho(Q, \mathcal{C})}{\rho^{\text{opt}}(\mathcal{C})} < \infty$$

(see Chapter 8).

When no co-observation is specified, it is quite common to use a heuristic: Determine $Q_{n_1}[f], Q_{n_2}[f], \ldots$ using different rules and stop when $\left|Q_{n_\nu}[f] - Q_{n_{\nu+1}}[f]\right|$ is within a prescribed bound; then, take the last computed estimate as the approximation. In order to keep the computational effort low, we should try to re-use function values that have already been computed. Methods that are frequently used in practice include those developed by Kronrod or Kronrod and Patterson (see Chapter 6) and those of Romberg (see Chapter 7). Obviously, any given method may produce completely wrong results for certain integrands f. But objecting that such functions f do not really come up in practice presents a logical dilemma: one cannot say simultaneously that we only know function values of f and that certain functions with these values cannot occur.

4.2. The Peano kernel theorem

Here and in what follows, we shall write, for $r = 0, 1, 2, \ldots$,

$$(u)_+^r := \begin{cases} 0 & \text{if } u \le 0, \\ u^r & \text{if } u > 0. \end{cases}$$

Let S be a continuous linear functional on $C[a, b]$ such that $S[\mathbb{P}_{r-1}] = \{0\}$ for some $r \in \{1, 2, \ldots\}$. The norms that we consider will be those of $C[a, b]$. The Peano kernel theorem is the identity

$$S[f] = \int_a^b S[\mathrm{tp}_{r-1}(\cdot - x)] f^{(r)}(x)\,dx, \quad \text{where } \mathrm{tp}_\nu(u) := \frac{(u)_+^\nu}{\nu!}, \tag{4.9}$$

which holds for all functionals S as above. (The symbol "tp" stands for "truncated power function".) See Peano (1913) and also Kowalewski (1932), Schmidt (1935) and von Mises (1936).

For the special case of $S = R$, where R is the error functional of a quadrature rule, the Peano kernel theorem will be an important tool in our investigations. In particular,

$$\big|R[f]\big| \le \|f^{(r)}\| \int_a^b \big|R[\mathrm{tp}_{r-1}(\cdot - x)]\big|\,dx$$

and

$$\rho(Q, \mathcal{C}_M^{(r)}) = M \int_a^b \big|R[\mathrm{tp}_{r-1}(\cdot - x)]\big|\,dx$$

are immediate consequences of the theorem and so we have, in principle, a simple method for determining $\rho(Q, \mathcal{C}_M^{(r)})$ whenever it exists (i.e. when $R[\mathbb{P}_{r-1}] = \{0\}$ holds). Below, our presentation follows that of Brass and Förster (1998).

When $r = 1$, there is a difficulty in (4.9): tp_0 is not continuous and therefore the meaning of $S[\mathrm{tp}_0(\cdot - x)]$ is unclear. We shall overcome this difficulty by a certain limiting process, which was first used by F. Riesz (see, e.g., Riesz and Sz.-Nagy 1956).

DEFINITION 4.2.1. Set

$$w_{\xi,\nu}(x) = \begin{cases} 0 & \text{if } x < \xi, \\ \nu(x - \xi) & \text{if } x \in [\xi,\ \xi + 1/\nu], \\ 1 & \text{if } x > \xi + 1/\nu. \end{cases}$$

The function K_1 defined by

$$K_1(\xi) = \begin{cases} 0 & \text{if } \xi = a, \\ \lim\limits_{\nu\to\infty} S[w_{\xi,\nu}] & \text{if } a < \xi \le b \end{cases}$$

is called the *first Peano kernel* of S.

We need to prove the following:

LEMMA 4.2.1. $\lim\limits_{\nu\to\infty} S[w_{\xi,\nu}]$ exists for all $a < \xi \le b$.

PROOF. We start with

$$S[w_{\xi,\nu}] = S[w_{\xi,1}] + \sum_{\mu=1}^{\nu-1} \big(S[w_{\xi,\mu+1}] - S[w_{\xi,\mu}]\big).$$

The lemma will be proved if we can establish that

$$\sum_{\mu=1}^{\infty} \big|S[w_{\xi,\mu+1}] - S[w_{\xi,\mu}]\big|$$

converges. This convergence follows from the boundedness of

$$\sum_{\mu=1}^{\lambda} \big|S[w_{\xi,\mu+1}] - S[w_{\xi,\mu}]\big| = S\left[\sum_{\mu=1}^{\lambda} \big(w_{\xi,\mu+1} - w_{\xi,\mu}\big) \operatorname{sgn} S[w_{\xi,\mu+1} - w_{\xi,\mu}]\right].$$

Note that the argument of S on the right-hand side is bounded by

$$\sum_{\mu=1}^{\lambda} \left|w_{\xi,\mu+1} - w_{\xi,\mu}\right| = w_{\xi,\lambda+1} - w_{\xi,1} \leq 1.$$

This completes the proof. □

Another result that we will be using is the next lemma.

LEMMA 4.2.2. Let $a = x_0 < x_1 < \cdots < x_n = b$ and set $\gamma := \sup_\nu |x_{\nu+1} - x_\nu|$. Then

$$\left|S[f] - \sum_{\mu=1}^{n-1} (f(x_{\mu+1}) - f(x_\mu)) K_1(x_\mu)\right| \leq \|S\| \, \omega(f\,;\gamma),$$

where $\omega(f\,;\,\cdot)$ denotes the modulus of continuity of f.

PROOF. Let

$$\overline{f}_\nu := f(x_1) + \sum_{\mu=1}^{n-1} [f(x_{\mu+1}) - f(x_\mu)]\, w_{x_\mu,\nu}.$$

For sufficiently large ν such that $x_\mu + 1/\nu < x_{\mu+1}$ for all μ, a short calculation shows that

$$|f(x) - \overline{f}_\nu(x)| \leq \omega(f\,;\gamma).$$

Therefore, using $S[\mathbb{P}_0] = \{0\}$, we obtain

$$\begin{aligned} &\left|S[f] - \sum_{\mu=1}^{n-1} (f(x_{\mu+1}) - f(x_\mu)) K_1(x_\mu)\right| \\ &\quad = \lim_{\nu\to\infty} \left|S[f] - \sum_{\mu=1}^{n-1} (f(x_{\mu+1}) - f(x_\mu)) S[w_{x_\mu,\nu}]\right| \\ &\quad = \lim_{\nu\to\infty} \left|S[f] - S[\overline{f}_\nu]\right| \\ &\quad \leq \lim_{\nu\to\infty} \|S\| \cdot \|f - \overline{f}_\nu\| \leq \|S\| \, \omega(f\,;\gamma) \end{aligned}$$

as asserted. □

THEOREM 4.2.1. K_1 is of bounded variation, and we have

$$\operatorname{Var} K_1 = \|S\|.$$

PROOF. Let $a = x_0 < x_1 < \cdots < x_n = b$. Then

$$\begin{aligned} &\sum_{\mu=0}^{n-1} \left|K_1(x_{\mu+1}) - K_1(x_\mu)\right| \\ &\quad = \lim_{\nu\to\infty} \left(\left|S[w_{x_1,\nu} - 1]\right| + \sum_{\mu=1}^{n-1} \left|S[w_{x_{\mu+1},\nu}] - S[w_{x_\mu,\nu}]\right| \right) \\ &\quad = \lim_{\nu\to\infty} S\Bigg[(w_{x_1,\nu} - 1) \operatorname{sgn} S[w_{x_1,\nu} - 1] \\ &\qquad\qquad + \sum_{\mu=1}^{n-1} \left(w_{x_{\mu+1},\nu} - w_{x_\mu,\nu}\right) \operatorname{sgn} S[w_{x_{\mu+1},\nu} - w_{x_\mu,\nu}] \Bigg]. \end{aligned}$$

For sufficiently large ν, the argument of S on the right-hand side is bounded by

$$|1 - w_{x_1,\nu}(x)| + \sum_{\mu=1}^{n-1} \big|w_{x_{\mu+1},\nu}(x) - w_{x_\mu,\nu}(x)\big| = 1.$$

This implies the inequality

$$\operatorname{Var} K_1 \le \|S\|.$$

To prove the reverse inequality, we shall use Lemma 4.2.2. By partial summation, we obtain

$$S[f] = \sum_{\mu=1}^{n} f(x_\mu)\big[K_1(x_{\mu-1}) - K_1(x_\mu)\big] + \delta \quad \text{where } |\delta| \le \|S\|\, \omega(f;\gamma).$$

Hence

$$\big|S[f]\big| \le \|f\| \operatorname{Var} K_1 + \delta.$$

Since this holds for every decomposition of the interval $[a, b]$, it follows that $\big|S[f]\big| \le \|f\| \operatorname{Var} K_1$, which yields $\operatorname{Var} K_1 \ge \|S\|$. □

The first main result is the following, which is Peano's theorem for the first derivative.

THEOREM 4.2.2. *Let f be absolutely continuous on $[a, b]$. Then*

$$S[f] = \int_a^b f'(u) K_1(u)\, du.$$

PROOF. For $a = x_0 < x_1 < \cdots < x_n = b$, we obtain

$$\begin{aligned}\sum_{\mu=0}^{n-1} \big(f(x_{\mu+1}) - f(x_\mu)\big) K_1(x_\mu) &= \sum_{\mu=0}^{n-1} K_1(x_\mu) \int_{x_\mu}^{x_{\mu+1}} f'(u)\, du \\ &= \int_a^b f'(u) K_1(u)\, du + \sum_{\mu=0}^{n-1} \int_{x_\mu}^{x_{\mu+1}} f'(u)\big(K_1(x_\mu) - K_1(u)\big)\, du.\end{aligned}$$

Estimating the second term on the right-hand side gives

$$\begin{aligned}&\sup_\lambda \int_{x_\lambda}^{x_{\lambda+1}} |f'(u)|\, du \sum_{\mu=0}^{n-1} \sup\big\{|K_1(x_\mu) - K_1(u)| \,:\, x_\mu \le u \le x_{\mu+1}\big\} \\ &\le \sup_\lambda \int_{x_\lambda}^{x_{\lambda+1}} |f'(u)|\, du \operatorname{Var} K_1.\end{aligned}$$

The first factor tends to zero if $\sup\limits_{\mu} |x_{\mu+1} - x_\mu|$ tends to zero. Therefore, the result follows from Lemma 4.2.2. □

A fundamental definition is the following:

DEFINITION 4.2.2. For $s = 2, 3, \ldots, r$, the function K_s defined by

$$K_s(x) := \int_x^b K_{s-1}(u)\, du$$

is called the *sth Peano kernel* of S.

We are now ready to prove the main theorem.

THEOREM 4.2.3 (Peano representation). For $s \in \{1, 2, \ldots, r\}$, let $f^{(s-1)}$ be absolutely continuous on $[a, b]$. Then

$$S[f] = \int_a^b f^{(s)}(u) K_s(u)\, du. \tag{4.10}$$

PROOF. The result can be proved by induction. For $s = 1$, we just have Theorem 4.2.2. Suppose that (4.10) has been proved for $s = \sigma$, and let $S[\mathbb{P}_\sigma] = \{0\}$. Then, using the notation $p_\sigma(u) := u^\sigma$, we have

$$K_{\sigma+1}(a) = \int_a^b K_\sigma(u)\, du = \frac{1}{\sigma!} \int_a^b p_\sigma^{(\sigma)}(u) K_\sigma(u)\, du = \frac{1}{\sigma!} S[p_\sigma] = 0.$$

Since $K_{\sigma+1}(b) = 0$, it follows from integration by parts that

$$S[f] = \int_a^b f^{(\sigma)}(u) K_\sigma(u)\, du = \int_a^b f^{(\sigma+1)}(u) K_{\sigma+1}(u)\, du,$$

which establishes the induction step. □

THEOREM 4.2.4.

$$K_s(x) = S\left[\frac{(\cdot - x)_+^{s-1}}{(s-1)!}\right] \quad \text{for } s > 1.$$

PROOF. Theorem 4.2.3 shows that

$$S\left[\frac{(\cdot - x)_+^{s-1}}{(s-1)!}\right] = \int_a^b (u - x)_+^0 K_{s-1}(u)\, du = \int_x^b K_{s-1}(u)\, du = K_s(x),$$

as claimed. □

We have now completed the proof of the Peano representation (4.9), and will proceed to specialize the results obtained to the case where S is the error of a quadrature rule.

THEOREM 4.2.5. Let

$$R[f] = \int_a^b f(x) w(x)\, dx - \sum_{\nu=1}^{n} a_\nu f(x_\nu)$$

and suppose that $R[\mathbb{P}_{r-1}] = \{0\}$. If $f^{(s-1)}$ is absolutely continuous for $s = 1, \ldots, r$, then

$$R[f] = \int_a^b f^{(s)}(u) K_s(u)\, du \tag{4.11}$$

with

$$K_s(x) = \frac{1}{(s-1)!}\left[\int_x^b (u - x)^{s-1} w(u)\, du - \sum_{\nu=\lambda+1}^{n} a_\nu (x_\nu - x)^{s-1}\right] \tag{4.12}$$

for $x \in\,]x_\lambda, x_{\lambda+1}[$, $\lambda = 0, 1, \ldots, n$ and $x_0 = a$, $x_{n+1} = b$. The Peano kernels K_s, $s \in \{1, 2, \ldots, r\}$, have the following properties:

(i) $K_s\big|_{]x_\lambda, x_{\lambda+1}[}(x) = \dfrac{(-1)^s}{(s-1)!}\left[\displaystyle\int_a^x (x - u)^{s-1} w(u)\, du - \sum_{\nu=1}^{\lambda} a_\nu (x - x_\nu)^{s-1}\right].$

(ii) $K_s(a) = K_s(b) = 0$.

(iii) $K_s(x) = \int_x^b K_{s-1}(x)$ for $s \geq 2$ and, in particular,

- $K_s \in C^{(s-2)}[a,b]$ for $2 \leq s \leq r$;
- $\int_a^b K_s(u)\,du = 0$ for $1 \leq s \leq r-1$.

(iv) In $]a,b[$, K_1 is continuous from the right.

(v) $K_1(x_\nu + 0) - K_1(x_\nu - 0) = a_\nu$, and K_1 is continuous for $x \in [a,b] \setminus \{x_1, \ldots, x_n\}$.

PROOF. Formula (4.12) is a special case of Definition 4.2.1 and Theorem 4.2.4. Property (i) is obtained from Theorem 4.2.4 with the help of the identity

$$(t-x)_+^{s-1} = (t-x)^{s-1} + (-1)^s (x-t)_+^{s-1}$$

(for $s = 1$, this holds for $t \neq x$). Properties (ii), (iii) and (iv) follow immediately, and (v) is a consequence of (i). □

EXAMPLE 4.2.1. In Example 2.2.1, we defined Simpson's rule by

$$Q_3^{\mathrm{Si}}[f] = \frac{b-a}{6}\left[f(a) + 4\,f\left(\frac{a+b}{2}\right) + f(b)\right]$$

and showed that $R_3^{\mathrm{Si}}[\mathbb{P}_3] = \{0\}$. Hence, there exist Peano kernels K_1, K_2, K_3 and K_4. A simple computation gives

$$K_2(a) = K_2(b) = 0, \qquad K_i(x) = (-1)^i K_i(b+a-x)$$

and, for $x \in\,]a,\,(a+b)/2[$,

$$\begin{aligned}
K_1(x) &= \frac{b+5a}{6} - x,\\
K_2(x) &= \frac{1}{6}(x-a)(3x-b-2a),\\
K_3(x) &= -\frac{1}{6}(x-a)^2\left(x - \frac{a+b}{2}\right),\\
K_4(x) &= \frac{1}{72}(x-a)^3(3x-a-2b).
\end{aligned}$$

If R has a Peano kernel K_j without sign changes, then

$$R[f] = f^{(j)}(\xi)\int_a^b K_j(x)\,dx \equiv \text{const}\cdot f^{(j)}(\xi) \tag{4.13}$$

with $\xi \in [a,b]$. For many applications, this representation is more helpful than a mere bound. Therefore, a study of error functionals that admit the representation (4.13) would be useful. We start with the following theorem.

THEOREM 4.2.6. Let R be an error functional of a quadrature rule on $[a,b]$. The following conditions are equivalent:

(i) The Peano kernel K_j of R has no sign changes.

(ii) If $f, g \in C^{(j)}[a,b]$ and $\inf_x g^{(j)}(x) > 0$, then there exists a $\xi \in [a,b]$ such that

$$R[f] = \frac{f^{(j)}(\xi)}{g^{(j)}(\xi)} R[g].$$

(iii) If $f \in C[a,b]$, then there exists a $\xi \in [a,b]$ such that

$$R[f] = \frac{f^{(j)}(\xi)}{j!} R[p_j] \quad \text{where } p_j(x) = x^j.$$

(iv) If $\inf_x f^{(j)}(x) > 0$, then $R[f] \neq 0$.

PROOF. (i) $\Longrightarrow$ (ii):

$$\begin{aligned} R[f] &= \int_a^b f^{(j)}(x) K_j(x)\,dx = \int_a^b \frac{f^{(j)}(x)}{g^{(j)}(x)} g^{(j)}(x) K_j(x)\,dx \\ &= \frac{f^{(j)}(\xi)}{g^{(j)}(\xi)} \int_a^b g^{(j)}(x) K_j(x)\,dx = \frac{f^{(j)}(\xi)}{g^{(j)}(\xi)} R[g]. \end{aligned}$$

(ii) $\Longrightarrow$ (iii): Note that (iii) is a special case of (ii), with $g = p_j$.

(iii) $\Longrightarrow$ (iv): R cannot vanish for all functions. So (iii) implies $R[p_j] \neq 0$ and hence (iv).

(iv) $\Longrightarrow$ (i): Define functions h_1 and h_2 by

$$h_1^{(j)}(x) := \begin{cases} 1 & \text{if } K_j(x) > 0, \\ 0 & \text{if } K_j(x) \leq 0 \end{cases}$$

and

$$h_2^{(j)}(x) := \begin{cases} 0 & \text{if } K_j(x) > 0, \\ 1 & \text{if } K_j(x) \leq 0. \end{cases}$$

Then set

$$h := h_2 R[h_1] - h_1 R[h_2].$$

If (i) were not true, i.e. if K_j were to have sign changes, we would get

$$R[h_2] < 0 \quad \text{and} \quad R[h_1] > 0$$

(using the fact that K_j is at least continuous from the right). We would then obtain $h^{(j)} > 0$ and $R[h] = 0$, which contradicts (iv). □

DEFINITION 4.2.3. R is said to be *definite of order j* if it satisfies (one of) the conditions in Theorem 4.2.6.

EXAMPLE 4.2.2. The generalized trapezoidal rule $Q^{\text{gen Tr}}$ (see Example 2.1.1) with $x_1 = a$ and $x_n = b$ is definite of order two. This can easily be seen from an application of Theorem 4.2.6(iv): $Q^{\text{gen Tr}}[f]$ is obtained by integrating the broken line interpolant s, and $f'' > 0$ gives the strict convexity of f. This implies that $f(x) < s(x)$ for $x \notin \{x_1, x_2, \ldots, x_n\}$, so that $I[f] < I[s] = Q^{\text{gen Tr}}[f]$.

In some cases (e.g. Theorem 6.7.1), the following result gives an improvement of the frequently used representation of definite functionals given in Theorem 4.2.6(iii).

THEOREM 4.2.7 (Brass 1975). Let Q be a quadrature rule on $[-1,1]$ whose associated R is definite of order j, and let $f \in C^{j+2}[-1,1]$. Then there exists $\eta \in [-1,1]$ such that

$$R[f] = \frac{f^{(j)}(0)}{j!}R[p_j] + \frac{f^{(j+1)}(0)}{(j+1)!}R[p_{j+1}] + \frac{f^{(j+2)}(\eta)}{(j+2)!}R[p_{j+2}], \quad \text{where } p_j(x) := x^j.$$

PROOF. Applying Taylor's theorem to $f^{(j)}$ yields

$$\begin{aligned} R[f] &= \int_{-1}^{1} f^{(j)}(x)K_j(x)\,dx \\ &= \int_{-1}^{1} \left(f^{(j)}(0) + xf^{(j+1)}(0) + \frac{1}{2}x^2 f^{(j+2)}(\xi) \right) K_j(x)\,dx \\ &= \frac{f^{(j)}(0)}{j!}R[p_j] + \frac{f^{(j+1)}(0)}{(j+1)!}R[p_{j+1}] + \frac{f^{(j+2)}(\eta)}{(j+2)!}R[p_{j+2}], \end{aligned}$$

where we have used the fact that K_j has no sign changes. □

Another feature of definiteness is that it allows the use of "reduced" functionals.

THEOREM 4.2.8 (Brass and Schmeisser 1981). Let $\eta \in [a,b]$ and suppose that an error functional R satisfying $R[\mathbb{P}_0] = \{0\}$ is given. Define the *reduced functional* $R^{(\eta)}$ by

$$R^{(\eta)}[f] := R[f_\eta] \quad \text{where } f_\eta(x) = f(x)(x-\eta).$$

If $R^{(\eta)}$ is definite of order $j-1$, then R is definite of order j.

PROOF. For $f \in C^j[a,b]$, set

$$g(\eta) := f'(\eta) \qquad \text{and} \qquad g(x) := \frac{f(x)-f(\eta)}{x-\eta} \quad \text{for } x \neq \eta.$$

We obtain inductively that

$$g^{(\nu-1)}(x) = \frac{1}{(x-\eta)^\nu} \int_\eta^x f^{(\nu)}(t)(t-\eta)^{\nu-1}\,dt \quad \text{for } \nu = 1,2,\ldots,j \tag{4.14}$$

(the existence of $g^{(\nu-1)}(\eta)$ follows from, for instance, l'Hôspital's rule).

Now we can use Theorem 4.2.6(iv). If $f^{(j)} > 0$, (4.14) yields $g^{(j-1)} > 0$ and therefore

$$R[f] = R[f - f(\eta)] = R^{(\eta)}[g] \neq 0,$$

which completes the proof. □

Here is an example to explain how reduced functionals are used. Further applications are given in Chapters 5 and 6.

THEOREM 4.2.9 (Steffensen 1921). The rule Q_n with nodes $x_1, \ldots, x_n$ is definite of order $n+1$ if:

(i) $R_n[\mathbb{P}_n] = \{0\}$; and
(ii) the function F defined by

$$F(x) := \int_x^b \prod_{\nu=1}^{n} (u - x_\nu) w(u)\,du$$

has no sign changes on $[a,b]$.

PROOF. The functional

$$R^{(x_1)(x_2)\dots(x_n)}[f] = \int_a^b f(u) \prod_{\nu=1}^{n} (u - x_\nu) w(u)\, du$$

is definite of order one. This follows from looking at its first Peano kernel

$$K_1(x) = \int_a^b (u-x)_+^0 \prod_{\nu=1}^{n} (u - x_\nu) w(u)\, du = \int_x^b \prod_{\nu=1}^{n} (u - x_\nu) w(u)\, du$$

and invoking condition (ii). According to Theorem 4.2.8, $R^{(x_1)(x_2)\dots(x_{n-1})}$ is then definite of order two. Repeating this argument proves the assertion. □

To see how these concepts can be developed further, the reader can refer to Köhler (1996).

The following "bracketing" property is one of the main reasons that rules with definite error functionals have engendered great interest.

Let R and R^* both be definite of order j, and suppose that $R[p_j] > 0$ and $R^*[p_j] < 0$. If $f^{(j)}$ is positive, then we have

$$Q[f] \leq I[f] \leq Q^*[f]. \tag{4.15}$$

This is an immediate consequence of Theorem 4.2.6(iii).

The inequalities (4.15) give a guaranteed inclusion of $I[f]$ without any actual error estimation. This is especially useful for functions that have complicated higher derivatives but which allow, at the same time, a simple proof of the sign-constancy of certain derivatives.

A well-known example of (4.15) is

$$Q_n^{\text{Mi}}[f] \leq I[f] \leq Q_n^{\text{Tr}}[f] \tag{4.16}$$

provided that $f'' > 0$. (For further discussion of this example see Section 7.7.)

Of course, the usefulness of (4.15) depends on how small the interval of uncertainty is. We are thus led to make the following definition.

DEFINITION 4.2.4. Denote by $D^*(n,j)$ the set of all rules $Q \in \mathbf{Q}_n$ that admit an error representation of the form

$$R[f] = c(Q) \cdot f^{(j)}(\xi) \quad \text{where } c(Q) > 0,\ \xi \in [a,b].$$

A rule $\overline{Q} \in D^*(n,j)$ is called an *optimal positive definite rule of order* j if

$$c(\overline{Q}) = \inf\{c(Q) : Q \in D^*(n,j)\}.$$

The analogue for negative definite rules is evident.

The existence, uniqueness and characterizing properties of optimal definite rules have been studied by Schmeisser (1972), Jetter (1976) and Lange (1977, 1979).

In the case of $j = 1$ and $w = 1$, the determination of the optimal definite rules is easy, giving the "rectangular" rules

$$Q_n^{\text{Re},a}[f] := h \sum_{\nu=0}^{n-1} f(a + \nu h) \quad \text{and} \quad Q_n^{\text{Re},b}[f] := h \sum_{\nu=1}^{n} f(a + \nu h)$$

where $h = \dfrac{b-a}{n}$, with

$$R_n^{\text{Re},a}[f] = \frac{(b-a)^2}{2n} f'(\xi) \quad \text{and} \quad R_n^{\text{Re},b}[f] = -\frac{(b-a)^2}{2n} f'(\eta).$$

Likewise, for $j = 2$ and $w = 1$ we obtain a simple result as well. In this case, the optimal positive definite rule is Q_n^{Mi} while the optimal negative definite rule is Q_n^{Tr}, and we have (again by using the Peano kernel theorem)

$$R_n^{\mathrm{Mi}}[f] = \frac{(b-a)^3}{24n^2} f''(\xi) \tag{4.17}$$

and

$$R_n^{\mathrm{Tr}}[f] = -\frac{(b-a)^3}{12(n-1)^2} f''(\xi).$$

To prove the optimality of Q_n^{Mi}, we start with any rule having a non-negative Peano kernel K_2:

$$\int_{x_\nu}^{x_{\nu+1}} K_2(x)\,dx \geq \int_{x_\nu}^{x_{\nu+1}} K_2(x) - K_2\left(\frac{x_\nu + x_{\nu+1}}{2}\right) dx = \frac{(x_{\nu+1}-x_\nu)^3}{24}.$$

Here we have used the fact that $K_2\big|_{[x_\nu, x_{\nu+1}]}$ is a polynomial of degree two with leading coefficient $\frac{1}{2}$ as well as (4.17) with $n = 1$. Therefore,

$$\begin{aligned}
\int_a^b K_2(x)\,dx &= \int_a^{x_1} K_2(x)\,dx + \sum_{\nu=1}^{n-1} \int_{x_\nu}^{x_{\nu+1}} K_2(x)\,dx + \int_{x_n}^b K_2(x)\,dx \\
&\geq \frac{1}{6}(x_1 - a)^3 + \frac{1}{24}\sum_{\nu=1}^{n-1}(x_{\nu+1} - x_\nu)^3 + \frac{1}{6}(b - x_n)^3 \\
&= \frac{n}{24}\left(\frac{1}{2n}(2x_1 - 2a)^3 + \sum_{\nu=1}^{n-1}\frac{1}{n}(x_{\nu+1} - x_\nu)^3 + \frac{1}{2n}(2b - 2x_n)^3\right) \\
&\geq \frac{(b-a)^3}{24n^2},
\end{aligned}$$

where we have applied Jensen's inequality in the last step. A comparison with (4.17) completes the proof.

For $j > 2$ the optimal definite rules are not known explicitly. Asymptotically optimal definite methods are presented in Schmeisser (1972), Lange (1977) and Nikolov (1996); for error bounds see Köhler and Nikolov (1995).

If we do not seek optimal rules, then the Gaussian and Lobatto rules can be used as bracketing rules when j is even, whereas for odd j the Radau rules can be used (see Theorem 6.4.4).

4.3. Bounded derivatives as co-observation

A function $f \in C[a, b]$ is said to be *absolutely continuous* if there is an integrable function g such that

$$f(x) = f(a) + \int_a^x g(u)\,du.$$

We say that g is a *generalized derivative* of f and write f' for g. For example, $g(x) = \operatorname{sgn}(x)$ is a generalized derivative of $f(x) = |x|$. We say, furthermore, that f has a generalized rth derivative if its (ordinary) $(r-1)$st derivative has a generalized derivative.

In the definition of the "classical" co-observation

$$\mathcal{C}_M^{(r)} := \{f \in C[a, b] \,:\, \sup\operatorname{ess}_{a \leq x \leq b} |f^{(r)}(x)| \leq M\},$$

we always take $f^{(r)}$ to mean the generalized rth derivative. As is shown in real analysis, we have

$$\mathcal{C}_M^{(r)} = \{f \in C[a,b] \,:\, f^{(r-1)} \in \operatorname{Lip}_M 1\}.$$

In particular, $\mathcal{C}_M^{(1)} = \operatorname{Lip}_M 1$.

As already mentioned, the most immediate consequence of Peano's representation is the determination of $\rho(Q, \mathcal{C}_M^{(r)})$.

THEOREM 4.3.1. Let Q be a quadrature rule with $R[\mathbb{P}_{r-1}] = \{0\}$, i.e. with rth Peano kernel K_r. Then

$$\rho(Q, \mathcal{C}_M^{(r)}) = M \int_a^b |K_r(x)|\, dx. \tag{4.18}$$

PROOF. From (4.11) we obtain that for all $f \in \mathcal{C}_M^{(r)}$,

$$R[f] \leq M \int_a^b |K_r(x)|\, dx.$$

Equality holds if $f^{(r)} = M \operatorname{sgn} K_r(x)$, which is a measurable and bounded (and therefore integrable) function. □

EXAMPLE 4.3.1 (a continuation of Example 4.2.1). An application of Theorem 4.3.1 leads to

$$\rho(Q_3^{\mathrm{Si}}, \mathcal{C}_M^{(1)}) = \frac{5}{36}(b-a)^2 M, \qquad \rho(Q_3^{\mathrm{Si}}, \mathcal{C}_M^{(2)}) = \frac{1}{81}(b-a)^3 M,$$

$$\rho(Q_3^{\mathrm{Si}}, \mathcal{C}_M^{(3)}) = \frac{1}{576}(b-a)^4 M, \qquad \rho(Q_3^{\mathrm{Si}}, \mathcal{C}_M^{(4)}) = \frac{1}{2880}(b-a)^5 M.$$

Simpson's rule (and generalizations of it—see Section 7.3) has been and still is quite popular. However, it is optimal in neither $\mathcal{C}_M^{(1)}$ nor $\mathcal{C}_M^{(2)}$. For the given observation and $\mathcal{C}_M^{(1)}$, the respective optimal rule is the trapezoidal rule Q_3^{Tr} (see Example 2.1.1); and, for $\mathcal{C}_M^{(2)}$, the corresponding optimal rule on $[-1,1]$ is the one given in Example 2.1.2, where

$$\rho^{\mathrm{opt}} = \frac{2-\sqrt{2}}{6} M = (0.0976\ldots)M < \frac{8}{81} M = (0.0987\ldots)M.$$

For $O[f] = \big(f(-1), f(0), f(1)\big)$, Q_3^{Si} is the only rule that has finite error in both $\mathcal{C}_M^{(3)}$ and $\mathcal{C}_M^{(4)}$; therefore it is optimal.

Saying that a rule is "optimal" means that it cannot be improved upon for a fixed observation. If we fix only the number of nodes used, i.e. the effort needed for the evaluation of a rule, it is possible to obtain a much larger set of estimation rules. Among those rules, some exist that are "better" than Simpson's rule, as shown in the next example.

EXAMPLE 4.3.2. Let

$$I[f] = \int_{-1}^{1} f(x)\, dx \qquad \text{and} \qquad Q[f] = f\left(-\frac{\sqrt{3}}{3}\right) + f\left(\frac{\sqrt{3}}{3}\right).$$

The quadrature rule thus defined is the "second Gaussian rule" Q_2^{G} (we will say more about Gaussian rules in Chapters 5 and 6). It is easily verified that $R[\mathbb{P}_3] = \{0\}$. From Theorems 4.2.5 and 4.3.1 together with some calculations, we get

$$\rho(Q_2^{\mathrm{G}}, \mathcal{C}_M^{(1)}) = \frac{5-2\sqrt{3}}{3}M, \qquad \rho(Q_2^{\mathrm{G}}, \mathcal{C}_M^{(2)}) = \frac{4}{9\sqrt{3}}(2\sqrt{3}-3)^{3/2}M,$$

$$\rho(Q_2^{\mathrm{G}}, \mathcal{C}_M^{(3)}) = \frac{9-4\sqrt{3}}{108}M, \qquad \rho(Q_2^{\mathrm{G}}, \mathcal{C}_M^{(4)}) = \frac{1}{135}M.$$

In all four classes we have $\rho(Q_2^{\mathrm{G}}) < \rho(Q_3^{\mathrm{Si}})$. In this respect, the Gaussian rule is better than Simpson's rule, especially since applying Q_2^{G} requires only two function values, compared with three for Q_3^{Si}.

The result of the above example raises the following question: What is the best quadrature rule in these situations? In $\mathcal{C}_M^{(1)}$ and with $w = 1$, we determined Q^{so} and ρ^{intr} for any observation in Example 2.1.1 and, in Example 2.4.1, obtained the midpoint rule as the best rule, with

$$\rho_n^{\mathrm{best}} = \frac{M(b-a)^2}{4n}.$$

Things are much more complicated for $\mathcal{C}_M^{(r)}$ with $r > 1$. In particular, in attempting to determine Q^{so}, ρ^{intr} and ρ_n^{best} we come across (up to now) insurmountable analytical difficulties. There is only one exception of some interest:

THEOREM 4.3.2 (Kuntzmann 1949). *In $\mathcal{C}_M^{(2)}$ we have that for $w = 1$,*

$$\rho_n^{\mathrm{best}} = \frac{M(b-a)h^2}{32} \quad \text{where } h = \frac{b-a}{n-1+\tau} \quad \text{and } \tau = \frac{\sqrt{3}}{2}.$$

The unique best rule is given by

$$x_\nu = a + \left(\nu - 1 + \frac{\tau}{2}\right)h, \qquad \nu = 1, \dots n,$$

and

$$a_1 = a_n = \frac{\tau+1}{2}h, \qquad a_2 = a_3 = \ldots = a_{n-1} = h.$$

This is a generalized trapezoidal rule (see Example 2.1.1).

PROOF. Let us denote the given quadrature rule by Q_n^{Ku}. After some calculations and with the help of Theorem 4.3.1, we obtain

$$\rho(Q_n^{\mathrm{Ku}}, \mathcal{C}_M^{(2)}) = \frac{M(b-a)h^2}{32}.$$

The remaining assertions will be proved in the next theorem. □

The following theorem gives lower bounds of the correct order and best possible bounds for $r = 2$.

THEOREM 4.3.3. *Let $w = 1$. Then*

$$\rho_n^{\mathrm{best}}(\mathcal{C}_M^{(s)}) \geq \frac{(b-a)^{s+1}M}{4^s s!\left(n + \frac{1}{2}\sqrt[s]{s+1} - 1\right)^s}.$$

PROOF. To prove this theorem, we shall use the following lemma (see, e.g., Cheney 1966: p. 222).

Solotarjeff–Korkin Lemma.

$$\inf\Big\{\int_{-1}^{1}|p(x)|\,dx \;:\; p\in\mathbb{P}_s,\; p^{(s)}(x)=s!\Big\}=2^{1-s}.$$

The restriction of the Peano kernel K_s to $]x_\nu, x_{\nu+1}[$ is a polynomial with leading coefficient $(s!)^{-1}$. Therefore, after an appropriate transformation, the lemma gives

$$\int_{x_\nu}^{x_{\nu+1}}|K_s(x)|\,dx \ge \frac{(x_{\nu+1}-x_\nu)^{s+1}}{2^{2s}s!} \quad \text{for } \nu=1,\dots,n-1.$$

Furthermore, the explicit representation of K_s on the boundary intervals gives

$$\int_a^{x_1}|K_s(x)|\,dx=\int_a^{x_1}\frac{(x-a)^s}{s!}\,dx=\frac{(x_1-a)^{s+1}}{(s+1)!},$$

$$\int_{x_n}^{b}|K_s(x)|\,dx=\int_{x_n}^{b}\frac{(b-x)^s}{s!}\,dx=\frac{(b-x_n)^{s+1}}{(s+1)!}.$$

So we have

$$\int_a^b|K_s(x)|\,dx\ge\frac{(x_1-a)^{s+1}}{(s+1)!}+\sum_{\nu=1}^{n-1}\frac{(x_{\nu+1}-x_\nu)^{s+1}}{2^{2s}s!}+\frac{(b-x_n)^{s+1}}{(s+1)!}.$$

In the notation

$$l_\nu:=x_{\nu+1}-x_\nu,\qquad x_0:=a,\qquad x_{n+1}:=b,$$

we have to minimize

$$\frac{l_0^{s+1}}{(s+1)!}+\sum_{\nu=1}^{n-1}\frac{l_\nu^{s+1}}{2^{2s}s!}+\frac{l_n^{s+1}}{(s+1)!}$$

subject to the condition $\sum_\nu l_\nu=b-a$. To this end, we use Hölder's inequality

$$\sum_{\nu=0}^{n}|c_\nu d_\nu|\le\Big(\sum_{\nu=0}^{n}|c_\nu|^p\Big)^{1/p}\Big(\sum_{\nu=0}^{n}|d_\nu|^q\Big)^{1/q},\quad\text{where } \frac{1}{p}+\frac{1}{q}=1,$$

with

$$p=s+1,\qquad q=\frac{s+1}{s},$$

$$c_0=\frac{2^{2s/(s+1)}l_0}{\sqrt[s+1]{s+1}},\qquad c_n=\frac{2^{2s/(s+1)}l_n}{\sqrt[s+1]{s+1}},\qquad c_\nu=l_\nu\quad\text{for } \nu=1,\dots,n-1$$

and

$$d_\nu=\frac{l_\nu}{c_\nu}.$$

This gives

$$b-a\le\left(\frac{l_0^{s+1}2^{2s}}{s+1}+\sum_{\nu=1}^{n-1}l_\nu^{s+1}+\frac{l_n^{s+1}2^{2s}}{s+1}\right)^{\frac{1}{s+1}}\left(\frac{\sqrt[s]{s+1}}{2^2}+n-1+\frac{\sqrt[s]{s+1}}{2^2}\right)^{\frac{s}{s+1}}$$

or

$$\frac{(b-a)^{s+1}}{2^{2s}s!\big(n+\frac12\sqrt[s]{s+1}-1\big)^s}\le\frac{l_0^{s+1}}{(s+1)!}+\sum_{\nu=1}^{n-1}\frac{l_\nu^{s+1}}{2^{2s}s!}+\frac{l_n^{s+1}}{(s+1)!}\le\int_a^b|K_s(x)|\,dx,$$

which proves the theorem. □

In the more general case where the weight function is positive on an interval, the method used above gives similar bounds (that are of the same order).

It is easy to obtain upper bounds for ρ_n^{best} in $\mathcal{C}_M^{(r)}$. There always exist quadrature methods $Q_{r+1}, Q_{r+2}, \ldots$ with

$$\rho(Q_n, \mathcal{C}_M^{(r)}) = O(n^{-r}). \tag{4.19}$$

This can be proved with the aid of the following important result.

THEOREM 4.3.4. Suppose $R[\mathbb{P}_{m-1}] = \{0\}$. Then, for any $f \in \mathcal{C}_M^{(r)}$,

$$\big|R[f]\big| \le \mathcal{K}_r \frac{(m-r)!}{m!} \left(\frac{b-a}{2}\right)^r \|R\|\, M \tag{4.20}$$

where $\mathcal{K}_r$ denotes the *Favard constant*, defined by

$$\mathcal{K}_r := \frac{4}{\pi} \sum_{\nu=0}^{\infty} \frac{(-1)^{\nu(r+1)}}{(2\nu+1)^{r+1}}.$$

Note that $\mathcal{K}_1 = \pi/2$, $\mathcal{K}_2 = \pi^2/8$, $\mathcal{K}_3 = \pi^3/24$ etc., and that $\mathcal{K}_r \in [\pi^2/8, \pi/2]$.

To deduce (4.19) from (4.20), we need the existence of a sequence Q_m, $m = 1, 2, \ldots$, of quadrature rules with $R_m[\mathbb{P}_{m-1}] = \{0\}$ and $\|R_m\| = O(1)$. This is a special case of Theorem 3.2.1 (the interpolation theorem). More explicit examples will be given in Chapters 5 and 6.

PROOF OF THEOREM 4.3.4. Combining Lebesgue's inequality from Theorem 3.1.1, namely

$$\big|R[f]\big| \le \|R\| \operatorname{dist}(f, \mathbb{P}_{m-1}) \quad \text{provided that } R[\mathbb{P}_{m-1}] = \{0\},$$

with the bound

$$\operatorname{dist}(f, \mathbb{P}_{m-1}) \le \mathcal{K}_r \frac{(m-r)!}{m!} \left(\frac{b-a}{2}\right)^r \|f^{(r)}\|_\infty \tag{4.21}$$

proved by Sinwel (1981) yields the result. □

The bound (4.21) is asymptotically the best possible for $m \to \infty$ and r fixed, but this is not true of (4.20), as we shall see in Chapter 5.

4.4. Bounded variation as co-observation

This section is devoted to the co-observation

$$\mathcal{C}_M^{\text{Var}} := \{f \in C[a,b] \,:\, \operatorname{Var} f \le M\}.$$

It can be regarded as a model co-observation, as it allows many of the hitherto introduced notions to be made concrete and a variety of explicit examples to be constructed.

THEOREM 4.4.1. In $\mathcal{C}_M^{\text{Var}}$, interpolation by natural splines of order one (see Example 2.3.1) is a reconstruction.

PROOF. Let g be the interpolating function; then

$$\operatorname{Var} g = \sum_{\nu=2}^{n} |g(x_\nu) - g(x_{\nu-1})| = \sum_{\nu=2}^{n} |f(x_\nu) - f(x_{\nu-1})| \le \operatorname{Var} f,$$

and this establishes the claim. □

We have shown that application of I to the interpolating natural spline of order one gives a very good estimation rule (with $\left|R[f]\right| \leq 2\rho^{\text{intr}}[f]$; see Theorem 2.2.4). For $w = 1$, the so-constructed rule is the generalized trapezoidal rule (see Example 2.1.1), which is moreover strongly optimal:

THEOREM 4.4.2 (Traub and Lee 1985). For $w \geq 0$ and $\mathcal{C} = \mathcal{C}_M^{\text{Var}}$, we have

$$Q^{\text{so}}[f] = \left(c_0 + \frac{c_1}{2}\right) f(x_1) + \sum_{\nu=2}^{n-1} \frac{c_\nu + c_{\nu-1}}{2} f(x_\nu) + \left(\frac{c_{n-1}}{2} + c_n\right) f(x_n),$$

where

$$c_i = \int_{x_i}^{x_{i+1}} w(x)\, dx \quad \text{for } i = 0, \ldots, n \quad \text{with } x_0 = a,\ x_{n+1} = b,$$

and

$$\rho^{\text{intr}}[f] = \frac{\Delta}{2}\left(M - \sum_{\nu=1}^{n-1} |f(x_\nu) - f(x_{\nu+1})|\right) + \frac{1}{2}\sum_{\nu=1}^{n-1} c_\nu |f(x_\nu) - f(x_{\nu+1})|$$

with

$$\Delta := \max\{2c_0, c_1, c_2, \ldots, c_{n-1}, 2c_n\}. \tag{4.22}$$

Furthermore,

$$\rho^{\text{opt}} = \frac{\Delta M}{2}.$$

PROOF. We write $f_i := f(x_i)$ for $i = 1, \ldots, n$ and set $f_0 := f_{n+1} := -\infty$. Let $g \in \text{Info}(f)$ and

$$u_i := \max\{g(x) \,:\, x_i \leq x \leq x_{i+1}\}.$$

Then

$$\begin{aligned} I[g] &\leq \sum_{\nu=0}^{n} u_\nu c_\nu \\ &= \sum_{\nu=0}^{n} (u_\nu - \max\{f_\nu, f_{\nu+1}\}) c_\nu + \sum_{\nu=0}^{n} \max\{f_\nu, f_{\nu+1}\} c_\nu \\ &\leq \frac{\Delta}{2}\left((u_0 - f_1) + 2\sum_{\nu=1}^{n-1}(u_\nu - \max\{f_\nu, f_{\nu+1}\}) + (u_n - f_n)\right) \\ &\qquad + \sum_{\nu=0}^{n} \max\{f_\nu, f_{\nu+1}\} c_\nu. \end{aligned}$$

Using the identity $2 \cdot \max\{s, t\} = s + t + |s - t|$, we obtain

$$\begin{aligned} I[g] &\leq \frac{\Delta}{2}\Bigg((u_0 - f_1) + \sum_{\nu=1}^{n-1}[2u_\nu - (f_\nu + f_{\nu+1})] \\ &\qquad + (u_n - f_n) - \sum_{\nu=1}^{n-1} |f_\nu - f_{\nu+1}|\Bigg) + \sum_{\nu=0}^{n} \max\{f_\nu, f_{\nu+1}\} c_\nu \\ &\leq \frac{\Delta}{2}\left(M - \sum_{\nu=1}^{n-1} |f_\nu - f_{\nu+1}|\right) + \sum_{\nu=0}^{n} \max\{f_\nu, f_{\nu+1}\} c_\nu. \end{aligned} \tag{4.23}$$

We have to show that this upper bound for $\hat{Q}[f]$ is sharp. For this purpose, we construct a step function $\hat{g}$ satisfying $\operatorname{Var}\hat{g} = M$ and $\hat{g}(x_\nu) = f(x_\nu)$, which can be approximated appropriately by a continuous function. Suppose that

$$\Delta = c_\mu \quad \text{for some } \mu \in \{1, 2, \ldots, n-1\}.$$

Define

$$\hat{g}(x) = \begin{cases} f_\nu & \text{if } x = x_\nu \text{ with } \nu = 1, \ldots, n, \\ f_1 & \text{if } x = x_0, \\ f_n & \text{if } x = x_{n+1}, \\ \max\{f_\nu, f_{\nu+1}\} & \text{if } x \in \,]x_\nu, x_{\nu+1}[\text{ with } \nu \in \{0, \ldots, n\} \setminus \{\mu\}, \\ \max\{f_\mu, f_{\mu+1}\} & \\ \quad + \frac{1}{2}\left(\sum_{\nu=1}^{n-1} |f_\nu - f_{\nu+1}|\right) & \text{if } x \in \,]x_\mu, x_{\mu+1}[. \end{cases}$$

If $2c_0 = \Delta$ or $2c_n = \Delta$, this construction needs to be modified slightly. Note that $I[\hat{g}]$ coincides with the right-hand side of (4.23) and hence with $\hat{Q}[f]$. Also, using (2.1), we obtain $\underset{\vee}{Q}[f]$ and hence $Q^{\text{so}}[f]$ and ρ^{intr}. Theorem 2.1.1 then gives ρ^{opt}. $\square$

We can now minimize (4.22) over all choices of nodes; that is, we can determine the best rule. Theorem 4.4.2 readily yields the following:

THEOREM 4.4.3. For $w \geq 0$ and $\mathcal{C} = \mathcal{C}_M^{\text{Var}}$, we have

$$\rho_n^{\text{best}} = \frac{M}{2n} \int_a^b w(x)\, dx.$$

A best rule is

$$Q_n^{\text{best}}[f] = c \sum_{\nu=1}^{n} f(x_\nu^{\text{best}}), \qquad c = \frac{1}{n} \int_a^b w(x)\, dx,$$

where the x_ν^{best} are chosen such that

$$\int_a^{x_\nu} w(x)\, dx = \left(\nu - \frac{1}{2}\right) c \quad \text{for } \nu = 1, \ldots, n.$$

For $w = 1$, we obtain the midpoint rule Q_n^{Mi} (see Zubrzycki 1963/64). A remarkable special case is where $[a, b] = [-1, 1]$ and $w(x) = \left(1 - x^2\right)^{-1/2}$. Theorem 4.4.3 gives

$$\rho_n^{\text{best}} = \frac{\pi M}{2n}$$

and

$$Q_n^{\text{best}}[f] = \frac{\pi}{n} \sum_{\nu=1}^{n} f\left(-\cos\frac{(2\nu - 1)\pi}{2n}\right).$$

This is the Gaussian quadrature rule for

$$I[f] := \int_{-1}^{1} \frac{f(x)}{\sqrt{1 - x^2}}\, dx \qquad \text{and} \qquad U = \mathbb{P}_{2n-1}$$

(see Section 3.4). It is called the Gauss–Chebyshev rule and will be investigated in more detail in Chapter 6. Of course, this result can be extended to general $[a, b]$ and $w(x) = 1/\sqrt{(x-a)(b-x)}$.

The determination of $\rho(Q, \mathcal{C}_M^{\text{Var}})$ for a given Q can be done systematically in a manner similar to that in $\mathcal{C}_M^{(1)}$. We may assume the existence of the first Peano kernel $\mathcal{K}_1$; otherwise, we would have $\rho(Q, \mathcal{C}_M^{\text{Var}}) = \infty$.

THEOREM 4.4.4. $\rho(Q, \mathcal{C}_M^{\text{Var}}) = \|K_1\| \cdot M$.

PROOF. We obtain

$$\big|R[f]\big| \leq \|K_1\| \cdot \operatorname{Var} f$$

as an easy consequence of Lemma 4.2.2, while

$$\sup\{\big|R[f]\big| \, : \, \operatorname{Var} f \leq 1\} \geq \|K_1\|$$

follows immediately from the definition of K_1. □

EXAMPLE 4.4.1. In Example 4.3.2 we compared Simpson's rule on $[-1, 1]$,

$$Q_3^{\text{Si}}[f] = \frac{1}{6}\left[f(-1) + 4f(0) + f(1)\right],$$

and the Gaussian rule

$$Q_2^{\text{G}}[f] = f\left(-\frac{\sqrt{3}}{3}\right) + f\left(\frac{\sqrt{3}}{3}\right)$$

with respect to the co-observations $\mathcal{C}_M^{(1)}, \ldots, \mathcal{C}_M^{(4)}$. Using Theorem 4.4.4, we can extend the comparison to $\mathcal{C}_M^{\text{Var}}$, obtaining

$$\rho(Q_3^{\text{Si}}, \mathcal{C}_M^{\text{Var}}) = \frac{2}{3}M \qquad \text{and} \qquad \rho(Q_2^{\text{G}}, \mathcal{C}_M^{\text{Var}}) = \frac{\sqrt{3}}{3}M.$$

Again, Q_2^{G} is better.

4.5. Error bounds using the modulus of continuity

In this section we study the classes $\mathcal{C}_\omega^{\text{mc}}$ defined in Section 4.1. The simplest special case is where $\omega(\delta) = M\delta$ with a fixed $M > 0$. The corresponding class is the Lipschitz class $\operatorname{Lip}_M 1$, which is identical to the $\mathcal{C}_M^{(1)}$ discussed in Section 4.3. For general ω, the determination of Q^{so} is difficult; nevertheless, we can obtain good quadrature rules with the method of reconstruction (see Theorem 2.2.4).

THEOREM 4.5.1. Let ω be concave and $\mathcal{C} = \mathcal{C}_\omega^{\text{mc}}$. Then, interpolation by natural splines of order one (see Example 2.3.1) is a reconstruction.

PROOF. Let s be the interpolating spline of f. We have to prove that

$$|s(x) - s(y)| \leq \omega(f; |x - y|) \quad \text{for all } x, y \in [a, b].$$

This is obvious if $x, y \in [x_\nu, x_{\nu+1}]$ for some ν. Write $x = \tau x_\nu + (1 - \tau)x_{\nu+1}$ and $y = \sigma x_\mu + (1 - \sigma)x_{\mu+1}$ with $\tau, \sigma \in [0, 1]$. Suppose without loss of generality that

$\nu < \mu$. Then

$$\begin{aligned}
&|s(x) - s(y)| \\
&\quad = \big|\tau f(x_\nu) + (1-\tau) f(x_{\nu+1}) - \sigma f(x_\mu) - (1-\sigma) f(x_{\mu+1})\big| \\
&\quad = \big|\tau\sigma[f(x_\nu) - f(x_\mu)] + \tau(1-\sigma)[f(x_\nu) - f(x_{\mu+1})] \\
&\qquad + (1-\tau)\sigma[f(x_{\nu+1}) - f(x_\mu)] + (1-\tau)(1-\sigma)[f(x_{\nu+1}) - f(x_{\mu+1})]\big| \\
&\quad \le \tau\big[\sigma\omega(f; x_\mu - x_\nu) + (1-\sigma)\omega(f; x_{\mu+1} - x_\nu)\big] \\
&\qquad + (1-\tau)\big[\sigma\omega(f; x_\mu - x_{\nu+1}) + (1-\sigma)\omega(f; x_{\mu+1} - x_{\nu+1})\big] \\
&\quad \le \tau\omega(f; \sigma x_\mu + (1-\sigma)x_{\mu+1} - x_\nu) + (1-\tau)\omega(f; \sigma x_\mu + (1-\sigma)x_{\mu+1} - x_{\nu+1}) \\
&\quad = \tau\omega(f; y - x_\nu) + (1-\tau)\omega(f; y - x_{\nu+1}) \\
&\quad \le \omega(f; y - \tau x_\nu - (1-\tau)x_{\nu+1}) \\
&\quad = \omega(f; y - x),
\end{aligned}$$

and the assertion is proved. □

For quadrature problems with positive weight functions, the optimal quadrature rules can be found explicitly:

THEOREM 4.5.2 (Lebed 1968). For $w \ge 0$, Q^{opt} in $\mathcal{C}_\omega^{\text{mc}}$ is given by

$$Q^{\text{opt}}[f] = \sum_{\nu=1}^{n} a_\nu f(x_\nu) \quad \text{where } a_\nu = \int_{z_{\nu-1}}^{z_\nu} w(x)\,dx$$

with

$$z_0 = a, \qquad z_n = b \quad \text{and} \quad z_\nu = \frac{x_{\nu+1} + x_\nu}{2} \quad \text{for } \nu = 1, \ldots, n-1.$$

Furthermore,

$$\rho^{\text{opt}}(\mathcal{C}_\omega^{\text{mc}}) = \int_a^b \omega\left(\inf_\nu |x - x_\nu|\right) w(x)\,dx.$$

Note that if $w = 1$, Q^{opt} is the generalized trapezoidal rule (see Example 2.1.1). To prove this theorem, we need two preparatory lemmas.

LEMMA 4.5.1. For $f_1, \ldots, f_k \in \mathcal{C}_\omega^{\text{mc}}$ and $f(x) := \min\{f_1(x), \ldots, f_k(x)\}$, we have $f \in \mathcal{C}_\omega^{\text{mc}}$.

PROOF. Choose $y \in [a, b]$, then there exists $j \in \{1, \ldots, k\}$ such that $f(y) = f_j(y)$, and therefore

$$f_j(x) - f_j(y) \le \omega(|x - y|), \quad \text{i.e.} \quad f_j(x) - f(y) \le \omega(|x - y|).$$

It follows from the definition of f that

$$f(x) - f(y) \le \omega(|x - y|).$$

Since this holds for all $x, y \in [a, b]$, we obtain

$$|f(x) - f(y)| \le \omega(|x - y|),$$

which proves the lemma. □

LEMMA 4.5.2. For

$$f(x) := \omega\left(\inf_\nu |x - x_\nu|\right),$$

we have $f \in \mathcal{C}_\omega^{\text{mc}}$.

PROOF. By the fact that

$$f(x) = \min\{\omega(|x - x_1|), \ldots, \omega(|x - x_n|)\}$$

together with Lemma 4.5.1, we need only show that

$$\omega(|x - x_\nu|) - \omega(|y - x_\nu|) \leq \omega(|x - y|).$$

This follows from the sub-additivity of ω. □

PROOF OF THEOREM 4.5.2. Let $f \in \mathcal{C}_\omega^{\mathrm{mc}} \cap \operatorname{Ker} O$. Then, for each ν,

$$|f(x)| \leq |f(x_\nu)| + \omega(|x - x_\nu|) = \omega(|x - x_\nu|)$$

so that

$$|f(x)| \leq \omega\left(\inf_\nu |x - x_\nu|\right).$$

Using Lemma 4.5.2 and Theorem 2.1.1, we obtain

$$\rho^{\mathrm{opt}}(\mathcal{C}_\omega^{\mathrm{mc}}) = \sup\{I[f] \,:\, f \in \mathcal{C}_\omega^{\mathrm{mc}} \cap \operatorname{Ker} O\} \leq I\left[\omega\left(\inf_\nu |\cdot - x_\nu|\right)\right].$$

It remains to show that the quadrature rule stated in Theorem 4.5.2 has the above error bound. This is done as follows:

$$\begin{aligned}
\left|\int_a^b f(x)w(x)\,dx - \sum_{\nu=1}^n a_\nu f(x_\nu)\right| &= \left|\sum_{\nu=1}^n \int_{z_{\nu-1}}^{z_\nu} [f(x) - f(x_\nu)]w(x)\,dx\right| \\
&\leq \sum_{\nu=1}^n \int_{z_{\nu-1}}^{z_\nu} \omega(|x - x_\nu|)w(x)\,dx \\
&= \int_a^b \omega\left(\inf_\nu |x - x_\nu|\right) w(x)\,dx,
\end{aligned}$$

so the theorem is proved. □

Lebed also obtained an expression for $\rho_n^{\mathrm{best}}(\mathcal{C}_\omega^{\mathrm{mc}})$ from Theorem 4.5.2. Here, we just point out the particularly simple result for $w = 1$:

THEOREM 4.5.3 (Lebed 1968). For $w = 1$ and $\mathcal{C}_\omega^{\mathrm{mc}}$, a best rule is the midpoint rule

$$Q_n^{\mathrm{Mi}}[f] = \frac{b-a}{n} \sum_{\nu=1}^n f\left(a + (\nu - \frac{1}{2})\frac{b-a}{n}\right),$$

and we have

$$\rho_n^{\mathrm{best}}(\mathcal{C}_\omega^{\mathrm{mc}}) = 2n \int_0^{(b-a)/(2n)} \omega(x)\,dx.$$

PROOF. Theorem 4.5.2 gives

$$\begin{aligned}
\rho^{\mathrm{opt}}(\mathcal{C}_\omega^{\mathrm{mc}}) &= \int_a^b \omega\left(\inf_\nu |x-x_\nu|\right) dx \\
&= \int_a^{x_1} \omega(x_1-x)\,dx + \int_{x_1}^{(x_1+x_2)/2} \omega(x-x_1)\,dx \\
&\quad + \int_{(x_1+x_2)/2}^{x_2} \omega(x_2-x)\,dx + \cdots + \int_{x_n}^{b} \omega(x-x_n)\,dx \\
&= \int_0^{x_1-a} \omega(y)\,dy + \int_0^{(x_2-x_1)/2} \omega(y)\,dy \\
&\quad + \int_0^{(x_2-x_1)/2} \omega(y)\,dy + \cdots + \int_0^{b-x_n} \omega(y)\,dy.
\end{aligned}$$

Upon setting $\Omega(x) := \int_0^x \omega(y)\,dy$, this reads

$$\rho^{\mathrm{opt}}(\mathcal{C}_\omega^{\mathrm{mc}}) = \Omega(x_1-a) + 2\sum_{\nu=1}^{n-1} \Omega\left(\frac{x_{\nu+1}-x_\nu}{2}\right) + \Omega(b-x_n).$$

Now, Ω is convex (since ω is increasing) and $\Omega(0)=0$. Hence, the defining inequality for convex functions gives

$$\begin{aligned}
\rho^{\mathrm{opt}}(\mathcal{C}_\omega^{\mathrm{mc}}) &= 2n\left[\frac{1}{2n}\Omega(x_1-a) + \sum_{\nu=1}^{n-1}\frac{1}{n}\Omega\left(\frac{x_{\nu+1}-x_\nu}{2}\right) + \frac{1}{2n}\Omega(b-x_n)\right] \\
&\geq 2n\,\Omega\left(\frac{(x_1-a)+(x_2-x_1)+\cdots+(b-x_n)}{2n}\right) = 2n\,\Omega\left(\frac{b-a}{2n}\right).
\end{aligned}$$

This bound is attained for $x_\nu = a + (\nu - \frac{1}{2})(b-a)/n$ and, by Theorem 4.5.2, the corresponding optimal rule is the midpoint rule. □

Explicit expressions for $\rho(Q, \mathcal{C}_\omega^{\mathrm{mc}})$ in the general case are not known. In the following two theorems, we establish upper bounds which are reasonable or even close to the true error in some important special cases. We can always assume the existence of the first Peano kernel K_1, because $\mathbb{P}_0 \subset \mathcal{C}_\omega^{\mathrm{mc}}$ and therefore $\rho(Q, \mathcal{C}_\omega^{\mathrm{mc}})$ would be infinite if $R[c] \neq 0$ for a constant function c.

Before stating the first theorem, we need another definition.

DEFINITION 4.5.1. $y \in [a,b]$ is called a *generalized zero* of K_1 if $K_1(y) = 0$ or $\operatorname{sgn} K_1(y+0)K_1(y-0) \leq 0$.

THEOREM 4.5.4 (Köhler 1994). Suppose that $a = \xi_0 < \xi_1 < \cdots < \xi_r = b$ are generalized zeros of K_1, and let

$$\delta = \sup\{|\xi_{\nu+1} - \xi_\nu| \,:\, \nu = 0, 1, \ldots, r\}.$$

Then

$$\rho(Q, \mathcal{C}_\omega^{\mathrm{mc}}) \leq \frac{1}{2}\|R\|\,\omega(\delta).$$

PROOF. Let $0 < 2\varepsilon < \min_\nu |\xi_{\nu+1} - \xi_\nu|$, and define

$$I_\nu := [\xi_\nu - \varepsilon,\ \xi_{\nu+1} + \varepsilon] \cap [a,b] \quad \text{and} \quad c_\nu := \frac{1}{2}\Big[\max_{x\in I_\nu} f(x) + \min_{x \in I_\nu} f(x)\Big].$$

It follows that

$$\sup_{x\in I_\nu} |f(x) - c_\nu| \le \frac{1}{2}\,\omega(\xi_{\nu+1} - \xi_\nu + 2\varepsilon) \le \frac{1}{2}\,\omega(\delta + 2\varepsilon). \tag{4.24}$$

For $\nu = 0,\dots,m$, let $S_{\nu,\varepsilon}$ be zero on $]-\infty,\ \xi_\nu - \varepsilon[$, monotonically increasing in $[\xi_\nu - \varepsilon,\ \xi_\nu + \varepsilon]$ and 1 in $]\xi_\nu + \varepsilon, \infty[$. Since ξ_ν is a generalized zero, this $S_{\nu,\varepsilon}$ can be chosen such that $\left|R[S_{\nu,\varepsilon}]\right| < \varepsilon/(2r+1)$. Setting

$$H_\varepsilon := c_0 S_{0,\varepsilon} + \sum_{\mu=1}^{r} (c_\mu - c_{\mu-1}) S_{\mu,\varepsilon},$$

we have

$$\left|f(x) - H_\varepsilon(x)\right| \le \frac{1}{2}\,\omega(\delta + 2\varepsilon).$$

In particular, $H_\varepsilon(x) = c_\mu$ for $x \in\ ]\xi_\mu + \varepsilon,\ \xi_{\mu+1} - \varepsilon[$, and the assertion follows from (4.24). If $x \in [\xi_\mu - \varepsilon,\ \xi_\mu + \varepsilon]$, we obtain from $x \in I_{\mu-1} \cap I_\mu$ and (4.24) that

$$\begin{aligned}
\left|f(x) - H_\varepsilon(x)\right| &= \left|f(x) - \big(1 - S_{\mu,\varepsilon}(x)\big)c_{\mu-1} - S_{\mu,\varepsilon}(x)c_\mu\right| \\
&\le \big(1 - S_{\mu,\varepsilon}(x)\big)\left|f(x) - c_{\mu-1}\right| + S_{\mu,\varepsilon}(x)\left|f(x) - c_\mu\right| \\
&\le \frac{1}{2}\,\omega(\delta + 2\varepsilon).
\end{aligned}$$

The cases where $x \in [\xi_0, \xi_0 + \varepsilon]$ and $x \in [\xi_r - \varepsilon, \xi_r]$ are similar. Consequently,

$$\begin{aligned}
\left|R[f]\right| &\le \left|R[f - H_\varepsilon]\right| + (2r+1)\max_\nu |c_\nu| \cdot \max_\mu \left|R[S_{\mu,\varepsilon}]\right| \\
&\le \frac{1}{2}\,\omega(\delta + 2\varepsilon)\|R\| + \varepsilon\|f\|
\end{aligned}$$

for every $\varepsilon > 0$. □

In the special case of $w = 1$, we obtain a result that is better in general:

THEOREM 4.5.5 (Proinov 1988). *If Q is a positive quadrature formula for $w = 1$ with $R[\mathbb{P}_0] = \{0\}$ and first Peano kernel K_1, then*

$$\rho(Q, \mathcal{C}_\omega^{\mathrm{mc}}) \le \int_a^b \omega(|K_1(x)|)\,dx.$$

PROOF. We have

$$a + a_1 + \cdots + a_i = x_i + K_1(x_i + 0) = x_{i+1} + K_1(x_{i+1} - 0)$$

and therefore

$$
\begin{aligned}
&\left|\int_a^b f(x)\,dx - \sum_{\nu=1}^n a_\nu f(x_\nu)\right| \\
&= \left|\sum_{\nu=1}^n \int_{a+a_1+\cdots+a_{\nu-1}}^{a+a_1+\cdots+a_\nu} [f(x)-f(x_\nu)]\,dx\right| \\
&\le \left|\sum_{\nu=1}^n \int_{a+a_1+\cdots+a_{\nu-1}}^{a+a_1+\cdots+a_\nu} \omega(|x-x_\nu|)\,dx\right| \qquad (4.25)\\
&= \sum_{\nu=1}^n \int_{K_1(x_\nu-0)}^{K_1(x_\nu+0)} \omega(|v|)\,dv \\
&= \sum_{\nu=1}^n \int_{K_1(x_\nu-0)}^{0} \omega(|v|)\,dv + \sum_{\nu=1}^n \int_0^{K_1(x_\nu+0)} \omega(|v|)\,dv \\
&= \left[\int_{K_1(x_1-0)}^{0} + \sum_{\nu=1}^{n-1}\left(\int_{K_1(x_{\nu+1}-0)}^{0} + \int_0^{K_1(x_\nu+0)}\right) + \int_0^{K_1(x_n+0)}\right]\omega(|v|)\,dv \\
&= \left(\int_{K_1(x_1-0)}^{0} + \sum_{\nu=1}^{n-1}\int_{K_1(x_{\nu+1}-0)}^{K_1(x_\nu+0)} + \int_0^{K_1(x_n+0)}\right)\omega(|v|)\,dv.
\end{aligned}
$$

Upon substituting $v = K_1(x)$, the right-hand side becomes

$$
\begin{aligned}
&\int_a^{x_1} \omega(|K_1(x)|)\,dx + \sum_{\nu=1}^{n-1}\int_{x_\nu}^{x_{\nu+1}} \omega(|K_1(x)|)\,dx + \int_{x_n}^b \omega(|K_1(x)|)\,dx \\
&= \int_a^b \omega(|K_1(x)|)\,dx
\end{aligned}
$$

and so the bound in the theorem is established. □

EXAMPLE 4.5.1. Again, we compare Simpson's rule Q_3^{Si} with the Gaussian rule Q_2^{G}. For $f \in \mathcal{C}_\omega^{\mathrm{mc}}$, Proinov's bound reads

$$\left|R_3^{\mathrm{Si}}[f]\right| \le 2\int_0^{1/3} \omega(x)\,dx + 2\int_0^{2/3}\omega(x)\,dx$$

for Simpson's rule and

$$\left|R_2^{\mathrm{G}}[f]\right| \le 2\int_0^{1/\sqrt{3}} \omega(x)\,dx + 2\int_0^{1-1/\sqrt{3}}\omega(x)\,dx$$

for the Gaussian rule. The Gaussian rule gives the better bound, since

$$F(t) := \int_0^t \omega(x)\,dx + \int_0^{1-t}\omega(x)\,dx$$

decreases for increasing $t < 1/2$ and we have $1/3 < 1/\sqrt{3}$. Furthermore, both inequalities are best possible, which can be seen as follows. In the proof of Proinov's bound, apart from (4.25) we have only equalities. Functions from $\mathcal{C}_\omega^{\mathrm{mc}}$ for which equality holds also in (4.25) are

$$f^{\mathrm{Si}}(x) = \begin{cases} \omega(|x|) & \text{for } |x| \le 2/3, \\ \omega(2/3) - \omega(1/3) + \omega(1-|x|) & \text{for } 2/3 < |x| \le 1 \end{cases}$$

for Simpson's rule and

$$f^{\mathrm{G}}(x) = \omega\big(\big||x| - 1/\sqrt{3}\big|\big)$$

for the Gaussian rule.

If ω is a concave function (as is usually the case), it is possible to derive a more manageable result from Proinov's theorem:

THEOREM 4.5.6 (Ligun 1976). Let $w = 1$ and let ω be concave. If Q is positive, then

$$\rho(Q, C_\omega^{\mathrm{mc}}) \leq (b-a)\,\omega\Big(\frac{1}{b-a}\int_a^b |K_1(x)|\,dx\Big).$$

PROOF. For any concave function f, Jensen's inequality says that

$$\frac{1}{b-a}\int_a^b f\big(g(x)\big)\,dx \leq f\Big(\frac{1}{b-a}\int_a^b g(x)\,dx\Big).$$

We prove the assertion by applying this inequality to Theorem 4.5.5 with $f = \omega$ and $g = |K_1|$. □

A general criterion for the sharpness of Proinov's bound is given in Petras (2005). There, it is also proved that Proinov's result is the best possible for all Gaussian rules Q_n^{G}.

Other, more general, bounds are proved in Köhler (1994).

Related to the co-observations discussed in this section are those using the modulus of continuity of derivatives; see Petras (1998b).

4.6. Derivatives of bounded variation

THEOREM 4.6.1. Let Q be a quadrature rule satisfying $R[\mathbb{P}_r] = \{0\}$ and having $(r+1)$st Peano kernel K_{r+1}. Then

$$\rho(Q, C_M^{\mathrm{Var},r}) = M\,\|K_{r+1}\|.$$

PROOF. According to a principal theorem from the theory of functions of bounded variation, the rth derivative of $f \in C_M^{\mathrm{Var},r}$ can be written as

$$f^{(r)} = g_1 - g_2 \quad \text{where } \operatorname{Var} g_1 + \operatorname{Var} g_2 = \operatorname{Var} f^{(r)}$$

and g_1 and g_2 are monotonically increasing. It is therefore sufficient to prove our theorem under the additional assumption of monotonicity of $f^{(r)}$. We shall use the second mean value theorem of integral calculus, which says the following: If u is integrable and v monotonic, then there is a $\xi \in [a,b]$ such that

$$\int_a^b u(x)v(x)\,dx = v(a)\int_a^\xi u(x)\,dx + v(b)\int_\xi^b u(x)\,dx.$$

Therefore, for $f \in C_M^{\mathrm{Var},r}$, Theorem 4.2.3 gives

$$\begin{aligned}
|R[f]| &= \Big|\int_a^b f^{(r)}(x)K_r(x)\,dx\Big| = \Big|f^{(r)}(a)\int_a^\xi K_r(x)\,dx + f^{(r)}(b)\int_\xi^b K_r(x)\,dx\Big| \\
&= \big|-f^{(r)}(a)K_{r+1}(\xi) + f^{(r)}(b)K_{r+1}(\xi)\big| \leq \operatorname{Var} f^{(r)} \cdot \|K_{r+1}\| \\
&\leq M \cdot \|K_{r+1}\|.
\end{aligned}$$

Equality holds for $f \in \mathcal{C}_M^{\mathrm{Var},r}$ such that

$$f^{(r)}(x) = \begin{cases} 0 & \text{if } x \leq \eta, \\ M & \text{if } x > \eta, \end{cases}$$

where η is chosen so that $|K_{r+1}(\eta)| = \|K_{r+1}\|$. □

As an application of this result, we again compare Q_3^{Si} and Q_2^{G} on $[-1,1]$.

EXAMPLE 4.6.1. Using Theorem 4.6.1, after some elementary calculations one obtains

$$\rho(Q_3^{\mathrm{Si}}, \mathcal{C}_M^{\mathrm{Var},1}) = \frac{1}{6}M, \qquad \rho(Q_2^{\mathrm{G}}, \mathcal{C}_M^{\mathrm{Var},1}) = \frac{2\sqrt{3}-3}{6}\,M,$$

$$\rho(Q_3^{\mathrm{Si}}, \mathcal{C}_M^{\mathrm{Var},2}) = \frac{2}{81}M, \qquad \rho(Q_2^{\mathrm{G}}, \mathcal{C}_M^{\mathrm{Var},2}) = \left(\frac{1}{\sqrt{3}} - \frac{1}{2}\right)\left(\sqrt{\frac{2}{\sqrt{3}}} - 1 - \frac{1}{3}\right)M,$$

$$\rho(Q_3^{\mathrm{Si}}, \mathcal{C}_M^{\mathrm{Var},3}) = \frac{1}{72}M, \qquad \rho(Q_2^{\mathrm{G}}, \mathcal{C}_M^{\mathrm{Var},3}) = \frac{9-4\sqrt{3}}{216}\,M.$$

In all cases, the Gaussian rule is better.

Now consider $w = 1$. In $\mathcal{C}_M^{\mathrm{Var}} \equiv \mathcal{C}_M^{\mathrm{Var},0}$, we obtained the midpoint method as the best rule (Theorem 4.4.3). In $\mathcal{C}_M^{\mathrm{Var},1}$, it is possible to obtain the best rule by following the strategy of the proof of Theorem 4.3.2. This leads to

$$\rho_n^{\mathrm{best}} = \frac{(b-a)^2}{16\left(n - 1 + \frac{1}{\sqrt{2}}\right)^2}\,M$$

(Stern 1967). This best rule looks similar to the one described in Theorem 4.3.2, except that here we set $\tau = 1/\sqrt{2}$. The best rules in $\mathcal{C}_M^{\mathrm{Var},2}$ were determined by Korneichuk and Luspai (1969) and, in simpler form, by Arnold (1977).

Theorem 4.6.1 in conjunction with a later result, Theorem 5.7.1, gives

$$R_n[f] = O(n^{-r-1}) \quad \text{if } f \in \mathcal{C}_M^{\mathrm{Var},r}$$

for a wide variety of quadrature methods. But, for any quadrature method, there exists $f \in \mathcal{C}_M^{\mathrm{Var},r} \cap C^{(r)}$ such that

$$\limsup_{n\to\infty} n^{r+1}\big|R_n[f]\big| > 0$$

(Petras 1994b).

4.7. Sard's co-observation

The Sard classes are defined by

$$\mathcal{C}_M^{\mathrm{Sa},r} := \Big\{f : \int_a^b \big[f^{(r)}(x)\big]^2\,dx \leq M^2\Big\}.$$

These were introduced in Sard (1949b, 1963) to enable the construction of Q^{opt} in a simpler manner than was possible in $\mathcal{C}_M^{(r)}$ or $\mathcal{C}_M^{\mathrm{Var},r}$. Sard used Peano's theorem and the Cauchy–Schwarz inequality to obtain

$$\big|R[f]\big|^2 \leq \int_a^b \big[f^{(r)}(x)\big]^2\,dx \cdot \int_a^b \big[K_r(x)\big]^2\,dx$$

and then minimized the second factor as a function of the coefficients. Schoenberg (1964) derived the same rules, but as projection rules corresponding to a space of natural splines. We begin with the definition of this space of splines.

DEFINITION 4.7.1. Let $r \in \{1, 2, \ldots\}$ and $x_1 < x_2 < \cdots < x_m$ be given. The real-valued function s defined on $]-\infty, \infty[$ is called a *natural spline function* of degree $2r-1$ with knots $x_1, \ldots, x_m$ if:

(i) $s \in C^{2r-2}$;
(ii) $s\big|_{[x_\nu, x_{\nu+1}]} \in \mathbb{P}_{2r-1}$ for $\nu = 1, 2, \ldots, m-1$;
(iii) $s\big|_{]-\infty, x_1]} \in \mathbb{P}_{r-1}$;
(iv) $s\big|_{[x_m, \infty]} \in \mathbb{P}_{r-1}$.

The space of all such functions is denoted by $\mathcal{S}^{\text{nat}}_{2r-1}(x_1, \ldots, x_m)$.

Recall that natural splines of degree one were defined already in Example 2.3.1. There, we obtained the special $r = 1$ case of the following result.

THEOREM 4.7.1 (Schoenberg 1964). In $C_M^{\text{Sa},r}$, the strongly optimal rule Q_n^{so} with $n \geq r$ is characterized by being exact on $\mathcal{S}^{\text{nat}}_{2r-1}(x_1, \ldots, x_n)$.

PROOF. This is a special case of Theorem 2.3.1 with the semi-definite inner product

$$\varphi(f, g) := \int_a^b f^{(r)}(x) g^{(r)}(x)\, dx.$$

We have to prove that $\mathcal{S}^{\text{nat}}_{2r-1}(x_1, \ldots, x_n)$ is the spline space in our situation, which is equivalent to showing that

$$\dim \mathcal{S}^{\text{nat}}_{2r-1}(x_1, \ldots, x_n) = n \tag{4.26}$$

(provided that the functions are restricted to $[a, b]$) and

$$\int_a^b f^{(r)}(x) s^{(r)}(x)\, dx = 0 \tag{4.27}$$

if $s \in \mathcal{S}^{\text{nat}}_{2r-1}(x_1, \ldots, x_n)$, $f^{(r-1)}$ is absolutely continuous and $f(x_\nu) = 0$ for $\nu = 1, \ldots, n$. Equation (4.27) can be proved via repeated integration by parts:

$$\int_a^b f^{(r)}(x) s^{(r)}(x)\, dx = \int_{x_1}^{x_n} f^{(r)}(x) s^{(r)}(x)\, dx = \cdots$$

$$\cdots = \int_{x_1}^{x_n} f''(x) s^{(2r-2)}(x)\, dx = \sum_{\nu=1}^{n-1} s^{(2r-1)}(x_\nu + 0) \int_{x_\nu}^{x_{\nu+1}} f'(x)\, dx = 0.$$

To verify (4.26), we use $n + 2r$ parameters $\alpha_1, \ldots, \alpha_n$ and $\beta_0, \ldots, \beta_{2r-1}$ to express $s \in \mathcal{S}^{\text{nat}}_{2r-1}(x_1, \ldots, x_n)$ as

$$s(x) = \sum_{\nu=1}^{n} \alpha_\nu (x - x_\nu)_+^{2r-1} + \sum_{\nu=0}^{2r-1} \beta_\nu x^\nu$$

with

$$s^{(\nu)}(x_1 - 0) = s^{(\nu)}(x_n + 0) = 0 \quad \text{for } \nu = r, r+1, \ldots, 2r-1.$$

The latter $2r$ conditions are easily seen to be independent, so (4.26) follows. □

EXAMPLE 4.7.1. Let $w = 1$. We want to determine Q^{so} in $\mathcal{C}_M^{\text{Sa},2}$ with the nodes $-1, \xi, 1$ on $[-1, 1]$. According to Theorem 4.7.1, we have to find a quadrature rule which is exact on a basis of $\mathcal{S}_3^{\text{nat}}(-1, \xi, 1)$, for example on

$$s_1(x) = 1, \qquad s_2(x) = x, \qquad s_3\big|_{[-1,1]}(x) = (1-\xi)(x+1)^3 - 2(x-\xi)_+^3.$$

The basis element s_3 can be extended to a natural spline, since $s_3''(-1+0) = s_3''(1-0) = 0$. A short calculation yields

$$Q_3[f] = \left(1 - \frac{5-\xi^2}{8(1+\xi)}\right) f(-1) + \frac{5-\xi^2}{4(1-\xi^2)} f(\xi) + \left(1 - \frac{5-\xi^2}{8(1-\xi)}\right) f(1). \tag{4.28}$$

The most interesting case is the one where $\xi = 0$:

$$Q_3[f] = \frac{3}{8} f(-1) + \frac{5}{4} f(0) + \frac{3}{8} f(1). \tag{4.29}$$

This quadrature rule was given in Sard (1949b) along with some other rules for equidistant nodes and for $r = 2, 3$ and 4. For (4.29), the determination of the refined error bound (2.35) is not too complicated and gives

$$\rho^{\text{intr}} = \left(1 - \frac{3}{2M^2}\left[f(-1) - 2f(0) + f(1)\right]^2\right)^{1/2} \rho^{\text{opt}},$$

where

$$\rho^{\text{opt}} = \frac{M}{\sqrt{160}}$$

is computed by using Peano's theorem.

Apart from optimality, there is a further motive for being interested in Sard's rules. We argued in Section 4.1 that in the absence of any co-observation, a reasonable method for obtaining an estimate of $I[f]$ entails computing $I[g]$, where among all of the functions interpolating f, g is the smoothest in the sense that

$$\int_a^b \left[g''(x)\right]^2 dx = \inf\left\{\int_a^b \left[h''(x)\right]^2 dx \; : \; h(x_\nu) = f(x_\nu) \text{ for } \nu = 1, \ldots, n\right\}. \tag{4.30}$$

This concept was first formulated by Holladay (1957). Schoenberg (1964) demonstrated its connection with Sard's rules:

THEOREM 4.7.2. *The quadrature rule $Q[f] = I[g]$, where g is determined by (4.30), coincides with Sard's rule for $r = 2$.*

PROOF. See Theorem 2.3.2. □

The method of Sard and Holladay gives useful rules (which we will refer to as SH rules) for any distribution of nodes. This is an important advantage over the more popular method of constructing rules $Q \in \mathbf{Q}_n$ by using the condition $R[\mathbb{P}_{n-1}] = \{0\}$, which can lead to quite bad rules (see Example 3.1.1 and Chapter 5).

By choosing $\xi = 2/5$ in (4.28), we can see that SH rules are not necessarily positive. Nevertheless, we have convergence. Obviously, a necessary condition for the convergence of any quadrature method based on the nodal system $x_{1,n} < \cdots < x_{n,n}$, $n = 1, 2, \ldots$, is

$$\lim_{n\to\infty} \sup_{\nu} |x_{\nu+1,n} - x_{\nu,n}| = 0. \tag{4.31}$$

Suppose, for simplicity, that $w = 1$, $x_1 = a$ and $x_n = b$; then the generalized trapezoidal rule has a second Peano kernel K_2. A short calculation gives

$$\int_a^b K_2^2(x)\,dx = \frac{1}{120}\sum_{\nu=1}^{n-1}(x_{\nu+1}-x_\nu)^5 \le \frac{b-a}{120}\bigl(\sup_\nu |x_{\nu+1}-x_\nu|\bigr)^4,$$

which shows that

$$|R[f]| \le \sqrt{\frac{b-a}{120}}\Bigl(\sup_\nu |x_{\nu+1}-x_\nu|\Bigr)^2 \cdot \Bigl\{\int_a^b [f''(x)]^2\,dx\Bigr\}^{1/2}. \tag{4.32}$$

Of course, this bound for the generalized trapezoidal rule also holds for the optimal SH rule (Köhler 1989a). Thus, we have proved convergence, provided that the necessary condition (4.31) is satisfied and

$$\int_a^b \bigl[f''(x)\bigr]^2\,dx < \infty.$$

If Q_n is any quadrature rule with $x_1 = a$, $x_n = b$ and second Peano kernel K_2, then we have

$$\begin{aligned}\int_a^b K_2^2(x)\,dx &= \sum_{\nu=1}^{n-1}\int_{x_\nu}^{x_{\nu+1}} K_2^2(x)\,dx \\ &\ge \sum_{\nu=1}^{n-1}\min_{\alpha_\nu,\beta_\nu}\int_{x_\nu}^{x_{\nu+1}}\Bigl(\frac{1}{2}x^2-\alpha_\nu x-\beta_\nu\Bigr)^2\,dx = \sum_{\nu=1}^{n-1}\frac{(x_{\nu+1}-x_\nu)^5}{720},\end{aligned}$$

where the minimization can be done easily via an affine transformation of

$$\min_{\alpha,\beta}\int_{-1}^{1}\Bigl(\frac{1}{2}x^2-\alpha x-\beta\Bigr)^2\,dx = \frac{8}{45},$$

which is a simple matter to check. So we lose at most a factor of $\sqrt{6}$ in the error bound by using the generalized trapezoidal rule instead of the SH rule. Moreover, for equidistant nodes, Kershaw (1972) proved that

$$\int_a^b \bigl[K_2^{\mathrm{SH}}(x)\bigr]^2\,dx = \frac{b-a}{720}\Bigl(\frac{b-a}{n-1}\Bigr)^4 + O(n^{-5}).$$

This small gain hardly justifies the drawback of using a method with a much more complicated structure.

For further relevant results, see Köhler (1988b, 1989b).

4.8. Co-observations of Davis type

Davis (1953) introduced a class of co-observations that has attracted a great deal of interest; see also Davis and Rabinowitz (1954) and Davis (1962). These co-observations are defined as balls in Hilbert spaces of analytic functions, where the inner product is given by a line or area integral. Using the theory of Section 2.3, we can construct optimal rules for such co-observations. Even more important is the possibility of obtaining "derivative-free" error bounds for any rule (in contrast to the co-observations of Section 4.3, which had been used almost exclusively up to that time).

Theorem 2.3.4 is fundamental to obtaining the pertinent error bounds. It requires knowledge of an orthonormal basis. We now give some concrete examples of

Hilbert spaces $\mathcal{H}$ together with their bases h_ν, $\nu = 0, 1, \dots$. In all cases, we take $[-1, 1]$ to be the fundamental interval for the quadrature rule and let $r > 1$ be fixed.

Special Case I (cf. Example 2.3.2)

$$\mathcal{H} := \{f : f \text{ is analytic for } |z| < r \text{ and } \int_{|z|=r} |f(z)|^2 \, ds < \infty\},$$

$$\varphi(f, g) := \int_{|z|=r} f(z)\overline{g(z)} \, ds, \qquad h_\nu(z) = \frac{z^\nu}{r^\nu \sqrt{2\pi r}}.$$

The integral is understood to mean $\lim\limits_{\rho \to r-0} \int_{|z|=\rho} \cdots$, and similarly in the following examples.

Special Case II

$$\mathcal{H} := \left\{f : f \text{ is analytic for } |z| < r \text{ and } \int_{|z|\le r} |f(z)|^2 \, dx \, dy < \infty\right\},$$

$$\varphi(f, g) := \int_{|z|\le r} f(z)\overline{g(z)} \, dx \, dy, \qquad h_\nu(z) = \frac{z^\nu \sqrt{\nu + 1}}{r^{\nu+1}\sqrt{\pi}}.$$

For the description of the next cases as well as in many future occasions, we will need the following special region in the complex plane:

DEFINITION 4.8.1. $\mathcal{E}_r := \{z : 2z = \zeta + \zeta^{-1},\ 1 \le |\zeta| < r\}$.

The boundary of $\mathcal{E}_r$ is an ellipse with foci ± 1 and semi-axes lengths $(r \pm r^{-1})/2$. We shall denote the boundary of $\mathcal{E}_r$ by $\partial\mathcal{E}_r$. Note that $\mathcal{E}_r \subset \mathcal{E}_{r'}$ if $r' > r$ and $\lim\limits_{r\to 1} \mathcal{E}_r = [-1, 1]$.

Special Case III

$$\mathcal{H} := \left\{f : f \text{ is analytic in } \mathcal{E}_r \text{ and } \int_{\mathcal{E}_r} |f(z)|^2 \, dx \, dy < \infty\right\},$$

$$\varphi(f, g) := \int_{\mathcal{E}_r} f(z)\overline{g(z)} \, dx \, dy, \qquad h_\nu(z) = \frac{2\sqrt{\nu+1}\, U_\nu(z)}{\sqrt{\pi}\left(r^{2\nu+2} - r^{-2\nu-2}\right)^{1/2}}.$$

(For the definition of U_ν, see Appendix A.)

Special Case IV

$$\mathcal{H} := \left\{f : f \text{ is analytic in } \mathcal{E}_r \text{ and } \int_{\partial\mathcal{E}_r} \frac{|f(z)|^2}{\left|1 - z^2\right|^{1/2}} \, ds < \infty\right\},$$

$$\varphi(f, g) := \int_{\partial\mathcal{E}_r} \frac{f(z)\overline{g(z)}}{\left|1 - z^2\right|^{1/2}} \, ds,$$

$$h_0(z)\frac{1}{\sqrt{2\pi}}, \qquad h_\nu(z) = \sqrt{\frac{2}{\pi}} \frac{T_\nu(z)}{\left(r^{2\nu} + r^{-2\nu}\right)^{1/2}} \quad \text{for } \nu = 1, 2, \dots.$$

(For the definition of T_ν, see Appendix A.)

Special Case V

$$\mathcal{H} := \left\{ f \,:\, f \text{ is analytic in } \mathcal{E}_r \text{ and } \int_{\partial\mathcal{E}_r} |f(z)|^2 \left|1-z^2\right|^{1/2} ds < \infty \right\},$$

$$\varphi(f,g) := \int_{\partial\mathcal{E}_r} f(z)\overline{g(z)} \left|1-z^2\right|^{1/2} ds,$$

$$h_\nu(z) = \sqrt{\frac{2}{\pi}} \frac{U_\nu(z)}{\left(r^{2\nu+2}+r^{-2\nu-2}\right)^{1/2}}.$$

For proof that these are indeed Hilbert spaces, see Davis (1963). With norms defined by $\|f\| = [\varphi(f,f)]^{1/2}$, in all of the above cases we have that

$$\mathcal{C}_M^{\mathrm{Da}} = \{f \,:\, \|f\| \le M\}$$

and

$$\rho(Q_n, \mathcal{C}_M^{\mathrm{Da}}) = M \left(\sum_{\nu=0}^{\infty} R_n^2[h_\nu] \right)^{1/2}. \tag{4.33}$$

For the computation of the sum on the right-hand side of (4.33), see, for instance, Stroud and Secrest (1966), Lo *et al.* (1965), Hämmerlin (1966) and Rabinowitz and Richter (1970a). Alternatively, good and simple bounds can be found, such as in Theorem 4.8.1 below. The main problem that arises in using (4.33) is the determination of $\|f\|$. If the Taylor coefficients $\alpha_\nu := f^{(\nu)}(0)/\nu!$ are known, we can use

$$\|f\|^2 = 2\pi r \sum_{\nu=0}^{\infty} |\alpha_\nu|^2 r^{2\nu} \quad \text{in Special Case I}$$

and

$$\|f\|^2 = \pi \sum_{\nu=0}^{\infty} \frac{|\alpha_\nu|^2 r^{2\nu+2}}{\nu+1} \quad \text{in Special Case II.}$$

In most cases, the computation of the (line or area) integral is more costly than solving our original problem. So, instead, we may use rough bounds such as

$$\|f\| \le \sqrt{2\pi r} \max_{|z|\le r} |f(z)| \quad \text{in Special Case I,}$$

$$\|f\| \le \sqrt{\pi}\, r \max_{|z|\le r} |f(z)| \quad \text{in Special Case II,}$$

$$\|f\| \le \frac{\sqrt{\pi}}{2} \left(r^2 - r^{-2}\right)^{1/2} \max_{z\in\mathcal{E}_r} |f(z)| \quad \text{in Special Case III,}$$

$$\|f\| \le \sqrt{2\pi} \max_{z\in\mathcal{E}_r} |f(z)| \quad \text{in Special Case IV}$$

and

$$\|f\| \le \sqrt{\frac{\pi}{2}} \left(r^2 + r^{-2}\right)^{1/2} \max_{z\in\mathcal{E}_r} |f(z)| \quad \text{in Special Case V.}$$

However, by using these bounds, we enter the realm of the co-observation $\max|f(z)|$ and, at least in principle, better bounds are possible; see the next section. On the other hand, there are usually fewer technical difficulties in using Davis-type co-observations.

Our aim now is to apply Davis's method in Special Case I to the trapezoidal rule

$$Q_{n+1}^{\mathrm{Tr}}[f] := \frac{1}{n}\Big(f(-1) + 2\sum_{\nu=1}^{n-1} f\Big(-1+\frac{2\nu}{n}\Big) + f(1)\Big).$$

We shall make use of the following results, which will be proved in Chapter 7:

$$\begin{aligned} R_{n+1}^{\mathrm{Tr}}[f] &= -\frac{2}{3n^2} f''(\xi), \\ R_{n+1}^{\mathrm{Tr}}[f] &= -\frac{1}{3n^2}\big(f'(1) - f'(-1)\big) + \frac{2}{45n^4} f^{(4)}(\eta) \end{aligned}$$

with some (unknown) $\xi, \eta \in [-1,1]$. Taking f to be $p_\nu(x) := x^\nu$, we obtain that for $\nu = 0, 1, \ldots,$

$$\big|R_{n+1}^{\mathrm{Tr}}[p_{2\nu}]\big| \le \frac{4\nu}{3n^2} \tag{4.34}$$

and

$$\lim_{n\to\infty} n^2 R_{n+1}^{\mathrm{Tr}}[p_{2\nu}] = \frac{4\nu}{3}. \tag{4.35}$$

Furthermore, it follows from the symmetry that $R_{n+1}^{\mathrm{Tr}}[p_{2\nu+1}] = 0$. Consequently, (4.33) yields

$$\rho^2(Q_{n+1}^{\mathrm{Tr}}, \mathcal{C}_M^{\mathrm{Da}}) \le \frac{M^2}{2\pi r}\,\frac{16}{9n^4}\sum_{\nu=0}^{\infty}\frac{\nu^2}{r^{4\nu}} = \frac{M^2}{2\pi r}\,\frac{16}{9n^4}\,\frac{r^8+r^4}{\left(r^4-1\right)^3}.$$

Equation (4.35) shows that this bound is asymptotically correct as $n \to \infty$. So we have proved the following:

THEOREM 4.8.1 (Hämmerlin 1965). *For the Davis-type co-observation in Special Case I, we have*

$$\rho(Q_{n+1}^{\mathrm{Tr}}, \mathcal{C}_M^{\mathrm{Da}}) \le \frac{4M}{3n^2r^2\sqrt{2\pi r}}\cdot\frac{1}{1-r^{-4}}\left(\frac{1+r^{-4}}{1-r^{-4}}\right)^{1/2}$$

and

$$\lim_{n\to\infty} n^2 \rho(Q_{n+1}^{\mathrm{Tr}}, \mathcal{C}_M^{\mathrm{Da}}) = \frac{4M}{3r^2\sqrt{2\pi r}}\cdot\frac{1}{1-r^{-4}}\left(\frac{1+r^{-4}}{1-r^{-4}}\right)^{1/2}.$$

We have already mentioned that this result is usually applied in the form

$$\big|R_{n+1}^{\mathrm{Tr}}[f]\big| \le \frac{4}{3n^2r^2}\cdot\frac{1}{1-r^{-4}}\left(\frac{1+r^{-4}}{1-r^{-4}}\right)^{1/2}\sup_{|z|\le r}|f(z)|.$$

In doing so, however, the asymptotic optimality as $n \to \infty$ is lost, for Schmeisser (1980) has shown that

$$\lim_{n\to\infty} n^2 \sup\Big\{\big|R_{n+1}^{\mathrm{Tr}}[f]\big| \;:\; \sup_{|z|\le r}|f(z)| \le M\Big\} = \frac{4}{3r^2}\cdot\frac{1}{1-r^{-4}}\cdot M.$$

On the other hand, we lose only a little when r is large, while an explicit bound is available.

In applying the same method to Special Case III, some new difficulties arise, but Hämmerlin (1966) found that these could be overcome. Subsequently, Rabinowitz and Richter (1970a) applied the method to Special Case IV. For a corresponding asymptotic result for increasing n, see Schmeisser (1980).

We have exhibited five co-observations of Davis type, and many more are possible. Comparing our five special cases, we remark that Cases III, IV and V have the advantage that any domain D of analyticity contains an ellipse $\mathcal{E}_r$; therefore these co-observations are suitable for any analytic function. This is of course not true for circular domains; on the other hand, computing bounds by hand is usually simpler on circles than on ellipses.

A further reason to be interested in $\mathcal{C}_M^{\mathrm{Da}}$ is the possibility it offers of obtaining optimal rules quite easily by solving a linear system. This is due to Hilbert space theory, as discussed in Section 2.3. We characterized Q^{so} by the property of being exact on the spline space $\mathcal{S}$ belonging to the underlying Hilbert space. According to the theory described in Section 2.3, a basis for $\mathcal{S}$ is given by

$$\widetilde{l}_\lambda(x) = \sum_{\nu=0}^{\infty} h_\nu(x) h_\nu(x_\lambda), \quad \lambda = 1, 2, \ldots, n.$$

In most cases, the resulting optimal rules are rather complicated; see Example 2.3.2. Moreover, they are very close to certain classical and well-studied rules when r is large. This is a consequence of the following result:

THEOREM 4.8.2 (Barnhill 1968). *Let $Q_n^{(r)}$ be an optimal rule for one of the aforementioned Davis-type co-observations with parameter r. Then*

$$\lim_{r\to\infty} Q_n^{(r)}$$

exists and is characterized by being exact for all $f \in \mathbb{P}_{n-1}$.

The rules with $R_n[\mathbb{P}_{n-1}] = \{0\}$ are not necessarily positive. In fact, even for small r, small n and, for example, equidistant nodes, we obtain negative coefficients for the optimal rules; see the tables in Engels and Mangold (1971).

PROOF OF THEOREM 4.8.2. Denote the spline space by $\mathcal{S}^{(r)}$ and let a basis for it be $\{g_\lambda^{(r)} : \lambda = 1, 2, \ldots, n\}$. The coefficients $a_\nu^{(r)}$ of the optimal rule are determined from the system

$$\sum_{\nu=1}^{n} a_\nu^{(r)} g_\lambda^{(r)}(x_\nu) = I[g_\lambda], \quad \lambda = 1, 2, \ldots, n.$$

The assertion will be proved if we can find a basis $\{g_\lambda^{(r)}\}$ such that

$$\lim_{r\to\infty} g_\lambda^{(r)} \in \mathbb{P}_{n-1}.$$

From the theory in Section 2.3 it follows that

$$\widetilde{l}_\sigma(x) := \sum_{\nu=0}^{\infty} h_\nu(x) h_\nu(x_\sigma), \quad \sigma = 1, 2, \ldots, n,$$

is a basis for $\mathcal{S}^{(r)}$. We construct a new basis $\{k_\lambda\}$ by setting

$$k_\lambda = \sum_{\sigma=1}^{n} c_{\sigma,\lambda} \widetilde{l}_\sigma, \quad \lambda = 1, 2, \ldots, n,$$

where the $c_{\sigma,\lambda}$ are determined as solutions of the system

$$\sum_{\sigma=1}^{n} c_{\sigma,\lambda} h_\nu(x_\sigma) = \begin{cases} 1 & \text{if } \nu = \lambda, \\ 0 & \text{otherwise,} \end{cases} \qquad \nu = 0, 1, \ldots, n-1, \tag{4.36}$$

which is regular in all of our special cases. We then have

$$k_\lambda = h_\lambda + \sum_{\nu \ge n} h_\nu \sum_{\sigma=1}^{n} c_{\sigma,\lambda} h_\nu(x_\sigma).$$

The remaining steps in the proof will be shown only for Special Case II; the other cases are similar. Now, the system (4.36) has the form

$$\sum_{\sigma=1}^{n} c_{\sigma,\lambda} x_\sigma^\nu = \begin{cases} \sqrt{\dfrac{\pi}{\nu+1}}\, r^{\lambda+1} & \text{if } \nu = \lambda, \\ 0 & \text{otherwise,} \end{cases} \qquad \nu = 0, 1, \ldots, n-1.$$

This implies that $c_{\sigma,\lambda} = O(r^{\lambda+1}) = O(r^n)$ as r increases. For $\nu \ge n$, we obtain

$$\sum_{\sigma=1}^{n} c_{\sigma,\lambda} h_\nu(x_\sigma) = O(r^{n-\nu-1}) = O(r^{-1})$$

and, finally,

$$\left| \sum_{\nu \ge n} h_\nu(x) \sum_{\sigma=1}^{n} c_{\sigma,\lambda} h_\nu(x_\sigma) \right| = O(r^{-n-2})$$

uniformly for $x \in [-1, 1]$. Therefore

$$\lim_{r \to \infty} r^{\lambda+1} k_\lambda(x) = \lim_{r \to \infty} r^{\lambda+1} h_\lambda(x) = \sqrt{\frac{\lambda+1}{\pi}}\, x^\lambda,$$

which proves our claim. □

Theorem 4.8.2 concerns optimal rules with large domains of holomorphy. In Rabinowitz and Richter (1970b), small domains of holomorphy were investigated. In this case, the result is (using the same notation as in Theorem 4.8.2)

$$\lim_{r \to 1} Q_n^{(r)} = 0$$

for Special Cases III and IV with $w = 1$ and $x_\nu \in \,]-1, 1[$.

Until now, we have assumed $r > 1$. When $r = 1$, things are radically different. To be sure, we can define a Hilbert space, as was done in Example 2.3.2. Now, however, not all elements are continuous functions on $[-1, 1]$; for example, $f(x) = \ln(1 - x)$ is in $\mathcal{C}_M^{\mathrm{Da}}$ (Special Case I) with $r = 1$ but has a singularity at $x = 1$. Stenger (1973) proved the somewhat surprising result that, in this class,

$$\rho_n = O\Big(\exp(-\frac{\pi}{2}\sqrt{n})\Big)$$

is possible with an appropriate set of nodes; moreover, he was able to show that the inequality

$$\rho_n \ge \exp(-c\sqrt{n}) \quad \text{with some positive constant } c$$

holds for any rule (Stenger 1978). The existence of best rules was proved by Barrar *et al.* (1974) and Paulik (1977). The determination of these rules is discussed in Engels and Eckhardt (1979), and a table is given in Engels and Eckhardt (1977).

4.9. Bounds in the complex plane as co-observations

In the previous section, we already considered a space of analytic functions. It was defined by an inner product, so the nice Hilbert space theory from Section 2.3 was applicable. On the other hand, bounds for analytic functions on a region of the complex plane might be more accessible (see Appendix C on the verification of co-observations). Let $\mathcal{G} \subset \mathbb{C}$ be open and simply connected with a rectifiable boundary, and set

$$\mathcal{C}_M(\mathcal{G}) := \{f : f \text{ is analytic in } \mathcal{G} \text{ and continuous on } \overline{\mathcal{G}} \\ \text{with } \|f\|_{\mathcal{G}} := \sup_{z\in\overline{\mathcal{G}}} |f(z)| \leq M\}.$$

We have an error theory for each such region $\mathcal{G}$. Sometimes it will be more convenient to map $\mathcal{G}$ onto a standard region—for instance a disc $D_r := \{z : |z| \leq r\}$ or an ellipse $\mathcal{E}_r$ as defined in the previous section—and keep in mind that the functional I will also be modified.

LEMMA 4.9.1 (cf. Bakhvalov 1967). Let m be a conformal mapping that maps the region $\mathcal{G}_1$ onto the region $\mathcal{G}_2$ and the interval $[\alpha_1, \beta_1] \subset \mathcal{G}_1$ onto the interval $[\alpha_2, \beta_2] \subset \mathcal{G}_2$. Let

$$Q^{[1]}[f] = \sum_{\nu=1}^{n} a_\nu f(x_\nu) \qquad \text{and} \qquad Q^{[2]}[f] = \sum_{\nu=1}^{n} a_\nu f(m(x_\nu))$$

be quadrature rules for

$$I^{[1]}[f] = \int_{\alpha_1}^{\beta_1} f(x) m'(x) w(m(x))\, dx \qquad \text{and} \qquad I^{[2]}[f] = \int_{\alpha_2}^{\beta_2} f(x) w(x)\, dx,$$

respectively. Then their errors in the classes $\mathcal{C}_M(\mathcal{G}_1)$ and $\mathcal{C}_M(\mathcal{G}_2)$, respectively, are the same. Optimal and best rules in the two settings are related in the same fashion.

PROOF. Define $g(z) = f(m(z))$. Clearly, $f \in \mathcal{C}_M(\mathcal{G}_2)$ is equivalent to $g \in \mathcal{C}_M(\mathcal{G}_1)$. Furthermore, $I^{[1]}[g] = I^{[2]}[f]$ and $Q^{[1]}[g] = Q^{[2]}[f]$. □

We now consider integration over $[a, b] \subset [-1, 1]$ along with the co-observation $\mathcal{C}_M(D_1)$. It is possible to say more about optimal quadrature rules of the type

$$Q^*[f] = \sum_{\nu=1}^{n} \left[a_\nu^{[0]} f(x_\nu) + a_\nu^{[1]} f'(x_\nu)\right]. \tag{4.37}$$

The main tool for obtaining many optimality results in $\mathcal{C}_M(D_1)$ is use of the Blaschke products

$$B(z) = B[x_1, \ldots, x_n](z) = \prod_{\nu=1}^{n} W(x_\nu, z) \quad \text{where } W(x, z) = \frac{z - x}{1 - z\overline{x}}; \tag{4.38}$$

see, e.g., Bojanov (1974a, 1974b), Loeb (1974) and Osipenko (1988, 1995). These functions are analytic in D_1 and satisfy $|B| = 1$ on ∂D_1 (which implies $|B| < 1$ in D_1). Furthermore, set

$$B_j(z) = \prod_{\substack{\nu=1 \\ \nu\neq j}}^{n} W(x_\nu, z) \quad \text{for } j = 1, \ldots, n.$$

The B_j are also Blaschke products. The following theory has its roots in the papers of Bojanov (1974a, 1974b) and Loeb (1974). We follow the ideas of Osipenko.

THEOREM 4.9.1 (Osipenko 1988). *Let $V = C[a,b]$, $O[f] = \big(f(x_1), \dots, f(x_n), f'(x_1), \dots, f'(x_n)\big)$ where $a \le x_1 < \dots < x_n \le b$, $I[f] = \int_a^b f(x)w(x)\,dx$ where $w(x) \ge 0$, and $\mathcal{C} = \mathcal{C}_M(D_1)$. Set*

$$h(x,z) = \frac{1}{B^2(z)(1-xz)(x-z)}.$$

Then, the quadrature rule Q^{opt} of the form (4.37) with

$$a_\nu^{[j]} = \int_a^b A_\nu^{[j]}(x)w(x)\,dx$$

and

$$A_\nu^{[j]}(x) = B(x)^2(1-x^2)\left(\frac{d}{dz}\right)^{(1-j)} \left[(z-x_\nu)^2 h(x,z)\right]\Big|_{z=x_\nu}$$

is optimal. Furthermore,

$$\rho^{\text{opt}} = M\int_a^b B^2(x)w(x)\,dx. \tag{4.39}$$

PROOF. Let $x \in (-1,1)$. We have

$$\begin{aligned} & f(x) - \sum_{\nu=1}^n \left[A_\nu^{[0]}(x)f(x_\nu) + A_\nu^{[1]}(x)f'(x_\nu)\right] \\ = \;& f(x) - B^2(x)(1-x^2)\sum_{\nu=1}^n \left[(z-x_\nu)^2 h(x,z)f(z)\right]'\Big|_{z=x_\nu}. \end{aligned}$$

This is the sum of all residues of $B^2(x)(1-x^2)h(x,\cdot)f$ in the unit circle. Hence, from the residue theorem, we obtain

$$f(x) - \sum_{\nu=1}^n \left[A_\nu^{[0]}(x)f(x_\nu) + A_\nu^{[1]}(x)f'(x_\nu)\right] = \frac{1}{2\pi i}\int_{|z|=1} \frac{B^2(x)(1-x^2)f(z)}{B^2(z)(1-xz)(z-x)}\,dz.$$

Since $|B^2| = 1$ on the unit circle, we have

$$\left|R^{\text{opt}}[f]\right| \le \frac{1}{2\pi}\int_a^b B^2(x)\|f\|\left[\int_0^{2\pi} \frac{(1-x^2)}{|e^{i\theta}-x|^2}\,d\theta\right] w(x)\,dx,$$

where

$$\int_0^{2\pi} \frac{(1-x^2)}{|e^{i\theta}-x|^2}\,d\theta = \int_0^{2\pi} \frac{(1-x^2)d\theta}{1-2x\cos\theta + x^2} = 2\pi.$$

We have therefore shown that

$$\rho(Q^{\text{opt}}) \le M\int_a^b B^2(x)w(x)\,dx.$$

Furthermore, $MB^2 \in \mathcal{C}_M(D_1)$ and $O[B^2] = (0,\dots,0)$, so that

$$\rho^{\text{opt}} \ge M\int_a^b B^2(x)w(x)\,dx$$

and the equality (4.39) is proved. □

EXAMPLE 4.9.1. Consider optimal quadrature rules of the form

$$Q^{\mathrm{opt}}[f] = a_1^{[0]} f(-y) + a_2^{[0]} f(y) + a_1^{[1]} f'(-y) + a_2^{[2]} f'(y)$$

for evaluation of the functional

$$I[f] = \int_{-a}^{a} f(x)\,dx, \qquad 0 < a < 1,$$

on the class $\mathcal{C}_M(D_1)$. We have explicit formulas for the coefficients, and obtain

$$\begin{aligned}
a_2^{[0]} &= a_1^{[0]} = \frac{(1-y^4)}{8y^3} \int_{-a}^{a} \frac{(1-x^2)(x+y)^2\left[4y(1+x^2)(1+y^4) - 2x\left(1+y^2\right)^3\right]}{(1+xy)^2(1-xy)^4}\,dx \\
&= \frac{(1-y^4)}{192y^9}\Bigg\{(35+15y^4+9y^8+5y^{12})\left(\frac{4ay(3-8a^2y^2+3a^4y^4)}{\left(1-a^2y^2\right)^3} - 6\ln\frac{1-ay}{1+ay}\right) \\
&\qquad - \frac{8ay}{\left(1-a^2y^2\right)^3}\Big[(7+3y^4)(15-40a^2y^2+24a^4y^4) \\
&\qquad\qquad + 3y^8 - 9y^{12} - 24a^6y^6(1+y^4)\Big]\Bigg\}
\end{aligned}$$

and

$$\begin{aligned}
a_2^{[1]} &= -a_1^{[1]} = \frac{(1-y^4)^2}{4y^2}\int_{-a}^{a} \frac{(1-x^2)(x+y)^2(x-y)}{(1+xy)^2(1-xy)^3}\,dx \\
&= \frac{(1-y^4)^2}{32y^8}\Bigg\{16ay + \frac{ay(1-y^4)[14+10y^4-6a^2y^2(3+y^4)]}{\left(1-a^2y^2\right)^2} \\
&\qquad + (15-2y^4+3y^8)\ln\frac{1-ay}{1+ay}\Bigg\}.
\end{aligned}$$

The error constant is

$$\begin{aligned}
\rho^{\mathrm{opt}} &= M\int_{-a}^{a} \frac{(x^2-y^2)^2}{\left(1-x^2y^2\right)^2}\,dx \\
&= \frac{2Ma}{y^4} + M\frac{1-y^4}{y^5}\left(\frac{ay(1-y^4)}{1-a^2y^2} + \frac{3+y^4}{2}\ln\frac{1-ay}{1+ay}\right).
\end{aligned}$$

If, for example, $a = 1/2$ and $y = 1/\sqrt{12} = 0.288675\ldots$, which is a node of the Gaussian quadrature formula Q_2^{G}, then we obtain

$$a_2^{[0]} = a_1^{[0]} = 0.4999648\ldots, \qquad a_2^{[1]} = -a_1^{[1]} = 0.00005347\ldots$$

and

$$\rho_2^{\mathrm{opt}} = M \cdot 0.005679983\ldots.$$

This optimal rule does not differ much from the Gaussian rule $Q_2^{\mathrm{G}}[f] = [f(y) + f(-y)]/2$. We remark that the coefficients $a_i^{[1]}$ are not always this close to zero; take $y = 1/2$, for instance, which gives $a_1^{[1]} = -a_2^{[1]} = 0.086\ldots$. On the other hand, there is a particular choice of y for which the $a_i^{[1]}$ vanish for the corresponding quadrature rule Q_2^*; this y-value is $0.288998\ldots$, which yields $a_1^{[0]} = a_2^{[0]} = 0.499964\ldots$ and $\rho_2^{\mathrm{opt}} = M \cdot 0.005679949\ldots$.

As we have seen, it can happen that all of the coefficients $a_\nu^{[1]}$ vanish. The following theorem tells us that such cases are of special interest.

THEOREM 4.9.2 (Osipenko 1988). Take a generalized SEF consisting of the SEF of Theorem 4.9.1 and

$$\begin{aligned} O_{2n} &= \{(o_1, \ldots, o_{2n}) : o_\nu[f] = f(x_\nu),\ o_{n+\nu}[f] = f'(x_\nu) \\ &\qquad \text{for } \nu = 1, \ldots, n,\ a \leq x_1 < \cdots < x_n \leq b\}. \end{aligned}$$

Then there exists a best rule, and all best rules Q_n^{best} are of the form

$$Q_n^{\text{best}}[f] = \sum_{\nu=1}^{n} a_\nu f(x_\nu)$$

where the nodes x_ν satisfy

$$\int_a^b B^2(x) \frac{1 - W(x_\nu, x)^2}{W(x_\nu, x)} w(x)\, dx = 0 \quad \text{for } \nu = 1, \ldots, n \tag{4.40}$$

(with B and W as defined in (4.38)) and the coefficients a_ν are given by

$$a_\nu = \int_a^b \frac{B_\nu^2(x)}{B_\nu^2(x_\nu)} \big[1 - W(x_\nu, x)^4\big] w(x)\, dx.$$

It is particularly surprising that the best rules are based on only half of the available information.

PROOF. First, we show that there exists a best rule. Consider

$$c(x_1, \ldots, x_n) := \int_a^b B[x_1, \ldots, x_n]^2(x)\, dx$$

on the compact set $\{(x_1, \ldots, x_n) : a \leq x_1 \leq \cdots \leq x_n \leq b\}$; $c(x_1, \ldots, x_n)$ represents ρ^{opt} if $a \leq x_1 < \cdots < x_n \leq b$. Being a continuous function, c attains its minimum, so we just need to show that $x_{\nu-1} < x_\nu$ for $\nu = 2, \ldots, n$. Suppose that $x_i = x_{i+1} = \cdots = x_{i+k}$. There is an $\varepsilon > 0$ such that

$$\int_{S_\varepsilon} B^2(x) w(x)\, dx = \gamma > 0,$$

where $S_\varepsilon = [a, b] \setminus \big([-1, \varepsilon - 1[\ \cup\ ]x_i - \varepsilon, x_i + \varepsilon[\ \cup\ ]1 - \varepsilon, 1]\big)$. Note that

$$\begin{aligned} &\frac{d}{dt} \left(\frac{x - (x_i - t)}{1 - x(x_i - t)}\right)^2 \left(\frac{x - (x_{i+k} + t)}{1 - x(x_{i+k} + t)}\right)^2 \\ &\qquad = -\frac{4t(1 - x^2)\big[(x - x_i)^2 - t^2\big](1 + x^2 - 2xx_i)}{[(1 - xx_i)^2 - t^2x^2]^3}. \end{aligned} \tag{4.41}$$

We replace x_i by $x_i - t$ and x_{i+k} by $x_{i+k} + t$, where $t > 0$, and denote the resulting Blaschke product by B_t. Then (4.41) implies that

$$B_t^2(x) \leq B^2(x) \quad \text{if } x(t + x_i) < 1 \text{ and } x \in [-1, x_i - t] \cup [x_i + t, 1].$$

For sufficiently small t, there is a constant $\alpha > 0$ such that

$$B_t^2(x) \leq (1 - \alpha t) B^2(x) \quad \text{for } x \in S_\varepsilon,$$

so we obtain

$$\begin{aligned}&\int_a^b B^2(x)w(t)\,dt - \int_a^b B_t^2(x)w(t)\,dt \\ &\quad > \gamma\alpha t - \int_{x_i-t}^{x_i+t}\left(\frac{x-(x_i-t)}{1-x(x_i-t)}\right)^2\left(\frac{x-(x_i+t)}{1-x(x_i+t)}\right)^2 w(t)\,dt = \gamma\alpha t + o(t^4),\end{aligned}$$

which is positive for sufficiently small t. If either $x_1 = a$ or $x_n = b$, we can reduce the integral (4.39) by shifting the end node to the interior of $[a,b]$. Hence, the minimal value is attained for $a < x_1 < x_2 < \cdots < x_n < b$. The best distribution of nodes has to satisfy

$$\begin{aligned}(4.42)\qquad 0 &= \frac{\partial}{\partial x_\nu}c(x_1,\ldots,x_n) \\ &= 2\int_a^b B_\nu^2(x)W(x_\nu,x)\left[\frac{d}{dx_\nu}W(x_\nu,x)\right]w(x)\,dx \\ &= 2\int_a^b \frac{B^2(x)}{W(x_\nu,x)}\cdot\frac{x^2-1}{1-xx_\nu}w(x)\,dx \\ (4.43)\qquad &= \frac{2}{x_\nu^2-1}\int_a^b \frac{B^2(x)}{W(x_\nu,x)}\left[1-W(x_\nu,x)^2\right]w(x)\,dx\end{aligned}$$

for $\nu = 1,\ldots,n$. This is (4.40). An explicit calculation yields that for the optimal rules from Theorem 4.9.1,

$$a_\nu^{[1]} = \int_a^b \frac{B_\nu^2(x)W(x_\nu,x)}{B_\nu^2(x_\nu)\frac{\partial}{\partial z}W(x_\nu,z)\big|_{z=x_\nu}}\left[1-W(x_\nu,x)^2\right]w(x)\,dx$$

and

$$a_\nu^{[0]} = \int_a^b \frac{B_\nu^2(x)}{B_\nu^2(x_\nu)}\left[1-W(x_\nu,x)^4\right]w(x)\,dx - \frac{B_\nu'(x_\nu)}{B_\nu(x_\nu)}a_\nu^{[1]}$$

for the optimal coefficients. By (4.42) and (4.43), the coefficients $a_\nu^{[1]}$ vanish. □

Originally (see Bojanov 1974a, 1974b), the special problem with $\mathcal{C} = \mathcal{C}_M(D_1)$ and $I[f] = \int_{-1}^1 f(x)\,dx$ was considered. The above results carry over to this particular setting, although we have assumed that $[a,b] \subset]-1,1[$. The reason is that the contributions of f to the functional I on the intervals $[-1,\varepsilon-1]$ and $[1-\varepsilon,1]$ are arbitrarily small for sufficiently small $\varepsilon > 0$.

EXAMPLE 4.9.1 (continued). Indeed, for $a = 1/2$, numerical minimization of the integral (4.39) over x_1 and x_2 indicates that the quadrature rule Q_2^* has the smallest error constant among all quadrature rules of the form

$$Q[f] = a_1^{[0]}f(x_1) + a_2^{[0]}f(x_2) + a_1^{[1]}f'(x_1) + a_2^{[1]}f'(x_2),$$

and hence that $\rho_2^{\text{best}} = 0.005679949\ldots.$

By using a suitable transformation mapping the unit circle conformally onto $\mathcal{E}_r$ and $[a,b]$ onto $[-1,1]$, and by choosing the correct weight function in the original setting, Osipenko (1988, 1995) was able to derive the following best rule for

$$I[f] = \int_{-1}^1 \frac{f(x)}{\sqrt{1-x^2}}\,dx$$

and $\mathcal{C} = \mathcal{C}_M(\mathcal{E}_r)$ in the case where $r \geq \sqrt{44 + n2^{4n-1}}$:

$$Q_n^{\text{best}}[f] = \frac{\pi}{n}\big[1 - d_n(r)\big] \sum_{\nu=1}^{n} f\Big(-\cos\Big(\frac{2\nu-1}{2n}\pi\Big)\Big), \tag{4.44}$$

where

$$d_n(r) = \frac{(\lambda^2+2)\Lambda - 2(\lambda^2+1)E}{3\lambda^2\Lambda} = 6r^{-4n} + O(r^{-8n}).$$

Here, Λ and E are complete elliptic integrals of the first and second kind, respectively (see Abramowitz and Stegun 1964), with modulus λ determined by the equation

$$\frac{\Lambda'}{\Lambda} = \frac{4n}{\pi}\ln r.$$

Furthermore,

$$\rho_n^{\text{best}} = \frac{\pi(\Lambda - E)}{\lambda\Lambda} = 2\pi r^{-2n} + O(r^{-6n}).$$

The quadrature rule is best even among all rules of the form (4.37). Upon omitting the assumption $r \geq \sqrt{44 + n2^{4n-1}}$, the rule is still at least optimal (among all rules using the same nodes). The proofs are very technical, so we will not present them here. Remarkably, the best rule for analytic, i.e. very smooth, functions is not even exact for the smoothest functions, namely constant ones. It is, however, very close to the corresponding Gaussian rule, which indicates that exactness for polynomials might again be a good construction principle. The Gaussian rule in this case is obtained by replacing d_n in (4.44) by 0. This means, in particular, that we lose less than the factor $1 + \pi d_n(r)/\rho_n^{\text{best}} = 1 + O(r^{-2n})$ in the error estimate if we use the Gaussian rule instead of the above rule.

Of course, such explicit examples are the exception. Therefore, in what follows, we show how to obtain good lower and upper bounds for ρ_n^{best}.

THEOREM 4.9.3. *Let Q be a quadrature rule on $C[a,b]$, with $[a,b] \subset \mathcal{G}$, and let $f \in \mathcal{C}_M(\mathcal{G})$. Then*

$$R[f] = \frac{1}{2\pi i}\int_{\partial\mathcal{G}} f(z)k(z)\,dz \quad \text{where } k(z) = R\left[\frac{1}{z - \cdot}\right], \tag{4.45}$$

and

$$\rho(Q) \leq \frac{M}{2\pi}\int_{\partial\mathcal{G}} |k(z)|\,ds. \tag{4.46}$$

PROOF. For each function $f \in \mathcal{C}_M(\mathcal{G})$, Cauchy's integral formula

$$f(x) = \frac{1}{2\pi i}\int_{\partial\mathcal{G}} \frac{f(z)\,dz}{z - x} \tag{4.47}$$

holds for all $x \in [a,b]$. Now, R can be applied to both sides of this equation and, furthermore, we may interchange R and the integration on the right-hand side, since $\partial\mathcal{G}$ and $[a,b]$ are a positive distance apart and Fubini's theorem is applicable. □

The error theory is more complicated than for differentiable functions. For instance, note that in general we do not have equality in (4.46). To be specific, we may subtract from k any analytic function on $\mathcal{G}$ and (4.45)–(4.46) will still be valid. Then, we minimize over all such functions g and obtain

$$\rho(Q) \leq \frac{M}{2\pi}\inf\Big\{\int_{\partial\mathcal{G}} |k(z) - g(z)|\,ds \;:\; g \text{ is analytic in } \mathcal{G}\Big\}. \tag{4.48}$$

At least for discs centered on the real axis, equality holds (see Duren 1970: Theorem 8.1). Nevertheless, the "kernel function" k and the bound (4.46) play an important role, since any improvement obtained via (4.48) is usually minor, as hinted at in the following example.

EXAMPLE 4.9.1 (continued). For the Gaussian rule

$$Q_2^{\mathrm{G}}[f] = \frac{1}{2}\left[f\left(-\frac{1}{\sqrt{12}}\right) + f\left(\frac{1}{\sqrt{12}}\right)\right]$$

for the evaluation of

$$I[f] = \int_{-1/2}^{1/2} f(x)\,dx,$$

we have

$$k(e^{i\varphi}) = -\frac{1}{2}\ln\left(\frac{9+16\sin^2\varphi}{(5+4\cos\varphi)^2}\right) + i\arctan\frac{4\sin\varphi}{3} - \frac{132\cos\varphi - 156i\sin\varphi}{145-24\cos(2\varphi)}.$$

Numerically calculating the integral in (4.46) with $\mathcal{G} = D_1$ gives $\rho(Q_2^{\mathrm{G}}, \mathcal{C}_M(D_1)) \le 0.005682$. Comparison with $\rho_2^{\mathrm{best}} = 0.005679949\ldots$ shows that the Gaussian rule is almost best and that, in this case, there is not much space for improvement by using the minimization in (4.48).

This example also shows that the kernel functions k are usually very complicated and do not allow the integration in (4.46) to be done exactly. A series expansion helps in estimating the error constant from above. For this purpose, we consider analytic functions on ellipses $\mathcal{E}_r$. Chebyshev polynomials T_ν (see Appendix A) play an important role in this context.

THEOREM 4.9.4.

$$\rho(Q, \mathcal{C}_M(\mathcal{E}_r)) \le \frac{M}{\pi}\int_0^{2\pi}\left|\tfrac{1}{2}R[T_0] + \sum_{\nu=1}^{\infty} r^{-\nu}e^{-i\nu t}R[T_\nu]\right|dt.$$

PROOF. We will need the identity

$$\frac{\zeta-\zeta^{-1}}{\zeta+\zeta^{-1}-2x} = 1 + 2\sum_{\nu=1}^{\infty}\zeta^{-\nu}T_\nu(x), \tag{4.49}$$

which holds for $|\zeta| > 1$ and $|x| \le 1$. To verify this identity, multiply through by $(\zeta + \zeta^{-1} - 2x)$ and use the relation

$$2xT_\nu(x) = T_{\nu+1}(x) + T_{\nu-1}(x).$$

We start with (4.46), using $\partial\mathcal{E}_r$ as the contour:

$$\rho(Q, \mathcal{C}_M(\mathcal{E}_r)) \le \frac{M}{2\pi}\int_{\partial\mathcal{E}_r}\left|R\left[\frac{1}{z-\cdot}\right]\right|ds.$$

The substitution $z = \left(re^{it} + r^{-1}e^{-it}\right)/2$ gives

$$\rho(Q, \mathcal{C}_M(\mathcal{E}_r)) \le \frac{M}{\pi}\int_0^{2\pi}\left|R\left[\frac{1}{re^{it}+r^{-1}e^{-it}-2\,\cdot}\right]\right|\cdot\left|re^{it}-r^{-1}e^{-it}\right|dt.$$

An application of (4.49) with $\zeta = re^{it}$ yields the result. □

Theorem 4.9.4 has many applications. The crude bound

$$\left|\sum_{\nu=m}^{\infty} r^{-\nu} e^{-i\nu t} R[T_\nu]\right| \le \sum_{\nu=m}^{\infty} r^{-\nu} \big|R[T_\nu]\big| \le \|R\| \cdot \sum_{\nu=m}^{\infty} r^{-\nu} = \|R\| \cdot \frac{r^{-m}}{1-r^{-1}}$$

leads to the following:

COROLLARY 4.9.1. For any quadrature rule Q with $R[\mathbb{P}_{m-1}] = \{0\}$, we have

$$\rho\big(Q, \mathcal{C}_M(\mathcal{E}_r)\big) \le \frac{2M\|R\|}{1-r^{-1}}\, r^{-m}.$$

An immediate improvement of this estimate is obtained in the case where w is even and Q is symmetric, because these conditions imply that $R[T_{2\nu+1}] = 0$.

Another interesting consequence of Theorem 4.9.4 comes from applying the Cauchy–Schwarz inequality:

$$\begin{aligned}
\int_0^{2\pi} \left|\sum_{\nu=m}^{\infty} r^{-\nu} e^{-i\nu t} R[T_\nu]\right| dt &\le \left(2\pi \int_0^{2\pi} \left|\sum_{\nu=m}^{\infty} r^{-\nu} e^{-i\nu t} R[T_\nu]\right|^2 dt\right)^{1/2} \\
&= \left(2\pi \int_0^{2\pi} \sum_{\nu=m}^{\infty} r^{-2\nu} R^2[T_\nu]\, dt\right)^{1/2} \\
&= 2\pi \left(\sum_{\nu=m}^{\infty} r^{-2\nu} R^2[T_\nu]\right)^{1/2}.
\end{aligned}$$

Thus, we have proved that for rules with $R[\mathbb{P}_{m-1}] = \{0\}$,

$$\rho\big(Q, \mathcal{C}_M(\mathcal{E}_r)\big) \le 2M \left(\sum_{\nu=m}^{\infty} \frac{R^2[T_\nu]}{r^{2\nu}}\right)^{1/2}. \tag{4.50}$$

With another method, we can obtain a slight improvement of this result at the expense of simplicity:

THEOREM 4.9.5. For any quadrature rule Q with $R[\mathbb{P}_{m-1}] = \{0\}$, we have

$$\frac{2M\big|R[T_m]\big|}{r^m + r^{-m}} \le \rho\big(Q, \mathcal{C}_M(\mathcal{E}_r)\big) \le 2M \left(\sum_{\nu=m}^{\infty} \frac{R^2[T_\nu]}{r^{2\nu} + r^{-2\nu}}\right)^{1/2}.$$

PROOF. The right-hand side comes from combining (4.33) in Special Case IV with the estimate

$$\|f\| \le \sqrt{2\pi}\, \max_{z \in \mathcal{E}_r} |f(z)|.$$

The left-hand side is a consequence of

$$\frac{2MT_m}{r^m + r^{-m}} \in \mathcal{C}_M(\mathcal{E}_r);$$

see Theorem A.9(vi). □

Using Theorem 4.9.5, we deduce the following result.

THEOREM 4.9.6. For any rule Q with $R[\mathbb{P}_{m-1}] = \{0\}$, we have

$$\lim_{r \to \infty} r^m \rho\big(Q, \mathcal{C}_M(\mathcal{E}_r)\big) = 2M\big|R[T_m]\big|.$$

Up to now we have used the representation (4.45) in combination with (4.46), but other bounds are possible, for example:

$$|R[f]| \leq \frac{l(\partial\mathcal{G})}{2\pi} \sup_{z\in\mathcal{G}} \left| R\left[\frac{1}{z-\cdot}\right] \right| \cdot \sup_{z\in\mathcal{G}} |f(z)|, \tag{4.51}$$

where $l(\partial\mathcal{G})$ denotes the length of $\partial\mathcal{G}$. It has been observed that, often, (4.51) is slightly better adapted to circular domains of holomorphy than to $\mathcal{G} = \mathcal{E}_r$; this is in contrast to (4.46). Some relevant papers are those by Gautschi and Varga (1983), Gautschi (1992), Schira (1997), and Scherer and Schira (2000).

For any specific rule, Theorem 4.9.5 gives a lower bound on the error constant. To obtain lower bounds that hold for all rules, we start with the next theorem.

THEOREM 4.9.7 (Petras 1998a). *Let w be a positive weight function, and let $q_0, \ldots, q_n$ be the orthonormal polynomials (see Appendix A) with respect to w. Then*

$$\rho_n^{\text{best}}(\mathcal{C}_1(\mathcal{G})) \geq \left(\sum_{\nu=0}^{n} \|q_\nu\|_{\mathcal{G}}^2\right)^{-1}.$$

PROOF. Let $p = \sum_{\nu=0}^{n} c_\nu q_\nu$ be such that $p(x_\mu) = 0$ for all μ. Then

$$\rho(Q_n, \mathcal{C}_1(\mathcal{G})) \geq \frac{|R_n[p^2]|}{\|p^2\|_{\mathcal{G}}} = \frac{I[p^2]}{\|p^2\|_{\mathcal{G}}} = \frac{\sum_{\nu=0}^{n} c_\nu^2}{\|p^2\|_{\mathcal{G}}} \geq \frac{\sum_{\nu=0}^{n} c_\nu^2}{\left(\sum_{\nu=0}^{n} |c_\nu| \, \|q_\nu\|_{\mathcal{G}}\right)^2}.$$

Applying the Cauchy–Schwarz inequality to the denominator in the right-hand side proves the theorem. □

In the proof, we used weighted integration of the square of a polynomial with zeros at the nodes. It is therefore clear that the lower bound thus derived is monotone with respect to the weight function. In order to derive lower bounds in the $w = 1$ case, we would usually do calculations with Legendre polynomials, but this requires some technical effort. A simpler way which produces results that are almost as good is to find a weight function smaller than $w = 1$ with particularly simple orthogonal polynomials. For $w(x) = \sqrt{1-x^2}$, the system of orthonormal polynomials is $\sqrt{2/\pi}\, U_n$, $n = 0, 1, \ldots$, where U_n is the Chebyshev polynomial of the second kind (see Definition A.3). It can be shown, using Theorem A.9 and Definition A.3, that

$$U_n\left(\frac{z^2+1}{2z}\right) = \frac{z}{z^2-1}\left(z^{n+1} - z^{-n-1}\right) = z^n + z^{n-2} + \cdots + z^{-n}.$$

This means that

$$\|U_n\|_{\mathcal{E}_r} \leq r^n + r^{n-2} + \cdots + r^{-n} \leq \frac{r^n}{1-r^{-2}}.$$

Upon squaring and summing these values, we obtain the following result, which is valid in particular for $w = 1$:

COROLLARY 4.9.2. *Let $w(x) \geq \left(1-x^2\right)^{1/2} dx$; then*

$$\rho_n^{\text{best}}(\mathcal{C}_1(\mathcal{E}_r)) \geq \frac{\pi(1-r^{-2})^3}{2r^{2n}}.$$

To find the asymptotic behaviour of the lower bound, we need to know the orthonormal polynomials only approximately. Szegö developed a theory that yields such approximations under rather general assumptions on the weight function:

DEFINITION 4.9.1. A weight function w is said to be of *Szegö type* if

$$\int_0^\pi \ln w(\cos x)\, dx$$

exists, and its associated *Szegö function* is given by

$$D(z) := \exp\left(\frac{1}{4\pi}\int_{-\pi}^{\pi} \frac{1+ze^{-it}}{1-ze^{-it}} \ln\left(w(\cos t)|\sin t|\right) dt\right).$$

It is known that for any $\varepsilon > 0$,

$$\sqrt{2\pi}\, D(z^{-1}) \lim_{n\to\infty} z^{-n} q_n\left(\frac{z^2+1}{2z}\right) = 1$$

holds uniformly for $|z| > 1+\varepsilon$ (see, e.g., Szegö 1939: Theorem 12.1.2). Note that v is on the boundary of $\mathcal{E}_r$ if and only if it is representable as $v = (z^2+1)/(2z)$ with $|z| = r$. Hence, ellipses $\partial\mathcal{E}_r$ are asymptotically level lines of orthonormal polynomials. The relationship between D and q_n yields the following:

COROLLARY 4.9.3 (Petras 1998a). Let w be a Szegö-type weight function. Then, for $\mathcal{C}_M(\mathcal{E}_r)$,

$$\liminf_{n\to\infty} r^{2n}\rho_n^{\text{best}} \geq 2\pi M(1-r^{-2}) \cdot \min_{|z|=r} \left|D(z^{-1})\right|^2.$$

Taking Corollary 4.9.1 together with Corollary 4.9.3, we get the next result:

COROLLARY 4.9.4 (Petras 1998a). Let w be a Szegö-type weight function. Then, for $\mathcal{C}_M(\mathcal{E}_r)$,

$$\lim_{n\to\infty} \sqrt[2n]{\rho_n^{\text{best}}} = \lim_{n\to\infty} \sqrt[2n]{\rho\left(Q_n^{\mathrm{G}}, \mathcal{C}_M(\mathcal{E}_r)\right)} = \frac{1}{r}.$$

Hence, for most of the interesting weight functions, the Gaussian method is almost best.

For $w = 1$, we combine Theorem 4.9.7 with Theorem A.17. A short calculation gives

$$\rho_n^{\text{best}}\left(\mathcal{C}_M(\mathcal{E}_r)\right) \geq \pi M r^{-2n}(1-r^{-2})^2\left(1+o(1)\right). \tag{4.52}$$

More precise results can be found in Petras (1998a).

The Szegö class includes all weight functions that are bounded from below by the square of a non-trivial polynomial (since functions with logarithmic singularities are integrable). For such functions, Bakhvalov (1967) gave a lower bound of order r^{-2n} for the corresponding error constants. His method, requiring a transformation of the ellipses onto an infinite strip and vice versa, is more complicated.

The most obvious class of analytic functions is $\mathcal{C}_M(D_r)$, and we are interested in proving inequalities similar to the one in Corollary 4.9.3. These can be derived by means of a conformal mapping that maps some $\mathcal{E}_\rho$ onto D_r and $[-1,1]$ onto itself. This fairly complicated mapping m was described in Kober (1957). We have

$$m(z) = \operatorname{sn}\left(\frac{2K}{\pi}\arcsin z\right), \qquad r = \frac{\vartheta_3(0,\rho^{-4})}{\vartheta_2(0,\rho^{-4})},$$

where sn is a Jacobian elliptic function (the "sine-amplitude"), $\vartheta_i(x,y)$ are the Jacobian theta functions, and

$$K = K(r^{-2}) = \int_0^{\pi/2} \frac{du}{\sqrt{1-r^{-4}\sin^2 u}}$$

is the complete elliptic integral of the first kind; see Abramowitz and Stegun (1964: Chapter 16) for more information on these functions. The weight function w for the disc and the weight function v for the corresponding ellipse are related via

$$w(x) = \frac{\pi}{2K} v\left(\sin\left(\frac{\pi}{2K}\operatorname{asn} x\right)\right) \frac{\cos\left(\frac{\pi}{2K}\operatorname{asn} x\right)}{\sqrt{(1-x^2)(1-r^{-4}x^2)}},$$

where asn is the inverse function of sn. If, for the disc, we start with a Szegö-type weight function w, then v will also be of Szegö type. We can therefore apply Lemma 4.9.1 and obtain, after some calculations, the following result:

COROLLARY 4.9.5 (Petras 1998a). *Let w be a Szegö-type weight function. Then, for $\mathcal{C}_M(D_r)$,*

$$\lim_{n\to\infty} \sqrt[2n]{\rho_n^{\text{best}}} = \exp\left(-\frac{\pi K(\sqrt{1-r^{-4}})}{4K(r^{-2})}\right).$$

For ellipses as class-defining regions of analyticity and a weight function of Szegö type, the Gaussian method is almost best. For a disc, the transformation of the Gaussian method is almost best. However, each disc requires a different transformation and therefore yields a different almost-best rule. Clearly, this situation is quite unsatisfactory and raises the question of what we would lose if we were to always take the Gaussian rule. To answer this question, we need to investigate error constants for circular regions of analyticity. Obviously, we have

$$\rho(Q_n, \mathcal{C}_M(\mathcal{G}_1)) \le \rho(Q_n, \mathcal{C}_M(\mathcal{G}_2)) \quad \text{if } \mathcal{G}_2 \subset \mathcal{G}_1.$$

Hence, an upper bound for $\rho(Q_n^{\mathrm{G}}, \mathcal{C}_M(D_r))$ is given by the error constant for $\rho(Q_n^{\mathrm{G}}, \mathcal{C}_M(\mathcal{E}_\rho))$, where $\mathcal{E}_\rho$ is the largest ellipse that is contained in D_r, while a lower bound is obtained from (4.52) with the smallest ellipse $\mathcal{E}_{r+\sqrt{r^2+1}}$ that contains the disc. For $R_n[\mathbb{P}_m] = \{0\}$ and $w = 1$, this technique yields

$$\begin{aligned}\frac{4\pi r^2 M}{\left(r+\sqrt{r^2+1}\right)^{2n+2}} \cdot \left(1+o(1)\right) &\le \rho(Q_n, \mathcal{C}_M(D_r)) \\ &\le \frac{2M\|R_n\|}{\left(r+\sqrt{r^2-1}\right)^m \left(r-1+\sqrt{r^2-1}\right)}.\end{aligned}$$

The larger r is, the closer together the bounds are.

That the upper bound is asymptotically correct for increasing n can be seen from a closer investigation of the error associated with the functions $f(x) = 1/(r+\varepsilon-x)$ with $\varepsilon > 0$. This can be done with the help of its Chebyshev series; see Petras (1998a). One finally obtains

$$\lim_{n\to\infty} \sqrt[2n]{\frac{\rho(Q_n^{\mathrm{G}}, \mathcal{C}_M(D_r))}{\rho_n^{\text{best}}}} = \frac{1}{r+\sqrt{r^2-1}} \exp\left(\frac{\pi K\sqrt{1-r^{-4}}}{4K(r^{-2})}\right).$$

Thus, for larger r we lose only a little, whereas for small r the loss is considerable. In practice, one is most interested in the number of nodes necessary to guarantee that the error will be less than some $\varepsilon > 0$. Upon setting

$$c_r^{\mathrm{G}} := \lim_{n\to\infty} \sqrt[2n]{\rho(Q_n, \mathcal{C}_M(D_r))} \qquad \text{and} \qquad c_r^{\text{best}} := \lim_{n\to\infty} \sqrt[2n]{\rho_n^{\text{best}}},$$

we see that the Gaussian method, for instance, requires about $\ln c_r^{\mathrm{G}}/(2\ln\varepsilon)$ nodes and hence about $\ln c_r^{\mathrm{G}}/\ln c_r^{\mathrm{best}}$ times as many nodes as a best method. In the following table, we show some examples of these values.

r	c_r^{G}	c_r^{best}	$\ln c_r^{\mathrm{G}}/\ln c_r^{\mathrm{best}}$
1.1	0.642	0.517	1.488
1.5	0.382	0.342	1.114
2.0	0.268	0.252	1.047

Therefore, if we can only guarantee $f \in \mathcal{C}_M(D_{1.1})$, the Gaussian method requires about one and a half times as many nodes as the best method to guarantee the same error bound, while for $r \geq 2$ this effect on the number of nodes almost vanishes.

4.10. Convex functions

Strictly speaking, non-trivial classes of convex functions are not co-observations, because $-f$ is not necessarily convex if f is convex. Nevertheless, all the definitions from the abstract framework—such as $\underset{\vee}{Q}$, $\hat{Q}$, Info, ρ^{intr}, ρ^{opt}, reconstruction, and optimal and strongly optimal rules—can be carried through unchanged, because the symmetry condition is never actually used in those contexts. One major difference in the convex-function setting is that there need not be linear rules among the optimal estimation rules. We will see, therefore, that the theorem of Smolyak is not valid in the asymmetric situation.

Convexity is a property of a function that is sometimes more easily detected than, say, a bound on the derivative. Not every convex function has a bounded derivative. Nonetheless, we will see that convexity can produce much tighter error bounds than those derived from boundedness of the first derivative.

We shall focus on the following "modified estimation framework":

(1) $V = C[-1,1]$;
(2) $I[f] = \int_{-1}^{1} f(x)w(x)\,dx$ where $w > 0$;
(3) $O[f] = \big(f(x_1),\ldots,f(x_n)\big)$ where $-1 \leq x_1 < x_2 < \cdots < x_n \leq 1$;
(4) $\mathcal{C}_M^{\mathrm{FP}} := \{f \in C[-1,1] \,:\, f \text{ is convex and } f(1) - 2f(0) + f(-1) \leq M\}$.

This is not an SEF because $\mathcal{C}_M^{\mathrm{FP}}$ is not symmetric for $M > 0$. In Förster and Petras (1990b), this framework was chosen because it allows the development of an elegant theory based on Peano kernels. Moreover, the easily accessible quantity $[f(1) - 2f(0) + f(-1)]/2$ gives the distance between $\big(0, f(0)\big)$ and the midpoint of the line from $\big(-1, f(-1)\big)$ to $\big(1, f(1)\big)$; it can therefore be considered as a measure of the convexity of a function.

In our situation, it is often easy to find a "good" estimation rule. For what follows, recall that the space $\mathcal{S}_r(\xi_1,\ldots,\xi_m)$ of spline functions was introduced after Definition 3.3.1, and note that $\mathcal{S}_1(x_2,\ldots,x_{n-1})$ for $n = 2$ is just $\mathbb{P}_1$.

THEOREM 4.10.1. Let $n \geq 2$. In our modified estimation framework, the projection P into $\mathcal{S}_1(x_2,\ldots,x_{n-1})$ based on $O[f] = \{f(x_1),\ldots,f(x_n)\}$ is a reconstruction. If $x_1 = -1$ and $x_n = 1$, we have

$$\hat{Q} = I \circ P.$$

PROOF. We have to show that $f \in \mathcal{C}_M^{\text{FP}}$ implies $s := P[f] \in \mathcal{C}_M^{\text{FP}}$. The convexity of s is obvious, so it suffices to prove

$$s(1) - 2s(0) + s(-1) \leq f(1) - 2f(0) + f(-1) =: l[f]. \tag{4.53}$$

For the proof, we use the fact that between any two points P_1 and P_2 on the graph of a convex function (with P_1 lying to the left of P_2), the function's graph lies below (or on) the line through P_1 and P_2, whereas on the left of P_1 and on the right of P_2, the graph lies above (or on) the line. This implies that $s(1) \leq f(1)$, $s(-1) \leq f(-1)$ and, in the case where $0 \in [x_1, x_n]$, $s(0) \geq f(0)$. In this case, the proof of the first statement is complete.

Now let $x_n < 0$. We cannot reuse the previous proof, because now $s(0) \leq f(0)$ and we may have strict inequality. Note that s restricted to $[x_{n-1}, 1]$ is a linear function which interpolates f at x_{n-1} and x_n. Let $s_1 \in \mathbb{P}_1$ interpolate f at x_n and 0, and denote by m_1 the slope of p and by m the slope of s at x_n. The convexity implies that $f(1) \geq s_1(1)$ and $m_1 \geq m$, so

$$\begin{aligned} f(1) - 2f(0) + f(-1) &\geq s_1(1) - 2s_1(0) + s(-1) = (1 + x_n)m_1 - f(x_n) + s(-1) \\ &\geq (1 + x_n)m - f(x_n) + s(-1) = s(1) - 2s(0) + s(-1) \end{aligned}$$

and we obtain the first assertion in this case as well.

The second part of the theorem follows immediately. □

For $n = 1$, we can take the projection onto $\mathbb{P}_0$ as the reconstruction. In the case of $n > 1$, $w = 1$ and $x_\nu = -1 + 2(\nu - 1)/(n - 1)$, we have

$$\hat{Q}[f] = Q_n^{\text{Tr}}[f].$$

It is much more difficult to find $\check{Q}$ for convex functions. If $w = 1$ and $\mathcal{C}_M^{\text{FP}}$ is replaced by the class of all continuous convex functions, Heindl (1982) determined strongly optimal estimation rules for up to nine equidistant nodes. Niessner (1985) presented an algorithm for the computation of $\check{Q}[f]$ for arbitrary f.

Next, we explain how the error of a quadrature rule can be derived systematically.

THEOREM 4.10.2 (Förster and Petras 1990b). *In the modified estimation framework, a quadrature rule Q_n with $\deg Q_n \geq 1$ has the worst-case error*

$$\rho(Q_n, \mathcal{C}_M^{\text{FP}}) = M \cdot \sup_{x \in]-1,1[} \frac{|K_2(x)|}{1 - |x|}. \tag{4.54}$$

If Q_n is not exact for $\mathbb{P}_1$, it has no finite error.

For the proof we shall use a lemma which, in fact, is not restricted to our special modified estimation framework.

LEMMA 4.10.1. *The error of a quadrature rule Q_n for a convex function is positive if Q_n is positive definite of order two and negative if Q_n is negative definite of order two.*

PROOF. The reason is that we can approximate a convex function f arbitrarily well by a twice-differentiable function: take the piecewise linear function $s \in \mathcal{S}_1(\xi_1, \ldots, \xi_{2m-1})$, where $\xi_\nu = -1 + \nu/m$, which satisfies $s(-1) = f(-1)$, $s(1) = f(1)$ and $s(\xi_\nu) = f(\xi_\nu)$ for $\nu = 1, \ldots, 2m - 1$. Then $\|f - s\|_\infty \leq \omega(f; 1/m)$, which can be arbitrarily close to zero. Now smooth s at its knots, thereby preserving convexity, and change s by an arbitrarily small amount. $\square$

PROOF OF THEOREM 4.10.2. Note that the functional given by

$$L[f] = f(-1) - 2f(0) + f(1)$$

has second Peano kernel

$$K_2^L[f] = 1 - |x|.$$

Let

$$\sup_{x \in]-1,1[} \frac{|K_2(x)|}{1-|x|} = c.$$

Then the quadrature rule $Q_*[f] = Q_n[f] + c \cdot L$ is negative definite of order two. According to the lemma, this means that

$$Q_n[f] + c \cdot L[f] \geq I[f] \quad \text{for convex } f.$$

Analogously, we can prove that $Q_n[f] - c \cdot L[f] \leq I[f]$. Thus we have shown that the left-hand side in (4.54) is less than or equal to the right-hand side. If $|K_2(x_0)|/(1-|x_0|) > c - \varepsilon$ where $\varepsilon > 0$, then the function $f \in \mathcal{C}_M^{\text{FP}}$ given by $f(x) = M \cdot (x - x_0)_+/(1-|x_0|)$ yields $\big|R[f]\big| > M \cdot (c - \varepsilon)$. Therefore, equality holds in (4.54).

The second assertion follows from $\mathbb{P}_1 \subset \mathcal{C}_M^{\text{FP}}$. $\square$

A geometric interpretation of this theorem is that the error is M times the height of the flattest hat function (a function that is zero outside $[-1,1]$, positive on $]-1,1[$ and symmetric) which fits above $|K_2|$.

We now establish a lower bound for the error; it implies that, for equidistant nodes, the error is no better than $O(n^{-1})$. There are better rules: see Theorem 4.10.4 and Corollary 6.5.1.

THEOREM 4.10.3. For $w = 1$ and $n \geq 3$, we have

$$\rho(Q_n, \mathcal{C}_M^{\text{FP}}) \geq \frac{M}{6+4\sqrt{2}} \cdot \max\{1+x_2, 1-x_{n-1}\}.$$

The proof is based on the following lemma, which will also be useful for proving the next theorem.

LEMMA 4.10.2. Let $\rho(Q_n, \mathcal{C}_M^{\text{FP}}) \leq c \cdot M$. Then $x_1 \leq 2c - 1$. Furthermore,

$$1 + x_{\nu+1} \leq \left(\sqrt{1+x_\nu} + 2\sqrt{c}\right)^2, \tag{4.55}$$

with equality if and only if

$$K_2\big|_{[x_\nu, x_{\nu+1}]}(x) = \frac{(x-\xi)^2}{2} - c(1+x), \qquad \xi = x_\nu + 2\sqrt{c(1+x_\nu)}. \tag{4.56}$$

PROOF. Theorem 4.10.2 implies that $|K_2(x)| \leq c(1+x)$. One consequence is that

$$K_2(x_1) = \frac{(x_1+1)^2}{2} \leq c(1+x_1) \quad \Longrightarrow \quad x_1 \leq 2c - 1.$$

The Peano kernel at $x \in [x_\nu, x_{\nu+1}]$ has to be between $-c(1+x)$ and $l(x) := c(1+x)$. The polynomial

$$p(x) = \frac{(x-\xi)^2}{2} - c(1+x)$$

has the properties that $p(x_\nu) = l(x_\nu)$, $p(\xi) = -l(\xi)$, $p'(\xi) = -l'(\xi)$ and $p(x_{\nu+1}) = l(x_{\nu+1})$; therefore it lies between $-l(x)$ and $l(x)$ on $[x_\nu, x_{\nu+1}]$. Any other quadratic polynomial $q(x)$ with leading coefficient $\frac{1}{2}$ satisfies either $q(x_\nu) > l(x_\nu)$, $q(\xi) < -l(\xi)$ or $q(x_{\nu+1}) > l(x_{\nu+1})$, because $p - q$ is a non-trivial linear function. Thus, the lemma is proved. □

PROOF OF THEOREM 4.10.3. The previous lemma tells us that

$$1 + x_2 \le \left(\sqrt{1+x_1} + 2\sqrt{c}\right)^2 \le \left(\sqrt{2c} + 2\sqrt{c}\right)^2 = \left(6 + 4\sqrt{2}\right)c;$$

that is, $\rho(Q_n, \mathcal{C}_M^{\mathrm{FP}}) \ge (1+x_2)\cdot M/(6+4\sqrt{2})$. From symmetry arguments, it follows that the same inequality holds with $1 + x_2$ in place of $1 - x_{n-1}$. □

In the next theorem, we shall see that we can find quadrature rules with $O(n^{-2})$ error.

THEOREM 4.10.4 (Petras 1993b). For $w(x) = 1$, we have

$$\inf_{Q_n \in \mathbf{Q}_n} \rho(Q_n, \mathcal{C}_M^{\mathrm{FP}}) = c_n \cdot M \quad \text{where } c_n := \frac{1}{(n - 1 + \sqrt{2}\,)^2}.$$

The unique quadrature rule for which the error attains this infimum is Q_n^{FP} with nodes

$$1 + x_\nu = 1 - x_{n+1-\nu} = \left[2(\nu - 1) + \sqrt{2}\right]^2 c_n \quad \text{for } \nu = 1, 2, \ldots, \left\lfloor \frac{n+1}{2} \right\rfloor$$

and coefficients

$$a_1 = a_n = \left(1 + \sqrt{2}\right)^2 c_n,$$

$$a_\nu = a_{n+1-\nu} = 4\left[2(\nu - 1) + \sqrt{2}\right] c_n \quad \text{for } \nu = 2, \ldots, \left\lfloor \frac{n}{2} \right\rfloor$$

and

$$a_{(n+1)/2} = \left[4n - 6 + 4\sqrt{2}\right] c_n \quad \text{for odd } n.$$

PROOF. Peano kernel theory and some calculations yield

$$\rho(Q_n^{\mathrm{FP}}, \mathcal{C}_M^{\mathrm{FP}}) = c_n \cdot M.$$

Next, note that symmetry and (4.55) imply that $x_n \ge 1 - 2c$ and

$$1 - x_{\nu-1} \le \left(\sqrt{1 - x_\nu} + 2\sqrt{c}\right)^2. \tag{4.57}$$

The nodes x_ν in the statement of the theorem satisfy (4.55) with equality for $c = c_n$ and $\nu = 1, 2, \ldots, \lfloor (n-1)/2 \rfloor$ as well as (4.57) with equality for $\nu = \lfloor (n+4)/2 \rfloor, \ldots, n$. Now suppose that $\rho(Q_n, \mathcal{C}_M^{\mathrm{FP}}) \le c_n \cdot M$ for

$$Q_n[f] = \sum_{\nu=1}^{n} b_\nu f(y_\nu).$$

The right-hand side of (4.55) is strictly increasing as a function of x_ν, while the right-hand side of (4.57) is strictly decreasing. This has the following two consequences:

- $y_\nu \le x_\nu$ for $\nu = 1, \ldots, \lfloor (n+1)/2 \rfloor$ and $y_\nu \ge x_\nu$ for $\nu = \lfloor (n+2)/2 \rfloor, \ldots, n$.
- $y_\nu < x_\nu$ implies $y_{\nu+1} < x_{\nu+1}$ as long as $x_{\nu+1} \le 0$, and $y_\nu > x_\nu$ implies $y_{\nu-1} > x_{\nu-1}$ as long as $x_{\nu+1} \ge 0$.

Let n be odd. If we had strict inequality in (4.55) or (4.57) for the nodes y_ν and some ν, we would obtain either $y_{(n+1)/2} < x_{(n+1)/2} = 0 \le y_{(n+1)/2}$ or $y_{(n+1)/2} \le x_{(n+1)/2} = 0 < y_{(n+1)/2}$, i.e. a contradiction. Therefore, the optimal nodal system is unique. Now the unique structure of the Peano kernel given in (4.56) gives $a_\nu = b_\nu$.

For even n we argue similarly. In this case, $y_\nu \ne x_\nu$ or $b_\nu \ne a_\nu$ implies either $K_2(R_n, 0-0) > 0 \ge K_2(R_n, 0+0)$ or $K_2(R_n, 0-0) \ge 0 > K_2(R_n, 0+0)$, although Q_n cannot have a node at 0. □

We have thus determined the best quadrature rule with n nodes; however, this is not the best estimation rule.

THEOREM 4.10.5. For the modified estimation framework and odd n, the estimation rule

$$\widehat{Q}[f] := Q_n[f] + \frac{M}{(n+1)^2}$$

with

$$Q_n[f] = \frac{4}{(n+1)^2}\left\{(n+1)f(0)+2\sum_{\nu=1}^{(n-1)/2}\left[f\left(-1+\frac{4\nu^2}{(n+1)^2}\right)+f\left(1-\frac{4\nu^2}{(n+1)^2}\right)\right]\right\}$$

satisfies

$$\rho(\widehat{Q}_n, \mathcal{C}_M^{\mathrm{FP}}) = \frac{M}{(n+1)^2}.$$

PROOF. Using Theorem 4.10.2, we see that

$$Q^*_{n+2}[f] = Q_n[f] + L[f] \quad \text{with } L[f] := \frac{1}{(n+1)^2}\Big(f(1) - 2f(0) + f(-1)\Big)$$

is a quadrature rule with error $M/(n+1)^2$. Hence

$$-L[f] \le R^*_{n+2}[f] \le L[f],$$

i.e.

$$0 \le R_n[f] \le 2L[f],$$

which shows that

$$-\frac{M}{(n+1)^2} \le \widehat{R}[f] \le 2L[f] - \frac{M}{(n+1)^2} \le \frac{M}{(n+1)^2}.$$

We have equality on the right-hand side with, for example, $f_0(x) = M \cdot x_+$, as can be seen by taking $R_n[f_0] = K_2(0)$ where K_2 is the second Peano kernel of R_n. □

This result shows that Smolyak's theorem does not hold in our modified estimation framework, since there is an estimation rule of the form

$$\widehat{Q}[f] = a_0 + \sum_{\nu=1}^{n} a_\nu f(x_\nu) \tag{4.58}$$

that has smaller error than any quadrature rule with n nodes. Rules of the form (4.58) are said to be *affine*. Sukharev (1986) showed that if we omit only the assumption of symmetry from the SEF definition, then there will always be an affine optimal estimation rule.

Investigations of various other classes of convex functions can also be found in the literature. Glinkin (1984) studied

$$\mathcal{C}_{uv}^{\mathrm{conv}} := \{f : f \text{ is convex on } [-1,1],\ f'(-1) \ge u \text{ and } f'(1) \le v\},$$

where derivatives should be understood as one-sided limits. Zwick (1988) showed that for $w(x) = 1$ and the above class of functions, the best estimation rule based on n nodes is

$$Q_n[f] = Q_n^{\mathrm{Mi}}[f] + \frac{(v-u)}{4n^2}.$$

The disadvantage of function classes of this type is that there exist convex functions which belong to none of the $\mathcal{C}_{uv}^{\mathrm{conv}}$. Furthermore, it may be difficult to obtain or estimate the derivatives of f to verify membership in a class.

Brass (1982) derived best quadrature rules for the class

$$\mathcal{C}_{M,\infty}^{\mathrm{conv}} := \{f : f \text{ is convex on } [-1,1] \text{ and } \|f\|_\infty \le M\}.$$

These are closely related to the affine rules of Theorem 4.10.5 (see Petras 1993b).

For $w(x) = 1$ and $\mathcal{C}_M^{\mathrm{FP}}$, the best quadrature rules have errors of the order $O(n^{-2})$. This is also true for rather general weight functions, but not for all of them—for example, it does not hold for

$$w(x) = \begin{cases} (1+x)^{-1}\ln^{-2}(1+x) & \text{if } -1 < x \le -1+e^{-2}, \\ e^2/4 & \text{otherwise;} \end{cases}$$

see Förster and Petras (1990b).

CHAPTER 5

Quadrature Rules of Interpolatory Type

This chapter focuses on a special class of quadrature rules that has attracted a great deal of interest since the beginnings of quadrature theory and for which a large body of results has been developed.

5.1. Recapitulation

Throughout this chapter, we assume that $w(x) \geq 0$ and $\int_a^b w(x)\,dx > 0$.

DEFINITION 5.1.1. $Q \in \mathbf{Q}_n$ is called a *quadrature rule of interpolatory type* if

$$R[\mathbb{P}_{n-1}] = \{0\}. \tag{5.1}$$

Of all the quadrature rules that have ever been recommended for practical use, a considerable proportion is of this type. We begin our discussion of such rules by specializing some theorems from Chapters 2 and 3.

THEOREM 5.1.1. For any given $x_1, x_2, \ldots, x_n \in [a, b]$, there is exactly one quadrature rule of interpolatory type with these points as nodes.

PROOF. See Theorem 2.2.2. □

COROLLARY 5.1.1. Suppose that $w(b-x) = w(a+x)$ for $x \in [0, b-a]$ and that $x_\nu - a = b - x_{n+1-\nu}$ for $\nu = 1, \ldots, n$. Then, the coefficients of the corresponding quadrature rule of interpolatory type satisfy $a_\nu = a_{n+1-\nu}$.

PROOF. Suppose that

$$Q[f] = \sum_{\nu=1}^{n} a_\nu f(x_\nu)$$

is of interpolatory type. For $f \in \mathbb{P}_{n-1}$, we have

$$Q'[f] := \sum_{\nu=1}^{n} a_\nu f(x_{n+1-\nu}) = Q[f(a+b-\cdot)] = I[f(a+b-\cdot)] = I[f]$$

in our situation. Hence, Q' is a rule of interpolatory type with the same nodes as Q and is therefore identical to Q. □

DEFINITION 5.1.2. A quadrature rule with $x_\nu - a = b - x_{n+1-\nu}$ and $a_\nu = a_{n+1-\nu}$ for $\nu = 1, \ldots, n$ is said to be *symmetric*.

The description of these quadrature rules as "interpolatory" is justified by the following theorem.

THEOREM 5.1.2. Let Q_n be the quadrature rule of interpolatory type with nodes $x_1, \ldots, x_n$. Then

$$Q_n[f] = \int_a^b \operatorname{intpol}(x_1, \ldots, x_n)[f](x) w(x)\, dx,$$

where $\operatorname{intpol}(x_1, \ldots, x_n)[f]$ denotes the polynomial that interpolates f at the nodes $x_1, \ldots, x_n$.

PROOF. See Theorem 2.2.3. □

Now, if

$$Q_n[f] = \sum_{\nu=1}^{n} a_\nu f(x_\nu)$$

is the rule from Theorem 5.1.2, then the Lagrange interpolation formula immediately yields

$$a_\nu = \int_a^b \prod_{\substack{k=1 \\ k\neq\nu}}^{n} \frac{x - x_k}{x_\nu - x_k}\, w(x)\, dx, \tag{5.2}$$

or

$$a_\nu = \int_a^b \frac{u(x)}{(x - x_\nu)u'(x_\nu)}\, w(x)\, dx \quad \text{with } u(x) = \prod_{k=1}^{n} (x - x_k).$$

It is possible that one or more of these a_ν are zero, in which case we would get a rule satisfying (5.1) with fewer than n nodes. A general result concerning this problem can be obtained from Theorem 2.2.1:

THEOREM 5.1.3. There exists a quadrature rule with nodes $x_1, \ldots, x_n$ and $R[\mathbb{P}_{n+s}] = \{0\}$ if and only if

$$\int_a^b p(x) \prod_{\nu=1}^{n} (x - x_\nu) w(x)\, dx = 0$$

for all $p \in \mathbb{P}_s$.

An easy consequence (upon setting $p(x) = \prod_{\nu=1}^{n}(x - x_\nu)$) is the following:

THEOREM 5.1.4. There is no $Q \in \mathbf{Q}_n$ with $R[\mathbb{P}_{2n}] = \{0\}$.

This theorem is particularly interesting because there exists $Q \in \mathbf{Q}_n$ with $R[\mathbb{P}_{2n-1}] = \{0\}$. The latter was proved in Theorem 3.3.3, and a different proof will be given in Chapter 6. Here, we add a statement of uniqueness.

THEOREM 5.1.5. There is a unique $Q \in \mathbf{Q}_n$ with $R[\mathbb{P}_{2n-1}] = \{0\}$.

DEFINITION 5.1.3. The rule Q from Theorem 5.1.5 is called the *nth Gaussian rule* and is denoted by Q_n^{G}. The sequence $\left(Q_n^{\mathrm{G}}\right)_{n\in\mathbb{N}}$ is called the Gaussian method.

These are the "classical" Gaussian rules, and we discussed a generalization of them in Section 3.4.

PROOF OF THEOREM 5.1.5. We prove only uniqueness here. Suppose that $Q_n^{(1)}$ and $Q_n^{(2)}$ are two rules such that $Q_n^{(i)} \in \mathbf{Q}_n$ and $R_n^{(i)}[\mathbb{P}_{2n-1}] = \{0\}$ for $i = 1, 2$. Then, for each α,

$$Q_{2n} := \alpha Q_n^{(1)} + (1 - \alpha) Q_n^{(2)}$$

is of interpolatory type, and all these rules have the same nodes. This contradicts the uniqueness given by Theorem 5.1.1. □

The polynomial degree of exactness will be important in many of our investigations. We shall write $\deg Q$, i.e. "degree of a quadrature rule", to mean

$$\deg Q := \sup \{\nu : R[\mathbb{P}_\nu] = \{0\}\}.$$

The statement

$$\|Q\| = \sum_{\nu=1}^{n} |a_\nu| \geq \sum_{\nu=1}^{n} a_\nu = I[1]$$

holds for any rule of interpolatory type. Therefore, we have the smallest norm if Q is positive.

If the nodes are chosen randomly, then positivity of the interpolatory-type rule generated would be an exception. Indeed, we would expect around half of the coefficients to be negative (see Section 5.2). A lower bound on the number of positive coefficients is given by the following.

THEOREM 5.1.6 (Steklov 1918). *Any quadrature rule Q with $\deg Q \geq m$ has at least $\lfloor (m+2)/2 \rfloor$ positive coefficients.*

PROOF. Let $\mathcal{M} := \{\nu : a_\nu > 0\}$. If $\mathcal{M}$ were to have fewer than $\lfloor (m+2)/2 \rfloor$ elements, then Q would be exact for

$$p(x) := \prod_{\nu \in \mathcal{M}} (x - x_\nu)^2.$$

So

$$\sum_{\nu \notin \mathcal{M}} a_\nu p(x_\nu) = Q[p] = \int_a^b p(x) w(x)\, dx > 0,$$

and not all a_ν with $\nu \notin \mathcal{M}$ can be less than or equal to zero. □

Theorem 5.1.6 says, in particular, that rules $Q \in \mathbf{Q}_n$ with $R[\mathbb{P}_{2n-2}] = \{0\}$ are positive and therefore good rules with respect to the norm criterion. This is not true of all rules $Q \in \mathbf{Q}_n$ with $\deg Q = 2n-3$, as seen in the following theorem.

THEOREM 5.1.7 (Micchelli and Rivlin 1973; Förster 1992). *Given any positive number g, there is a rule $Q \in \mathbf{Q}_n$ with $\deg Q = 2n-3$ and $\|Q\| > g$.*

PROOF. We apply Theorem 3.3.3 to the strictly positive linear functional I^* defined by

$$I^*[f] := I[f] + g \cdot f(z),$$

where I is the integral under consideration and $z \in [a,b]$ is any number distinct from all the nodes of Q_{n-1}^{G} (the Gaussian rule for I). There is an $(n-1)$-point Gaussian formula for the functional I^*, which means that there are x_ν and a_ν such that

$$I[f] + g \cdot f(z) = \sum_{\nu=1}^{n-1} a_\nu f(x_\nu) \quad \text{for all } f \in \mathbb{P}_{2n-3},$$

i.e.

$$I[f] = \sum_{\nu=1}^{n-1} a_\nu f(x_\nu) - g \cdot f(z) =: Q[f] \quad \text{for all } f \in \mathbb{P}_{2n-3}.$$

We have $z \notin \{x_1, \ldots, x_{n-1}\}$, because otherwise Q would be a quadrature rule with $n-1$ nodes and $\deg Q \geq 2n-3$, i.e. $Q = Q_{n-1}^{\mathrm{G}}$, and this would contradict the choice of z. □

By specializing Theorem 3.4.1, we can obtain an instructive necessary condition for positivity:

THEOREM 5.1.8 (Bernstein 1936). *Let $Q \neq Q_n^{\mathrm{G}}$ be a positive quadrature rule with $R[\mathbb{P}_{2n-1}] = \{0\}$, and denote the nodes of Q_n^{G} by $x_1^{\mathrm{G}} < x_2^{\mathrm{G}} < \cdots < x_n^{\mathrm{G}}$. Then Q has a node in each of the intervals $[a, x_1^{\mathrm{G}}[,\ ]x_1^{\mathrm{G}}, x_2^{\mathrm{G}}[,\ ]x_2^{\mathrm{G}}, x_3^{\mathrm{G}}[, \ldots,\ ]x_n^{\mathrm{G}}, b]$.*

We shall discuss the distribution of nodes of Q_n^{G} in the next chapter. The main point is that for most w of interest (including all $w > 0$), this distribution is far from being equidistant. For example, two consecutive nodes near the boundary of $[a, b]$ have distance $O(n^{-2})$.

One of the most useful applications of the notion of norm is Lebesgue's inequality (Theorem 3.1.1),

$$|R_n[f]| \leq \|R_n\| \operatorname{dist}(f, \mathbb{P}_{n-1}), \tag{5.3}$$

which leads to a wide variety of error bounds.

Let us summarize various results from approximation theory concerning the degree of approximation by polynomials that are applicable here. For simplicity, we consider the fundamental interval $[-1, 1]$.

$$\operatorname{dist}(f, \mathbb{P}_n) \leq \omega\left(\frac{\pi}{n+1}\right) \qquad \text{if } f \in C_\omega^{\mathrm{mc}}, \tag{5.4}$$

$$\operatorname{dist}(f, \mathbb{P}_n) \leq \frac{1}{2}\omega\left(\frac{\pi}{n+1}\right) \qquad \text{if } f \in C_\omega^{\mathrm{mc}} \text{ and } \omega \text{ is concave}, \tag{5.5}$$

$$\operatorname{dist}(f, \mathbb{P}_n) \leq \mathcal{K}_r \frac{(n+1-r)!}{(n+1)!} M \qquad \text{if } f \in C_M^{(r)} \tag{5.6}$$

(where $\mathcal{K}_r$ is the Favard constant defined in Theorem 4.3.4),

$$\operatorname{dist}(f, \mathbb{P}_n) \leq \frac{1}{r!\, 2^{r-1}} \left(\frac{(2r)!(n+1-r)!}{(n+r)!(2r-1)}\right)^{1/2} M \qquad \text{if } f \in C_M^{(r)}, \tag{5.7}$$

$$\operatorname{dist}(f, \mathbb{P}_n) \leq \frac{M}{(n+1)!\, 2^n} \qquad \text{if } f \in C_M^{(n+1)}, \tag{5.8}$$

$$\operatorname{dist}(f, \mathbb{P}_n) \leq \frac{2M}{r^n(r-1)} \qquad \text{if } f \in C_M(\mathcal{E}_r), \tag{5.9}$$

$$\operatorname{dist}(f, \mathbb{P}_n) \leq \frac{8}{\pi} \frac{M}{r^{n+1} + r^{-n-1}} \qquad \text{if } f \in C_M(\mathcal{E}_r). \tag{5.10}$$

The bounds (5.4) and (5.5) were proved by Korneichuk (1991). For (5.6) see Sinwel (1981), and for (5.7) see Brass and Förster (1987). Inequalities (5.8) and (5.9) are classic results of Bernstein, presented in most books on approximation theory. Inequality (5.10) is a result of Achieser (1956); see also Schönhage (1971).

It should be noted that proofs of important theorems via (5.3) tend to be quite simple, because the technical difficulties are transferred to approximation theory.

A quadrature method of interpolatory type is a sequence $(Q_n)_{n \in \mathbb{N}}$ of quadrature rules of interpolatory type. By specializing Theorem 3.1.2 and using the Weierstrass approximation theorem, we obtain the following:

THEOREM 5.1.9. A quadrature method $(Q_n)_{n\in\mathbb{N}}$ of interpolatory type is convergent (for all $f \in C[a,b]$) if and only if $(\|Q_n\|)_{n\in\mathbb{N}}$ is bounded. In particular, all positive methods of interpolatory type are convergent.

Thus, the Gaussian method is always convergent. The existence of convergent interpolatory quadrature methods is remarkable in view of a theorem of Faber (see, e.g., Cheney 1966: p. 215) which says that for any triangular scheme $(x_{\nu,n})$ with $\nu = 1,\dots,n$, $n = 1,2,\dots$ and $x_{\nu,n} \in [a,b]$, there exists an $f \in C[a,b]$ such that

$$\limsup_{n\to\infty} \|\operatorname{intpol}(x_{1,n},\dots,x_{n,n})[f]\| = \infty.$$

We end this section with an example of an interpolatory quadrature method which is unique in terms of its simplicity:

EXAMPLE 5.1.1. The Gauss–Chebyshev (also known as the Hermite or Mehler) method is the interpolatory quadrature method for

$$I[f] = \int_{-1}^{1} \frac{f(x)}{\sqrt{1-x^2}}\,dx$$

with the nodes

$$x_{\nu,n} := -\cos\Big(\frac{2\nu-1}{2n}\pi\Big), \qquad \nu = 1,\dots,n.$$

Its explicit form is

$$Q_n[f] = \frac{\pi}{n}\sum_{\nu=1}^{n} f\Big(-\cos\Big(\frac{2\nu-1}{2n}\pi\Big)\Big). \tag{5.11}$$

In fact, Q_n is even a Gaussian rule, which means that

$$Q_n[p] = I[p] \tag{5.12}$$

for all $p \in \mathbb{P}_{2n-1}$. Equation (5.12) has to be proved for all p taken from a basis of $\mathbb{P}_{2n-1}$. A very convenient basis is the one consisting of Chebyshev polynomials (see Appendix A): $T_k(x) = \cos(k\arccos x)$ for $k = 0,\dots,2n-1$. We readily see that $Q_n[T_0] = \pi = I[T_0]$ and $I[T_k] = 0$ for $k > 0$. For $k > 0$, we have

$$\begin{aligned}
&\frac{2n}{\pi}\sin\Big(\frac{k\pi}{2n}\Big)Q_n[T_k]\\
&\quad= \sum_{\nu=1}^{n} 2\sin\Big(\frac{k\pi}{2n}\Big)T_k\Big(-\cos\Big(\frac{2\nu-1}{2n}\pi\Big)\Big)\\
&\quad= \sum_{\nu=1}^{n} 2\sin\Big(\frac{k\pi}{2n}\Big)\cos\Big(k\arccos\Big(\cos\Big(\frac{2n-2\nu+1}{2n}\pi\Big)\Big)\Big)\\
&\quad= \sum_{\nu=1}^{n} 2\sin\Big(\frac{k\pi}{2n}\Big)\cos\Big(k\,\frac{2n-2\nu+1}{2n}\,\pi\Big)\\
&\quad= \sum_{\nu=1}^{n}\Big(\sin\Big(k\,\frac{2n-2\nu+2}{2n}\,\pi\Big) - \sin\Big(k\,\frac{2n-2\nu}{2n}\,\pi\Big)\Big) = \sin k\pi - \sin 0 = 0.
\end{aligned}$$

Now, $Q_n[T_k] = 0$ for $k = 1,\dots,2n-1$ follows from the fact that $\sin\frac{k\pi}{2n} \neq 0$.

5.2. The Newton–Cotes method

DEFINITION 5.2.1. The quadrature rule of interpolatory type with nodes

$$x_\nu = a + (\nu - 1)h, \qquad h = \frac{b-a}{n-1}, \quad \nu = 1, \dots, n$$

and $w = 1$ is called the nth Newton–Cotes rule and is denoted by Q_n^{NC}. The sequence $\left(Q_n^{\mathrm{NC}}\right)_{n=2,3,\dots}$ is called the Newton–Cotes method.

Newton–Cotes rules use a rather obvious choice of nodes; we shall see that, at least when n is not very small, they have some undesirable properties. In fact, the Newton–Cotes method provides an important example illustrating what can go wrong if the nodes of interpolatory quadrature rules are not chosen with great care.

We begin our discussion with a simple remark concerning the interval of integration. The Newton–Cotes rule on the interval $[a, b]$,

$$\sum_{\nu=1}^{n} a_\nu f\big(a + (\nu - 1)h\big) \quad \text{with } h = \frac{b-a}{n-1},$$

and the Newton–Cotes rule on the interval $\left[\widetilde{a}, \widetilde{b}\,\right]$,

$$\sum_{\nu=1}^{n} \widetilde{a}_\nu f\big(\widetilde{a} + (\nu - 1)\widetilde{h}\big) \quad \text{with } \widetilde{h} = \frac{\widetilde{b}-\widetilde{a}}{n-1},$$

are related through

$$\widetilde{a}_\nu = \frac{\widetilde{b}-\widetilde{a}}{b-a} \cdot a_\nu. \tag{5.13}$$

This can be shown by means of an affine transformation of the intervals, which does not change the characteristic property $R[\mathbb{P}_{n-1}] = \{0\}$. Owing to (5.13), we can restrict our investigation to the fundamental interval $[-1, 1]$ without loss of generality. The same holds for any rule of interpolatory type with $w = 1$.

EXAMPLE 5.2.1. For small n, the Newton–Cotes rules on $[-1, 1]$ are:

$$\begin{aligned}
Q_2^{\mathrm{NC}}[f] &= f(-1) + f(1),\\
Q_3^{\mathrm{NC}}[f] &= \tfrac{1}{3} f(-1) + \tfrac{4}{3} f(0) + \tfrac{1}{3} f(1),\\
Q_4^{\mathrm{NC}}[f] &= \tfrac{1}{4} f(-1) + \tfrac{3}{4} f\left(-\tfrac{1}{3}\right) + \tfrac{3}{4} f\left(\tfrac{1}{3}\right) + \tfrac{1}{4} f(1),\\
Q_5^{\mathrm{NC}}[f] &= \tfrac{7}{45} f(-1) + \tfrac{32}{45} f\left(-\tfrac{1}{2}\right) + \tfrac{12}{45} f(0) + \tfrac{32}{45} f\left(\tfrac{1}{2}\right) + \tfrac{7}{45} f(1).
\end{aligned}$$

Note that Q_2^{NC} coincides with Q_2^{Tr} and Q_3^{NC} with Q_3^{Si} (see Example 2.2.1).

The main drawback of the Newton–Cotes method is that the corresponding norms increase exponentially:

THEOREM 5.2.1. For even n,

$$\|Q_{n+1}^{\mathrm{NC}}\| \geq \frac{b-a}{n} \left(\frac{2^{n+1}}{3n(n+1)} - 2 \right).$$

A similar, but slightly more involved, argument yields the analogous result for odd n.

The proof is based on the following expression for the coefficients, which also has many other applications.

THEOREM 5.2.2.

$$a_{\nu+1,n+1}^{\mathrm{NC}} = \frac{b-a}{n}\Big(1 - c_{n,\nu} - c_{n,n-\nu}\Big) \tag{5.14}$$

with

$$c_{n,\nu} := (-1)^{\nu} \sum_{\lambda=0}^{n} \binom{\lambda}{\nu} L_{\lambda+1},$$

where

$$L_\lambda := (-1)^{\lambda+1} \int_0^1 \binom{x}{\lambda}\, dx. \tag{5.15}$$

The numbers L_λ, sometimes called Laplace coefficients, were tabulated by Lowan and Salzer (1943). Their asymptotic properties were investigated in Steffensen (1924). The following rough lower bound will be sufficient for our purposes:

$$\begin{aligned} L_\lambda &= \frac{(-1)^{\lambda+1}}{\lambda!} \int_0^1 x(x-1)(x-2)(x-3)\cdots(x-\lambda+1)\, dx \\ &= \frac{1}{\lambda!} \int_0^1 x(1-x)(2-x)(3-x)\cdots(\lambda-1-x)\, dx \\ &\geq \frac{1}{\lambda!} \int_0^1 x(1-x)\cdot 1\cdot 2\cdots(\lambda-2)\, dx, \end{aligned}$$

so that

$$L_\lambda \geq \frac{1}{6\lambda(\lambda-1)} \quad \text{for } \lambda \geq 2. \tag{5.16}$$

PROOF OF THEOREM 5.2.1. We use (5.14) and (5.16) to obtain

$$\begin{aligned} \|Q_{n+1}^{\mathrm{NC}}\| &= \sum_{\nu=0}^{n} |a_{\nu+1,n+1}^{\mathrm{NC}}| \\ &\geq \sum_{\nu=0}^{n} (-1)^{\nu+1} a_{\nu+1,n+1}^{\mathrm{NC}} \\ &= \frac{b-a}{n} \sum_{\nu=0}^{n} \big((-1)^{\nu+1} + |c_{n,\nu}| + |c_{n,n-\nu}|\big) = \frac{b-a}{n}\Big(-1 + 2\sum_{\nu=0}^{n} |c_{n,\nu}|\Big) \\ &= \frac{b-a}{n}\Big(-1 + 2\sum_{\nu=0}^{n}\sum_{\lambda=0}^{n} \binom{\lambda}{\nu} L_{\lambda+1}\Big) = \frac{b-a}{n}\Big(-1 + 2\sum_{\lambda=0}^{n} 2^\lambda L_{\lambda+1}\Big) \\ &\geq \frac{b-a}{n}\Big(-1 + \frac{1}{3n(n+1)}\sum_{\lambda=0}^{n} 2^\lambda\Big) = \frac{b-a}{n}\Big(-1 + \frac{2^{n+1}-1}{3n(n+1)}\Big), \end{aligned}$$

which gives the desired result. □

Before proving the explicit representation of the coefficients given above, we draw another conclusion. Theorem 5.2.1 shows that not all coefficients of $Q_n^{\rm NC}$ can be positive. Indeed, a much stronger statement holds; the next theorem says that almost half of the coefficients will be negative.

THEOREM 5.2.3 (Uspensky 1935). For sufficiently large even n,

$$\operatorname{sgn} a_{\nu,n+1}^{\rm NC} = (-1)^\nu \quad \text{for } \nu = 2, 3, \ldots, n.$$

By including a more careful estimation in the following proof, it is possible to show that "sufficiently large" means $n \geq 10$.

PROOF OF THEOREM 5.2.3. Starting with (5.14), we use (5.16) to get

$$\begin{aligned}(-1)^{\nu+1} a_{\nu+1,n+1}^{\rm NC} &= \frac{b-a}{n}\left((-1)^{\nu+1} + |c_{n,\nu}| + |c_{n,n-\nu}|\right)\\ &\geq \frac{b-a}{n}\left(|c_{n,\nu}| - 1\right)\\ &= \frac{b-a}{n}\left(\sum_{\lambda=0}^{n}\binom{\lambda}{\nu} L_{\lambda+1} - 1\right)\\ &\geq \frac{b-a}{n}\left(\sum_{\lambda=0}^{n}\binom{\lambda}{\nu}\frac{1}{6\lambda(\lambda+1)} - 1\right).\end{aligned}$$

If $\nu \notin \{0, n-2, n-1, n\}$, this is easily seen to be positive if n is large. The cases of $\nu = n-2$ and $\nu = n-1$ follow from the symmetry. □

We have still not proved Theorem 5.2.2. A crucial step in the proof is the following lemma, which will also be very useful in the discussion of Gregory rules in Section 7.5.

LEMMA 5.2.1. For every $p \in \mathbb{P}_n$, we have

$$\int_0^n p(x)\,dx = \sum_{k=0}^{n} p(k) - \sum_{\lambda=0}^{n} L_{\lambda+1}\left[\Delta^\lambda p(n-\lambda) + (-1)^\lambda \Delta^\lambda p(0)\right], \tag{5.17}$$

where L_λ is defined by (5.15) and the Δ-operator is given (as usual) by

$$\Delta^\lambda f = \sum_{\mu=0}^{\lambda} (-1)^{\lambda-\mu}\binom{\lambda}{\mu} f(\,\cdot + \mu).$$

PROOF. For $h \in \mathbb{P}_{n+1}$ defined uniquely by the interpolation condition

$$h(\nu) = -\sum_{k=\nu}^{n} p(k), \qquad \nu = 0, 1, \ldots, n+1,$$

we obtain

$$p(x) = h(x+1) - h(x).$$

(Both sides are polynomials of degree n or less; since equality holds for $x = 0, 1, \ldots, n$, it must hold for all x.) This yields $\Delta^\lambda h = \Delta^{\lambda-1} p$ and

$$\int_0^n p(x)\,dx = \int_1^{n+1} h(x)\,dx - \int_0^n h(x)\,dx = \int_0^1 \left[h(n+1-x) - h(x)\right] dx. \tag{5.18}$$

We apply Newton's interpolation formula

$$f(x) = \sum_{\lambda=0}^{n+1} \binom{x}{\lambda} \Delta^\lambda f(0) \quad \text{for } f \in \mathbb{P}_{n+1}$$

first to $f = h$, obtaining

$$h(x) = h(0) + \sum_{\lambda=1}^{n+1} \binom{x}{\lambda} \Delta^{\lambda-1} p(0) = -\sum_{k=0}^{n} p(k) + \sum_{\lambda=1}^{n+1} \binom{x}{\lambda} \Delta^{\lambda-1} p(0),$$

and then to $f = h(n+1-\cdot\,)$, to get

$$h(n+1-x) = \sum_{\lambda=0}^{n+1} \binom{x}{\lambda} (-1)^\lambda \Delta^\lambda h(n+1-\lambda) = \sum_{\lambda=1}^{n+1} \binom{x}{\lambda} (-1)^\lambda \Delta^{\lambda-1} p(n+1-\lambda).$$

Plugging these identities into the right-hand side of (5.18) yields the lemma. □

PROOF OF THEOREM 5.2.2. For p in Lemma 5.2.1 we choose the interpolation polynomial of f with nodes $0, 1, \ldots, n$. Then $p(k) = f(k)$, $\Delta^\lambda p(n-\lambda) = \Delta^\lambda f(n-\lambda)$ and $\Delta^\lambda p(0) = \Delta^\lambda f(0)$. Therefore, the right-hand side of (5.17) is the Newton–Cotes rule for the interval $[0, n]$, and Theorem 5.2.2 is an immediate consequence. □

A very important implication of Theorem 5.2.1 is the existence of an $f \in C[a,b]$ such that

$$\limsup_{n\to\infty} \left| Q_n^{\mathrm{NC}}[f] \right| = \infty.$$

The explicit determination of an f with this property can be done easily with the help of Lemma 5.2.1 as follows: Replace n by $2m$ and choose p to be the interpolation polynomial of

$$g(x) := |x - m|$$

with nodes $0, 1, \ldots, 2m$. Using

$$\Delta^1 g(l) = \begin{cases} -1 & \text{if } l = 0, 1, \ldots, m-1, \\ 1 & \text{if } l = m, m+1, \ldots, 2m-1 \end{cases}$$

and, for $k \geq 2$,

$$\Delta^k g(l) = 2(-1)^{k+l+m+1} \binom{k-2}{m-l-1}$$

(which can be proved by induction), we obtain

$$\int_0^{2m} p(x)\, dx = m^2 - \frac{1}{6} + 4(-1)^m \sum_{k=m+1}^{2m} L_{k+1} \binom{k-2}{m-1}.$$

This means that for the Newton–Cotes rule $Q_{2m+1}^{\mathrm{NC},[0,2m]}$ on $[0, 2m]$, we have

$$R_{2m+1}^{\mathrm{NC},[0,2m]}[g] = \frac{1}{6} - 4(-1)^m \sum_{k=m+1}^{2m} L_{k+1} \binom{k-2}{m-1}.$$

Now let R_{2m+1}^{NC} be the error functional of the Newton–Cotes rule on $[-1, 1]$ for

$$f(x) := |x|.$$

A simple transformation shows that

$$R_{2m+1}^{\mathrm{NC}}[f] = \frac{1}{m^2} R_{2m+1}^{\mathrm{NC},[0,2m]}[g]$$

and thus

$$\left|R_{2m+1}^{\mathrm{NC}}[f]\right| = \frac{1}{m^2}\left|4\sum_{k=m+1}^{2m} L_{k+1}\binom{k-2}{m-1} - \frac{(-1)^m}{6}\right|.$$

Finally, using (5.16), we find that

$$\lim_{n\to\infty}\left|R_{2m+1}^{\mathrm{NC}}[f]\right| = \infty.$$

It is not so important to understand the error of Newton–Cotes rules, because typically these rules are not used unless n is small and we can use Peano kernel methods without difficulty. The definiteness of R_n^{NC}, however, comes up frequently and is quite a remarkable property; therefore, we will give a proof of it.

THEOREM 5.2.4 (Steffensen 1921, 1925, 1933). For even n and $f \in C^{(n)}[a,b]$, there is a $\xi \in [a,b]$ with

$$R_n^{\mathrm{NC}}[f] = -2h^{n+1}L_{n+1}f^{(n)}(\xi).$$

For odd n and $f \in C^{(n+1)}[a,b]$, there is a $\xi \in [a,b]$ with

$$R_n^{\mathrm{NC}}[f] = -h^{n+2}(2L_{n+2} - L_{n+1})f^{(n+1)}(\xi). \tag{5.19}$$

In both cases, the coefficients of $f^{(n)}(\xi)$ and $f^{(n+1)}(\xi)$ are negative.

Equation (5.19) shows that $R_n^{\mathrm{NC}}[\mathbb{P}_n] = \{0\}$ if n is odd. This property of the Newton–Cotes rules follows immediately from Theorem 5.1.3 and the symmetry of the nodes.

PROOF. (i) Let n be even. According to Theorem 4.2.8, definiteness of order n will follow if we can prove definiteness of the $(n-1)$-fold reduced functional

$$R_n^{\mathrm{NC}(x_1)(x_2)\cdots(x_{n-1})}[f] = \int_a^b f(x)\prod_{\nu=1}^{n-1}(x-x_\nu)\,dx - a_n f(x_n)\prod_{\nu=1}^{n-1}(x_n - x_\nu)\,dx.$$

Applying (5.2) yields

$$R_n^{\mathrm{NC}(x_1)(x_2)\cdots(x_{n-1})}[f] = \int_a^b [f(x)-f(x_n)]\prod_{\nu=1}^{n-1}(x-x_\nu)\,dx,$$

and an integration by parts leads to

$$R_n^{\mathrm{NC}(x_1)(x_2)\cdots(x_{n-1})}[f] = -\int_a^b f'(x)h_1(x)\,dx$$

with

$$h_1(x) := \int_a^x \prod_{\nu=1}^{n-1}(u-x_\nu)\,du.$$

If h_1 has no sign changes on $[a,b]$, then

$$R_n^{\mathrm{NC}(x_1)(x_2)\cdots(x_{n-1})}[f] = -f'(\xi)\int_a^b h_1(x)\,dx$$

with $\xi \in [a,b]$, and definiteness is proved.

(ii) Let $n \geq 3$ be odd. We apply Theorem 4.2.9 and consequently have to prove that

$$h_2(x) := \int_x^b \prod_{\nu=1}^{n}(u-x_\nu)\,du = -\int_a^x \prod_{\nu=1}^{n}(u-x_\nu)\,du$$

has no sign changes on $[a, b]$.

(iii) An affine transformation of $[a, b]$ onto $[0, n]$ shows that the conditions on h_1 and h_2 are satisfied if

$$G_n(x) := \int_0^x \prod_{\mu=0}^{n-1} (y - \mu)\, dy$$

does not change sign on $[0, n]$ when $n \geq 3$. Since for odd n the integrand, and therefore G_n, is positive on $[0, 1]$, we have to show non-negativity throughout $[0, n]$. Now, the integrand is also positive for $y \geq n - 1$, so proving non-negativity of G_n on $[0, n-1]$ would be sufficient. We shall prove by induction that

$$\begin{aligned} G_{2n-1}(x) &\geq 0 \quad \text{for } x \in [0, 2n-2], \\ G_{2n}(x) &\leq 0 \quad \text{for } x \in [0, 2n-2]. \end{aligned}$$

First, note that $G_3(x) = x^2(x-2)^2/4 \geq 0$. The induction step is based on the following identity that comes from integration by parts:

$$G_n(x) = [x - (n-1)]G_{n-1}(x) - \int_0^x G_{n-1}(y)\, dy.$$

The induction hypothesis concerning G_{2n-1} immediately yields the assertion about G_{2n}. The same is true for the step from G_{2n} to G_{2n+1} but only on the interval $[0, 2n-2]$. The symmetry $G_{2n+1}(x) = G_{2n+1}(2n - x)$ completes the proof.

(iv) Let n be even. The definiteness of R_n^{NC} means that

$$R_n^{\mathrm{NC}}[f] = \frac{R_n^{\mathrm{NC}}[p_n]}{n!} f^{(n)}(\xi) \quad \text{where } p_n(x) = x^n. \tag{5.20}$$

Using $R_n^{\mathrm{NC}}[\mathbb{P}_{n-1}] = \{0\}$ together with an affine transformation, we get

$$R_n^{\mathrm{NC}}[p_n] = \int_a^b \prod_{\nu=1}^{n} (x - x_\nu)\, dx = h^{n+1} \int_0^{n-1} \prod_{\nu=0}^{n-1} (y - \nu)\, dy.$$

Now,

$$\begin{aligned} R_n^{\mathrm{NC}}[p_n] &= \frac{h^{n+1}}{n+1} \int_0^{n-1} \left(\prod_{\nu=0}^{n} (y + 1 - \nu) - \prod_{\nu=0}^{n} (y - \nu) \right) dy \\ &= \frac{h^{n+1}}{n+1} \left(\int_1^n - \int_0^{n-1} \right) \prod_{\nu=0}^{n} (y - \nu)\, dy \\ &= \frac{h^{n+1}}{n+1} \int_0^1 \left(\prod_{\nu=0}^{n} (n - \nu - x) - \prod_{\nu=0}^{n} (x - \nu) \right) dx \\ &= -\frac{2h^{n+1}}{n+1} \int_0^1 \prod_{\nu=0}^{n} (x - \nu)\, dx = -2h^{n+1} n! \int_0^1 \binom{x}{n+1} dx. \end{aligned}$$

Plugging this result into (5.20) establishes the error representation in this case.

(v) Let n be odd. Starting with

$$R_n^{\mathrm{NC}}[p_{n+1}] = \int_a^b \prod_{\nu=1}^{n+1} (x - x_\nu)\, dx \quad \text{where } x_{n+1} := b + h$$

and using the same methods as above, we obtain

$$
\begin{aligned}
R_n^{\mathrm{NC}}[p_{n+1}] &= \frac{h^{n+2}}{n+2} \int_0^1 \left(-\prod_{\nu=-1}^{n} (x-\nu) - \prod_{\nu=0}^{n+1} (x-\nu) \right) dx \\
&= -\frac{h^{n+2}}{n+2} \int_0^1 (2x-n) \prod_{\nu=0}^{n} (x-\nu)\, dx \qquad (5.21) \\
&= -\frac{h^{n+2}}{n+2} \int_0^1 \left(2\prod_{\nu=0}^{n+1} (x-\nu) + (n+2) \prod_{\nu=0}^{n} (x-\nu) \right) dx \\
&= -h^{n+2}(n+1)!\,\big(2L_{n+2} - L_{n+1}\big).
\end{aligned}
$$

From the expression (5.21) it can be seen that this number is negative. □

To close this section, we mention some works that deal with the Newton–Cotes method: Ouspensky (1925) described asymptotics of the coefficients; Polya (1933), Brass (1968), Brass (1977b), and Mills and Smith (1991) each contains explicit examples of divergence; Barrett (1964) determined the smallest region of holomorphy that guarantees convergence; Uspensky (1935) gave an expansion of the error functional; further discussions of the error can be found in Kambo (1970); and Locher (1973b) gave a short proof of $\|R_n^{\mathrm{NC}}\| \to \infty$.

5.3. A theorem of Sloan and Smith

In the previous section, we saw that a "natural" choice of nodes can lead to a method that is quite unsatisfactory. So let us consider some other choices now. Owing our experience with the Newton–Cotes method, and bearing in mind Lebesgue's inequality (Theorem 3.1.1), we will be particularly interested in rules with small norms. The smallest possible norm for a rule of interpolatory type is $I[1]$, so we want to look at methods with

$$\|Q_n\| = I[1] = \|I\| \quad \text{for } n = 1, 2, \ldots. \qquad (5.22)$$

Note that (5.22) holds for the Gaussian rule, i.e. a rule with nodes depending on I. The main result of this section is that there is a fixed system of nodes such that (5.22) is asymptotically satisfied for virtually any quadrature problem.

We restrict ourselves again to the interval $[-1, 1]$, since all the essential results carry over to arbitrary compact intervals via an affine transformation.

DEFINITION 5.3.1 (Clenshaw and Curtis 1960). The interpolatory quadrature method on $[-1, 1]$ whose nth formula Q_n^{CC} $(n = 2, 3, \ldots)$ is based on the nodes

$$x_{\nu,n}^{\mathrm{CC}} := -\cos\left(\frac{\nu-1}{n-1}\pi\right), \quad \nu = 1, \ldots, n,$$

is called the Clenshaw–Curtis method.

EXAMPLE 5.3.1. The Clenshaw–Curtis method for the weight function $w(x) = (1-x^2)^{-1/2}$ takes a simple form:

$$Q_{n+1}^{*}[f] = \frac{\pi}{n} \sum_{\nu=0}^{n}{}'' f\left(-\cos\frac{\nu\pi}{n}\right),$$

where $\sum''$ means that the first and last summands have to be halved. To establish this formula, we will show that

$$Q_{n+1}^*[T_k] = \int_{-1}^{1} \frac{T_k(x)}{\sqrt{1-x^2}}\,dx = \begin{cases} \pi & \text{for } k = 0, \\ 0 & \text{for } k = 1, 2, \ldots, 2n-1. \end{cases} \tag{5.23}$$

We emphasize that this is a stronger statement than is required for a rule of interpolatory type and implies that Q_{n+1}^* is a Lobatto rule as in Definition 3.4.4.

For $k > 0$ (the $k = 0$ case is trivial), formula (5.23) is equivalent to

$$\sum_{\nu=0}^{n}{}'' \cos\frac{k\nu\pi}{n} = 0 \quad \text{for } k = 1, 2, \ldots, 2n-1.$$

This equality can be proved by the same method as in Example 5.1.1.

Our next theorem is the important result of Sloan and Smith (1978) that is the title of this section.

THEOREM 5.3.1. Let

$$\int_{-1}^{1} w^p(x)\,dx < \infty \quad \text{for some } p > 1. \tag{5.24}$$

Then

$$\lim_{n\to\infty} \|Q_n^{\mathrm{CC}}\| = \|I\|.$$

This theorem implies, in particular (see Theorem 5.1.9), that

$$\lim_{n\to\infty} Q_n^{\mathrm{CC}}[f] = I[f].$$

In the proof, we will write Q_{n+1}^{CC} and I for the case of a general weight function, and Q_{n+1}^* and I^* for the special case where $w(x) = \left(1-x^2\right)^{-1/2}$. We need four lemmas, as follows.

LEMMA 5.3.1. We have

$$Q_{n+1}^{\mathrm{CC}}[f] = Q_{n+1}^*[fS_n]$$

where

$$S_n := \sum_{\nu=0}^{n}{}'' \alpha_\nu T_\nu \quad \text{with } \alpha_\nu = \frac{2}{\pi}\int_{-1}^{1} w(x)T_\nu(x)\,dx.$$

Hence, S_n is the nth partial sum with halved final summand of the Chebyshev expansion of $w(x)\sqrt{1-x^2}$.

PROOF. We have to show that

$$Q_{n+1}^*[T_\lambda S_n] = I[T_\lambda] \quad \text{for } \lambda = 0, 1, \ldots, n.$$

Since Q_{n+1}^* is a Lobatto rule, we obtain from the orthogonality of the Chebyshev polynomials that

$$\begin{aligned} Q_{n+1}^*[T_\lambda S_n] &= \sum_{\nu=0}^{n}{}'' \alpha_\nu Q_{n+1}^*[T_\lambda T_\nu] = \sum_{\nu=0}^{n}{}'' \alpha_\nu I^*[T_\lambda T_\nu] \\ &= \alpha_\lambda \underbrace{I^*[T_\lambda^2]}_{=\pi/2} = \int_{-1}^{1} w(x)T_\lambda(x)\,dx = I[T_\lambda] \end{aligned}$$

for $\lambda = 1, \ldots, n-1$. For the remaining cases, we use $Q^*_{n+1}[T_n^2] = Q^*_{n+1}[T_0^2] = I^*[T_0^2] = \pi$. $\square$

LEMMA 5.3.2. For $x, t \in [-1, 1]$, we have

$$\sum_{\nu=0}^{m}{}'(m-\nu)T_\nu(x)T_\nu(t) \ge 0,$$

where $\sum'$ means that the first summand has to be halved.

PROOF. We have

$$\begin{aligned}\sum_{\nu=0}^{m}{}'(m-\nu)T_\nu(\cos\xi)T_\nu(\cos\tau) &= \sum_{\nu=0}^{m}{}'(m-\nu)\cos\nu\xi\,\cos\nu\tau \\ &= \frac{1}{2}\sum_{\nu=0}^{m}{}'(m-\nu)\Big(\cos\nu(\xi+\tau)+\cos\nu(\xi-\tau)\Big).\end{aligned}$$

The non-negativity of this expression follows directly from

$$\sum_{\nu=0}^{m}{}'(m-\nu)\cos\nu\eta = \frac{1-\cos m\eta}{2(1-\cos\eta)},$$

which can be proved by multiplying both sides by $1-\cos\eta$. $\square$

LEMMA 5.3.3. For $p \in \mathbb{P}_n$, we have

$$Q^*_{n+1}\big[|p|\big] \le 3I^*\big[|p|\big].$$

PROOF. Orthogonality yields

$$p = \frac{2}{\pi}\int_{-1}^{1} p(t)\left[\frac{1}{n}\sum_{\nu=0}^{2n}{}'(2n-\nu)T_\nu T_\nu(t) - \frac{1}{n}\sum_{\nu=0}^{n}{}'(n-\nu)T_\nu T_\nu(t)\right]\frac{dt}{\sqrt{1-t^2}}.$$

Together with Lemma 5.3.2, this gives

$$\begin{aligned}|p(x)| \le \frac{2}{\pi}\int_{-1}^{1}|p(t)|\left[\frac{1}{n}\sum_{\nu=0}^{2n}{}'(2n-\nu)T_\nu(x)T_\nu(t)\right]\frac{dt}{\sqrt{1-t^2}} \\ +\frac{2}{\pi}\int_{-1}^{1}|p(t)|\left[\frac{1}{n}\sum_{\nu=0}^{n}{}'(n-\nu)T_\nu(x)T_\nu(t)\right]\frac{dt}{\sqrt{1-t^2}},\end{aligned}$$

so that

$$\begin{aligned}Q^*_{n+1}\big[|p|\big] &\le \frac{2}{\pi}\int_{-1}^{1}|p(t)|\left[\frac{1}{n}\sum_{\nu=0}^{2n}{}'(2n-\nu)Q^*_{n+1}[T_\nu]T_\nu(t)\right]\frac{dt}{\sqrt{1-t^2}} \\ &\quad+\frac{2}{\pi}\int_{-1}^{1}|p(t)|\left[\frac{1}{n}\sum_{\nu=0}^{n}{}'(n-\nu)Q^*_{n+1}[T_\nu]T_\nu(t)\right]\frac{dt}{\sqrt{1-t^2}} \\ &= \frac{2}{\pi}\int_{-1}^{1}|p(t)|\,I^*[T_0]\,T_0(t)\,\frac{dt}{\sqrt{1-t^2}} + \frac{1}{\pi}\int_{-1}^{1}|p(t)|\,I^*[T_0]\,T_0(t)\,\frac{dt}{\sqrt{1-t^2}} \\ &= 3\int_{-1}^{1}|p(t)|\,\frac{dt}{\sqrt{1-t^2}}\end{aligned}$$

as claimed. $\square$

LEMMA 5.3.4. With the notation from Lemma 5.3.1 and under the assumption (5.24), we have

$$\lim_{n\to\infty}\int_{-1}^{1}\frac{\left|w(x)\sqrt{1-x^2}\,-S_n(x)\right|}{\sqrt{1-x^2}}\,dx=0.$$

PROOF. We have

$$S_n(\cos t)=\sum_{\nu=0}^{n}{}''\alpha_\nu\cos\nu t\quad\text{where }\alpha_\nu=\frac{1}{\pi}\int_{-\pi}^{\pi}w(\cos t)\,|\sin t|\,\cos\nu t\,dt.$$

This is a modified partial sum of the Fourier series of $W:=w\big(\cos(\cdot)\big)\,\big|\sin(\cdot)\big|$, and the assertion of the lemma can also be written as

$$\lim_{n\to\infty}\int_{-\pi}^{\pi}\big|W(t)-S_n(\cos t)\big|\,dt. \tag{5.25}$$

Since $\alpha_n\to 0$ (proven in the theory of Fourier series), we may replace $\sum''$ by $\sum'$. Therefore, (5.25) is a L_1 convergence result for Fourier series. Now, if L_p convergence were known, then Hölder's inequality would give L_1 convergence as well. But L_p convergence is known from the theory of Fourier series (see, e.g., Zygmund 1977: vol. 1, p. 266 or Bary 1964: vol. 2, p. 138), so we are done. □

PROOF OF THEOREM 5.3.1. Lemma 5.3.1 yields $\|Q_{n+1}^{\mathrm{CC}}\|=Q_{n+1}^*[|S_n|]$. Hence, for each $m<n$, we have

$$\begin{aligned}\Big|\|I\|-\|Q_{n+1}^{\mathrm{CC}}\|\Big|&=\left|\int_{-1}^{1}w(x)\,dx-Q_{n+1}^*\big[|S_n|\big]\right|\\&=\bigg|\Big(\int_{-1}^{1}w(x)\,dx-I^*\big[|S_m|\big]\Big)\\&\qquad+\Big(I^*\big[|S_m|\big]-Q_{n+1}^*\big[|S_m|\big]\Big)+\Big(Q_{n+1}^*\big[|S_m|\big]-Q_{n+1}^*\big[|S_n|\big]\Big)\bigg|\\&\le I^*\Big[\big|w\sqrt{1-(\cdot)^2}-S_m\big|\Big]+\Big|I^*\big[|S_m|\big]-Q_{n+1}^*\big[|S_m|\big]\Big|+Q_{n+1}^*\big[|S_m-S_n|\big].\end{aligned}$$

The last summand can be estimated by using Lemma 5.3.3:

$$\begin{aligned}Q_{n+1}^*\big[|S_m-S_n|\big]&\le 3I^*\big[|S_m-S_n|\big]\\&\le 3I^*\Big[\big|w\sqrt{1-(\cdot)^2}-S_n\big|\Big]+3I^*\Big[\big|w\sqrt{1-(\cdot)^2}-S_m\big|\Big].\end{aligned}$$

Now we have

$$\begin{aligned}\Big|\|I\|-\|Q_{n+1}^{\mathrm{CC}}\|\Big|&\le 4I^*\Big[\big|w\sqrt{1-(\cdot)^2}-S_m\big|\Big]+3I^*\Big[\big|w\sqrt{1-(\cdot)^2}-S_n\big|\Big]\\&\qquad+\Big|I^*\big[|S_m|\big]-Q_{n+1}^*\big[|S_m|\big]\Big|.\end{aligned} \tag{5.26}$$

By Lemma 5.3.4, for a given $\varepsilon>0$ we can choose m such that

$$I^*\Big[\big|w\sqrt{1-(\cdot)^2}-S_\nu\big|\Big]<\frac{\varepsilon}{8}\quad\text{for all }\nu\ge m.$$

Finally, choose $n_0>m$ such that for all $n\ge n_0$, the last summand in the right-hand side of (5.26) is less than $\varepsilon/8$. This is possible due to the positivity of Q_{n+1}^*. Hence

$$\big|\,\|I\|-\|Q_{n+1}^{\mathrm{CC}}\|\,\big|<\varepsilon\quad\text{for all }n\ge n_0$$

and the theorem is proved. □

We remark that the proof of the theorem of Sloan and Smith is not elementary, since it relies on the deep theorem on L_p convergence (for $p > 1$) of Fourier series. Should one be content with just $w \in L_p$ for $p \geq 2$, then by a short and elementary proof one can obtain a result which goes beyond Theorem 5.3.1 in that it gives a bound for $\|Q_{n+1}^{\mathrm{CC}}\|$.

THEOREM 5.3.2. Suppose $\int_{-1}^{1} w^2(x)\sqrt{1-x^2}\,dx < \infty$. Then

$$\|Q_{n+1}^{\mathrm{CC}}\| \leq \left[\pi \int_{-1}^{1} w^2(x)\sqrt{1-x^2}\,dx\right]^{1/2}.$$

PROOF. Denote by x_ν the nodes and a_ν the coefficients of Q_{n+1}^{CC}. Apply Lemma 5.3.1 to a function f with

$$f(x_1) = 2a_1, \quad f(x_{n+1}) = 2a_{n+1} \quad \text{and} \quad f(x_\nu) = a_\nu \text{ for } \nu = 2, \ldots, n$$

to obtain

$$\begin{aligned} 2a_1^2 + \sum_{\nu=2}^{n} a_\nu^2 + 2a_{n+1}^2 = Q_{n+1}^{\mathrm{CC}}[f] &= \frac{\pi}{n}\sum_{\nu=1}^{n+1} a_\nu S_n(x_\nu) = \frac{\pi}{n} I[S_n] \\ &= \frac{\pi}{n}\sum_{\nu=0}^{n}{}'' \alpha_\nu I[T_\nu] = \frac{2}{n}\sum_{\nu=0}^{n}{}'' I^2[T_\nu] \\ &\leq \frac{\pi}{n}\int_{-1}^{1} w^2(x)\sqrt{1-x^2}\,dx, \end{aligned}$$

where we have used Bessel's inequality. Finally, use the Cauchy–Schwarz inequality to get

$$\|Q_{n+1}^{\mathrm{CC}}\|^2 = \left(\sum_{\nu=1}^{n+1} |a_\nu|\right)^2 \leq n\left(2a_1^2 + \sum_{\nu=2}^{n} a_\nu^2 + 2a_{n+1}^2\right),$$

which proves the assertion. □

5.4. Error bounds for the Clenshaw–Curtis method

Error bounds of the type

$$\left|R_n^{\mathrm{CC}}[f]\right| \leq c_r(n)\|f^{(r)}\|$$

can be obtained for any fixed w, n and $r \leq n$ by an application of the Peano kernel theorem. It is even possible to find $\rho\big(Q_n^{\mathrm{CC}}, C_1^{(r)}\big)$, i.e. the best possible value of $c_r(n)$. The disadvantage of using this procedure is the complicated structure of the Peano kernel, which impedes the derivation of useful general results in most cases. One exception should be mentioned: Consider $w(x) = \left(1-x^2\right)^{-1/2}$ and $r = 1$. Then

$$K_1\big|_{]x_\lambda, x_{\lambda+1}[} = \arccos x - \frac{2n - 2\lambda + 1}{2n}\pi$$

and from this, after some calculations, one finds that

$$\rho\big(Q_n^{\mathrm{CC}}, C_1^{(r)}\big) = \int_{-1}^{1} |K_1(x)|\,dx = \frac{\pi}{n} - 2\tan\frac{\pi}{4n}.$$

For w of fairly general form, fixed r and $n \to \infty$, bounds of order $O(n^{-r})$ can be obtained easily by using Theorem 4.3.4. For special weight functions, improvements are possible. Let us start with $w = 1$. From Lemma 5.3.1 and the fact that

$$\begin{aligned} \alpha_\nu &= \frac{2}{\pi}\int_{-1}^{1} T_\nu(x)\,dx \\ &= \frac{2}{\pi}\int_0^\pi \cos\nu t\,\sin t\,dt = \begin{cases} 0 & \text{for odd } \nu, \\ \dfrac{2}{\pi}\left(\dfrac{1}{\nu+1} - \dfrac{1}{\nu-1}\right) & \text{for even } \nu, \end{cases} \end{aligned} \tag{5.27}$$

we obtain the following:

Coefficients of the Clenshaw–Curtis rule Q_n^{CC} for $w = 1$:

$$a_1 = a_n = \begin{cases} \dfrac{1}{(n-1)^2} & \textit{for even } n, \\ \dfrac{1}{n(n-2)} & \textit{for odd } n, \end{cases}$$

$$a_k = \frac{2}{n-1}\left(1 - \sum_{\mu=1}^{\lfloor (n-1)/2 \rfloor}{}^{*} \frac{2}{4\mu^2 - 1}\cos\left(2\mu\frac{k-1}{n-1}\pi\right)\right) \quad \text{for } k = 2, 3, \ldots, n-1,$$

where $\sum^*$ means that the last summand should be halved if n is odd.

The most important conclusion is this next theorem.

THEOREM 5.4.1 (Imhof 1963). The Clenshaw–Curtis method for $w = 1$ is a positive method.

PROOF. For $2 \le k \le n-1$,

$$\frac{n-1}{2}\cdot a_k \ge 1 - \sum_{\mu=1}^{\lfloor (n-1)/2 \rfloor}{}^{*}\left(\frac{1}{2\mu-1} - \frac{1}{2\mu+1}\right) = \begin{cases} \dfrac{1}{2}\left(\dfrac{1}{n} + \dfrac{1}{n-2}\right) & \text{for odd } n, \\ \dfrac{1}{n-1} & \text{for even } n, \end{cases}$$

and so all coefficients are positive. □

Let us continue studying the weight function $w = 1$. The error bound in $\mathcal{C}_M^{(r)}$, obtained by specializing Theorem 4.3.4, is of the correct order for $n \to \infty$ and fixed r. For the extensively studied bound involving the derivative of highest possible order, an application of Lebesgue's inequality gives

$$\left|R_n^{CC}[f]\right| \le \frac{\|f^{(m)}\|}{m!\,2^{m-3}} \quad \text{where } m = 2\left\lceil \frac{n}{2} \right\rceil \tag{5.28}$$

(see Example 3.1.2). Here, considerable improvement is possible. We will mainly be interested in the asymptotic behaviour of $\rho(Q_n^{CC}, \mathcal{C}_1^{(m)})$. For strict bounds, we refer to our remarks at the beginning of this section.

We shall use results on the degree of approximation by polynomials, as in Lebesgue's inequality, but in a somewhat trickier manner. Our main tool is the

Chebyshev expansion

$$f = \frac{1}{2}\alpha_0[f]T_0 + \sum_{\nu=1}^{\infty} \alpha_\nu[f]T_\nu \quad \text{where } \alpha_\nu[f] = \frac{2}{\pi}\int_{-1}^{1} \frac{f(x)T_\nu(x)}{\sqrt{1-x^2}}\,dx.$$

We have uniform convergence for all continuous functions of bounded variation (this follows immediately from the analogous theorem concerning Fourier series).

LEMMA 5.4.1. *Let $R[\mathbb{P}_{m-1}] = \{0\}$. Then there exists a $\xi \in [a,b]$ such that*

$$R[f] = \frac{f^{(m)}(\xi)}{m!\,2^{m-1}} R[T_m] + \delta,$$

where

$$|\delta| \le \frac{8B_m}{\pi}\operatorname{dist}(f,\mathbb{P}_m) + \|R\|\operatorname{dist}(f,\mathbb{P}_{m+\lambda}) \quad \text{with } B_m = \sum_{\nu=m}^{m+\lambda} |R[T_\nu]|$$

for any $\lambda \in \{0,1,\ldots\}$.

Usually, applications of this result are based on the inequality

$$\operatorname{dist}(f,\mathbb{P}_{m+\lambda}) \le \frac{1}{m!\,2^{m-1}}\frac{\sqrt{(\lambda+1)!}}{(2m-1)^{(\lambda+1)/2}}\|f^{(m)}\|, \tag{5.29}$$

which is a coarsened version of (5.7).

PROOF. We start by proving

$$\alpha_\nu[f] = \frac{f^{(\nu)}(\eta)}{\nu!\,2^{\nu-1}} \quad \text{with } \eta \in [-1,1]. \tag{5.30}$$

To this end, we use the interpolation polynomial $\operatorname{intpol}(y_1,\ldots,y_\nu)[f]$, where the nodes $y_1,\ldots,y_\nu$ are the zeros of T_ν. The well-known representation

$$f(x) - \operatorname{intpol}(y_1,\ldots,y_\nu)[f](x) = \frac{f^{(\nu)}(\xi_x)}{\nu!\,2^{\nu-1}}T_\nu(x), \quad x,\xi_x \in [-1,1],$$

yields

$$\begin{aligned}\alpha_\nu[f] &= \frac{2}{\pi}\int_{-1}^{1}\{f(x) - \operatorname{intpol}(y_1,\ldots,y_\nu)[f](x)\}\frac{T_\nu(x)}{\sqrt{1-x^2}}\,dx \\ &= \frac{2}{\pi}\int_{-1}^{1}\frac{f^{(\nu)}(\xi_x)}{\nu!\,2^{\nu-1}}\frac{T_\nu^2(x)}{\sqrt{1-x^2}}\,dx = \frac{2}{\pi}\frac{f^{(\nu)}(\eta)}{\nu!\,2^{\nu-1}}\int_{-1}^{1}\frac{T_\nu^2(x)}{\sqrt{1-x^2}}\,dx = \frac{f^{(\nu)}(\eta)}{\nu!\,2^{\nu-1}}.\end{aligned}$$

We furthermore need a bound for $|\alpha_\nu[f]|$, which can be obtained immediately from an application of Lebesgue's inequality (Theorem 3.1.1):

$$|\alpha_\nu[f]| \le \|\alpha_\nu\|\operatorname{dist}(f,\mathbb{P}_{\nu-1}) = \frac{4}{\pi}\operatorname{dist}(f,\mathbb{P}_{\nu-1}). \tag{5.31}$$

After these preliminary steps, we turn to the main part of the proof:

$$\begin{aligned}R[f] &= R\left[f - \sum_{\nu=0}^{m+\lambda}{}'\alpha_\nu[f]\,T_\nu\right] + R\left[\sum_{\nu=0}^{m+\lambda}{}'\alpha_\nu[f]\,T_\nu\right] \\ &= \alpha_m[f]R[T_m] + \sum_{\nu=m+1}^{m+\lambda}\alpha_\nu[f]\,R[T_\nu] + R\left[f - \sum_{\nu=0}^{m+\lambda}{}'\alpha_\nu[f]\,T_\nu\right].\end{aligned}$$

We use (5.30) for the first term, while (5.31) yields, for the second term,

$$\left|\sum_{\nu=m+1}^{m+\lambda} \alpha_\nu[f]\, R[T_\nu]\right| \le \frac{4}{\pi} \sum_{\nu=m+1}^{m+\lambda} \operatorname{dist}(f, \mathbb{P}_{\nu-1})\big|R[T_\nu]\big| \le \frac{4}{\pi} \operatorname{dist}(f, \mathbb{P}_m) B_m.$$

Finally, we have

$$p = \sum_{\nu=0}^{m+\lambda}{}' \alpha_\nu[p] T_\nu \quad \text{for all } p \in \mathbb{P}_{m+\lambda}$$

so that

$$\begin{aligned}\left|R\left[f - \sum_{\nu=0}^{m+\lambda}{}' \alpha_\nu[f]\, T_\nu\right]\right| &= \left|R\left[(f-p) - \sum_{\nu=0}^{m+\lambda}{}' \alpha_\nu[f-p]\, T_\nu\right]\right| \\ &\le \|R\|\, \|f-p\| + \sum_{\nu=m}^{m+\lambda} \|\alpha_\nu\|\, \|f-p\|\, \big|R[T_\nu]\big| \\ &\le \|R\|\, \|f-p\| + \frac{4}{\pi} B_m \|f-p\| \\ &\le \|R\| \operatorname{dist}(f, \mathbb{P}_{m+\lambda}) + \frac{4}{\pi} B_m \operatorname{dist}(f, \mathbb{P}_m).\end{aligned}$$

Upon combining the bounds obtained above, the assertion follows. □

Application of this lemma to the Clenshaw–Curtis method starts with the observation that

$$T_{2\mu(n-1)\pm\lambda}(x_\nu) = T_\lambda(x_\nu) \quad \text{for } x_\nu = -\cos\left(\frac{\nu-1}{n-1}\pi\right).$$

This means that (for arbitrary w)

$$Q_n^{\mathrm{CC}}[T_{2\mu(n-1)\pm\lambda}] = Q_n^{\mathrm{CC}}[T_\lambda] \tag{5.32}$$

and, in particular (upon taking $\mu = 1$ and $\lambda = n-k-2$),

$$R_n^{\mathrm{CC}}[T_{n+k}] = I[T_{n+k}] - I[T_{n-k-2}] \quad \text{for } k \le n-2. \tag{5.33}$$

For $w = 1$, we obtain by using (5.27) that

$$R_n^{\mathrm{CC}}[T_n] = \frac{8}{(n+1)(n-1)(n-3)} \quad \text{for even } n, \tag{5.34}$$

$$R_n^{\mathrm{CC}}[T_{n+1}] = \frac{16(n-1)}{(n+2)n(n-2)(n-4)} \quad \text{for odd } n, \tag{5.35}$$

$$R_n^{\mathrm{CC}}[T_{n+s}] = O(n^{-3}) \quad \text{for fixed } s \text{ and } n \to \infty.$$

So Lemma 5.4.1 with $\lambda = 6$ yields the following:

THEOREM 5.4.2. *If $m = 2\lceil n/2 \rceil$ and $w = 1$, then*

$$\rho(Q_n^{\mathrm{CC}}, C_1^{(m)}) = \frac{|R_n^{\mathrm{CC}}[T_m]|}{m!\, 2^{m-1}}\left(1 + O(n^{-1/2})\right),$$

where $R_n^{\mathrm{CC}}[T_m]$ is as in (5.34) and (5.35).

This theorem gives an improvement over (5.28) by a factor of order n^{-3}. The actual value of the $O(n^{-1/2})$ term is quite small; by a different method of proof, Fiedler (1987) was able to replace it by $O(n^{-n/2})$. In Brass and Förster (1987), it was mentioned that an upper bound can be obtained by replacing $O(n^{-1/2})$ with 10^{-6} for $n > 8$. We cannot omit this term entirely, however, as Akrivis and Förster (1984) have shown. Their argument is as follows: Omitting the $O(n^{-1/2})$ term would mean that

$$\int_{-1}^{1} |K_m(x)|\, dx = \left| R_n \left[\frac{T_m}{m!\, 2^{m-1}} \right] \right| = \left| \int_{-1}^{1} K_m(x)\, dx \right|,$$

i.e. K_m would have no sign changes. But $K_m < 0$ contradicts (5.34) or (5.35), whereas $K_m > 0$ is impossible since we have $K_m(x) < 0$ for x near -1, as can be seen from the explicit expression for K_m (Theorem 4.2.5(i)) together with $a_1 > 0$.

With some additional work, we may replace the asymptotic result of Theorem 5.4.2 by a bound, such as

$$|R_n^{\mathrm{CC}}[f]| \le \frac{\|f^{(n+1)}\|}{(n+1)!(n+1)(n-1)(n-3)2^{n-5}}$$

for odd n; see Brass (1973), Brass and Schmeisser (1981) and Kütz (1984b).

Let us now turn to more general weight functions w. We always have

$$|R_n^{\mathrm{CC}}[f]| \le \|R\| \frac{\|f^{(n)}\|}{n!\, 2^{n-1}} \tag{5.36}$$

(see Example 3.1.2). Depending on what properties are assumed of w, Lemma 5.4.1 may allow us to improve this bound. For example, if we assume monotonicity of w, then the second mean value theorem gives

$$I[T_n] = \int_{-1}^{1} T_n(x) w(x)\, dx = w(-1) \int_{-1}^{\xi} T_n(x)\, dx + w(1) \int_{\xi}^{1} T_n(x)\, dx = O(n^{-1}),$$

and from (5.33) we get

$$R_n^{\mathrm{CC}}[T_{n+s}] = O(n^{-1})$$

for fixed s. So, in this case, the lemma yields an improvement upon (5.36) by a factor of order n^{-1}. Evidently, the same would be true if w is assumed to be of bounded variation. For further and more precise results in this direction, see Petras (1994a). Obtaining a considerable improvement on the simple bound (5.36) is possible only for a small class of weight functions. We define "considerable improvement" by the condition

$$\limsup_{n\to\infty} \left(n!\, 2^n \rho(Q_n, C_1^{(n)}) \right)^{1/n} < 1. \tag{5.37}$$

THEOREM 5.4.3 (Petras 1994a). *The quadrature method Q_n^{CC} satisfies (5.37) if and only if the weight function w has the form*

$$w(x) = \frac{h(x)}{\sqrt{1-x^2}}$$

where h is analytic on $[-1,1]$.

PROOF. Note that (5.33) implies that

$$
\begin{aligned}
R_n^{\mathrm{CC}}[T_{n+k}] &= \int_{-1}^{1} h(x)\frac{T_{n+k}(x)}{\sqrt{1-x^2}}\,dx - \int_{-1}^{1} h(x)\frac{T_{n-k-2}(x)}{\sqrt{1-x^2}}\,dx \\
&= \frac{\pi}{2}\big(\alpha_{n+k}[h] - \alpha_{n-k-2}[h]\big).
\end{aligned}
\tag{5.38}
$$

First, we prove that analyticity of h is necessary. It is trivially true that

$$n!\,2^{n-1}\rho(Q_n, C_1^{(n)}) \geq \big|R_n[T_n]\big| = \frac{\pi}{2}\,\big|\alpha_n[h] - \alpha_{n-2}[h]\big|.$$

Assuming (5.37), we obtain

$$\big|\alpha_n[h] - \alpha_{n-2}[h]\big| \leq c\cdot q^n \quad \text{for all } n,$$

with some constants $q < 1$ and c. This implies that

$$\big|\alpha_n[h]\big| = \left|\sum_{\nu=0}^{\infty}\Big(\alpha_{n+2\nu}[h] - \alpha_{n+2\nu+2}[h]\Big)\right| \leq c\sum_{\nu=0}^{\infty} q^{n+2\nu+2} = \frac{cq^2}{1-q^2}\cdot q^n.$$

The expansion

$$h = \frac{\alpha_0[h]}{2}T_0 + \sum_{\nu=1}^{\infty}\alpha_\nu[h]T_\nu$$

holds on $[-1,1]$, since the series is uniformly convergent. This series can be continued into the interior of the ellipse $\mathcal{E}_r$ for some $r < 1/q$ (see Section 4.9), and it converges since $\sup\{\,|T_\nu(z)| : z \in \mathcal{E}_r\} \leq r^\nu$ (see Appendix A). Therefore h is analytic.

To prove the sufficiency, we use Lemma 5.4.1. Upon combining it with the bound

$$\operatorname{dist}(f, \mathbb{P}_m) \leq \frac{\|f^{(m)}\|}{m!\,2^{m-1}}$$

(see (3.4)), we find that

$$\big|R[f]\big| \leq \|R\|\operatorname{dist}(f, \mathbb{P}_{m+\lambda}) + \left(\frac{8B_m}{\pi} + \big|R[T_m]\big|\right)\frac{\|f^{(m)}\|}{m!\,2^{m-1}}.$$

Set $\lambda = \lfloor m/2 \rfloor$; then it remains to show that

$$\operatorname{dist}(f, \mathbb{P}_{m+\lfloor m/2\rfloor}) \leq r_1^{-m}\frac{\|f^{(m)}\|}{m!\,2^{m-1}} \tag{5.39}$$

and

$$R[T_\nu] = O(r_2^{-m}) \quad \text{for } m \leq \nu \leq m + \lfloor m/2 \rfloor\,, \tag{5.40}$$

for certain numbers $r_1, r_2 > 1$.

The inequality (5.39) follows from (5.29) with the help of Stirling's formula. To prove (5.40), we start with (5.38), for $k \leq m/2$, and use (5.31):

$$\big|R_m^{\mathrm{CC}}[T_{m+k}]\big| = \frac{\pi}{2}\,\big|\alpha_{m+k}[h] - \alpha_{m-k-2}[h]\big| \leq 4\operatorname{dist}(h, \mathbb{P}_{m-k-3}).$$

According to (5.9), $h \in \mathcal{C}_M(\mathcal{E}_r)$ implies $\operatorname{dist}(h, \mathbb{P}_\nu) = O(r^{-\nu})$ and therefore

$$\big|R_m^{\mathrm{CC}}[T_{m+k}]\big| = O(r^{-m+k}) = O\left(\big(\sqrt[3]{r}\big)^{-(m+k)}\right),$$

so (5.40) is established. $\square$

Next, we investigate error bounds in $\mathcal{C}_M(\mathcal{E}_r)$. For all weight functions, we obtain from (5.9) that

$$\left|R_n^{\mathrm{CC}}[f]\right| \leq \frac{2M}{r^{n-1}(r-1)}\|R_n^{\mathrm{CC}}\|.$$

Again, this result can be improved for special forms of w. We shall consider the asymptotics for the $w = 1$ case, using a method which allows strengthening and generalization.

THEOREM 5.4.4. Let $w = 1$. Then there exist positive numbers c_1 and c_2, independent of n, such that

$$c_1 \leq r^n n^3 \rho\left(Q_n^{\mathrm{CC}}, \mathcal{C}_1(\mathcal{E}_r)\right) \leq c_2.$$

PROOF. Suppose that n is even; the case of odd n is similar, with no additional difficulties.

From $2T_n/(r^n + r^{-n}) \in \mathcal{C}_1(\mathcal{E}_r)$ (see Theorem A.9(vi)) we obtain

$$\rho\left(Q_n^{\mathrm{CC}}, \mathcal{C}_1(\mathcal{E}_r)\right) \geq \frac{2\left|R_n^{\mathrm{CC}}[T_n]\right|}{r^n + r^{-n}},$$

so that (5.34) yields the lower bound in the theorem.

To establish the upper bound, we apply R to the Chebyshev expansion of f:

$$\begin{aligned} \left|R_n^{\mathrm{CC}}[f]\right| &= \left|\sum_{\nu=n}^{\infty} \alpha_\nu[f]\, R_n^{\mathrm{CC}}[T_\nu]\right| \\ &\leq \sum_{\nu=0}^{s} \left|\alpha_{n+2\nu}[f]\, R_n^{\mathrm{CC}}[T_{n+2\nu}]\right| + \|R_n^{\mathrm{CC}}\| \sum_{\nu=n+2s}^{\infty} \left|\alpha_\nu[f]\right|. \end{aligned} \tag{5.41}$$

Choose $s = \lfloor n/4 \rfloor$; then, for the second sum on the right-hand side, (5.31) and (5.9) give

$$\sum_{\nu=n+2s}^{\infty} \left|\alpha_\nu[f]\right| = O(r^{-n-2s}) = O(r^{-n}n^{-3}).$$

The first sum on the right-hand side of (5.41) requires more careful estimation. According to (5.33), we have

$$\left|R_n^{\mathrm{CC}}[T_{n+2\nu}]\right| = \left|\frac{2}{1-(n+2\nu)^2} - \frac{2}{1-(n-2\nu-2)^2}\right| \leq \frac{(2\nu+1)d_1}{n^3},$$

where d_1 is independent of n if $\nu \leq s$ and $n > 2$. We again apply $\alpha_k[f] = O(r^{-k})$ and obtain

$$\sum_{\nu=0}^{s} \left|\alpha_{n+2\nu}[f]\, R_n^{\mathrm{CC}}[T_{n+2\nu}]\right| \leq \frac{d_2}{r^n n^3} \sum_{\nu=0}^{s} (2\nu+1) r^{-2\nu} = O(r^{-n}n^{-3}).$$

Thus we have proved the upper bound. □

5.5. Relatives of the Clenshaw–Curtis method

Several variants of the Clenshaw-Curtis method have been proposed. A seemingly small modification, which nevertheless deserves attention, involves deletion of the boundary nodes.

DEFINITION 5.5.1 (Filippi 1964). The interpolatory quadrature method on $[-1, 1]$ whose nth formula Q_n^{Fi} $(n = 1, 2, \ldots)$ is based on the nodes

$$x_{\nu,n}^{\text{Fi}} := -\cos\left(\frac{\nu}{n+1}\pi\right), \qquad \nu = 1, \ldots, n,$$

is called the *Filippi method.*

EXAMPLE 5.5.1. A simple explicit example of a Filippi method can be obtained by modifying Example 5.3.1. We proved that

$$\int_{-1}^{1} \frac{f(x)}{\sqrt{1-x^2}}\, dx = \frac{\pi}{n} \sum_{\nu=0}^{n}{}'' f\left(-\cos\frac{\nu\pi}{n}\right)$$

for all $f \in \mathbb{P}_{2n-1}$. Setting $f(x) = (1-x^2)g(x)$ with $g \in \mathbb{P}_{2n-3}$ gives the Filippi rule for $w(x) = \sqrt{1-x^2}$, since

$$\int_{-1}^{1} g(x)\sqrt{1-x^2}\, dx = \frac{\pi}{n} \sum_{\nu=1}^{n-1} \sin^2\frac{\nu\pi}{n} \cdot g\left(-\cos\frac{\nu\pi}{n}\right).$$

The nodes of Q_n are the zeros of U_n, the Chebyshev polynomial of the second kind (see Definition A.3). We express the connection between the Clenshaw–Curtis and Filippi methods in terms of divided differences.

DEFINITION 5.5.2. Let $a \leq x_1 < x_2 < \cdots < x_n \leq b$. The divided difference at the points $x_1, \ldots, x_n$, denoted by $\operatorname{dvd}(x_1, \ldots, x_n)$, is a functional on $C[a, b]$ such that

$$\operatorname{dvd}(x_1, \ldots, x_n)[f] = \sum_{\nu=1}^{n} \alpha_\nu f(x_\nu)$$

with

$$\begin{aligned}
&\operatorname{dvd}(x_1, \ldots, x_n)[p_\nu] = 0 \quad \text{for } \nu = 0, 1 \ldots, n-2,\\
&\operatorname{dvd}(x_1, \ldots, x_n)[p_{n-1}] = 1,
\end{aligned}$$

where $p_\nu(x) = x^\nu$.

It is easy to prove that $\operatorname{dvd}(x_1, \ldots, x_n)$ exists and is uniquely determined, and that

$$\alpha_\nu = \prod_{\substack{\mu=1\\ \mu\neq\nu}}^{n} (x_\nu - x_\mu)^{-1}. \tag{5.42}$$

We can now establish the following result:

THEOREM 5.5.1. Let

$$x_\nu := -\cos\left(\frac{\nu-1}{n+1}\pi\right) \ (= x_{\nu,n+2}^{\text{CC}} = x_{\nu-1,n}^{\text{Fi}}).$$

Then

(5.43) $$Q_{n+2}^{\mathrm{CC}}[f]-Q_n^{\mathrm{Fi}}[f] = \frac{c_1^{(n)}}{2^{n+1}}\,\mathrm{dvd}(x_2,\ldots,x_{n+2})[f]+\frac{c_2^{(n)}}{2^{n+1}}\,\mathrm{dvd}(x_1,\ldots,x_{n+1})[f],$$

where

$$c_1^{(n)} = \int_{-1}^{1}(x+1)U_n(x)w(x)\,dx, \qquad c_2^{(n)} = -\int_{-1}^{1}(x-1)U_n(x)w(x)\,dx.$$

PROOF. Both sides of the asserted equality are linear combinations of the $f(x_\nu)$. The coefficient of $f(x_1)$ on the left-hand side is

$$a_1^{\mathrm{CC}} = \int_{-1}^{1}\frac{(x-1)U_n(x)}{-2U_n(-1)}\,w(x)\,dx,$$

and that on the right-hand side is

$$\frac{c_2^{(n)}}{2^{n+1}}\cdot\frac{2^n}{U_n(-1)}$$

(using (5.42) and $U_n(x) = 2^n x^n + \cdots$). Hence, $c_2^{(n)}$ as stated above is the right choice. The same argument can be applied to the coefficient of $f(x_{n+2})$.

We conclude that the difference between the two sides is a linear combination of $f(x_2),\ldots,f(x_{n+1})$ which is zero for $f\in\mathbb{P}_{n-1}$. Choosing

$$f(x) = f_k(x) = \prod_{\nu\in\{2,\ldots,n+1\}\setminus\{k\}}(x-x_\nu)$$

implies that all coefficients in this linear combination have to vanish. □

Our next aim is to obtain a bound on the difference $Q_{n+2}^{\mathrm{CC}}[f]-Q_n^{\mathrm{Fi}}[f]$. To this end, we estimate divided differences by using Lebesgue's inequality (with $I\equiv 0$ and $Q=\mathrm{dvd}(\cdots)$) as follows:

$$\left|\mathrm{dvd}(y_1,\ldots,y_n)[f]\right| \le \|\,\mathrm{dvd}(y_1,\ldots,y_n)\|\,\mathrm{dist}(f,\mathbb{P}_{n-2}).$$

From (5.42), we see that the signs of $\alpha_1,\alpha_2,\ldots$ are alternating. Since $T_{n+1}(x_\nu) = (-1)^\nu$, we have, for both divided differences in (5.43),

$$\begin{aligned}\|\,\mathrm{dvd}(\cdots)\| &= \left|\mathrm{dvd}(\cdots)(T_{n+1})\right|\\ &= \left|\mathrm{dvd}(\cdots)(T_{n+1}\pm U_n)\mp\mathrm{dvd}(\cdots)(U_n)\right| = \left|\mathrm{dvd}(\cdots)(U_n)\right| = 2^n,\end{aligned}$$

where we have used

$$T_{n+1}\pm U_n = 2^n p_{n+1}\pm 2^n p_n = (p_1\pm 1)U_n \pmod{\mathbb{P}_{n-1}} \quad\text{and}\quad p_\nu(x)=x^\nu.$$

Now, Theorem 5.5.1 gives

(5.44) $$\left|Q_{n+2}^{\mathrm{CC}}[f]-Q_n^{\mathrm{Fi}}[f]\right| \le \frac{1}{2}\left(|c_1^{(n)}|+|c_2^{(n)}|\right)\mathrm{dist}(f,\mathbb{P}_{n-1}).$$

An important consequence of this inequality is the following:

THEOREM 5.5.2. Q_n^{Fi} is a convergent method if w satisfies the Sloan–Smith assumption

$$\int_{-1}^{1} w^p(x)\,dx<\infty \quad \text{for some } p>1$$

and, additionally,

$$\int_{-1}^{1} w(x)\left(1-x^2\right)^{-1/2}dx<\infty.$$

PROOF. Owing to the fact that

$$|U_n(x)| = \left|\frac{\sin(n+1)\arccos x}{\sqrt{1-x^2}}\right| \le \frac{1}{\sqrt{1-x^2}},$$

we have

$$\left|c_i^{(n)}\right| = \left|\int_{-1}^{1} (1-(-1)^i x) U_n(x) w(x)\, dx\right| \le 2\int_{-1}^{1} w(x)\bigl(1-x^2\bigr)^{-1/2}\, dx$$

for $i = 1, 2$. Inequality (5.44) and the Weierstrass approximation theorem give

$$\left|Q_{n+2}^{\mathrm{CC}}[f] - Q_n^{\mathrm{Fi}}[f]\right| \to 0,$$

and the assertion then follows from Theorem 5.3.1. □

The Sloan–Smith assumption alone is not sufficient to give convergence. This can be seen from

$$\|R_n^{\mathrm{Fi}}\| \ge \left|R_n^{\mathrm{Fi}}[T_n]\right| = \frac{1}{2}\left|R_n^{\mathrm{Fi}}[U_n]\right| = \frac{1}{2}\left|\int_{-1}^{1} U_n(x) w(x)\, dx\right|.$$

This sequence of numbers is not bounded for $w(x) = \bigl(1-x^2\bigr)^{-\alpha}$, $\alpha \in \,]\frac{1}{2}, 1[$. Specifically, if

$$\left|\int_{-1}^{1} U_k(x)\bigl(1-x^2\bigr)^{-\alpha}\, dx\right| = \left|\int_0^{\pi} \frac{\sin(k+1)t}{\sin^{2\alpha} t}\, dt\right| < B$$

were to hold for $k = 1, 2, \dots$, then the well-known identity

$$\sum_{k=0}^{n-1} \sin\,(2k+1)t = \frac{\sin^2 nt}{\sin t}$$

would yield

$$\frac{1}{n}\int_0^{\pi} \frac{\sin^2 nt}{\sin^{2\alpha+1} t}\, dt \le B.$$

But this contradicts

$$\begin{aligned}\frac{1}{n}\int_0^{\pi} \frac{\sin^2 nt}{\sin^{2\alpha+1} t}\, dt &> \frac{1}{n}\int_{\pi/(4n)}^{3\pi/(4n)} \frac{\frac{1}{2}}{\sin^{2\alpha+1}\frac{3\pi}{4n}}\, dt \\ &\ge \frac{1}{n}\cdot\frac{\pi}{2n}\cdot\frac{1}{2}\cdot\left(\frac{3\pi}{4n}\right)^{-2\alpha-1} = \frac{\pi}{4}\left(\frac{4}{3\pi}\right)^{2\alpha+1}\cdot n^{2\alpha-1} \to \infty.\end{aligned}$$

The assertion $R_n^{\mathrm{Fi}}[f] \to 0$ in Theorem 5.5.2 does not imply

$$\|Q_n^{\mathrm{Fi}}\| \to \|I\|, \tag{5.45}$$

which would be the exact counterpart of Theorem 5.3.1. In Smith and Sloan (1980), (5.45) is proved under the assumption that

$$\int_{-1}^{1} \left[w(x)\bigl(1-x^2\bigr)^{-1/2}\right]^p dx < \infty$$

for some $p > 1$.

We now turn to the $w = 1$ case. Here we have $c_i^{(n)} = o(1)$, so that (5.44) gives

$$\left|Q_{n+2}^{\mathrm{CC}}[f] - Q_n^{\mathrm{Fi}}[f]\right| = o\bigl(\mathrm{dist}(f, \mathbb{P}_{n-1})\bigr).$$

Using (5.6) and Theorem 4.3.3, we find that the error bounds for the Filippi and Clenshaw–Curtis methods are quite similar in the class $\mathcal{C}_M^{(r)}$. This is not the case for error bounds that utilize the highest possible derivative or holomorphy. The crucial point is the behaviour of $R_n[T_m]$ for $m = 2\lceil n/2\rceil$. Integration by parts shows that

$$R_n^{\mathrm{Fi}}[T_n] = \frac{1}{2}R_n^{\mathrm{Fi}}[U_n] = \frac{1}{2}\int_{-1}^{1} U_n(x)\,dx = \frac{1}{n+1} \quad \text{for even } n$$

and

$$R_n^{\mathrm{Fi}}[T_{n+1}] = \int_{-1}^{1} x\,U_n(x)\,dx = \frac{2(n+1)}{n(n+2)} \quad \text{for odd } n.$$

Furthermore, from $Q_n^{\mathrm{Fi}}[T_{2(n+1)\mu\pm\lambda}] = Q_n^{\mathrm{Fi}}[T_\lambda]$ one sees that

$$R_n^{\mathrm{Fi}}[T_{n+\lambda}] = O(n^{-1}) \quad \text{for } \lambda \le n/2.$$

Lemma 5.4.1, together with (5.29), yields

$$\rho\big(Q_n^{\mathrm{Fi}}, \mathcal{C}_1^{(m)}\big) = \frac{R_n^{\mathrm{Fi}}[T_m]}{m!\,2^{m-1}}\Big(1 + O(n^{-1/2})\Big) \quad \text{for } m = 2\left\lceil \frac{n}{2}\right\rceil.$$

Compared with the Clenshaw–Curtis rules, we lose a factor of order n^{-2}. The method of proving Theorem 5.4.4 carries over to $\rho\big(Q_n^{\mathrm{Fi}}, \mathcal{C}_1(\mathcal{E}_r)\big)$ and leads to the same loss. Another difference between Q_n^{CC} and Q_n^{Fi} is the definiteness of the latter. We prove this only for odd n; for the case of even n, which is more involved, see Brass and Schmeisser (1979).

THEOREM 5.5.3 (Locher 1969). *Suppose that n is odd and $w = 1$. Then for every $f \in C^{(n+1)}[-1,1]$ there exists $\xi \in [-1,1]$ such that*

$$R_n^{\mathrm{Fi}}[f] = \frac{n+1}{2^{n-1}(n+2)n}\cdot\frac{f^{(n+1)}(\xi)}{(n+1)!}.$$

PROOF. The function g defined by

$$g(x) = \int_x^1 U_n(t)\,dt = \frac{T_{n+1}(1) - T_{n+1}(x)}{n+1} = \frac{1 - T_{n+1}(x)}{n+1}$$

does not change sign on $[-1,1]$. Hence, definiteness follows from Theorem 4.2.9. This means that (see Theorem 4.2.6)

$$R_n^{\mathrm{Fi}}[f] = \frac{f^{(n+1)}(\xi)}{(n+1)!}R[p_{n+1}] \quad \text{where } p_{n+1}(x) := x^{n+1}$$

with

$$R_n^{\mathrm{Fi}}[p_{n+1}] = \frac{1}{2^n}\int_{-1}^{1} x\,U_n(x)\,dx = \frac{1}{2^n}\cdot\frac{2(n+1)}{n(n+2)},$$

and so the theorem is proved. □

Another remarkable property of Q_n^{Fi} is the possibility of determining a (good) bound using $f(-1), f(x_1^{\mathrm{Fi}}), \ldots, f(x_n^{\mathrm{Fi}}), f(1)$ only. Certainly, this requires some kind of assumption on f. We shall assume $f^{(n+1)} \ge 0$, a property that is sometimes more easily satisfied than a bound on $\|f^{(n+1)}\|$.

THEOREM 5.5.4. *Suppose that $w = 1$, n is odd and $f^{(n+1)} \ge 0$. Then*

$$0 \le R_n^{\mathrm{Fi}}[f] \le \frac{1}{2^{n-2}(n+1)}\,\mathrm{dvd}(-1, x_1^{\mathrm{Fi}}, \ldots, x_n^{\mathrm{Fi}}, 1)[f].$$

PROOF. The first inequality $R_n^{\mathrm{Fi}}[f] \geq 0$ follows from Theorem 5.5.3. We now prove that the linear functional R defined by

$$R[f] := R_n^{\mathrm{Fi}}[f] - \frac{1}{2^{n-2}(n+1)} \operatorname{dvd}(-1, x_1^{\mathrm{Fi}}, \ldots, x_n^{\mathrm{Fi}}, 1)[f]$$

is negative definite. To this end, we apply Theorem 4.2.8. For the reduced functional

$$\begin{aligned} R^{(x_1^{\mathrm{Fi}})\cdots(x_n^{\mathrm{Fi}})}[f] &= \frac{1}{2^n} \int_{-1}^{1} f(x) U_n(x)\, dx \\ &\quad - \frac{1}{2^{n-2}(n+1)} \frac{2^{n-1}}{U_n(1)} \left(f(-1) \frac{U_n(-1)}{2^n} + f(1) \frac{U_n(1)}{2^n} \right), \end{aligned}$$

we prove definiteness of order one by looking at its Peano kernel:

$$\begin{aligned} K_1(x) &= \frac{1}{2^n} \int_x^1 U_n(x)\, dx - \frac{1}{2^{n-1}(n+1)} \\ &= \frac{1}{2^n} \left(\frac{T_{n+1}(1) - T_{n+1}(x)}{n+1} - \frac{2}{n+1} \right) = \frac{1}{2^n} \cdot \frac{-1 - T_{n+1}(x)}{n+1}, \end{aligned}$$

and so we are done. □

So far, we lack simple expressions for the coefficients of Q_n^{Fi}. These could be obtained from the coefficients of Q_n^{CC} via Theorem 5.5.1, but we prefer to use a method of wider applicability.

THEOREM 5.5.5 (Szegö 1933/34). *Let $q_0, q_1, \ldots$ be the system of orthonormal polynomials associated with the weight function W. Denote by δ_n the leading coefficient and by $x_1 \ldots, x_n$ the zeros of q_n. Then, the coefficients of the quadrature rule of interpolatory type with respect to the weight function w and based on the nodes x_k, $k = 1, \ldots, n$, are given by*

$$a_k = -\frac{\delta_{n+1}}{\delta_n} \cdot \frac{1}{q_n'(x_k) q_{n+1}(x_k)} \sum_{\mu=0}^{n-1} q_\mu(x_k) \int_a^b q_\mu(x) w(x)\, dx.$$

PROOF. Using (5.2) and $q_n(x_k) = 0$, we obtain

$$a_k = \frac{1}{q_n'(x_k) q_{n+1}(x_k)} \int_a^b \frac{q_n(x) q_{n+1}(x_k) - q_n(x_k) q_{n+1}(x)}{x - x_k} w(x)\, dx.$$

Now, the Christoffel–Darboux formula (see Appendix A) gives the assertion. □

In the special case where $[a, b] = [-1, 1]$, $W(x) = \sqrt{1 - x^2}$ and $q_n = \sqrt{\pi/2} \cdot U_n$, i.e. $\delta_n = \sqrt{\pi/2} \cdot 2^n$ and $w = 1$, a short calculation leads to the following expressions:

Coefficients of the Filippi rule Q_n^{Fi} for $w = 1$:

$$\begin{aligned} a_k &= \frac{4}{n+1} \sin \frac{k\pi}{n+1} \sum_{\mu=0}^{m} \frac{1}{2\mu+1} \sin\left((2\mu+1) \frac{k\pi}{n+1} \right) \\ &= \frac{2}{n+1} \left(1 - \sum_{\mu=1}^{m} \frac{2}{4\mu^2 - 1} \cos\left(2\mu \frac{k\pi}{n+1} \right) - \frac{1}{2m+1} \cos\left((2m+2) \frac{k\pi}{n+1} \right) \right) \end{aligned} \tag{5.46}$$

where $m := \lfloor (n-1)/2 \rfloor$.

A similar calculation as in the proof of Theorem 5.4.1 yields the following positivity result.

THEOREM 5.5.6 (Féjer 1933). The Filippi method for $w = 1$ is a positive method.

A comparison of (5.46) with the Fourier series

$$|\sin x| = \frac{2}{\pi}\left(1 - \sum_{\mu=1}^{\infty} \frac{2\cos 2\mu x}{4\mu^2 - 1}\right)$$

proves the simple formula

$$a_k = \frac{\pi}{n+1} \sin \frac{k\pi}{n+1} + \frac{4\theta}{(n+1)^2}, \qquad \theta \in [-1, 1]. \tag{5.47}$$

Let us now consider a different system of nodes.

DEFINITION 5.5.3. The interpolatory quadrature method on $[-1, 1]$ whose nth formula Q_n^{Po} $(n = 1, 2, \ldots)$ is based on the nodes

$$x_{\nu,n}^{\mathrm{Po}} := -\cos\left(\frac{2\nu - 1}{2n}\pi\right), \qquad \nu = 1, \ldots, n,$$

is called the *Polya method.*

The nodes of Q_n^{Po} are the zeros of T_n, the nth Chebyshev polynomial of the first kind. The name "Polya method" comes from Polya's proof of the convergence of Q_n^{Po} for $w = 1$ (see Fejér 1933). Until this proof appeared, the only interesting quadrature methods of interpolatory type being studied were the methods of Gauss and Newton–Cotes, as well as their relatives. Sloan (1978) (see also Sloan and Smith 1978) went beyond Polya's result to prove convergence and show that, moreover,

$$\lim_{n\to\infty} \|Q_n^{\mathrm{Po}}\| = \|I\|$$

under the assumption

$$\int_{-1}^{1} w^p(x)\,dx < \infty \quad \text{for some } p > 1.$$

The proof is a minor modification of the proof of Theorem 5.3.1. Under more restrictive assumptions on w, fairly simple convergence proofs can be constructed. Here we present one, which is of independent interest, based on an asymptotic expression for the coefficients.

THEOREM 5.5.7. For continuous w of bounded variation, we have

$$a_{\nu,n}^{\mathrm{Po}} = \frac{\pi}{n} w\left(-\cos\left(\frac{2\nu - 1}{2n}\pi\right)\right) \sin\left(\frac{2\nu - 1}{2n}\pi\right) + o(n^{-1})$$

uniformly in ν.

The convergence proof is a simple application of the following result:

$$\begin{aligned}\|Q_n^{\mathrm{Po}}\| = \sum_{\nu=1}^{n} |a_{\nu,n}^{\mathrm{Po}}| &= \sum_{\nu=1}^{n} \frac{\pi}{n} w\left(-\cos\left(\frac{2\nu - 1}{2n}\pi\right)\right) \sin\left(\frac{2\nu - 1}{2n}\pi\right) + o(1) \\ &= \int_0^{\pi} w(-\cos t)\sin t\,dt + o(1) = \int_{-1}^{1} w(x)\,dx + o(1).\end{aligned}$$

PROOF OF THEOREM 5.5.7. We apply Theorem 5.5.5 with

$$W(x) = \left(1 - x^2\right)^{-1/2}$$

and

$$q_0(x) = \sqrt{\frac{1}{\pi}}\, T_0(x), \qquad q_k(x) = \sqrt{\frac{2}{\pi}}\, T_k(x) \quad \text{for } k = 1, 2, \ldots.$$

After a short calculation we obtain

$$a_{\nu,n}^{\text{Po}} = \frac{\pi}{n} S_{n-1}[g](x_{\nu,n}^{\text{Po}}) \quad \text{with } g(x) = w(x)\sqrt{1 - x^2},$$

where $S_{n-1}[g]$ means the $(n-1)$st partial sum of the Chebyshev expansion of g. The assumptions on g guarantee uniform convergence of this expansion, and the assertion follows. □

Using the same method, we can obtain an explicit formula for the coefficients in the $w = 1$ case:

$$a_{\nu,n}^{\text{Po}} = \frac{2}{n}\left(1 - \sum_{\mu=1}^{\lfloor (n-1)/2 \rfloor} \frac{2}{4\mu^2 - 1} \cos\left(2\mu \frac{2\nu - 1}{2n}\pi\right)\right).$$

As in Theorems 5.4.1 and 5.5.6, we get positivity as a corollary.

THEOREM 5.5.8 (Féjer 1933). *The Polya method for $w = 1$ is a positive method.*

For error estimates, we can use the Peano kernel method or, again, Lebesgue's inequality.

EXAMPLE 5.5.2. The Polya method for $w(x) = \left(1 - x^2\right)^{-1/2}$, which was already discussed in Example 5.1.1, permits rather precise determination of the error constants, owing to the relatively simple structure of the Peano kernel. Specifically, we have

$$K_1(x) = \arccos x - \frac{\pi}{n}(n - \lambda) \quad \text{for } x \in \left]x_{\lambda,n}^{\text{Po}}, x_{\lambda+1,n}^{\text{Po}}\right[,$$

which implies that

$$\int_{-1}^{1} |K_1(x)|\, dx = 2 \tan \frac{\pi}{4n}$$

(see Ponomarenko 1983) and

$$\|K_1\| = \frac{\pi}{2n}.$$

Thus, we have obtained error bounds in $\mathcal{C}_M^{(1)}$ and $\mathcal{C}_M^{\text{Var}}$ that are impossible to improve upon (see also the remarks after Theorem 4.4.3). Moreover, we can apply Theorem 4.5.4 and obtain that

$$\left|R_n[f]\right| \le \pi\, \omega\Big(f; \sin \frac{\pi}{2n}\Big).$$

Using the Bernoulli function B_1^* (see Appendix B), we can find a more compact expression for K_1, namely

$$K_1(x) = \frac{\pi}{n} B_1^* \left(\frac{n \arccos x}{\pi} + \frac{1}{2}\right).$$

Now, note that

$$-\frac{\pi^2}{n^2} B_2^* \left(\frac{n \arccos x}{\pi} + \frac{1}{2}\right)$$

is a primitive of

$$\frac{\pi}{n} B_1^*\left(\frac{n \arccos x}{\pi}+\frac{1}{2}\right)\left(1-x^2\right)^{-1/2}.$$

Therefore, integration by parts gives

$$\begin{aligned} K_2(x) &= \int_x^1 K_1(u)\,du = \int_x^1 \left[K_1(u)\left(1-u^2\right)^{-1/2}\right]\sqrt{1-u^2}\,du \\ &= \frac{\pi^2}{n^2} B_2^*\left(\frac{n \arccos x}{\pi}+\frac{1}{2}\right)\sqrt{1-x^2} \\ &\quad - \int_x^1 \frac{\pi^2}{n^2} B_2^*\left(\frac{n \arccos u}{\pi}+\frac{1}{2}\right)\frac{u}{\sqrt{1-u^2}}\,du. \end{aligned}$$

Upon taking $B_3^*(1/2)=0$ into account, a further integration by parts yields

$$K_2(x) = \frac{\pi^2}{n^2}\sqrt{1-x^2}\, B_2^*\left(\frac{n \arccos x}{\pi}+\frac{1}{2}\right) + O(n^{-3}).$$

Similarly, we can establish that

$$K_s(x) = \left(\frac{\pi}{n}\right)^s \left(1-x^2\right)^{(s-1)/2} B_s^*\left(\frac{n \arccos x}{\pi}+\frac{1}{2}\right) + O(n^{-s-1}).$$

This shows that (cf. Polya's lemma in Section 7.1)

$$\lim_{n\to\infty} n^s \int_{-1}^1 |K_s(x)|\,dx = \pi^s \int_0^\pi \sin^s(x)\,dx \cdot \int_0^1 |B_s(u)|\,du.$$

Using a method presented in Section 5.8, it is possible to show that $\rho(Q_n^{\mathrm{Fi}}, \mathcal{C}_M^{(r)})$ and $\rho(Q_n^{\mathrm{Po}}, \mathcal{C}_M^{(r)})$ have the same asymptotic behaviour for fixed r as $n\to\infty$. Again, this is not true of bounds based on the highest possible derivative. The situation is similar to that in Theorem 5.4.2; the only difference in the proof lies in the calculation of $R_n^{\mathrm{Po}}[T_\mu]$. We have

$$T_{2kn\pm\ell}(x_{\nu,n}^{\mathrm{Po}}) = (-1)^k T_\ell(x_{\nu,n}^{\mathrm{Po}})$$

and hence

$$R_n^{\mathrm{Po}}[T_{2kn\pm\ell}] = I[T_{2kn\pm\ell}] - \begin{cases} (-1)^k I[T_\ell] & \text{for } 0\le \ell < n, \\ 0 & \text{for } \ell = n. \end{cases}$$

For $w=1$ we have, in particular, that

$$R_n^{\mathrm{Po}}[T_n] = \frac{2}{1-n^2} \quad \text{for even } n$$

$$\text{and} \quad R_n^{\mathrm{Po}}[T_{n+1}] = \frac{4}{4-n^2} \quad \text{for odd } n.$$

Thus, we obtain that

$$\|R_n[f]\| \le c_n \frac{\|f^{(m)}\|}{m!\,2^{m-1}} \quad \text{with } m := 2\left\lceil \frac{n}{2}\right\rceil, \tag{5.48}$$

where c_n is of best possible order, i.e. n^{-1}, n^{-2} and n^{-3} for Q_n^{Fi}, Q_n^{Po} and Q_n^{CC}, respectively.

Similar bounds can be derived for weight functions of quite general form. For this purpose, we use the well-known representation

$$f(x) - \operatorname{intpol}(x_1, \dots, x_n)[f](x) = \frac{f^{(n)}(\xi)}{n!} \prod_{\nu=1}^{n} (x - x_\nu).$$

For any quadrature rule of interpolatory type, this yields

$$\begin{aligned} |R[f]| &= \big|I\big[f - \operatorname{intpol}(x_1, \dots, x_n)[f]\big]\big| \le \|I\| \cdot \|f - \operatorname{intpol}(x_1, \dots, x_n)[f]\| \\ &\le \|I\| \frac{\|f^{(n)}\|}{n!} \sup_x \Big| \prod_{\nu=1}^{n} (x - x_\nu) \Big|. \end{aligned}$$

The supremum on the right-hand side is minimal on $[-1, 1]$ if $x_\nu = x_{\nu,n}^{\mathrm{Po}}$ (see, e.g., Cheney 1966: p. 222). In this case, we obtain

$$\rho(Q_n^{\mathrm{Po}}, \mathcal{C}_1^{(n)}) \le \frac{\|I\|}{n!\, 2^{n-1}}.$$

For this method, nothing needs to be known about $\|Q_n\|$.

Note that Q_n^{Po} and Q_n^{Fi} are based on the zeros of the Chebyshev polynomials of first and second kind, respectively; these are both Jacobi polynomials $P_n^{(\alpha,\beta)}$, i.e. orthogonal polynomials with respect to the weight function $W_{\alpha,\beta}(x) = (1 - x)^\alpha (1 + x)^\beta$. The nodes of the Clenshaw–Curtis, Lobatto and Radau methods (see Chapter 6) which lie in $]-1, 1[$ are likewise zeros of certain Jacobi polynomials. It therefore seems reasonable to consider quadrature rules of interpolatory type that are based on zeros of $P_n^{(\alpha,\beta)}$. The main difficulty is the lack of simple expressions for these zeros, except when

$$(\alpha, \beta) \in \Big\{ \Big(\frac{1}{2}, \frac{1}{2}\Big), \Big(-\frac{1}{2}, \frac{1}{2}\Big), \Big(\frac{1}{2}, -\frac{1}{2}\Big), \Big(-\frac{1}{2}, -\frac{1}{2}\Big) \Big\}. \tag{5.49}$$

The choices of $(-\frac{1}{2}, -\frac{1}{2})$ and $(\frac{1}{2}, \frac{1}{2})$ for (α, β) correspond to Q_n^{Po} and Q_n^{Fi}. For the other two parameter-pairs, it can be proved that

$$P_n^{(1/2,-1/2)}(\cos t) = \mathrm{const}_n \frac{\sin (n + 1/2)t}{\sin(t/2)}$$

and

$$P_n^{(-1/2,1/2)}(\cos t) = \mathrm{const}_n \frac{\cos (n + 1/2)t}{\cos(t/2)},$$

giving

$$x_{\nu,n}^{(1/2,-1/2)} = -\cos\Big(\frac{2\nu - 1}{2n + 1}\pi\Big) \quad \text{and} \quad x_{\nu,n}^{(-1/2,1/2)} = -\cos\Big(\frac{2\nu}{2n + 1}\pi\Big).$$

Although the methods based on these nodes are not symmetric, their behaviour resembles that of Q_n^{Po} and Q_n^{Fi}. Adding an endpoint (or both endpoints) to the set of nodes associated with the parameter choices in (5.49)—see Notaris (1997), Mason and Venturino (1997) or Notaris (2001)—does not lead to substantially new properties. The relationship between Q_n^{CC} and Q_n^{Fi} is typical. If we want to retain symmetry, the only possibility is to add ± 1 to the nodes of Q_n^{Po}. This method (see Riess and Johnson 1972), while not positive, is convergent. Using the same method as for the Clenshaw–Curtis rules, we can prove that $c_n = O(n^{-4})$ for the constants in (5.48); see Brass and Schmeisser (1981).

If (α, β) is not one of the possibilities in (5.49), then any deeper examination of rules based on the zeros of $P_n^{(\alpha,\beta)}$ necessitates a study of these Jacobi polynomials; such investigations belong to the realm of special functions rather than to quadrature theory. Some relevant works are Askey and Fitch (1968), Askey (1972, 1979), Lether *et al.* (1979), Micchelli (1980), Smith and Sloan (1980), Sloan and Smith (1982), Sottas (1982), Kütz (1982, 1986), Rabinowitz (1986a, 1987), Sottas (1989), and Notaris (2002, 2003). Szegö (1939) proved that $\max\{\alpha, \beta\} \leq 3/2$ is necessary and sufficient for convergence if $w = 1$ (cf. Locher 1973b). For $\max\{\alpha, \beta\} < 1$, a simpler way of proving convergence is based on the following result of Erdős and Turán (1937):

THEOREM 5.5.9. Let $(Q_n)_{n\in\mathbb{N}}$ be the quadrature method of interpolatory type based on the zeros of the orthogonal polynomials corresponding to the weight function W. The method is convergent if

$$\int_{-1}^{1} \frac{w^2(x)}{W(x)}\, dx < \infty.$$

PROOF. We shall prove that $\sup_n \|Q_n\| < \infty$. To this end, define $p_n \in \mathbb{P}_{n-1}$ by

$$p_n(x_{\nu,n}) = \operatorname{sgn} a_{\nu,n}.$$

Then

$$\|Q_n\| = Q_n[p_n] = \int_{-1}^{1} p_n(x)w(x)\, dx \leq \left(\int_{-1}^{1} p_n^2(x)W(x)\, dx \cdot \int_{-1}^{1} \frac{w^2(x)}{W(x)}\, dx\right)^{1/2},$$

and application of the Gaussian rule Q_n^{G} shows boundedness of the first integral. □

5.6. The distribution of nodes

All the "good" quadrature rules of interpolatory type that have been discussed so far have a rather similar distribution of nodes. To make this more precise, we introduce the notion of arcsine distribution.

DEFINITION 5.6.1. Let $A := \big\{\{x_{\nu,n} : \nu = 1, 2, \ldots, n\},\ n = 1, 2, \ldots\big\}$ be a system of points in $[-1, 1]$. Denote by $N_n(c, d)$ the number of points from $x_{1,n}, \ldots x_{n,n}$ that lie in $[c, d]$. We say that A has an *arcsine distribution* if

$$\lim_{n\to\infty} \frac{N_n[c, d]}{n} = \frac{\arcsin d - \arcsin c}{\pi}.$$

Evidently, the nodal systems of the Clenshaw–Curtis and related methods have an arcsine distribution. Later we shall prove that the same is true of Gaussian methods.

In order to exhibit examples with different behaviour, let us regard $Q^{\mathrm{G}}_{\lceil (n+1)/2\rceil}$ as a rule with n nodes, of which $n - \lceil (n + 1)/2\rceil$ are arbitrarily chosen and have corresponding zero coefficients. Hence, there are convergent methods of interpolatory type such that about half of the nodes have an arcsine distribution while the other half may be arbitrary. The following remarkable theorem, originally proved under slightly more general assumptions, shows that this is the case in general.

THEOREM 5.6.1 (Bloom *et al.* 1993a, 1993b). Suppose $w \geq \text{const} > 0$ on $[-1,1]$ and let $(Q_n)_{n\in\mathbb{N}}$ be a quadrature method of interpolatory type such that

$$\lim_{n\to\infty} \|Q_n\|^{1/n} = 1. \tag{5.50}$$

Then

$$\liminf_{n\to\infty} \frac{N_n(c,d)}{n} \geq \frac{1}{2\pi}(\arcsin d - \arcsin c).$$

The interval $[\sqrt{0.75}, 1] = [\sin(\pi/3), \sin(\pi/2)]$, for instance, has to contain at least $[\pi/2 - \pi/3]/(2\pi) = 1/12$ of all the nodes asymptotically. If the nodes are equidistant, the proportion of nodes in this interval must be $(1 - \sqrt{0.75}\,)/2 < 1/12$. Therefore, (5.50) cannot hold and the quadrature method of interpolatory type is necessarily divergent.

LEMMA 5.6.1. For any set $\{x_1, \ldots, x_n\} \subset [-1,1]$ and $s \in \{2, 3, \ldots, n\}$, it is possible to find a subset $\{y_1, \ldots, y_s\}$ such that

$$h(x) := \prod_{\nu=1}^{s} (x - y_\nu)$$

satisfies

$$\sup_\nu |h(x_\nu)| \leq (s+1)2^{1-s}.$$

PROOF. Find y_ν so that

$$\prod_{1\leq i<j\leq s} |y_i - y_j| = \sup\left\{ \prod_{1\leq i<j\leq s} |u_i - u_j| \,:\, \{u_1, \ldots, u_s\} \subset \{x_1, \ldots, x_n\} \right\}. \tag{5.51}$$

Then the cardinal functions for Lagrange interpolation,

$$\ell_i(x) := \prod_{\substack{\nu=1\\ \nu\neq i}}^{s} \frac{x - y_\nu}{y_i - y_\nu},$$

satisfy

$$|\ell_i(x_j)| \leq 1 \quad \text{for } i = 1, \ldots, s \text{ and } j = 1, \ldots, n, \tag{5.52}$$

or else we would have

$$\prod_{\substack{\nu=1\\ \nu\neq i}}^{s} |x_j - y_\nu| > \prod_{\substack{\nu=1\\ \nu\neq i}}^{s} |y_i - y_\nu|$$

and could replace y_i by x_j, increasing the supremum in (5.51).

Lagrange's interpolation formula yields

$$2^{1-s}T_s - h = \sum_{\nu=1}^{s} 2^{1-s}T_s(y_\nu)\ell_\nu.$$

By (5.52), we get

$$|2^{1-s}T_s(x_k) - h(x_k)| \leq \sum_{\nu=1}^{s} 2^{1-s}|T_s(y_\nu)| \leq s2^{1-s}$$

and the lemma follows. □

PROOF OF THEOREM 5.6.1. Set $s := \lfloor (n-1)/2 \rfloor$ and determine h_n according to Lemma 5.6.1, where the x_ν are the nodes of Q_n. Then

$$\begin{aligned}\int_{-1}^{1} h_n^2(x)w(x)\,dx &= Q_n[h_n^2] \le \|Q_n\| \sup_\nu |h_n(x_\nu)|^2 \\ &\le \|Q_n\|(s+1)^2 2^{2-2s} \le \|Q_n\|(n+1)^2 2^{2-n}.\end{aligned}$$

Using (5.50), we obtain

$$\limsup_{n\to\infty} \left(\int_{-1}^{1} h_n^2(x)w(x)\,dx \right)^{1/n} \le \frac{1}{2}. \tag{5.53}$$

Our next aim is to prove that

$$\lim_{n\to\infty} \|h_n\|^{2/n} = \frac{1}{2}. \tag{5.54}$$

We argue as follows: Take $p \in \mathbb{P}_m$ and let x_0 be such that $|p(x_0)| = \|p\|$; then

$$|p(x)| \ge \|p\| - |x - x_0|\, \|p'\|.$$

Markov's inequality (see, e.g., DeVore and Lorentz 1993: p. 27) says that $\|p'\| \le m^2\|p\|$, so

$$|p(x)| \ge \|p\|(1 - |x - x_0|m^2)$$

and therefore

$$|p(x)| > \frac{1}{2}\|p\| \quad \text{for } x \in J_n,$$

where $J_n \subset [-1,1]$ is an interval of length $1/(2m^2)$. In particular, we have

$$\int_{-1}^{1} h_n^2(x)w(x)\,dx \ge \frac{1}{2}\|h_n^2\| \int_{J_n} w(x)\,dx \ge \|h_n^2\| \frac{\text{const}}{4n^2}.$$

Passing to the limit as $n \to \infty$ shows that

$$\limsup_{n\to\infty} \left(\int_{-1}^{1} h_n^2(x)w(x)\,dx \right)^{1/n} \ge \limsup_{n\to\infty} \|h_n\|^{2/n},$$

giving (see (5.53))

$$\limsup_{n\to\infty} \|h_n\|^{2/n} \le \frac{1}{2}. \tag{5.55}$$

On the other hand, it is well known that for any monic $p \in \mathbb{P}_m$ we have

$$\|p\| \ge 2^{1-m},$$

so (5.54) follows from (5.55).

The final step is to use a theorem of Fekete (see, e.g., Erdős and Turán 1940a), which says that property (5.54) for all monic polynomials $h_n \in \mathbb{P}_{\lfloor (n-1)/2 \rfloor}$ with all their zeros in $[-1,1]$ implies the arcsine distribution of these zeros. Obviously, this is equivalent to the assertion of the theorem. □

COROLLARY 5.6.1. The set of nodes of a Gaussian method with respect to a strictly positive weight function has an arcsine distribution.

PROOF. For each n, add n nodes that have zero coefficients outside $[c,d]$ to Q_n^{G}. This gives a new quadrature method $\left(\widetilde{Q}_{2n}\right)_{n\in\mathbb{N}}$ which has property (5.50) and the same number of nodes in $[c,d]$, i.e. $\widetilde{N}_{2n}[c,d] = N_n[c,d]$. Hence, from Theorem 5.6.1 we obtain

$$\liminf_{n\to\infty} \frac{N_n[c,d]}{n} = 2\liminf_{n\to\infty} \frac{\widetilde{N}_{2n}[c,d]}{2n} \geq \frac{1}{\pi}(\arcsin d - \arcsin c).$$

We may apply the same argument to $[-1,c]$ and $[d,1]$, obtaining

$$\begin{aligned}\limsup_{n\to\infty} \frac{N_n[c,d]}{n} &= \limsup_{n\to\infty} \frac{n - N_n[-1,c] - N_n[d,1]}{n} \\ &\leq 1 - \liminf_{n\to\infty} \frac{N_n[-1,c]}{n} - \liminf_{n\to\infty} \frac{N_n[d,1]}{n} \\ &\leq 1 - \frac{1}{\pi}\big(\arcsin 1 - \arcsin d + \arcsin c - \arcsin(-1)\big) \\ &= \frac{1}{\pi}(\arcsin d - \arcsin c).\end{aligned}$$

Therefore, the condition for an arcsine distribution is satisfied. □

Theorem 5.6.1 gives a necessary condition for convergence. A further necessary condition rules out over-large gaps between consecutive nodes. More precisely, we have the following statement.

THEOREM 5.6.2. Let $w \geq \text{const} > 0$ and let $\left(Q_n\right)_{n\in\mathbb{N}}$ be a convergent quadrature method of interpolatory type. Then

$$\sup_{\nu} \left| \arccos x_{\nu,n} - \arccos x_{\nu+1,n} \right| < \frac{4\ln n}{n}$$

for sufficiently large n.

In particular, this means that for convergent methods, gaps near the boundary are of order $O(n^{-2}\ln^2 n)$. For Q_n^{CC} and Q_n^{G}, they are even of order $O(n^{-2})$, but this asserted order cannot be relaxed; see Brass (1977a: p. 98) or Erdős and Turán (1938).

The main tool used in the proof is a lemma that ensures the existence of a polynomial g with the following property: $\int_{-1}^{1} g(x)\,dx$ is large even though $|g(x)|$ is small, except for $x \in J$ where J is a given interval. One special case is given in the following lemma.

LEMMA 5.6.2. For $n \in \mathbb{N}$ and $\delta \in [0,\pi/2]$, there is an $h \in \mathbb{P}_n$ with the following properties:

(i) $h(x) \geq -1$ for $x \in [-1,1]$;
(ii) $h(x) \geq 0$ for $x \in [\cos\delta, 1]$;
(iii) $|h(x)| \leq 1$ for $x \in [-1,\cos\delta]$;
(iv) $\displaystyle\int_{\cos\delta}^{1} h(x)\,dx \geq \frac{1}{2n}\left(\sin\frac{\delta}{2} - \frac{1}{2n}\right)\exp\left(2n\sin\frac{\delta}{2}\right).$

PROOF. Let

$$h(x) = T_n\left(\frac{2x+1-\cos\delta}{1+\cos\delta}\right).$$

Then properties (i), (ii) and (iii) follow immediately from simple properties of the Chebyshev polynomials. The proof of (iv) starts with the substitution $x = (y + \cos\delta)/(1-y)$, which gives

$$\begin{aligned} \int_{\cos\delta}^{1} h(x)\,dx &= \int_0^{\sin^2\delta/2} T_n\left(\frac{1+y}{1-y}\right)\frac{1+\cos\delta}{(1-y)^2}\,dy \\ &\geq \int_0^{\sin^2\delta/2} T_n\left(\frac{1+y}{1-y}\right)dy. \end{aligned} \tag{5.56}$$

Now,

$$2T_n(x) = \left(x+\sqrt{x^2-1}\right)^n + \left(x-\sqrt{x^2-1}\right)^n \geq \left(x+\sqrt{x^2-1}\right)^n \quad \text{for } x \geq 1$$

(see Theorem A.9(vii)), and the Taylor expansion

$$\ln\frac{1+z}{1-z} = 2\left(z+\frac{z^3}{3}+\frac{z^5}{5}+\cdots\right) > 2z \quad \text{for } z > 0$$

shows that

$$2T_n\left(\frac{1+y}{1-y}\right) \geq \left(\frac{1+y+\sqrt{4y}}{1-y}\right)^n = \left(\frac{1+\sqrt{y}}{1-\sqrt{y}}\right)^n = \exp\left(n\ln\frac{1+\sqrt{y}}{1-\sqrt{y}}\right) \geq e^{2n\sqrt{y}}.$$

This estimate for the integrand on the right-hand side of (5.56) yields

$$\begin{aligned} \int_{\cos\delta}^{1} h(x)\,dx &\geq \frac{1}{2}\int_0^{\sin^2\delta/2} e^{2n\sqrt{y}}\,dy = \int_0^{\sin\delta/2} ze^{2nz}\,dz \\ &= \left(\frac{\sin\delta/2}{2n}-\frac{1}{4n^2}\right)\exp\left(2n\sin\frac{\delta}{2}\right)+\frac{1}{4n^2}, \end{aligned}$$

so that (iv) is established. □

LEMMA 5.6.3. Let α and δ be positive numbers such that $\delta \leq \alpha \leq \pi-\delta$. For any $n \in \mathbb{N}$, there exists a $g \in \mathbb{P}_n$ with the following properties:

(i) $g(x) \geq -1$ for $x \in [-1,1]$;
(ii) $|g(x)| \leq 1$ for $x \in [-1,1] \setminus [\cos(\alpha+\delta), \cos(\alpha-\delta)]$;
(iii) $\displaystyle\int_{-1}^{1} g(x)\,dx \geq \cos\delta \int_{\cos\delta}^{1} h(x)\,dx - 2$, where h is as in Lemma 5.6.2.

PROOF. Define

$$\begin{aligned} t(x) &:= h(\cos x), \\ t^*(x) &:= \tfrac{1}{2}t(x+\alpha)+\tfrac{1}{2}t(x-\alpha), \\ g(x) &:= t^*(\arccos x). \end{aligned}$$

Since t^* is an even trigonometric polynomial of order n, we have $g \in \mathbb{P}_n$. Properties (i) and (ii) are obvious. By arguing that

$$\begin{aligned}
\int_{-1}^{1} g(x)\,dx &= \int_0^\pi t^*(y)\sin y\,dy \\
&= \frac{1}{2}\int_0^\pi t(y-\alpha)\sin y\,dy + \frac{1}{2}\int_0^\pi t(y+\alpha)\sin y\,dy \\
&\geq \frac{1}{2}\int_0^\pi t(y-\alpha)\sin y\,dy - 1 = \frac{1}{2}\int_{-\alpha}^{\pi-\alpha} t(z)\sin(z+\alpha)\,dz - 1 \\
&\geq \frac{1}{2}\int_{-\delta}^{\delta} t(z)\sin(z+\alpha)\,dz - 2 = \frac{\sin\alpha}{2}\int_{-\delta}^{\delta} t(z)\cos(z)\,dz - 2 \\
&= \sin\alpha \int_0^\delta t(z)\cos(z)\,dz - 2 = \sin\alpha \int_{\cos\delta}^{1} h(u)\frac{u}{\sqrt{1-u^2}}\,du - 2 \\
&\geq \frac{\sin\alpha}{\sin\delta}\cos\delta \int_{\cos\delta}^{1} h(u)\,du - 2 \geq \cos\delta \int_{\cos\delta}^{1} h(u)\,du - 2,
\end{aligned}$$

we prove (iii). □

PROOF OF THEOREM 5.6.2. Construct g according to Lemma 5.6.3 with

$$\begin{aligned}
\alpha &= \frac{1}{2}\left(\arccos x_{\nu,n+1} + \arccos x_{\nu+1,n+1}\right), \\
\delta &= \frac{1}{2}\left(\arccos x_{\nu,n+1} - \arccos x_{\nu+1,n+1}\right).
\end{aligned}$$

Then we have

$$\begin{aligned}
\|Q_{n+1}\| &\geq \left|Q_{n+1}[g]\right| = I[g] = \int_{-1}^{1} [g(x)+1]w(x)\,dx - \int_{-1}^{1} w(x)\,dx \\
&\geq \text{const} \int_{-1}^{1} [g(x)+1]\,dx - \int_{-1}^{1} w(x)\,dx \\
&\geq \text{const} \int_{-1}^{1} g(x)\,dx - \int_{-1}^{1} w(x)\,dx.
\end{aligned}$$

For

$$\arccos x_{\nu,n+1} - \arccos x_{\nu+1,n+1} \geq 4\,\frac{\ln n}{n},$$

we obtain $\delta \geq 2\frac{\ln n}{n}$ and therefore $\|Q_{n+1}\| \to \infty$. □

Finally, it should be remarked that for positive quadrature rules of interpolatory type, improved estimates for the distances between nodes are possible; these follow from Theorem 5.1.8 and estimates for the distances between zeros of orthogonal polynomials.

5.7. Bounds for the norms of Peano kernels of interpolatory rules

In this section, we shall always take

$$[a,b] = [-1,1].$$

Our first main result is the following:

THEOREM 5.7.1. For any positive quadrature rule with $R[\mathbb{P}_m] = \{0\}$,

$$\sup_{-1\le x\le 1} |K_s(x)| \le \frac{\pi^s(m-s+1)!}{(m+1)!} \sup_{-1\le x\le 1} w(x)\sqrt{1-x^2}.$$

This is a coarser version of a theorem in Brass (1993a), whose proof is rather long. In the special case where $w(x) = \left(1-x^2\right)^{-1/2}$ and $s = 1$, Theorem 5.7.1 says that

$$\sup_{-1\le x\le 1} |K_1(x)| \le \frac{\pi}{m+1},$$

and Example 5.5.2 shows that this bound cannot be improved. To prove the theorem, we need one-sided approximations of the function t_x given by

$$t_x(u) := \begin{cases} 0 & \text{if } u < x, \\ 1 & \text{if } u \ge x. \end{cases}$$

LEMMA 5.7.1. Assume that

$$\sup_{-1\le x\le 1} w(x)\sqrt{1-x^2} < \infty.$$

Then, for each $m \in \mathbb{N}$ and $x \in [-1, 1]$, there is a $q_x \in \mathbb{P}_m$ such that

$$q_x(u) \ge t_x(u) \quad \text{for all } u \in [-1, 1]$$

and

$$\int_{-1}^{1} \left[t_x(u) - q_x(u)\right] w(u)\, du \ge -\frac{\pi}{m+1} \sup_{-1\le x\le 1} w(x)\sqrt{1-x^2}.$$

We will prove this lemma a little later. For now, just note that upon assuming the truth of the lemma, the theorem can be proved easily:

PROOF OF THEOREM 5.7.1. The proof is by induction over s. First, consider $s = 1$ and use the polynomial q_x from Lemma 5.7.1:

$$\begin{aligned} K_1(x-0) = R[t_x] = R[t_x - q_x] &= I[t_x - q_x] - Q[t_x - q_x] \\ &\ge I[t_x - q_x] \\ &\ge -\frac{\pi}{m+1} \sup_{-1\le x\le 1} w(x)\sqrt{1-x^2}. \end{aligned}$$

In an analogous fashion, we argue with

$$v_x(u) := \begin{cases} 0 & \text{if } u \le x, \\ 1 & \text{otherwise} \end{cases}$$

and approximation from below. From $v_x(u) = 1 - t_{-x}(-u)$ we obtain

$$K_1(x+0) \le \frac{\pi}{m+1} \sup_{-1\le x\le 1} w(x)\sqrt{1-x^2}$$

and hence the assertion for $s = 1$. We start the inductive step as follows:

$$K_{s+1}(x) = \int_x^1 K_s(u)\, du = \int_{-1}^{1} K_s(u) t_x(u)\, du = \int_{-1}^{1} K_s(u)\left[t_x(u) - p(u)\right] du,$$

with an arbitrary $p \in \mathbb{P}_{m-s}$. In Lemma 5.7.1, replace m by $m - s$ and then choose $p = q_x$. This gives

$$|K_{s+1}(x)| \le \sup_u |K_s(u)| \cdot \int_{-1}^{1} \left|t_x(u) - p(u)\right| du \le \sup_u |K_s(u)| \cdot \frac{\pi}{m-s+1}.$$

Thus the inductive step, and hence the theorem, is established. □

It remains to prove Lemma 5.7.1. To do this, we will use a result about one-sided trigonometric approximation of

$$B(u) := \frac{u}{2\pi} - \left\lfloor \frac{u}{2\pi} \right\rfloor .$$

In the following, we write

$$\mathbb{T}_m = \Big\{ t \,:\, t(x) = a_0 + \sum_{\nu=1}^{m} \big(a_\nu \cos \nu x + b_\nu \sin \nu x\big) \Big\}.$$

LEMMA 5.7.2 (Ganelius 1956). There is a $G \in \mathbb{T}_m$ satisfying

$$G(u) \le B(u) \quad \text{for all } u \in \mathbb{R} \tag{5.57}$$

and

$$G\left(\frac{2\nu\pi}{m+1}\right) = B\left(\frac{2\nu\pi}{m+1}\right) \quad \text{for } \nu = 0, 1, \dots. \tag{5.58}$$

PROOF. Set $h = 2\pi/(m+1)$; G is constructed from the Hermitian interpolation condition (5.58) together with the condition

$$G'(\nu h) = \frac{1}{2\pi} \quad \text{for } \nu = 1, 2, \dots, m.$$

First we have to show that these equations are actually solvable. Solvability is equivalent to:

$$\begin{aligned} t \in \mathbb{T}_m \quad &\text{with} \quad t(\nu h) = 0 \quad (\nu = 0, \dots, m) \\ &\text{and} \quad t'(\nu h) = 0 \quad (\nu = 1, \dots, m) \\ \Longrightarrow \quad & t(x) \equiv 0. \end{aligned}$$

For any t with such (double) zeros, $t' \in \mathbb{T}_m$ must have at least $2m+1$ zeros in $[0, 2\pi[$, which implies $t'(x) \equiv 0$ and hence $t(x) \equiv 0$.

Now, it suffices to prove (5.57) for $u \in [0, 2\pi[$. Suppose that the statement is false; then $G - B$ would have more than $2m+1$ zeros (counting multiplicity). This yields at least $2m+1$ zeros for $G' - B'$, which contradicts $G' - B'\big|_{[0,2\pi[} \in \mathbb{T}_m$ and $G'(x) \not\equiv B'$. □

Now we are in a position to prove Lemma 5.7.1.

PROOF OF LEMMA 5.7.1. We start with

$$t_x(\cos v) = 1 + \frac{\arccos x}{\pi} - B(\arccos x - v) - B(\arccos x + v) \quad \text{for } x > -1,$$

which follows from the definition of B. Upon taking

$$q_x(\cos v) = 1 + \frac{\arccos x}{\pi} - G(\arccos x - v) - G(\arccos x + v),$$

where G is the function defined in Lemma 5.7.2, we have $q_x \in \mathbb{P}_m$, $t_x - q_x \le 0$ and

$$\begin{aligned} \int_{-1}^{1} \big[t_x(u) - q_x(u)\big] w(u)\, du &= \int_0^{\pi} \big[t_x(\cos v) - q_x(\cos v)\big] w(\cos v) \sin v\, dv \\ &\ge \sup_\tau w(\cos\tau)\,|\sin\tau| \cdot \int_0^{\pi} \big[t_x(\cos v) - q_x(\cos v)\big]\, dv. \end{aligned}$$

From the definitions of t_x and q_x we obtain

$$\begin{aligned}
\int_0^\pi &\big[t_x(\cos v) - q_x(\cos v)\big]\,dv \\
&= \int_0^\pi \big[G(\arccos x - v) - B(\arccos x - v)\big]\,dv \\
&\quad + \int_0^\pi \big[G(\arccos x + v) - B(\arccos x + v)\big]\,dv \\
&= \int_{\arccos x - \pi}^{\arccos x} \big[G(u) - B(u)\big]\,du + \int_{\arccos x}^{\arccos x + \pi} \big[G(u) - B(u)\big]\,du \\
&= \int_0^{2\pi} \big[G(u) - B(u)\big]\,du,
\end{aligned}$$

where the last equality follows from the 2π-periodicity. Setting

$$B^*(v) := \begin{cases} B(v) & \text{if } v \in [0, 2\pi[, \\ 1 & \text{if } v = 2\pi, \end{cases}$$

we have $B^*\big|_{[0,2\pi]} \in \mathbb{P}_1$ and $G \in \mathbb{T}_m$. Integrals of such functions between 0 and 2π can be calculated exactly by using the trapezoidal formula with $m+2$ nodes (see Chapter 7); therefore

$$\begin{aligned}
\int_0^{2\pi} &\big[G(v) - B(v)\big]\,dv = \int_0^{2\pi} \big[G(v) - B^*(v)\big]\,dv \\
&= \frac{\pi}{m+1}\Big\{\underbrace{G(0) - B^*(0)}_{=0} + 2\sum_{\nu=1}^{m}\big[\underbrace{G(\nu h) - B^*(\nu h)}_{=0}\big] + \underbrace{G(2\pi)}_{=0} - \underbrace{B^*(2\pi)}_{=1}\Big\} \\
&= -\frac{\pi}{m+1},
\end{aligned}$$

and the assertion of the lemma is proved. □

Now we present a localized estimate for the first Peano kernel.

THEOREM 5.7.2. For any positive quadrature rule with $R[\mathbb{P}_m] = \{0\}$ and a continuous weight function w, we have

$$|K_1(x)| \le \frac{\pi}{m}\, w(x)\sqrt{1-x^2} + o(m^{-1})$$

uniformly in x.

The proof of this theorem is based on a more detailed investigation of the one-sided approximation G of B.

LEMMA 5.7.3 (Ganelius 1956). The function $G_m := G$ from Lemma 5.7.2 has the representation

$$G_m(x) = \frac{m}{2m+2} + \sum_{\nu=1}^{m} \big(\alpha_\nu \cos \nu x + \beta_\nu \sin \nu x\big), \tag{5.59}$$

where

$$\alpha_\nu = \frac{\nu - m - 1}{(m+1)^2} \qquad \text{and} \qquad \beta_\nu = \frac{\nu - m - 1}{(m+1)^2}\cot\frac{\nu\pi}{m+1} - \frac{1}{\pi(m+1)}.$$

PROOF. Denote the right-hand side of (5.59) by $g(x)$. This is a trigonometric polynomial of degree at most m and is therefore defined uniquely if the function values at $z_k = 2k\pi/(m+1)$, $k = 0, \ldots, m$, and the derivatives at z_k, $k = 1, \ldots, m$, are given. Hence we have to verify that $g(z_k) = k/(m+1)$ for $k = 0, \ldots, m$ and $g'(z_k) = 1/(2\pi)$ for $k = 1, \ldots, m$. In what follows, we shall say that a sum $S = \sum_{\nu=1}^{m} c_m$ vanishes due to symmetry if $c_{m+1-\nu} = -c_\nu$ and therefore $S = 0$. We have

$$g(z_k) = \frac{m}{2m+2} + \sum_{\nu=1}^{m} \frac{\nu - m - 1}{(m+1)^2} \cdot \frac{\sin \frac{(2k+1)\nu\pi}{m+1}}{\sin \frac{\nu\pi}{m+1}} - \frac{1}{\pi(m+1)} \sum_{\nu=1}^{m} \sin \frac{2k\nu\pi}{m+1}.$$

The second sum on the right-hand side vanishes due to symmetry. Trigonometric identities show that

$$g(z_k) - g(z_{k-1}) = 2 \sum_{\nu=1}^{m} \frac{\nu - m - 1}{(m+1)^2} \cos \frac{2k\nu\pi}{m+1},$$

which can be simplified by using the formulas

$$\sum_{\nu=1}^{m} \cos \frac{2k\nu\pi}{m+1} = -1 \quad \text{and} \quad \sum_{\nu=1}^{m} \nu \cos \frac{2k\nu\pi}{m+1} = -\frac{m+1}{2} \tag{5.60}$$

for $0 < k < m+1$. We therefore get

$$g(z_k) - g(z_{k-1}) = \frac{1}{m+1}.$$

Since it is easy to see that $g(0) = 0$, we have proved that $g(z_k) = k/(m+1)$ for $k = 0, \ldots, m$. Next, note that

$$g'(z_k) = \sum_{\nu=1}^{m} \frac{\nu(\nu - m - 1)}{(m+1)^2} \cdot \frac{\cos \frac{(2k+1)\nu\pi}{m+1}}{\sin \frac{\nu\pi}{m+1}} - \frac{1}{\pi(m+1)} \sum_{\nu=1}^{m} \nu \cos \frac{2k\nu\pi}{m+1}.$$

Symmetry now shows that the first sum on the right-hand side vanishes. For the second sum, use (5.60). □

LEMMA 5.7.4. Let G_m be as defined in Lemma 5.7.3. Then, for every continuous f,

$$\lim_{m \to \infty} H_m[f] = f$$

where

$$H_m[f](t) = \frac{m+1}{\pi} \int_{-\pi}^{\pi} \left[B(u) - G_m(u)\right] f(t-u)\, du.$$

PROOF. H_m as defined above is a positive operator. By Korovkin's theorem (see DeVore and Lorentz 1993: p. 8), the assertion of the lemma is true if it holds for $f \in \{1, \sin, \cos\}$. We use

$$B(x) = \frac{1}{2} - \sum_{\mu=1}^{\infty} \frac{\sin \mu x}{\pi\mu}.$$

In the notation of Lemma 5.7.3, we have

$$\begin{aligned}
H_m[1] &= 1,\\
H_m[\sin](t) &= \frac{m+1}{\pi}\big[\cos t - \pi(\alpha_1 \sin t - \beta_1 \cos t)\big]\\
&= \frac{m}{m+1}\sin t + \left(\frac{m}{\pi} - \frac{m}{m+1}\cot\frac{\pi}{m+1}\right)\cos t,\\
H_m[\cos](t) &= \frac{m+1}{\pi}\big[\sin t - \pi(\alpha_1 \cos t + \beta_1 \sin t)\big]\\
&= \frac{m}{m+1}\cos t - \left(\frac{m}{\pi} - \frac{m}{m+1}\cot\frac{\pi}{m+1}\right)\sin t.
\end{aligned}$$

The assertion now follows from the fact that $\cot x = 1/x + O(x)$ for $|x| \le \pi/2$. □

PROOF OF THEOREM 5.7.2. The following bound was already used in the proof of Theorem 5.7.1:

$$I[t_x - q_x] \le K_1(x-0).$$

We have

$$\begin{aligned}
&\int_{-1}^{1} \big[t_x(u) - q_x(u)\big] w(u)\, du\\
&\qquad = \int_0^{\pi} \big[t_x(\cos v) - q_x(\cos v)\big] w(\cos v) \sin v\, dv\\
&\qquad = \int_0^{\pi} \Big\{\big[G_m(\arccos x - v) - B(\arccos x - v)\big]\\
&\qquad\qquad + \big[G_m(\arccos x + v) - B(\arccos x + v)\big]\Big\} w(\cos v)\sin v\, dv\\
&\qquad = \int_{\arccos x - \pi}^{\arccos x} \big[G_m(u) - B(u)\big] w\big(\cos(\arccos x - u)\big)\sin(\arccos x - u)\, du\\
&\qquad\quad + \int_{\arccos x}^{\arccos x + \pi} \big[G_m(u) - B(u)\big] w\big(\cos(u - \arccos x)\big)\sin(\arccos x - u)\, du\\
&\qquad \ge \int_{\arccos x - \pi}^{\arccos x + \pi} \big[G_m(u) - B(u)\big] w\big(\cos(\arccos x - u)\big)\big|\sin(\arccos x - u)\big|\, du.
\end{aligned}$$

The integrand on the right-hand side is 2π-periodic, so we may as well integrate over $[-\pi, \pi]$. Lemma 5.7.4 then tells us that $(m+1)/\pi$ times the right-hand side tends to $-w(x)\sqrt{1-x^2}$.

Arguing analogously for approximation from below leads to the same asymptotic bound for $K_1(x+0)$, and the theorem follows. □

We now turn to discussion of $\int_{-1}^{1} |K_s(x)|\, dx$. For a concrete quadrature rule, good bounds can be found for these quantities with the help of a computer; for example, we could apply the trapezoidal rule Q^{Tr}_{m+1}, along with its error bound $(b-a)M/(2m)$, to $|K_s|$ (here M is a bound for the total variation of the integrand). We have

$$\operatorname{Var}|K_s| \le \operatorname{Var} K_s, \qquad \operatorname{Var} K_1 = \|R\|, \qquad \operatorname{Var} K_s \le (b-a)\|K_{s-1}\| \quad \text{for } s \ge 2$$

and may use the bounds for $\|K_{s-1}\|$ from Theorem 5.7.1.

In the investigations that follow, our purpose will be to derive bounds valid for all positive quadrature rules of interpolatory type. Such bounds can of course be obtained immediately from Theorem 5.7.1 or Theorem 4.3.4. Owing to the importance of estimates of the type

$$\left|R[f]\right| \leq \operatorname{const} \sup_x \left|f^{(s)}(x)\right|,$$

good bounds are particularly desirable. The method below (Brass and Petras 2003) yields good bounds with relatively little effort; its starting point is the following theorem.

THEOREM 5.7.3 (Petras 1996b). Let K_r be the Peano kernel of a linear functional S that satisfies $S[\mathbb{P}_m] = \{0\}$. Then

$$\int_{-1}^{1} \frac{K_r^2(x)}{\left(1-x^2\right)^{r-1/2}}\,dx = \frac{2}{\pi} \sum_{\mu=m+1}^{\infty} \left[\prod_{\lambda=0}^{r-1} (\mu^2 - \lambda^2)\right]^{-1} S[T_\mu]^2. \tag{5.61}$$

Denote by $q_{\nu,\alpha}$ the νth orthonormal polynomial with respect to the weight function $W(x) = \left(1-x^2\right)^\alpha$ on $[-1,1]$. We shall use the following lemma in the proof of Theorem 5.7.3.

LEMMA 5.7.5.

$$q'_{\nu,\alpha} = \sqrt{\nu(\nu+1+2\alpha)}\, q_{\nu-1,\alpha+1}.$$

PROOF. Let $p \in \mathbb{P}_{\nu-1}$. Integration by parts gives

$$\begin{aligned}
&\int_{-1}^{1} q'_{\nu,\alpha}(x) p(x) \left(1-x^2\right)^{\alpha+1} dx \\
&\quad = \int_{-1}^{1} q_{\nu,\alpha}(x) p(x) (\alpha+1) 2x \left(1-x^2\right)^\alpha dx \\
&\quad\quad - \int_{-1}^{1} q_{\nu,\alpha}(x) p'(x) (1-x^2) \left(1-x^2\right)^\alpha dx.
\end{aligned} \tag{5.62}$$

By orthogonality, the right-hand side of (5.62) vanishes if $p \in \mathbb{P}_{\nu-2}$. Hence

$$q'_{\nu,\alpha} = c(\nu,\alpha) \cdot q_{\nu-1,\alpha+1}, \tag{5.63}$$

with some constant $c(\nu,\alpha)$. Denote by $k_{\nu,\alpha}$ the leading coefficient of $q_{\nu,\alpha}$; then (5.63) shows that

$$\nu \cdot k_{\nu,\alpha} = c(\nu,\alpha) \cdot k_{\nu-1,\alpha+1}. \tag{5.64}$$

Upon setting $p = q_{\nu-1,\alpha+1}$ in (5.62) and using (5.63), we obtain the relation

$$c(\nu,\alpha) = (2\alpha+2)\frac{k_{\nu-1,\alpha+1}}{k_{\nu,\alpha}} + (\nu-1)\frac{k_{\nu-1,\alpha+1}}{k_{\nu,\alpha}} = (\nu+2\alpha+1)\frac{k_{\nu-1,\alpha+1}}{k_{\nu,\alpha}}.$$

Taking (5.64) into account yields

$$c(\nu,\alpha) = (\nu+2\alpha+1)\frac{\nu}{c(\nu,\alpha)}$$

and hence the lemma. □

PROOF OF THEOREM 5.7.3. The existence of the integral for $r > 1$ follows from the fact that K_r has zeros at ± 1 of multiplicity $r-1$ or greater. Therefore, Parseval's identity for the polynomials $q_{\nu,r-1/2}$ yields

$$\begin{aligned}\int_{-1}^{1} \frac{K_r^2(x)}{\left(1-x^2\right)^{r-1/2}}\,dx &= \int_{-1}^{1}\left(\frac{K_r(x)}{\left(1-x^2\right)^{r-1/2}}\right)^2 \left(1-x^2\right)^{r-1/2}\,dx \\ &= \sum_{\nu=0}^{\infty}\left(\int_{-1}^{1} K_r(x) q_{\nu,r-1/2}(x)\,dx\right)^2.\end{aligned}$$

We integrate by parts $r-1$ times, applying Lemma 5.7.5 each time, to get

$$\begin{aligned}\left(\int_{-1}^{1} K_r(x) q_{\nu,r-1/2}(x)\,dx\right)^2 &= \frac{\nu!\,(\nu+r)}{(\nu+2r-1)!}\left(\int_{-1}^{1} K_1(x) q_{\nu+r-1,1/2}(x)\,dx\right)^2 \\ &= \frac{\nu!\,(\nu+r)}{(\nu+2r-1)!}\cdot\frac{1}{(\nu+r)^2}\,S[q_{\nu+r,-1/2}]^2.\end{aligned}$$

We then use

$$q_{\mu,-1/2} = \sqrt{\frac{2}{\pi}}\,T_\mu \quad \text{for } \mu \geq 1$$

(see Appendix A) to obtain

$$\begin{aligned}\int_{-1}^{1} \frac{K_r^2(x)}{\left(1-x^2\right)^{r-1/2}}\,dx &= \frac{2}{\pi}\sum_{\nu=0}^{\infty}\frac{\nu!}{(\nu+2r-1)!\,(\nu+r)}\,S[T_{\nu+r}]^2 \\ &= \frac{2}{\pi}\sum_{\mu=m+1}^{\infty}\frac{(\mu-r)!}{(\mu+r-1)!\,\mu}\,S[T_\mu]^2,\end{aligned}$$

and the assertion of the theorem follows. □

A first consequence of Theorem 5.7.3 is the following:

THEOREM 5.7.4. *Under the assumption of Theorem 5.7.3, we have*

$$\int_{-1}^{1} \frac{K_r^2(x)}{\left(1-x^2\right)^{r-1/2}}\,dx \leq \frac{(m-r+1)!\,(m+1)}{(m+r)!}\int_{-1}^{1}\frac{K_1^2(x)}{\sqrt{1-x^2}}\,dx.$$

PROOF.

$$\begin{aligned}\int_{-1}^{1} \frac{K_r^2(x)}{\left(1-x^2\right)^{r-1/2}}\,dx &= \frac{2}{\pi}\sum_{\mu=m+1}^{\infty}\left[\prod_{\lambda=0}^{r-1}(\mu^2-\lambda^2)\right]^{-1} S[T_\mu]^2 \\ &\leq \left[\prod_{\lambda=1}^{r-1}((m+1)^2-\lambda^2)\right]^{-1}\cdot\frac{2}{\pi}\sum_{\mu=m+1}^{\infty}\frac{S[T_\mu]^2}{\mu^2} \\ &= \left[\prod_{\lambda=1}^{r-1}((m+1)^2-\lambda^2)\right]^{-1}\cdot\int_{-1}^{1}\frac{K_1^2(x)}{\sqrt{1-x^2}}\,dx,\end{aligned}$$

which is what we needed to show. □

The next result is our main application.

THEOREM 5.7.5. Let K_r be the rth Peano kernel of a positive quadrature rule that has $R[\mathbb{P}_m] = \{0\}$. Then

$$\int_{-1}^{1} |K_r(x)|\,dx \le \frac{7.5}{\sqrt[4]{r}} \left(\frac{(m+1-r)!}{(m+r)!\,(m+1)} \right)^{1/2} \sup_x w(x)\sqrt{1-x^2}.$$

PROOF.

$$\begin{aligned}\int_{-1}^{1} |K_r(x)|\,dx &\le \left(\int_{-1}^{1} \frac{K_r^2(x)}{(1-x^2)^{r-1/2}}\,dx \right)^{1/2} \cdot \left(\int_{-1}^{1} (1-x^2)^{r-1/2}\,dx \right)^{1/2} \\ &= \left(\int_{-1}^{1} \frac{K_r^2(x)}{(1-x^2)^{r-1/2}}\,dx \right)^{1/2} \cdot \left(\pi \frac{(2r)!}{(r!)^2\,4^r} \right)^{1/2} \\ &\le \sqrt[4]{\frac{\pi}{r}} \left(\int_{-1}^{1} \frac{K_r^2(x)}{(1-x^2)^{r-1/2}}\,dx \right)^{1/2}.\end{aligned}$$

For the last line we used Wallis's product. Combining this estimate with Theorem 5.7.4 and the inequality

$$\int_{-1}^{1} \frac{K_1^2(x)}{\sqrt{1-x^2}}\,dx \le \sup_x |K_1^2(x)| \cdot \int_{-1}^{1} \frac{dx}{\sqrt{1-x^2}} = \pi \sup_x |K_1^2(x)|$$

yields

$$\int_{-1}^{1} |K_r(x)|\,dx \le \sqrt[4]{\frac{\pi^3}{r}} \left(\frac{(m+1-r)!\,(m+1)}{(m+r)!} \right)^{1/2} \sup_x |K_1(x)|$$

(which holds for arbitrary functionals). The theorem follows upon specializing to positive quadrature rules and applying Theorem 5.7.1. □

From a comparison with Theorem 4.3.4, it is evident that the bounds derived here are better for $w = 1$ and $r \ge 5$, for instance. For smaller r, the same method with refined arguments can also be used to improve Theorem 4.3.4 slightly.

It is possible to improve the bounds in Theorem 5.7.5 further, at least for $w = 1$; see Brass and Fischer (1999b), where asymptotically optimal bounds for $m \to \infty$ are given. The asymptotic behaviour of

$$\rho_{r,m} := \sup\left\{ \int_{-1}^{1} |K_r(x)|\,dx \;:\; R[\mathbb{P}_m] = \{0\} \text{ and } Q \text{ is positive} \right\}$$

is given by

$$\lim_{m\to\infty} m^r \rho_{r,m} = (2\pi)^r \int_0^1 |B_r(x)|\,dx \cdot I[h^r], \quad \text{with } h(x) := \sqrt{1-x^2},$$

for a large class of weight functions including $w = 1$ (Brass 1993b).

This means that for all positive quadrature rules of interpolatory type, we have

$$\rho\big(Q_n, \mathcal{C}_1^{(1)}\big) \le \frac{\pi^2}{4n}\big(1 + o(1)\big) < \frac{2.47}{n}\big(1 + o(1)\big)$$

if $w = 1$. On the other hand, we have

$$\rho\big(Q_n, \mathcal{C}_1^{(1)}\big) \ge \frac{1}{n}$$

(see Example 2.4.1). Hence, all positive quadrature rules of interpolatory type are of similar quality with respect to the co-observation $\mathcal{C}_M^{(1)}$.

If $w = 1$, we may obtain bounds that hold for any $\int_{-1}^{1} |K_r(x)|\,dx$ of a positive rule:

$$\int_{-1}^{1} |K_{2s}(x)|\,dx \le \sup_{m\ge 2s-1} \rho_{2s,m} = \rho_{2s,2s-1} = \frac{s+1}{s(2s+1)}\frac{(s!)^4}{[(2s)!]^3}\,2^{2s+1}$$

(see Theorems 6.5.1 and 6.5.2) and

$$\int_{-1}^{1} |K_{2s-1}(x)|\,dx \le \sup_{m\ge 2s-2} \rho_{2s-1,m} = \rho_{2s-1,2s-2} = \frac{(s!)^4}{[(2s)!]^3}\,2^{2s+2}$$

(Brass 1985).

5.8. Asymptotic behaviour of a class of Peano kernels

Many quadrature rules of interest, including Krein-type rules (such as the Gaussian, Radau and Lobatto rules) and Clenshaw–Curtis rules and their relatives, have similar properties; specifically, the coefficients are related in some way to the distances between consecutive nodes. Together with certain other properties that are often easy to verify, this implies a particular type of behaviour of the Peano kernels. We will show that on an interval between consecutive nodes, each K_s coincides asymptotically with a multiple of the Bernoulli polynomial B_s transformed to that interval.

THEOREM 5.8.1 (Petras 1988). *Suppose that the quadrature rules Q_n, $n = 1, 2, \ldots$, with respect to the weight function w have sth Peano kernels $K_s := K_{s,n}$ for $s = 1, 2, \ldots r$, and suppose that for some $\varepsilon > 0$, w is continuous on the interval $[c-\varepsilon, d+\varepsilon] \cap [a, b]$. Define*

$$k_s\big|_{[x_{\nu-1},x_\nu[}(x) = (x_{\nu-1} - x_\nu)^s w(z_\nu) B_s\left(\frac{x - x_{\nu-1}}{x_\nu - x_{\nu-1}}\right) \quad \text{for } \nu = 1, \ldots, n+1,$$

where

$$z_\nu = \frac{x_\nu + x_{\nu-1}}{2} \quad \text{and} \quad -x_0 := x_{n+1} := 1.$$

Assume that the following hold:

(1) $\sup_{x\in[c,d]} |K_s(x)| = O(n^{-s})$ *for* $s = 1, \ldots, r$;

(2) *for some $A, B > 0$ and for all n,*

$$\frac{A}{n} < x_\nu - x_{\nu-1} < \frac{B}{n} \quad \text{if } [x_{\nu-1}, x_\nu] \cap [c, d] \ne \emptyset;$$

(3) $\max_{[x_{\nu-1},x_{\nu+1}]\cap[c,d]\ne\emptyset} \big|(x_{\nu+1} - x_\nu) - (x_\nu - x_{\nu-1})\big| = o(n^{-1})$;

(4) $\max_{[x_{\nu-1},x_{\nu+1}]\cap[c,d]\ne\emptyset} \left|a_\nu - \int_{z_\nu}^{z_{\nu+1}} w(x)\,dx\right| = o(n^{-1})$.

Then

$$\sup_{x\in[c,d]} |K_s(x) - k_s(x)| = o(n^{-s}) \quad \text{for } s = 1, 2, \ldots, r-1.$$

As a special example, we mention the asymptotic representation of Peano kernels of Q_{n+1}^{CC} with $w = 1$; see Petras (1988) and Brass (1992b). Here we have

$$K_s(x) = \left(\frac{\pi}{n}\right)^s B_s^*\left(\frac{n \arccos x}{\pi}\right)(1 - x^2)^{s/2} + o(n^{-s})$$

uniformly for $x \in [-1, 1]$.

The following lemma is useful for an inductive proof of the theorem. Think of F_n as a multiple of a Peano kernel and think of G_n as its approximation.

LEMMA 5.8.1. Let F_n and G_n be defined on $[c,d]$, with

$$\sup_{x,y\in[c,d]}\left|\int_x^y F_n(t)\,dt\right| \le C\cdot n^{-2} \quad\text{and}\quad \sup_{x,y\in[c,d]}\left|\int_x^y G_n(t)\,dt\right| \le C\cdot n^{-2} \tag{5.65}$$

for some constant C. Furthermore, suppose that there exists a null sequence (C_n) such that $\Delta_n := F_n - G_n$ satisfies

$$\bigl|\Delta_n(x) - \Delta_n(y)\bigr| \le C_n\left(\frac{1}{n} + |x-y|\right) \quad \text{for all } x,y\in[c,d].$$

Then

$$F_n(x) - G_n(x) = o(n^{-1}) \quad \text{uniformly for } x\in[c,d].$$

PROOF. Suppose without loss of generality that $x < (c+d)/2$ and that $n^2\cdot C_n$ tends to infinity. Setting $\varepsilon = 2n^{-1}\sqrt{C/C_n} = o(1)$, we have $x+\varepsilon < d$ for sufficiently large n, so that

$$\begin{aligned}
\frac{2C}{n^2} &\ge \left|\int_x^{x+\varepsilon}\Delta_n(y)\,dy\right| = \left|\int_x^{x+\varepsilon}\Delta_n(x) - \bigl(\Delta_n(x)-\Delta_n(y)\bigr)\,dy\right| \\
&\ge \left|\int_x^{x+\varepsilon}\Delta_n(x)\,dy\right| - \int_x^{x+\varepsilon}\bigl|\Delta_n(x)-\Delta_n(y)\bigr|\,dy \\
&\ge \varepsilon\,|\Delta_n(x)| - \frac{\varepsilon C_n}{n} - \frac{C_n\varepsilon^2}{2}.
\end{aligned}$$

Hence

$$|\Delta_n(x)| \le \frac{2C}{n^2\varepsilon} + \frac{C_n}{n} + \frac{C_n\varepsilon}{2} = \frac{C_n + 2\sqrt{C_n\cdot C}}{n} = o(n^{-1})$$

as claimed. □

PROOF OF THEOREM 5.8.1. In this proof, let σ and ρ always be defined in such a way that $x\in[x_{\rho-1},x_\rho[$ and $y\in[x_{\sigma-1},x_\sigma[$; moreover, suppose without loss of generality that $\rho\le\sigma$. In particular, this assumption means that

$$y - x \ge (\sigma-\rho-1)\frac{A}{n}, \quad \text{i.e. } \sigma-\rho \le 1 + \frac{n(y-x)}{A}.$$

We shall furthermore always take the modulus of continuity of w to be a function on $[c-\varepsilon, d+\varepsilon]$.

First, we prove the statement for $s=1$ by showing that the assumptions of Lemma 5.8.1 are satisfied with $F_n = K_1$ and $G_n = k_1$. The conditions (5.65) are verified as follows:

$$\left|\int_x^y F_n(t)\,dt\right| = \bigl|K_2(x) - K_2(y)\bigr| \le |K_2(x)| + |K_2(y)| = O(n^{-2}).$$

The integral over G_n vanishes on each interval $[x_\nu, x_{\nu+1}]$, so we only have to integrate over $[x,x_\rho]$ and $[x_{\sigma-1},y]$. Hence,

$$\left|\int_x^y G_n(t)\,dt\right| \le 2\max_{[x_\nu,x_{\nu+1}]\cap[c,d]\neq\emptyset}\int_{x_\nu}^{x_{\nu+1}}|k_1(t)|\,dt \le \frac{B^2}{2n^2}\cdot\max_{x\in[c-\frac{B}{n},d+\frac{B}{n}]}|w(x)|.$$

Now,

$$\begin{aligned}
&\left|\Delta_n(x) - \Delta_n(y)\right| \\
&\quad = \left|\{K_1(x) - K_1(y)\} - \{k_1(x) - k_1(y)\}\right| \\
&\quad = \Bigg|\Big\{ - \int_x^{z_\rho} w(t)\,dt + \sum_{\nu=\rho}^{\sigma-1} \Big(a_\nu - \int_{z_\nu}^{z_{\nu+1}} w(t)\,dt\Big) - \int_{z_\sigma}^{y} w(t)\,dt\Big\} \\
&\qquad\quad - \Big\{w(z_\rho)(z_\rho - x) + w(z_\sigma)(z_\sigma - y)\Big\}\Bigg| \\
&\quad \le (\sigma-\rho)o(n^{-1}) + \left|\int_x^{z_\rho} \left|w(t) - w(z_\rho)\right| dt\right| + \left|\int_{z_\sigma}^{y} \left|w(t) - w(z_\sigma)\right| dt\right| \\
&\quad \le (\sigma-\rho)o(n^{-1}) + \frac{B}{n}\,\omega\Big(w; \frac{B}{2n}\Big) \\
&\quad \le \left(1 + \frac{n(y-x)}{A}\right) o(n^{-1}) + o(n^{-1}).
\end{aligned}$$

The assumptions of Lemma 5.8.1 have been verified and so the theorem is proved for $s = 1$.

For the inductive step, suppose that the statement has already been proved for K_s, and set $F_n = n^s K_{s+1}$ and $G_n = n^s k_{s+1}$. The conditions (5.65) are checked in an analogous fashion to the K_1 case. Furthermore, observe that G_n is an antiderivative of $-n^s k_s$ plus a jump function with possible jumps as follows (note that $B_{s+1}(0) = B_{s+1}(1)$):

$$\begin{aligned}
&\left|G_n(x_\nu + 0) - G_n(x_\nu - 0)\right| \\
&\quad = n^s\left|B_{s+1}(0)\right| \cdot \left|w(z_{\nu+1})(x_{\nu+1} - x_\nu)^{s+1} - w(z_\nu)(x_\nu - x_{\nu-1})^{s+1}\right| \\
&\quad = n^s\left|B_{s+1}(0)\right| \cdot \Bigg|\Big[w(z_{\nu+1}) - w(z_\nu)\Big](x_{\nu+1} - x_\nu)^{s+1} \\
&\qquad\qquad + w(z_\nu)\cdot\Big[(x_{\nu+1} - x_\nu) - (x_\nu - x_{\nu-1})\Big] \sum_{i=0}^{s} (x_{\nu+1} - x_\nu)^i (x_\nu - x_{\nu-1})^{s-i}\Bigg| \\
&\quad \le n^s\left|B_{s+1}(0)\right| \cdot \Bigg(\omega\Big(w; \frac{B}{n}\Big)\cdot\Big(\frac{B}{n}\Big)^{s+1} \\
&\qquad\qquad + \sup_{x\in[c-\varepsilon,\, d+\varepsilon]} w(x)\cdot o(n^{-1})\cdot(s+1)\Big(\frac{B}{n}\Big)^{s}\Bigg) \\
&\quad = o(n^{-1}).
\end{aligned}$$

We obtain

$$\begin{aligned}
\left|\Delta_n(x) - \Delta_n(y)\right| &\le n^s \int_x^y \left|k_s(t) - K_s(t)\right| dt + \sum_{\nu=\rho}^{\sigma-1} \left|G_n(x_\nu - 0) - G_n(x_\nu + 0)\right| dt \\
&\le n^s|x-y| \max_{t\in[c,d]} \left|k_s(t) - K_s(t)\right| + (\sigma-\rho)\cdot o(n^{-1}) \\
&\le |x-y|\cdot o(1) + \left(1 + \frac{n|x-y|}{A}\right)\cdot o(n^{-1}).
\end{aligned}$$

Thus, all the assumptions of Lemma 5.8.1 are satisfied and we can conclude that $|K_{s+1} - k_{s+1}| = o(n^{-s-1})$ on $[c, d]$. □

This result is most frequently applied to the situation where the assumptions of the theorem are satisfied on intervals that "almost exhaust $[a,b]$".

COROLLARY 5.8.1. Let $(Q)_{n\in\mathbb{N}}$ be a quadrature method with $\|K_s\| = O(n^{-s})$ and $x_{\nu+1} - x_\nu = O(n^{-1})$, and suppose that the assumptions of Theorem 5.8.1 are satisfied for all $[c,d] \subset]a,b[$. Then

$$\int_a^b |K_s(x)|\,dx = \int_0^1 |B_s(y)|\,dy \cdot \sum_{\nu=0}^{n} w(z_{\nu+1})(x_{\nu+1} - x_\nu)^{s+1} + o(n^{-s}),$$

where $x_{n+1} := 1$ and $x_0 := -1$.

PROOF. For all $\varepsilon > 0$ we have

$$\begin{aligned}\left|\int_a^b |K_s(x)|\,dx - \int_a^b |k_s(x)|\,dx\right| \\ \leq \left(\int_a^{a+\varepsilon} + \int_{b-\varepsilon}^b\right) |K_s(x)| + |k_s(x)|\,dx + \int_{a+\varepsilon}^{b-\varepsilon} |K_s(x) - k_s(x)|\,dx.\end{aligned}$$

The first summand on the right-hand side is bounded by $2\varepsilon\big(\|K_s\| + \|k_s\|\big) = \varepsilon \cdot O(n^{-s})$, while the second is of order $o(n^{-s})$. Upon noting that

$$\int_a^b |k_s(x)|\,dx = \int_0^1 |B_s(y)|\,dy \cdot \sum_{\nu=0}^{n} w(z_{\nu+1})(x_{\nu+1} - x_\nu)^{s+1},$$

the proof is complete. □

EXAMPLE 5.8.1. We demonstrate the application of this corollary by looking at the Filippi rule for $w = 1$. Filippi rules are positive (see Theorem 5.5.6), so K_s is of order n^{-s} according to Theorem 5.7.1. Assumptions (2) and (4) of Theorem 5.8.1 are easily proved with trigonometric identities, where (4) requires (5.47). Therefore the corollary is applicable, and so

$$\begin{aligned}\int_{-1}^1 |K_s(x)|\,dx &= \int_0^1 |B_s(y)|\,dy \cdot \sum_{\nu=1}^{n-1} (x_{\nu+1} - x_\nu)^{s+1} + o(n^{-s}) \\ &= \int_0^1 |B_s(y)|\,dy \cdot \sum_{\nu=1}^{n-1} 2^{s+1} \left(\sin\frac{\pi}{2n+2}\right)^{s+1} \left(\sin\frac{(2\nu+1)\pi}{2n+2}\right)^{s+1} \\ &\quad + o(n^{-s}).\end{aligned}$$

We obtain

$$\lim_{n\to\infty} n^s \int_{-1}^1 |K_s(x)|\,dx = \pi^s \int_0^\pi \sin^{s+1} x\,dx \cdot \int_0^1 |B_s(y)|\,dy.$$

All of the arguments, and hence the result, also hold for the Clenshaw–Curtis and related methods. In particular, we have

$$\rho(Q_n^{\mathrm{Fi}}, \mathcal{C}_M^{(1)}) = M \cdot \frac{\pi^2}{8n} + o(n^{-1})$$

and, for comparison,

$$\rho_n^{\mathrm{best}}(\mathcal{C}_M^{(1)}) = \rho(Q_n^{\mathrm{Mi}}, \mathcal{C}_M^{(1)}) = M \cdot \frac{1}{n}.$$

Hence, we lose a factor of about 1.2337 compared with the best method.

We shall see in the next chapter that results like those in the preceding example hold for the Gaussian and similar rules.

CHAPTER 6

Gaussian Quadrature

The Gaussian method is the most extensively studied quadrature method in the literature, owing to both its theoretical importance and its practical advantages. On the theoretical side, Q_n^{G} occupies an extremal or almost extremal position in regard to many interesting properties. One of the practical advantages is that $\rho(Q_n^{\mathrm{G}}, \mathcal{C})$ is only slightly larger than $\rho_n^{\mathrm{best}}(\mathcal{C})$ for most realistic co-observations. The main tool that will be used in this chapter is the theory of orthogonal polynomials, of which an overview is given in Appendix A.

6.1. Rules of high degree and orthogonal polynomials

Throughout this chapter, we consider a non-negative weight function w on $[a, b]$ which is positive on a set of positive measure. Hence, the functional I will be strictly positive on $\mathbb{P}_n$ for all n.

We denote by q_n the nth orthonormal polynomial with respect to the weight function w, and we let $\delta_n > 0$ denote the leading coefficient of q_n.

In terms of orthogonal polynomials, Theorem 5.1.3 can be restated in the following form, which will be fundamental for our future considerations.

THEOREM 6.1.1. There exists a quadrature rule with nodes $x_1, \dots, x_n$ and $R_n[\mathbb{P}_{n+s}] = \{0\}$ if and only if there exist numbers $\alpha_{s+1}, \alpha_{s+2}, \dots, \alpha_n$ such that

$$\prod_{\nu=1}^{n} (\cdot - x_\nu) = \sum_{\mu=s+1}^{n} \alpha_\mu q_\mu.$$

The simplest case is that where $s = n - 1$, described in the following theorem.

THEOREM 6.1.2. The quadrature rule Q_n^{G} of interpolatory type whose nodes are the zeros of q_n satisfies $R_n^{\mathrm{G}}[\mathbb{P}_{2n-1}] = \{0\}$ and is positive.

In Definition 5.1.3 as well as in Definition 3.4.4, with U chosen to be $\mathbb{P}_{2n-1}$, we already called Q_n^{G} a "Gaussian rule". Its uniqueness among all $Q \in \mathbf{Q}_n$ was proved in Theorem 5.1.5, and its positivity follows from Theorem 5.1.6.

Before studying Gaussian rules in more detail, we mention the survey articles by Gautschi (1981) and Brass *et al.* (1996), which the reader may wish to refer to.

EXAMPLE 6.1.1. Take $w = 1$ and $[a, b] = [-1, 1]$; then Q_n^{G} is usually called the Gauss–Legendre quadrature rule. We have

$$q_0(x) = \sqrt{\frac{1}{2}}, \qquad q_1(x) = \sqrt{\frac{3}{2}}\, x, \qquad q_2(x) = \sqrt{\frac{5}{2}} \left(\frac{3}{2}x^2 - \frac{1}{2}\right),$$

$$q_3(x) = \sqrt{\frac{7}{2}} \left(\frac{5}{2}x^3 - \frac{3}{2}\right), \qquad q_4(x) = \sqrt{\frac{9}{2}} \left(\frac{35}{8}x^3 - \frac{15}{4}x^2 + \frac{3}{8}\right).$$

Upon calculating the corresponding zeros and applying (5.2), we get

$$Q_1^{\mathrm{G}}[f] = 2f(0),$$

$$Q_2^{\mathrm{G}}[f] = f\left(-\frac{1}{\sqrt{3}}\right) + f\left(\frac{1}{\sqrt{3}}\right),$$

$$Q_3^{\mathrm{G}}[f] = \frac{5}{9} f\left(-\sqrt{\frac{3}{5}}\right) + \frac{8}{9} f(0) + \frac{5}{9} f\left(\sqrt{\frac{3}{5}}\right),$$

$$Q_4^{\mathrm{G}}[f] = \frac{18-\sqrt{30}}{36}\left[f\left(-\sqrt{\frac{15+2\sqrt{30}}{35}}\right) + f\left(\sqrt{\frac{15+2\sqrt{30}}{35}}\right)\right] + \frac{18+\sqrt{30}}{36}\left[f\left(-\sqrt{\frac{15-2\sqrt{30}}{35}}\right) + f\left(\sqrt{\frac{15-2\sqrt{30}}{35}}\right)\right].$$

For larger n, tables listing the nodes and coefficients are given in, for instance, Abramowitz and Stegun (1964) or Stroud and Secrest (1966). Algorithms for computing these values can be found in Lether (1978), Lether and Wenston (1995), Petras (1999a) or Swarztrauber (2002).

Numerical problems may arise in the computation of nodes and coefficients for general weight functions w; see Gautschi (1983). Suitable algorithms are discussed in Gautschi (1968, 1979) and Golub and Welsch (1969); see also Laurie (1999).

Calculating coefficients via (5.2) requires an integration. It is important to also have expressions that do not involve calculating integrals.

THEOREM 6.1.3. *The coefficients a_ν^{G} of Q_n^{G} with nodes x_ν^{G} satisfy*

$$a_\nu^{\mathrm{G}} = -\frac{\delta_{n+1}}{\delta_n} \cdot \frac{1}{q_{n+1}(x_\nu^{\mathrm{G}})q_n'(x_\nu^{\mathrm{G}})} = \frac{\delta_n}{\delta_{n-1}} \cdot \frac{1}{q_{n-1}(x_\nu^{\mathrm{G}})q_n'(x_\nu^{\mathrm{G}})} \tag{6.1}$$

and

$$a_\nu^{\mathrm{G}} = \left(\sum_{\mu=0}^{n} \left[q_\mu(x_\nu^{\mathrm{G}})\right]^2\right)^{-1}. \tag{6.2}$$

PROOF. The first equality in (6.1) is a special case of Theorem 5.5.5; the second equality follows from the three-term recursion relation for orthogonal polynomials (Theorem A.3). From (6.1), we obtain (6.2) by applying Theorem A.6. □

COROLLARY 6.1.1. *The nodes and coefficients of the Gauss–Legendre rules satisfy*

$$a_\nu = -\frac{2}{(n+1)P_n'(x_\nu)P_{n+1}(x_\nu)} = \frac{2}{nP_n'(x_\nu)P_{n-1}(x_\nu)} \tag{6.3}$$

$$= \frac{2}{(1-x_\nu^2)\left[P_n'(x_\nu)\right]^2} \tag{6.4}$$

(where P_k is the Legendre polynomial of degree k; see Definition A.4).

Note that with

$$q_k = \sqrt{\frac{2k+1}{2}} \cdot P_k(x) \quad \text{and} \quad \delta_k = \frac{(2k)!}{2^k \cdot (k!)^2}\sqrt{\frac{2k+1}{2}}, \tag{6.5}$$

(6.3) is a special case of (6.1). Equality (6.4) follows from (6.3) upon applying Theorem A.14(ii).

The explicit representations (6.1) and (6.2) do not give much immediate information about the coefficients; a more detailed investigation will be undertaken in the next section. Here we shall consider some special cases of Theorem 6.1.1. The second simplest case is where $s = n-2$; this is of particular interest because, according to Theorem 5.1.6, all the associated rules are positive. The same is not true for smaller s. Theorem 6.1.1 implies that for every Q_n with

$$R_n[\mathbb{P}_{2n-2}] = \{0\}, \tag{6.6}$$

there exists α such that the nodes x_ν satisfy

$$q_n(x_\nu) - \alpha\, q_{n-1}(x_\nu) = 0 \quad \text{for } \nu = 1, \ldots, n.$$

On the other hand, the zeros of $q_n - \alpha q_{n-1}$ are nodes of a quadrature rule Q_n satisfying (6.6) if these zeros are all simple and contained in $[a,b]$. However, these conditions do not always hold; in fact, we have the following:

THEOREM 6.1.4. *$q_n - \alpha q_{n-1}$ has exactly n zeros in $[a,b]$ if and only if*

$$\alpha \in \left[\frac{q_n(a)}{q_{n-1}(a)}, \frac{q_n(b)}{q_{n-1}(b)}\right].$$

PROOF. We denote the zeros of q_n by x_ν and those of q_{n-1} by y_ν. Then

$$a < x_1 < y_1 < x_2 < y_2 < \cdots < y_{n-1} < x_n < b.$$

With $L = Q_n^{\mathrm{G}} - Q_{n-1}^{\mathrm{G}}$, $U = \mathbb{P}_{2n-3}$ and $p = 2n-1$, this follows from Lemma 3.4.3. The positivity of the leading coefficient gives $q_n(b) > 0$ and therefore

$$\operatorname{sgn} q_n(x) = (-1)^{n-\mu} \quad \text{for } x \in\,]x_\mu, x_{\mu+1}[.$$

This last relation holds, in particular, for $x = y_\mu$, which means that $q_n - \alpha q_{n-1}$ has at least $n-2$ sign changes on $[y_1, y_{n-1}]$.

Now, $q_n(y_{n-1}) < 0$, so another zero of $q_n - \alpha q_{n-1}$ lies in $[y_{n-1}, b]$ if and only if

$$q_n(b) - \alpha\, q_{n-1}(b) \geq 0, \quad \text{i.e. } \alpha \leq \frac{q_n(b)}{q_{n-1}(b)}.$$

In an analogous fashion, we can show that $\alpha \geq q_n(a)/q_{n-1}(a)$ by considering $[a, y_1]$ and using the fact that $\operatorname{sgn} q_n(a) = (-1)^n$. □

Simple special cases of Theorem 6.1.4 are obtained by taking

$$\alpha = \frac{q_n(a)}{q_{n-1}(a)} \qquad \text{or} \qquad \alpha = \frac{q_n(b)}{q_{n-1}(b)},$$

and we get the following result.

THEOREM 6.1.5. *There is exactly one quadrature rule $Q_n^{\mathrm{Ra},a} \in \mathbf{Q}_n$ with a as a node and $R_n^{\mathrm{Ra},a}[\mathbb{P}_{2n-2}] = \{0\}$, namely the quadrature rule of interpolatory type with respect to the zeros $x_\nu^{\mathrm{Ra},a}$ of $q_{n-1}(a)q_n - q_n(a)q_{n-1}$. It is positive, and its coefficients satisfy*

$$a_\nu^{\mathrm{Ra},a} = \frac{\delta_n}{\delta_{n-1}} \cdot \frac{q_n(a)}{q_n(x_\nu^{\mathrm{Ra},a})} \cdot \frac{1}{q_{n-1}(a)q_n'(x_\nu^{\mathrm{Ra},a}) - q_n(a)q_{n-1}'(x_\nu^{\mathrm{Ra},a})} \tag{6.7}$$

and

$$a_\nu^{\mathrm{Ra},a} = \frac{\delta_n}{\delta_{n-1}} \cdot \frac{1}{q_{n-1}(x_\nu^{\mathrm{Ra},a})q_n'(x_\nu^{\mathrm{Ra},a}) - q_n(x_\nu^{\mathrm{Ra},a})q_{n-1}'(x_\nu^{\mathrm{Ra},a})}. \tag{6.8}$$

Analogous results hold with b in place of a everywhere.

The quadrature rules described in this theorem are the Radau rules $Q_n^{\mathrm{Ra},a}$ and $Q_n^{\mathrm{Ra},b}$. It is easy to see that they coincide with the "Radau rules" introduced in Definition 3.4.5 if one sets $U = \mathbb{P}_{2n-2}$.

PROOF. We still have to prove (6.7) and (6.8). Starting from (5.2) and using Theorem A.5 together with the relation

$$q_{n-1}(a)q_n(x_\nu^{\mathrm{Ra},a}) = q_n(a)q_{n-1}(x_\nu^{\mathrm{Ra},a}) \tag{6.9}$$

yields (6.7). On applying (6.9) again, we get (6.8). □

COROLLARY 6.1.2. For $Q_n^{\mathrm{Ra}} \in \{Q_n^{\mathrm{Ra},a}, Q_n^{\mathrm{Ra},b}\}$ and $I[f] = \int_{-1}^{1} f(x)\,dx$, we have

$$a_\nu^{\mathrm{Ra}} = \frac{1 \mp x_\nu^{\mathrm{Ra}}}{n^2 P_n^2(x_\nu^{\mathrm{Ra}})},$$

where the nodes x_ν^{Ra} are zeros of $P_n \pm P_{n-1}$. For $Q_n^{\mathrm{Ra},a}$ we choose the upper sign and for $Q_n^{\mathrm{Ra},b}$ the lower one.

PROOF. Formula (6.7) directly gives

$$a_\nu^{\mathrm{Ra}} = \mp\frac{2}{n} \cdot \frac{1}{P_n(x_\nu^{\mathrm{Ra}})\left[P_n'(x_\nu^{\mathrm{Ra}}) \pm P_{n-1}'(x_\nu^{\mathrm{Ra}})\right]}$$

for these special cases. We then apply the identity

$$(1 \mp x)[P_n'(x) \pm P_{n-1}'(x)] = \mp n[P_n(x) \mp P_{n-1}(x)]$$

(see Theorem A.14(vii)) and the relation $\mp P_{n-1}(x_\nu^{\mathrm{Ra}}) = P_n(x_\nu^{\mathrm{Ra}})$ to obtain the result. □

It is sometimes desirable that a quadrature rule have some nodes which are prescribed, for example when function values at these points are already known or easily accessible. One principle for choosing the other nodes is to maximize the polynomial degree of exactness. If, for instance, only $x_1 = a$ (or only $x_n = b$) is specified, then the Radau rules have maximal degree. Things become slightly more complicated if a node from $]a, b[$ is prescribed.

THEOREM 6.1.6. There is exactly one $Q \in \mathbf{Q}_n$ with η as a node and $R[\mathbb{P}_{2n-2}] = \{0\}$ if and only if

$$\eta \in \bigcup_{\nu=1}^{n} [x_\nu^{\mathrm{Ra},a}, x_\nu^{\mathrm{Ra},b}] \tag{6.10}$$

Upon lowering the degree requirement to $2n-3$ and prescribing η, there is just one trivial solution, namely Q_{n-1}^{G}. Less trivial are the Krein rules in this case; for $\mathbb{P}_{2n-2}$, the Krein rules are described in Theorem 6.1.6.

PROOF. The positivity of δ_n gives

$$\operatorname{sgn}\left(q_n(x) - \frac{q_n(a)}{q_{n-1}(a)} q_{n-1}(x)\right) = (-1)^{n-\mu} \quad \text{for } x \in \left]x_\mu^{\mathrm{Ra},a}, x_{\mu+1}^{\mathrm{Ra},a}\right[$$

and

$$\operatorname{sgn}\left(q_n(x) - \frac{q_n(b)}{q_{n-1}(b)} q_{n-1}(x)\right) = (-1)^{n-\mu} \quad \text{for } x \in \left]x_\mu^{\mathrm{Ra},b}, x_{\mu+1}^{\mathrm{Ra},b}\right[.$$

According to Theorem 3.4.1, the nodes of $Q_n^{\mathrm{Ra},a}$ and $Q_n^{\mathrm{Ra},b}$ are strictly interlacing, which implies that

$$\left(q_n(\eta) - \frac{q_n(a)}{q_{n-1}(a)} q_{n-1}(\eta)\right) \cdot \left(q_n(\eta) - \frac{q_n(b)}{q_{n-1}(b)} q_{n-1}(\eta)\right)^{-1} < 0 \quad \text{for } \eta \in \left]x_\nu^{\mathrm{Ra},a}, x_\nu^{\mathrm{Ra},b}\right[.$$

Combining this with

$$\frac{q_n(b)}{q_{n-1}(b)} > 0 > \frac{q_n(a)}{q_{n-1}(a)},$$

we obtain

$$\frac{q_n(\eta)}{q_{n-1}(\eta)} \in \left]\frac{q_n(a)}{q_{n-1}(a)}, \frac{q_n(b)}{q_{n-1}(b)}\right[.$$

If there exists any quadrature rule with the desired properties, its nodes have to be the zeros of $q_n - [q_n(\eta)/q_{n-1}(\eta)] \cdot q_{n-1}$. Then, Theorem 6.1.4 is applicable and the sufficiency of (6.10) is proved once we realize that the statement is trivial if η is a node of $Q_n^{\mathrm{Ra},a}$ or $Q_n^{\mathrm{Ra},b}$.

To prove necessity, suppose that $\eta \in \left]x_\nu^{\mathrm{Ra},b}, x_{\nu+1}^{\mathrm{Ra},a}\right[$, where we set $x_{n+1}^{\mathrm{Ra},a} = b$. As above, we can show that the condition of Theorem 6.1.4 is not satisfied. □

Before leaving behind the discussion of Radau rules, we point out that tables of their nodes and coefficients can be found in Stroud and Secrest (1966).

The next special case of Theorem 6.1.1 that we shall consider is the one where $s = n-3$. The set of such rules is discussed in detail in Micchelli and Rivlin (1973) and Förster (1992). We shall deal with only one particular rule, which is extremal in a certain sense (see, e.g., Theorem 6.5.2).

THEOREM 6.1.7. *There is exactly one quadrature rule $Q_n^{\mathrm{Lo}} \in \mathbf{Q}_n$ with a and b as nodes and $R_n^{\mathrm{Lo}}[\mathbb{P}_{2n-3}] = \{0\}$, namely the quadrature rule of interpolatory type with respect to the zeros of S_n, where*

$$S_n(x) := \det \begin{pmatrix} q_n(x) & q_n(a) & q_n(b) \\ q_{n-1}(x) & q_{n-1}(a) & q_{n-1}(b) \\ q_{n-2}(x) & q_{n-2}(a) & q_{n-2}(b) \end{pmatrix}.$$

This rule is positive and its coefficients satisfy

$$a_\nu^{\mathrm{Lo}} = \frac{\delta_n}{\delta_{n-2}} \cdot \frac{(b-a) q_{n-1}(a) q_{n-1}(b)}{S_n'(x_\nu^{\mathrm{Lo}}) q_{n-1}(x_\nu^{\mathrm{Lo}})}. \tag{6.11}$$

The rules described in this theorem are the so-called Lobatto rules. They coincide with the "Lobatto rules" from Definition 3.4.4 if we choose $U = \mathbb{P}_{2n-3}$ there.

PROOF. We have to prove that:

(i) S_n has n different zeros in $[a, b]$;

(ii) the coefficients are positive;
(iii) equation (6.11) holds.

(i) It is easily verified that $(x-a)^{-1}(b-x)^{-1}S_n(x)$ is an orthogonal polynomial of degree $n-2$ with respect to the weight function $(\cdot - a)(b - \cdot)w$. Theorem A.2 then yields $n-2$ inner zeros and hence (i).

(ii) We define the polynomial

$$l(x) := (x-a)\prod_{\substack{\mu=2\\ \mu\neq\nu}}^{n-1}\left(x-x_\mu^{\mathrm{Lo}}\right)^2(b-x)$$

of degree $2n-4$ and obtain

$$0 < I[l] = Q_n^{\mathrm{Lo}}[l] = a_\nu l(x_\nu).$$

The positivity of the coefficients corresponding to interior nodes follows. For $\nu \in \{1, n\}$ we proceed similarly.

(iii) We start with

$$\begin{aligned}
&q_{n-1}(x_\nu^{\mathrm{Lo}})S_n(x)\\
&\quad= q_{n-1}(x_\nu^{\mathrm{Lo}})S_n(x) - q_{n-1}(x)S_n(x_\nu^{\mathrm{Lo}})\\
&\quad= \det\begin{pmatrix} q_n(x)q_{n-1}(x_\nu^{\mathrm{Lo}}) - q_{n-1}(x)q_n(x_\nu^{\mathrm{Lo}}) & q_n(a) & q_n(b)\\ 0 & q_{n-1}(a) & q_{n-1}(b)\\ q_{n-2}(x)q_{n-1}(x_\nu^{\mathrm{Lo}}) - q_{n-1}(x)q_{n-2}(x_\nu^{\mathrm{Lo}}) & q_{n-2}(a) & q_{n-2}(b)\end{pmatrix}.
\end{aligned}$$

Dividing by $x - x_\nu^{\mathrm{Lo}}$ and applying Theorem A.5 yields

$$\int_a^b \frac{q_{n-1}(x_\nu^{\mathrm{Lo}})S_n(x)}{x-x_\nu^{\mathrm{Lo}}}w(x)\,dx = \det\begin{pmatrix} \delta_n/\delta_{n-1} & q_n(a) & q_n(b)\\ 0 & q_{n-1}(a) & q_{n-1}(b)\\ -\delta_{n-1}/\delta_{n-2} & q_{n-2}(a) & q_{n-2}(b)\end{pmatrix}.$$

Row operations using the three-term recurrence relations of Theorem A.3 give

$$\int_a^b \frac{q_{n-1}(x_\nu^{\mathrm{Lo}})S_n(x)}{x-x_\nu^{\mathrm{Lo}}}w(x)\,dx = \frac{\delta_n}{\delta_{n-2}}(b-a)q_{n-1}(a)q_{n-1}(b).$$

Now (6.11) follows with the aid of (5.2). □

COROLLARY 6.1.3. In the case of $I[f] = \int_{-1}^1 f(x)\,dx$, the coefficients of Q_n^{Lo} satisfy

$$a_\nu^{\mathrm{Lo}} = \frac{2}{n(n-1)P_{n-1}^2(x_\nu^{\mathrm{Lo}})}$$

where x_ν^{Lo} are the zeros of $P_n - P_{n-2}$.

This corollary is derived from the relations

$$S_n = (-1)^{n-1}\sqrt{\frac{(2n+1)(2n-1)(2n-3)}{2}}\,(P_n - P_{n-2})$$

and

$$P_n' - P_{n-2}' = (2n-1)P_{n-1}$$

(see Theorem A.14(v)).

We mention that $Q_3^{\mathrm{Lo}} = Q_3^{\mathrm{Si}}$, the three-point Simpson's rule. In general, Lobatto rules are so complicated that we can only calculate the nodes and coefficients numerically or read them off from tables; see Stroud and Secrest (1966). There is, however, a remarkable exception—a case where the Gauss, Radau and Lobatto rules are particularly simple.

EXAMPLE 6.1.2. For

$$I[f] = \int_{-1}^{1} \frac{f(x)}{\sqrt{1-x^2}}\, dx,$$

we have

$$Q_n^{\mathrm{G}}[f] = \frac{\pi}{n} \sum_{\nu=1}^{n} f\left(-\cos\Big(\frac{2\nu-1}{2n}\pi\Big)\right),$$

$$Q_n^{\mathrm{Ra},-1}[f] = \frac{\pi}{2n-1} f(-1) + \frac{2\pi}{2n-1} \sum_{\nu=2}^{n} f\left(-\cos\Big(\frac{2\nu-2}{2n-1}\pi\Big)\right),$$

$$Q_n^{\mathrm{Ra},1}[f] = \frac{2\pi}{2n-1} \sum_{\nu=1}^{n-1} f\left(-\cos\Big(\frac{2\nu-1}{2n-1}\pi\Big)\right) + \frac{\pi}{2n-1} f(1),$$

$$Q_n^{\mathrm{Lo}}[f] = \frac{\pi}{2n-2} f(-1) + \frac{\pi}{n-1} \sum_{\nu=2}^{n-1} f\left(-\cos\Big(\frac{\nu-1}{n-1}\pi\Big)\right) + \frac{\pi}{2n-2} f(1).$$

These formulas can be obtained by specializing Theorems 6.1.3, 6.1.5 and 6.1.7, or one can verify them as in Example 5.1.1.

As l increases, the investigation of rules with $R_n[\mathbb{P}_{2n-l}] = \{0\}$ becomes more and more complicated. Research in this area has focused mainly on seeking criteria that guarantee positivity of the rules. For the $l = 4$ case, see Nikolov (2001); for arbitrary l, see Peherstorfer (1981, 1984, 2008), Sottas and Wanner (1982), Schmid (1989) and Xu (1994). A particularly elegant criterion can be obtained from Peherstorfer (1993a) by letting α_ν and β_ν be the coefficients in the following form of the three-term recurrence relation:

$$\frac{q_\nu(x)}{\delta_\nu} = (x - \alpha_\nu)\frac{q_{\nu-1}(x)}{\delta_{\nu-1}} - \beta_\nu \frac{q_{\nu-2}(x)}{\delta_{\nu-2}}.$$

Then, we have a surprisingly simple result: The quadrature rule Q_n on $[-1, 1]$ with $R_n[\mathbb{P}_{n+s}] = \{0\}$ and nodes in $]-1, 1[$ is positive if and only if its nodes are the zeros of a polynomial $p_n \in \mathbb{P}_n$ constructed recursively by taking $p_0 = 1$, $p_1 = q_1/\delta_1$, and

$$p_\nu(x) = (x - \widetilde{\alpha}_\nu)p_{\nu-1}(x) - \widetilde{\beta}_\nu p_{\nu-2}(x) \quad \text{for } \nu = 2, 3, \ldots, n,$$

where

$$\widetilde{\alpha}_\nu = \alpha_\nu \quad \text{for } \nu \le \frac{n+s+1}{2}, \qquad \widetilde{\beta}_\nu = \beta_\nu \quad \text{for } \nu \le \frac{n+s+2}{2}$$

and

$$\operatorname{sgn} p_\nu(\pm 1) = (\pm 1)^\nu \quad \text{for } \nu = 1, \ldots, n.$$

Among quadrature methods with $R_n[\mathbb{P}_{2n-4}] = \{0\}$, one that has attracted some interest is the method of Ralston; see Ralston (1959), Filippi (1966) and Schneider (1988). Ralston's motivation for imposing the conditions

$$R_n[\mathbb{P}_{2n-4}] = \{0\}, \qquad x_1 = a, \qquad x_n = b, \qquad 0 < a_1 = -a_n$$

was to construct compound rules with favourable properties.

Finally, we present another framework in which the Gaussian rule comes up in a natural way. First, note that if we have a quadrature rule of interpolatory type with nodes $x_1, \ldots, x_n$, and this rule satisfies $R_n[\mathbb{P}_{n+s}] = \{0\}$, then $Q[f] = I[p]$ holds for every $p \in \mathbb{P}_{n+s}$ with $p(x_\nu) = f(x_\nu)$ for $\nu = 1, \ldots, n$. In the Gaussian case, we may choose p as follows:

$$p \in \mathbb{P}_{2n-1}, \qquad p(x_\nu^{\mathrm{G}}) = f(x_\nu^{\mathrm{G}}) \quad \text{and} \quad p'(x_\nu^{\mathrm{G}}) = f'(x_\nu^{\mathrm{G}}) \quad \text{for } \nu = 1, \ldots, n \tag{6.12}$$

(i.e. p is the Hermitian interpolation polynomial). We then obtain the following result.

THEOREM 6.1.8.

$$Q_n^{\mathrm{G}}[f] = \int_a^b p(x)w(x)\,dx,$$

where p is as defined in (6.12).

It is quite surprising that the values $f'(x_\nu)$ necessary for the construction of p do not actually appear in the quadrature rule. This could also be taken as a defining property of the Gaussian rules.

If we were to prescribe derivatives only at interior nodes, then Theorem 6.1.8 would also hold for the Radau and Lobatto rules. Another relationship between Gaussian rules and Hermite interpolation is discussed in Scherer and Zeller (1979).

6.2. Coefficients and nodes for general weights

To establish bounds for the coefficients of the Gaussian rule, we examine the function

$$\begin{aligned} C_m(\eta) &= \sup\{A(\eta, Q) \,:\, Q \in \mathbf{Q}^+(I, \mathbb{P}_m)\} \\ &= \inf\{I[p] \,:\, p \in \mathbb{P}_m,\ p > 0 \text{ on } [a,b],\ p(\eta) = 1\} \end{aligned} \tag{6.13}$$

(for the definition of A, see Theorem 3.3.5), where the second equality comes from Theorem 3.3.5. Obviously,

$$C_{2r}(\eta) \le \Phi_r(\eta) := \inf\{I[p^2] \,:\, p \in \mathbb{P}_r,\ p(\eta) = 1\}.$$

As in the previous section, we let q_ν denote the orthonormal polynomial of degree ν with respect to the weight function w.

LEMMA 6.2.1.

$$\Phi_r(\eta) = \left(\sum_{\nu=0}^{r} q_\nu^2(\eta)\right)^{-1}.$$

PROOF. We use the orthonormal expansion of a polynomial $p \in \mathbb{P}_r$ along with the Cauchy–Schwarz inequality and Parseval's identity to obtain

$$1 = p(\eta) = \sum_{\nu=0}^{r} I[pq_\nu]q_\nu(\eta) \le \left(\sum_{\nu=0}^{r} I^2[pq_\nu] \sum_{\nu=0}^{r} q_\nu^2(\eta)\right)^{1/2} = I[p^2]^{1/2}\left(\sum_{\nu=0}^{r} q_\nu^2(\eta)\right)^{1/2}.$$

Therefore

$$I[p^2] \ge \left(\sum_{\nu=0}^{r} q_\nu^2(\eta)\right)^{-1},$$

and equality holds for

$$p = \left(\sum_{\nu=0}^{r} q_\nu^2(\eta)\right)^{-1} \cdot \sum_{\nu=0}^{r} q_\nu(\eta)q_\nu,$$

which is what we wanted to prove. □

The function Φ_r is known as the Christoffel function and has been investigated intensively in the theory of orthogonal polynomials. Although we have obtained it only through rough estimation, we shall see that it is of help in describing our functions C_m completely.

THEOREM 6.2.1 (Krein 1951). Let q_ν, $q_{\nu;a}$, $q_{\nu;b}$ and $q_{\nu;a,b}$ be the orthonormal polynomials with respect to the weight functions w, $(\cdot - a)w$, $(b - \cdot)w$ and $(b - \cdot)(\cdot - a)w$, respectively. Then, the functions defined in (6.13) satisfy

$$C_{2r}(\eta) = \min\left\{\left(\sum_{\nu=0}^{r} q_\nu^2(\eta)\right)^{-1}, \left((b-\eta)(\eta-a)\sum_{\nu=0}^{r-1} q_{\nu;a,b}^2(\eta)\right)^{-1}\right\}$$

and

$$C_{2r+1}(\eta) = \min\left\{\left((\eta-a)\sum_{\nu=0}^{r} q_{\nu;a}^2(\eta)\right)^{-1}, \left((b-\eta)\sum_{\nu=0}^{r} q_{\nu;b}^2(\eta)\right)^{-1}\right\}.$$

PROOF. We prove only the formula for C_{2r}; the one for C_{2r+1} can be shown analogously. For any $u \in \mathbb{P}_{2r}$ we have $I[u] = Q^\eta[u]$, where Q^η is the Krein rule for the space $U = \mathbb{P}_{2r}$ with a node at $\eta \in\,]a,b[$. Since u has to be positive in order to contribute to the minimum, the value $Q^\eta[u]$ is minimal if u vanishes at all nodes of Q^η other than η. This occurs when u has double zeros at the interior nodes of Q^η and zeros at the boundary nodes of Q^η. Each Krein rule of index $2r+2$ for $\mathbb{P}_{2r}$ has either $r+1$ interior nodes and no nodes at the boundary or r interior nodes and two boundary nodes. Hence, the minimizing polynomials are either of the form $u(x) = p^2(x)$ with $p \in \mathbb{P}_r$, or of the form $u(x) = (b-x)(x-a)p^2(x)$ with $p \in \mathbb{P}_{r-1}$. We therefore need only minimize over such polynomials and can apply Lemma 6.2.1 with either q_ν or $q_{\nu;a,b}$. For Krein rules of index $2r+1$ and $\eta \in \{a,b\}$, the same result can be obtained by continuity arguments. □

The simplest example involves the weight function $w(x) = \left(1-x^2\right)^{-1/2}$ on $[-1,1]$, which gives rise to the orthogonal polynomial systems

$$q_0(x) = \frac{1}{\sqrt{\pi}}, \qquad q_\nu(x) = \sqrt{\frac{2}{\pi}}\,T_\nu(x) \quad \text{for } \nu \ge 1,$$

$$q_{\nu;-1,1}(x) = \sqrt{\frac{2}{\pi}}\,U_\nu(x), \qquad q_{\nu;1}(x) = (-1)^\nu q_{\nu;-1}(-x) = \frac{1}{\sqrt{\pi}}\,\frac{T_{\nu+1}(x) - T_\nu(x)}{x-1}.$$

After some calculations using the formulas

$$\sum_{\nu=1}^{n} \sin^2 \nu x = \frac{1}{4}\left(2n+1 - \frac{\sin(2n+1)x}{\sin x}\right),$$

$$\sum_{\nu=1}^{n} \cos^2 \nu x = \frac{1}{4}\left(2n-1 + \frac{\sin(2n+1)x}{\sin x}\right),$$

we obtain the following result.

THEOREM 6.2.2 (Achieser and Krein 1938; Erdős and Turan 1940; see also Krein 1951). For the weight function $w(x) = \left(1-x^2\right)^{-1/2}$ on $[-1,1]$, we have

$$C_m(\eta) = \frac{2\pi}{m+1+|U_m(\eta)|}.$$

The characterization of C_m as an infimum in (6.13) shows that it increases as the weight function increases. This yields the next theorem.

THEOREM 6.2.3. The coefficients a_ν of a positive quadrature rule on $[-1,1]$ with $R[\mathbb{P}_m] = \{0\}$ satisfy

$$a_\nu \le \frac{2\pi}{m+1+|U_m(x_\nu)|} \cdot \sup_{x\in[-1,1]} w(x)\sqrt{1-x^2}.$$

The application of Theorem 6.2.1 to $w = 1$ requires more work. We restrict our attention to C_{2r-1}; for the C_{2r} case and further pertinent results, see Förster (1995). The first main results in this area are due to Bernstein (1937c).

THEOREM 6.2.4. Let $w = 1$ and $[a,b] = [-1,1]$. Then

$$C_{2r-1}(x) = \frac{2}{r(r+1)P_r^2(x) + (1-x^2)[P_r'(x)]^2 - xP_r(x)P_r'(x) + |P_r(x)P_r'(x)|}.$$

PROOF. It is easily verified that the orthonormal polynomials with respect to $w(x) = 1 \pm x$ are

$$q_\nu(x) = \sqrt{\frac{\nu+1}{2}} \cdot \frac{P_{\nu+1}(x) \pm P_\nu(x)}{x \pm 1}.$$

Together with Theorem A.6, this gives

$$(1 \pm x)\sum_{\nu=0}^{r-1} q_\nu^2(x) = \frac{r(r+1)}{4r+2} \cdot \frac{\big(P_{r+1}'(x) \pm P_r'(x)\big)\big(P_r(x) \pm P_{r-1}(x)\big) - \big(P_r'(x) \pm P_{r-1}'(x)\big)\big(P_{r+1}(x) \pm P_r(x)\big)}{1 \pm x}.$$

Using Theorem A.14(vii), we see that this equals

$$\frac{r(r+1)}{2} \cdot \frac{P_r^2(x) - P_{r-1}(x)P_{r+1}(x)}{1-x^2} \pm \frac{1}{2}P_r(x)P_r'(x).$$

The assertion is then proved by plugging in

$$-P_{r-1}(x)P_{r+1}(x) = -x^2P_r^2(x) + \frac{(1-x^2)^2[P_r'(x)]^2 - x(1-x^2)P_r(x)P_r'(x)}{r(r+1)}.$$

Proving this latter equation is straightforward if we use the relations

$$xP_r(x) = \frac{r+1}{2r+1}P_{r+1}(x) + \frac{r}{2r+1}P_{r-1}(x)$$

and

$$(1-x^2)P_r'(x) = \frac{r(r+1)}{2r+1}\big(P_{r-1}(x) - P_{r+1}(x)\big)$$

(Theorem A.13 and Theorem A.14(iv), respectively). □

An important consequence is the following:

COROLLARY 6.2.1. For $w = 1$ and $[a,b] = [-1,1]$, we have

$$\sup_{-1\le x\le 1} C_{2r-1}(x) = \begin{cases} a^{\mathrm{G}}_{(r+1)/2,r} = \dfrac{\pi}{r}\big(1+o(1)\big) & \text{if } r \text{ is odd,} \\ a^{\mathrm{Lo}}_{(r+2)/2,r+1} = \dfrac{\pi}{r}\big(1+o(1)\big) & \text{if } r \text{ is even.} \end{cases}$$

PROOF. From Theorem 6.2.4 and $-xP_r(x)P_r'(x)+|P_r(x)P_r'(x)|\geq 0$, we obtain

$$C_{2r-1}(x)\leq\frac{2}{r(r+1)P_r^2(x)+(1-x^2)[P_r'(x)]^2}.$$

The denominator on the right-hand side is increasing on $[0,1]$ because its derivative is

$$\begin{aligned}&2r(r+1)P_r(x)P_r'(x)-2x[P_r'(x)]^2+2(1-x^2)P_r'(x)P_r''(x)\\&=2P_r'(x)\big[(1-x^2)P_r''(x)-xP_r'(x)+r(r+1)P_r(x)\big]\\&=2x\big[P_r'(x)\big]^2\geq 0,\end{aligned}$$

where we have used Theorem A.14(i) in the second step. Furthermore, C_{2r-1} is an even function. We have thus shown that

$$\sup_{-1\leq x\leq 1}C_{2r-1}(x)\leq\frac{2}{r(r+1)P_r^2(0)+[P_r'(0)]^2}.$$

The right-hand side is $a^{\mathrm{G}}_{(r+1)/2,r}$ if r is odd (see Corollary 6.1.1) and $a^{\mathrm{Lo}}_{(r+2)/2,r+1}$ if r is even (see Corollary 6.1.3). The asymptotic values of these coefficients are given in Theorem 6.2.6. □

Now we come to the asymptotic behaviour of the function C_m.

Assumption P. *Let $[-1,1]$ be the the underlying fundamental interval. Assume that $w(x)\sqrt{1-x^2}$ is continuous and that there exists a polynomial p such that*

$$0\leq p(x)\leq w(x)\sqrt{1-x^2}\quad\textit{for } x\in[-1,1]$$

and

$$p(x)>0\quad\textit{for } x\in[c,d]\subset\,]-1,1[.$$

THEOREM 6.2.5. Under Assumption P, we have

$$\lim_{m\to\infty}m\cdot C_m(\eta)=2\pi w(\eta)\sqrt{1-\eta^2}$$

uniformly for $\eta\in[c,d]$.

PROOF. (i) For every $\varepsilon>0$, we shall prove the existence of a polynomial q satisfying

$$0\leq q(x)\leq w(x)\sqrt{1-x^2}\quad\text{for } x\in[-1,1]\tag{6.14}$$

and

$$q(x)\geq(1-\varepsilon)w(x)\sqrt{1-x^2}\quad\text{for } x\in[c,d]\subset[-1,1].\tag{6.15}$$

Choose $\delta>0$ such that $p(x)>0$ for $x\in[c-\delta,d+\delta]$. Then, define the function α by

$$\alpha(x)=\begin{cases}0 & \text{if } x\in[-1,c-\delta]\cup[d+\delta,1],\\ 1 & \text{if } x\in[c,d],\\ \text{linear} & \text{if } x\in[c-\delta,c] \text{ or } x\in[d,d+\delta].\end{cases}$$

The Weierstrass approximation theorem implies the existence of a polynomial r with

$$(1-\varepsilon)\frac{\alpha(x)w(x)\sqrt{1-x^2}}{p(x)}\leq r(x)\leq\frac{\alpha(x)w(x)\sqrt{1-x^2}}{p(x)}+1-\alpha(x)$$

(note that the difference between the left-hand side and the right-hand side is positive for $x \in [-1, 1]$). Now, $q := r \cdot p$ has the desired property.

(ii) Consider the functionals

$$I[f] = \int_{-1}^{1} f(x)w(x)\,dx,$$
$$I_1[f] = \int_{-1}^{1} f(x)\frac{q(x)}{\sqrt{1-x^2}}\,dx,$$
$$I_2[f] = \int_{-1}^{1} \frac{f(x)}{\sqrt{1-x^2}}\,dx,$$

with the q from part (i), and denote by s the degree of q. Using (6.13), we obtain

$$\begin{aligned} C_m(\eta) &= \inf\left\{I\left[\frac{p}{p(\eta)}\right] : p > 0,\ p \in \mathbb{P}_m\right\} \\ &\geq \inf\left\{I_1\left[\frac{p}{p(\eta)}\right] : p > 0,\ p \in \mathbb{P}_m\right\} \\ &= q(\eta)\cdot \inf\left\{I_2\left[\frac{pq}{p(\eta)q(\eta)}\right] : p > 0,\ pq \in \mathbb{P}_{m+s}\right\} \\ &\geq q(\eta)\cdot \inf\left\{I_2\left[\frac{p}{p(\eta)}\right] : p > 0,\ p \in \mathbb{P}_{m+s}\right\} \\ &= q(\eta)\cdot \sup\left\{A(\eta, Q) : Q \in \mathbf{Q}^+(I_2, \mathbb{P}_{m+s})\right\}. \end{aligned}$$

Inequality (6.15) and the formula from Theorem 6.2.2 yield

$$\sup\left\{A(\eta, Q) : Q \in \mathbf{Q}^+(I_2, \mathbb{P}_k)\right\} \geq (1-\varepsilon)w(\eta)\sqrt{1-\eta^2}\cdot\frac{2\pi}{k+1+|U_k(\eta)|}.$$

We have $|U_k(\eta)| \leq 1/\sqrt{1-\eta^2}$, which is bounded since $\eta \in [c, d] \subset\,]-1, 1[$. Hence we have shown that for every $\varepsilon_1 > 0$ there is an m_0, independent of $\eta \in [c, d]$, such that for all $m > m_0$,

$$mC_m(\eta) \geq 2\pi w(\eta)\sqrt{1-\eta^2} - \varepsilon_1.$$

(iii) Derivation of the upper bound is similar. For $\varepsilon > 0$, choose a polynomial q (of some degree s) such that

$$\frac{1}{w(x)\sqrt{1-x^2}+2\varepsilon} \leq q(x) \leq \frac{1}{w(x)\sqrt{1-x^2}+\varepsilon} \qquad \text{for } x \in [-1, 1].$$

This means, in particular, that

$$w(x) \leq \frac{1}{q(x)\sqrt{1-x^2}}.$$

Now, choose functionals I and I_2 as in part (ii) and let

$$I_1[f] = \int_{-1}^{1} \frac{f(x)}{q(x)\sqrt{1-x^2}}\,dx.$$

This time, we get

$$\begin{aligned} C_m(\eta) &\le \inf\left\{ I_1\left[\frac{p}{p(\eta)}\right] : p > 0,\ p \in \mathbb{P}_m \right\} \\ &\le \frac{1}{q(\eta)} \inf\left\{ I_2\left[\frac{p}{p(\eta)}\right] : p > 0,\ p \in \mathbb{P}_{m-s} \right\} \\ &= \frac{1}{q(\eta)} \cdot \sup\left\{ A(\eta, Q) : Q \in \mathbf{Q}^+(I_2, \mathbb{P}_{m-s}) \right\} \\ &\le \left[w(\eta)\sqrt{1-\eta^2} + 2\varepsilon \right] \cdot \frac{2\pi}{m - s + |U_{m-s}(\eta)|} \end{aligned}$$

and therefore

$$m \cdot C_m(\eta) \le 2\pi w(\eta)\sqrt{1-\eta^2} + \varepsilon_1,$$

for arbitrary $\varepsilon_1 > 0$ and all $m > m_1(\varepsilon_1)$.

Taking parts (ii) and (iii) together, we get the assertion of the theorem. □

Theorem 6.2.5 is, in fact, a statement about the coefficients of Krein-type rules with $m \approx 2n$, since the function values of C_m are attained by precisely such quadrature rules.

THEOREM 6.2.6 (Freud 1969). *For weight functions that satisfy Assumption P, the coefficients a_ν associated with the nodes x_ν of a Krein-type rule with n nodes are such that*

$$a_\nu = \frac{\pi}{n} w(x_\nu)\sqrt{1 - x_\nu^2} + o(n^{-1})$$

uniformly for $x_\nu \in [c, d]$.

As can be seen in the following theorem, even more is true.

THEOREM 6.2.7. *Suppose that w satisfies Assumption P, and let Q_n be a Krein-type rule with n nodes. Then, for all ν with $[x_{\nu-1}, x_{\nu+1}] \subset [c, d]$, the following asymptotic relations hold uniformly:*

$$x_{\nu+1} - x_\nu = \frac{\pi}{n}\sqrt{1 - x_\nu^2} + o(n^{-1}), \tag{6.16}$$

$$x_{\nu+1} - 2x_\nu + x_{\nu-1} = o(n^{-1}), \tag{6.17}$$

$$a_\nu = \int_{(x_\nu + x_{\nu-1})/2}^{(x_{\nu+1}+x_\nu)/2} w(x)\,dx + o(n^{-1}). \tag{6.18}$$

PROOF. Denote by K_1 the first Peano kernel of the rule.

(i) From $w(x) > \text{const} > 0$ for $x \in [c, d]$ we deduce that

$$\text{const} \cdot (x_{\nu+1} - x_\nu) \le \int_{x_\nu}^{x_{\nu+1}} w(x)\,dx = K_1(x_\nu + 0) - K_1(x_{\nu+1} - 0) \le 2\,\|K_1\|,$$

and Theorem 5.7.2 gives

$$x_{\nu+1} - x_\nu = O(n^{-1}). \tag{6.19}$$

(ii) The two relations

$$K_1(x_\nu + 0) - K_1(x_\nu - 0)\ (= a_\nu) = \frac{\pi}{n} w(x_\nu)\sqrt{1 - x_\nu^2} + o(n^{-1})$$

(i.e. Theorem 6.2.6) and

$$|K_1(x_\nu \pm 0)| \le \frac{\pi}{2n} w(x_\nu)\sqrt{1 - x_\nu^2} + o(n^{-1})$$

(i.e. Theorem 5.7.2) can both hold only if

$$K_1(x_\nu \pm 0) = \pm\frac{\pi}{2n}w(x_\nu)\sqrt{1-x_\nu^2} + o(n^{-1}).$$

(iii) Using (i) and (ii), we get

$$\begin{aligned}(x_{\nu+1}-x_\nu)w(x_\nu) &= \int_{x_\nu}^{x_{\nu+1}} w(x)\,dx + o(n^{-1})\\ &= K_1(x_\nu+0) - K_1(x_{\nu+1}-0) + o(n^{-1})\\ &= \frac{\pi}{2n}\left(w(x_\nu)\sqrt{1-x_\nu^2} + w(x_{\nu+1})\sqrt{1-x_{\nu+1}^2}\right) + o(n^{-1})\\ &= \frac{\pi}{n}w(x_\nu)\sqrt{1-x_\nu^2} + o(n^{-1}).\end{aligned}$$

Relation (6.16) follows immediately from $w(x_\nu) \geq \text{const} > 0$, and (6.17) is also a direct consequence once (6.19) is taken into account. Now, we have

$$\begin{aligned}\int_{(x_\nu+x_{\nu-1})/2}^{(x_{\nu+1}+x_\nu)/2} w(x)\,dx &= \frac{x_{\nu+1}-x_{\nu-1}}{2}\,w(x_\nu) + o(n^{-1})\\ &= \frac{\pi}{n}\sqrt{1-x_\nu^2}\,w(x_\nu) + o(n^{-1}) = a_\nu + o(n^{-1}),\end{aligned}$$

which is (6.18) □

6.3. Nodes and coefficients for $w = 1$

For constant weight functions, one can obtain much more precise results on the nodes and coefficients of Gaussian quadrature rules than is possible in the general case. The tool that is available in this special case, but not in general, is the differential equation[1] for Legendre polynomials. All the results of this section are based on this differential equation.

Without loss of generality, we consider the weight function $w = 1$ on $[-1, 1]$. A solution of the Legendre differential equation

$$(1-x^2)y'' - 2xy' + n(n+1)y = 0$$

is the Legendre polynomial P_n (see Theorem A.14). In what follows we shall need a transformed version of P_n. Defining

$$z(\theta) = \sqrt{\sin\theta}\,P_n(\cos\theta), \tag{6.20}$$

one can verify that z satisfies the differential equation

$$z'' + g(\theta)z = 0 \tag{6.21}$$

where

$$g(\theta) = \left(n+\frac{1}{2}\right)^2 + \frac{1}{4\sin^2\theta}. \tag{6.22}$$

Let us write the zeros of z as

$$(0=)\,\theta_0 < \theta_1 < \theta_2 < \cdots < \theta_n < \theta_{n+1}\,(=\pi).$$

[1] Certain classes of weight functions have differential equations associated with them. However, looking at each of these cases in turn yields no further insight over what can be gained by investigating just the constant weight function.

These zeros are related to the zeros of P_n via

$$x_\nu = -\cos\theta_\nu \quad \text{for } \nu = 1, 2, \ldots, n.$$

Following Szegö (1936), we derive results for z by applying Sturm's comparison theorem. The following version of the theorem is suitable for our purposes.

THEOREM 6.3.1 (Sturm's comparison theorem). Let f_1 and f_2 be continuous functions on $]c, d[$, and let y_1 and y_2 be non-trivial solutions of

$$y_i'' + f_i y_i = 0, \qquad y_i(c+0) = 0.$$

Assume $f_1 \le f_2$ and that $f_1 = f_2$ does not hold on any non-degenerate interval. Assume, furthermore, that

$$\lim_{x \to c+0} \left[y_1(x) y_2'(x) - y_2(x) y_1'(x) \right] = 0. \tag{6.23}$$

Then, if y_1 has a zero on $]c, d]$, y_2 must have a sign change on $]c, d]$.

PROOF. Without loss of generality, we may assume that y_1 and y_2 are positive on some interval $]c, c+\varepsilon]$, with $\varepsilon > 0$. Let ξ be the smallest zero of y_1 on $]c, d]$, and set

$$v := y_1 y_2' - y_2 y_1'.$$

Then

$$v' := y_1 y_2'' - y_2 y_1'' = -f_2 y_1 y_2 + f_1 y_2 y_1 = (f_1 - f_2) y_1 y_2.$$

Suppose that y_2 does not change sign. Then v is decreasing on $]c, \xi]$, and therefore

$$v(c+0) = 0 > v(\xi) = -y_2(\xi) y_1'(\xi);$$

but this contradicts $y_2(\xi) \ge 0$ and $y_1'(\xi) < 0$. □

We apply Sturm's comparison theorem to the differential equation (6.21) with solution (6.20).

LEMMA 6.3.1. We have

$$\theta_{\nu+1} - \theta_\nu > \frac{\pi}{\sqrt{g(\theta_\nu)}} \qquad \text{for } \nu = 1, 2, \ldots, \left\lfloor \frac{n}{2} \right\rfloor \tag{6.24}$$

and

$$\theta_{\nu+1} - \theta_\nu < \frac{\pi}{\sqrt{g(\theta_{\nu+1})}} \qquad \text{for } \nu = 0, 1, \ldots, \left\lfloor \frac{n-1}{2} \right\rfloor. \tag{6.25}$$

PROOF. In Theorem 6.3.1, choose $]c, d] =]\theta_\nu, \theta_{\nu+1}]$ and

$$f_1 = g, \qquad f_2(x) = g(\theta_\nu), \qquad y_1 = z, \qquad y_2(\theta) = \sin\left(\sqrt{g(\theta_\nu)}(\theta - \theta_\nu)\right).$$

Sturm's comparison theorem then yields a sign change for y_2; that is,

$$\theta_\nu + \frac{\pi}{\sqrt{g(\theta_\nu)}} \in]\theta_\nu, \theta_{\nu+1}],$$

which is (6.24). Next, choose

$$f_1 = g(\theta_{\nu+1}), \qquad f_2 = g, \qquad y_1(\theta) = \sin\left(\sqrt{g(\theta_{\nu+1})}(\theta - \theta_\nu)\right), \qquad y_2 = z.$$

In the $\nu = 0$ case we have to verify assumption (6.23), which is no problem. Now, y_2 does not change sign on $]\theta_\nu, \theta_{\nu+1}]$, so y_1 has no zero on this interval. Therefore,

$$\theta_\nu + \frac{\pi}{\sqrt{g(\theta_{\nu+1})}} > \theta_{\nu+1},$$

which is (6.25). □

Upon taking $\theta_\nu = \pi - \theta_{n+1-\nu}$ into account, we gain some essential information about the locations of the zeros of P_n.

THEOREM 6.3.2.
(i) $\theta_{\nu+1} - \theta_\nu < \dfrac{\pi}{n+1/2}$ for $\nu = 0, 1, \ldots, n$;
(ii) $\dfrac{\nu - 1/4}{n+1/2}\pi \le \theta_\nu \le \dfrac{\nu}{n+1/2}\pi$ for $\nu = 0, 1, \ldots, \left\lfloor \dfrac{n+1}{2} \right\rfloor$;
(iii) $\theta_\nu - \theta_{\nu-1} < \theta_{\nu+1} - \theta_\nu$ for $\nu = 1, 2, \ldots, \left\lfloor \dfrac{n}{2} \right\rfloor$.

PROOF. Assertion (i) follows immediately from (6.24) and $g(\theta) > (n+1/2)^2$. It implies that

$$\theta_m - \theta_k = (\theta_m - \theta_{m-1}) + (\theta_{m-1} - \theta_{m-2}) + \cdots + (\theta_{k+1} - \theta_k) < \frac{(m-k)\pi}{n+1/2}.$$

Taking $m = \nu$ and $k = 0$ then gives the second inequality in (ii). With $m = n+1-\nu$ and $k = \nu$, we get

$$2\theta_\nu = \theta_\nu + (\pi - \theta_{n+1-\nu}) \ge \pi - (n+1-2\nu) \cdot \frac{\pi}{n+1/2} = \frac{2\nu - 1/2}{n+1/2}\pi,$$

which is the left-hand inequality in (ii). Comparing (6.24) and (6.25) gives (iii). □

Many variants of this theorem, as well as sharper results, have been proved; see Brass *et al.* (1996) and the references therein. However, for our central topic, namely the quadrature error, they are not as helpful as one might expect. For example, direct application of the Peano kernel representation in Theorem 4.2.5(i) fails, because we have a difference of two almost-equal sums—this is a problem not only for theoretical investigations but also for numerical computations; see Stroud and Secrest (1966: p. 66). Nevertheless, in Theorem 6.6.2, we will obtain useful estimates with our simple inequalities by using a different representation of K_1. A simple, but quite remarkable, result on the distribution of nodes of Gauss–Legendre rules is the following:

THEOREM 6.3.3. $x_{\nu+1} - x_\nu > x_\nu - x_{\nu-1}$ for $\nu = 1, 2, \ldots, \lfloor n/2 \rfloor$.

PROOF. Theorem 6.3.2(iii) and monotonicity of the cosine give

$$\begin{aligned}
&\cos\theta_{\nu+1} - 2\cos\theta_\nu + \cos\theta_{\nu-1}\\
&\qquad = \cos[\theta_\nu + (\theta_{\nu+1} - \theta_\nu)] - 2\cos\theta_\nu + \cos[\theta_\nu - (\theta_\nu - \theta_{\nu-1})]\\
&\qquad < \cos[\theta_\nu + (\theta_\nu - \theta_{\nu-1})] - 2\cos\theta_\nu + \cos[\theta_\nu - (\theta_\nu - \theta_{\nu-1})]\\
&\qquad = 2\cos\theta_\nu \cos(\theta_\nu - \theta_{\nu-1}) - 2\cos\theta_\nu\\
&\qquad < 0.
\end{aligned}$$

With $x_\nu = -\cos\theta_\nu$, this is just a reformulation of the assertion. □

Now, let us turn to the coefficients $a_\nu = a_{\nu,n}$ of Q_n^{G}. We have the almost-explicit representations of Corollary 6.1.1, but the following theorem gives a clearer idea as to the size of a_ν.

THEOREM 6.3.4 (Förster and Petras 1990a). We have

$$a_\nu = \vartheta_\nu \cdot \frac{\pi}{n+1/2}\sin\theta_\nu, \quad \text{where } 1 - \frac{1}{(2n+1)^2 \sin^2\theta_\nu} < \vartheta_\nu < 1.$$

Very sharp asymptotic results of similar character are given in Petras (1996a). For results concerning coefficients that correspond to nodes near the boundary of $[-1, 1]$, where the theorem is less useful, see Gatteschi (1988).

Another illustrative result is the following:

THEOREM 6.3.5 (Förster and Petras 1990a). *The sequence*

$$\frac{a_\nu}{\sqrt{1-x_\nu^2}}, \qquad \nu = 1, 2, \ldots, \left\lfloor \tfrac{n+1}{2} \right\rfloor,$$

is strictly increasing.

Here are two consequences of the theorem:

COROLLARY 6.3.1 (Winston 1934).

$$a_1 < a_2 < \cdots < a_{\lfloor \frac{n+1}{2} \rfloor}.$$

COROLLARY 6.3.2.

$$\sup_\nu a_\nu < \frac{\pi}{n+1/2} \qquad \text{and} \qquad \lim_{n\to\infty} n \cdot \sup_\nu a_{\nu,n} = \pi.$$

PROOF OF THEOREMS 6.3.4 AND 6.3.5. On $]0, \pi/2]$ set

$$v_1 := gz^2 + (z')^2 \qquad \text{and} \qquad v_2 := z^2 + \frac{1}{g}(z')^2,$$

with g and z as in (6.20) and (6.22). Then

$$v_1' = g'z^2 + 2gzz' + 2z'z'' = g'z^2 + 2z'(z'' + gz) = g'z^2$$

and

$$v_2' = 2zz' - \frac{g'}{g^2}(z')^2 + \frac{2}{g}z'z'' = \frac{2}{g}z'(z'' + gz) - \frac{g'}{g^2}(z')^2 = -\frac{g'}{g^2}(z')^2.$$

Since $g' < 0$ on $]0, \pi/2[$, we have that v_1 is decreasing and v_2 is increasing.

The third representation in Corollary 6.1.1 gives

$$v_1(\theta_\nu) = \left[z'(\theta_\nu)\right]^2 = \sin^3\theta_\nu \left[P_n'(\cos\theta_\nu)\right]^2 = \frac{2\sqrt{1-x_\nu^2}}{a_\nu},$$

which proves Theorem 6.3.5. In an analogous fashion, we obtain

$$v_2(\theta_\nu) = \frac{1}{g(\theta_\nu)}\left[z'(\theta_\nu)\right]^2 = \frac{2\sqrt{1-x_\nu^2}}{g(\theta_\nu)a_\nu}.$$

We will show that

$$v_1(x) > \frac{2}{\pi}\left(n + \frac{1}{2}\right) \qquad \text{and} \qquad v_2(x) \le \frac{2}{\pi}\left(n + \frac{1}{2}\right)^{-1}, \tag{6.26}$$

which would imply

$$\frac{2\sqrt{1-x_\nu^2}}{a_\nu} > \frac{2}{\pi}\left(n + \frac{1}{2}\right) \qquad \text{and} \qquad \frac{2\sqrt{1-x_\nu^2}}{g(\theta_\nu)a_\nu} \le \frac{2}{\pi}\left(n + \frac{1}{2}\right)^{-1},$$

or equivalently

$$\frac{\pi}{n+1/2}\sqrt{1-x_\nu^2} > a_\nu \ge \frac{\pi}{n+1/2}\sqrt{1-x_\nu^2} \cdot \frac{(n+1/2)^2}{g(\theta_\nu)}.$$

Then we would have

$$\frac{(n+1/2)^2}{g(\theta_\nu)} = \left(1+\frac{1}{(2n+1)^2\sin^2\theta_\nu}\right)^{-1} > 1-\frac{1}{(2n+1)^2\sin^2\theta_\nu}.$$

So it remains to prove (6.26). We shall prove the first inequality; the second can be verified with similar techniques.

Theorem A.14(ii) gives $P_n'(0) = nP_{n-1}(0)$, which we will use to show that

$$\begin{aligned} v_1(x) \geq v_1\left(\frac{\pi}{2}\right) &= \left[\left(n+\frac{1}{2}\right)^2+\frac{1}{4}\right] P_n^2(0) + [P_n'(0)]^2 \\ &= \left[\left(n+\frac{1}{2}\right)^2+\frac{1}{4}\right] P_n^2(0) + n^2P_{n-1}^2(0). \end{aligned}$$

For odd n, we set $n-1=2k$ and then have to check that

$$b_k := \frac{(2k+1)^2}{2k+3/2}\, P_{2k}^2(0) > \frac{2}{\pi}. \tag{6.27}$$

The recurrence relation for Legendre polynomials (Theorem A.13) gives

$$P_{2k}(0) = -\frac{2k-1}{2k}\, P_{2k-2}(0).$$

Hence

$$\frac{b_k}{b_{k-1}} = \frac{(2k+1)^2(2k-1/2)}{(2k)^2(2k+3/2)} = 1-\frac{1}{16k^3+12k^2} < 1.$$

Therefore the sequence of b_k is monotonically decreasing, and Wallis's formula shows that it converges to $2/\pi$. Thus, (6.27) is proved. In the case of even n, we proceed analogously. □

Finally, we remark that all orthogonal polynomials with respect to $w(x) = (1-x)^\alpha(1+x)^\beta$, for $\alpha, \beta > -1$, on $[-1,1]$ also satisfy some differential equation that is a generalization of the Legendre equation. Therefore, results similar to those presented in this section can also be established for these weight functions.

6.4. Peano kernels for general weights

In this section we analyze the structure of Peano kernels of Gauss-type rules. Their sign changes play a central role. Sometimes, we may find that a Peano kernel is, for example, positive for $x < y$, zero on the interval $[y, z]$, and negative for $x > z$. In such a situation, we wish to be able to speak of one and only one sign change. We will therefore consider the location of the sign change as being exactly in the middle of the interval on which the kernel is zero. This motivates the following definition, which might seem a bit strange at first glance.

DEFINITION 6.4.1. We say that x_0 is a *sign change* of f if there exists an $l \geq 0$ such that f vanishes in the interval $]x_0-l,\, x_0+l[$ and

$$f(x_0-l-\varepsilon)f(x_0+l+\varepsilon) < 0$$

for all sufficiently small $\varepsilon > 0$.

LEMMA 6.4.1. *Let $S \neq 0$ be a linear functional on $C[a,b]$ satisfying $S[\mathbb{P}_m] = \{0\}$. Then the Peano kernel K_r of S has at least $m-r+1$ sign changes.*

PROOF. Suppose that K_r changes sign only at $z_1, \dots, z_{m-r}$. Then, the non-vanishing function $p \cdot K_r$ with

$$p(x) = \prod_{\nu=1}^{m-r} (x - z_\nu)$$

has no sign changes. For $P^{(r)} = p$, i.e. $P \in \mathbb{P}_m$, we obtain

$$S[P] = \int_a^b p(x) K_r(x)\, dx \neq 0,$$

which contradicts the assumption that $S[\mathbb{P}_m] = \{0\}$. □

This lemma leads to the following important theorem.

THEOREM 6.4.1 (Markoff 1884; Stieltjes 1884). Let Q^η be the Krein rule with node $\eta \in]a, b[$ for the space $\mathbb{P}_{m-1}$, and let $Q \neq Q^\eta$ be any positive rule with $R[\mathbb{P}_{m-1}] = \{0\}$. Denoting the first Peano kernels of Q^η and Q by K_1^η and K_1, respectively, we have

$$K_1^\eta(\eta - 0) < K_1(\eta - 0) \leq K_1(\eta + 0) < K_1^\eta(\eta + 0). \tag{6.28}$$

PROOF. We apply Lemma 6.4.1 to

$$S := Q - Q^\eta = R^\eta - R$$

and count the sign changes of the first Peano kernel K_1^S of S. Note that K_1^S is a piecewise constant function that has jumps in the positive direction at nodes of Q^η and jumps in the negative direction at nodes of Q (at least when they are not common nodes). We refer to a sign change as "up" if it goes from negative to positive, and "down" otherwise. From the structure of K_1^S it is clear that ups are possible only at interior nodes of Q^η. Let l be the number of ups. If Q^η has k boundary nodes, then there are at most $l-1+k$ downs. This can be seen as follows: If there is no node of Q^η at a, then we have $K_1^S(a+0) = -K_1(a+0) \leq 0$ owing to the positivity of Q. Hence there is no down to the left of the first up. Furthermore, we have a down between two ups. Thus, there can be at most $2l - 1 + k$ sign changes. We now use Lemma 6.4.1 to obtain

$$2l - 1 + k \geq m - 1.$$

Suppose that Q^η has r interior nodes. Then

$$2(l - r) + 2r + k \geq m, \quad \text{i.e.} \quad 2(l - r) \geq m - \operatorname{ind} Q^\eta.$$

Since $\operatorname{ind} Q^\eta \in \{m, m+1\}$, it follows that $l \geq r$. We have thus shown that there is an up at each interior node of Q^η, which is what the theorem asserts. □

Before drawing further conclusions from this result, we point out that it has an alternative, equivalent formulation:

THEOREM 6.4.1 (alternative formulation). Under the same assumptions as in the original Theorem 6.4.1 above, let x_ν and a_ν be, respectively, the zeros and coefficients of Q, and let x_ν^η and a_ν^η be those of Q^η. Then

$$\sum_{x_\nu^\eta < \eta} a_\nu^\eta < \sum_{x_\nu < \eta} a_\nu < \sum_{x_\nu^\eta \leq \eta} a_\nu^\eta. \tag{6.29}$$

In formula (6.28), Q may be chosen arbitrarily; in particular, it can be chosen so that $\|K_1\|$ is arbitrarily close to 0. This yields

$$K_1^\eta(\eta - 0) < 0 < K_1^\eta(\eta + 0),$$

or

$$-\int_a^{x_\nu^\eta} w(u)\,du + \sum_{\mu=1}^{\nu-1} a_\mu^\eta < 0 \qquad \text{and} \qquad -\int_a^{x_\nu^\eta} w(u)\,du + \sum_{\mu=1}^{\nu} a_\mu^\eta > 0,$$

and this also holds for $x_\nu^\eta \in \{a, b\}$ (with equality instead of inequality in the trivial cases, where either none or all of the coefficients are involved). We conclude that

$$\int_a^{x_\nu^\eta} w(u)\,du \le \sum_{\mu=1}^{\nu} a_\mu^\eta \le \int_a^{x_{\nu+1}^\eta} w(u)\,du \quad \text{for } \nu = 1, \ldots, n \tag{6.30}$$

(where $x_{n+1} = b$), with strict inequality if an interior node is the limit of integration.

Now, we can easily show the following:

THEOREM 6.4.2. *Let Q_n be a Krein-type rule with nodes x_ν and coefficients a_ν. Then there exist numbers y_ν with*

$$a = y_0 \le x_1 \le y_1 \le x_2 \le \cdots \le x_n \le y_n = b \tag{6.31}$$

and

$$a_\nu = \int_{y_{\nu-1}}^{y_\nu} w(u)\,du.$$

This means that Q_n is a Riemann–Stieltjes sum, i.e.

$$Q_n[f] = \sum_{\nu=1}^{n} f(x_\nu) \int_{y_{\nu-1}}^{y_\nu} w(u)\,du.$$

PROOF. Choose y_ν so that

$$\sum_{\mu=1}^{\nu} a_\mu = \int_a^{y_\nu} w(u)\,du,$$

which is possible because $w > 0$ and Q_n is exact for constant functions (y_ν is not necessarily unique if w is zero on certain intervals). Now (6.31) is equivalent to (6.30), and the other assertions are trivial consequences. □

A simple implication of the above theorem is that

$$a_\nu \le \int_{x_{\nu-1}}^{x_{\nu+1}} w(u)\,du, \tag{6.32}$$

a fact which is sometimes useful. To study Peano kernels further, we need a counterpart of Lemma 6.4.1.

LEMMA 6.4.2. *The Peano kernel K_r of a quadrature rule with l nodes in $]a, b[$ and k nodes in $\{a, b\}$ has at most $2l + k - r$ sign changes.*

PROOF. First, consider $r = 1$. Between any two consecutive nodes, K_1 is decreasing and so has at most one sign change. Hence, away from nodes, K_1 can have at most $l + k - 1$ sign changes. Additional sign changes may occur only at the l interior nodes. Therefore, the number of sign changes of K_1 is at most $2l + k - 1$.

Now consider $r \ge 2$, and suppose that K_r has $2l + k - r + s$ sign changes. Then, owing to the multiple zeros at the boundary, K_r has $2l + r + s$ zeros on $[a, b]$.

According to Rolle's theorem, K_2 must have at least $2l+2+s$ zeros on $[a,b]$, with $2l+s+k-2$ of them in $]a,b[$. Restricted to $[x_\nu, x_{\nu+1}] \subset\,]a,b[$, K_2 is convex and therefore has at most two sign changes. Additionally, K_2 can have at most k sign changes on the boundary intervals; hence it has at most $2(l-1)+k$ sign changes, so that $s \le 0$, which is what we needed to prove. □

A comparison of Lemmas 6.4.1 and 6.4.2 gives the following result.

THEOREM 6.4.3. The Peano kernel K_r has

(i) $2n-r$ sign changes for the Gaussian rule Q_n^{G};
(ii) $2n-r-1$ sign changes for the Radau rules $Q_n^{\mathrm{Ra},a}$ and $Q_n^{\mathrm{Ra},b}$;
(iii) $2n-r-2$ sign changes for the Lobatto rule Q_n^{Lo}.

The theorem says that the Peano kernel of maximal order for the above-mentioned rules has no sign change, i.e. the rules are definite. From this, we derive the following classical error representations.

THEOREM 6.4.4. We have

$$R_n^{\mathrm{G}}[f] = \frac{1}{\delta_n^2} \cdot \frac{f^{(2n)}(\xi)}{(2n)!},$$

$$R_n^{\mathrm{Ra},a}[f] = -\frac{q_n(a)}{\delta_n \delta_{n-1} q_{n-1}(a)} \cdot \frac{f^{(2n-1)}(\xi)}{(2n-1)!},$$

$$R_n^{\mathrm{Ra},b}[f] = -\frac{q_n(b)}{\delta_n \delta_{n-1} q_{n-1}(b)} \cdot \frac{f^{(2n-1)}(\xi)}{(2n-1)!}$$

and

$$R_n^{\mathrm{Lo}}[f] = \frac{q_n(a)q_{n-1}(b) - q_n(b)q_{n-1}(a)}{q_{n-1}(a)q_{n-2}(b) - q_{n-1}(b)q_{n-2}(a)} \cdot \frac{1}{\delta_n \delta_{n-2}} \cdot \frac{f^{(2n-2)}(\xi)}{(2n-2)!}$$

for some $\xi \in [a,b]$ (which is unknown and generally different for the different quadrature rules). The coefficient of the derivative is positive for Q_n^{G} and $Q_n^{\mathrm{Ra},a}$ and negative for $Q_n^{\mathrm{Ra},b}$ and Q_n^{Lo}.

PROOF. According to Theorem 4.2.6, a quadrature rule which is definite of order j satisfies

$$R[f] = R[g] \cdot \frac{f^{(j)}(\xi)}{j!},$$

where $g \in \mathbb{P}_j$ has leading coefficient 1. To calculate $R[g]$, we choose a g with $Q[g] = 0$. In the Gaussian case take $g = q_n^2/\delta_n^2$; in the case of $Q_n^{\mathrm{Ra},a}$ take

$$g(x) = \frac{[q_{n-1}(a)q_n(x) - q_n(a)q_{n-1}(x)]^2}{x-a} \cdot \frac{1}{\delta_n^2 q_{n-1}^2(a)},$$

and in the case of $Q_n^{\mathrm{Ra},b}$ take the analogue; finally, for the Lobatto case, choose

$$g(x) = \frac{S_n^2(x)}{(x-a)(x-b)} \cdot \frac{1}{\delta_n^2} \cdot \frac{1}{[q_{n-1}(a)q_{n-2}(b) - q_{n-2}(a)q_{n-1}(b)]^2},$$

where S_n is as defined in Theorem 6.1.7. The Christoffel–Darboux formula (A.4) can be used to calculate $I[g]$. In the Lobatto case, the calculations in Theorem 6.1.7 should be taken into account. The details are left to the reader. The assertion regarding the signs of the coefficients follows from the relation $(-1)^n \operatorname{sgn} q_n(a) = \operatorname{sgn} q_n(b) = 1$. □

Another major result on Peano kernels is the following:

THEOREM 6.4.5 (Markoff's theorem). Let K_m be the Peano kernel of a positive quadrature rule Q with $R[\mathbb{P}_{m-1}] = \{0\}$. For $m = 2p$,

$$K_m^{\mathrm{Lo}}(x) \le K_m(x) \le K_m^{\mathrm{G}}(x) \tag{6.33}$$

where K_m^{Lo} and K_m^{G} are the Peano kernels of R_{p+1}^{Lo} and R_p^{G}, respectively. For $m = 2p-1$,

$$K_m^{\mathrm{Ra},b}(x) \le K_m(x) \le K_m^{\mathrm{Ra},a}(x) \tag{6.34}$$

where $K_m^{\mathrm{Ra},b}$ and $K_m^{\mathrm{Ra},a}$ are the Peano kernels of $R_p^{\mathrm{Ra},b}$ and $R_p^{\mathrm{Ra},a}$, respectively.

PROOF. Let $m = 2p$. If $f^{(m)}$ is strictly positive, then $\mathbb{P}_{m-1} \cup \{f\}$ spans a Haar space. In particular, if $f+p$ were to have $m+1$ zeros, Rolle's theorem would yield a zero of $(f+p)^{(m)} = f^{(m)}$, which does not exist. Therefore, we may apply Theorem 3.4.4 and obtain that for $Q \notin \{Q_p^{\mathrm{G}}, Q_{p+1}^{\mathrm{Lo}}\}$,

$$\min\left\{Q_p^{\mathrm{G}}[f], Q_{p+1}^{\mathrm{Lo}}[f]\right\} < Q[f] < \max\left\{Q_p^{\mathrm{G}}[f], Q_{p+1}^{\mathrm{Lo}}[f]\right\},$$

that is,

$$\max\left\{R_p^{\mathrm{G}}[f], R_{p+1}^{\mathrm{Lo}}[f]\right\} > R[f] > \min\left\{R_p^{\mathrm{G}}[f], R_{p+1}^{\mathrm{Lo}}[f]\right\}.$$

According to the sign considerations in Theorem 6.4.4, this means that

$$R_p^{\mathrm{G}}[f] > R[f] > R_{p+1}^{\mathrm{Lo}}[f],$$

or

$$R_p^{\mathrm{G}}[f] - R[f] > 0 > R_{p+1}^{\mathrm{Lo}}[f] - R[f].$$

Theorem 4.2.6(iv) then gives the definiteness, so that Theorem 4.2.6(i) shows (6.33). The inequalities (6.34) can be proved analogously. □

Upon applying (6.33) to $Q = Q_q^{\mathrm{G}}$ with $q > p$, we obtain

$$R_q^{\mathrm{G}}[f] \le R_p^{\mathrm{G}}[f]$$

if $f^{(2p)} \ge 0$. A consequence of this is the following:

THEOREM 6.4.6 (Stenger 1966; Brass 1978b). If $f^{(2n)} \ge 0$ for all n, then the sequence of $R_n^{\mathrm{G}}[f]$ is decreasing.

Strauss (1984) proved similar results for Gauss-type rules corresponding to subspaces of spline functions.

Before turning to results that concern all positive quadrature rules, we state an asymptotic result which combines Theorems 5.8.1 and 6.2.7.

THEOREM 6.4.7 (Petras 1988). Suppose that w satisfies Assumption P (stated just before Theorem 6.2.5) on all intervals $[c,d] \subset]{-1},1[$, and let (Q_n) be a sequence of Krein-type rules. Then

$$K_s\big|_{]x_\nu, x_{\nu+1}[} = (x_\nu - x_{\nu+1})^s w(z_\nu) B_s\left(\frac{x - x_\nu}{x_{\nu+1} - x_\nu}\right) + o(n^{-s}),$$

where $z_\nu = \dfrac{x_\nu + x_{\nu+1}}{2}$.

This result does not merely give us information about Krein rules; in conjunction with the results that follow, it will also provide asymptotic information about the maximal size of Peano kernels of rules from class $\mathbf{Q}^+(I, \mathbb{P}_m)$ (the class of positive quadrature rules exact on $\mathbb{P}_m$). We have already seen from Theorem 6.4.5 that all the $(m+1)$st Peano kernels of rules from $Q \in \mathbf{Q}^+(I, \mathbb{P}_m)$ are enclosed by those of the Lobatto and Gaussian rules or of the Radau rules. Furthermore, the local suprema of the first Peano kernels of Krein-type rules give an enclosure for the rules from $Q \in \mathbf{Q}^+(I, \mathbb{P}_m)$ (see the Markoff–Stieltjes theorem, Theorem 6.4.1). These results can be generalized to a certain extent. For second Peano kernels, in particular, we get a rather explicit result.

THEOREM 6.4.8 (Förster 1982c). *Let Q^η be a quadrature rule of Krein type with node η for the space $\mathbb{P}_{2m-1}$. Denote by K_2^η the second Peano kernel of Q^η and by K_2 the second Peano kernel of $Q \in \mathbf{Q}^+(I, \mathbb{P}_{2m-1})$. Then*

$$K_2(\eta) \le K_2^\eta(\eta).$$

PROOF. In the proof of Theorem 6.4.1, we saw that the first Peano kernel K_1^S of the functional

$$S := Q - Q^\eta = R^\eta - R$$

changes sign at η from negative to positive. It is, moreover, monotonically decreasing between any two nodes of Q_η, where we consider -1 and 1 to be nodes also (potentially with zero coefficient). Hence $K_2^S = K_2^\eta - K_2$ (which is negative the antiderivative of K_1^S) is convex between each pair of nodes. This shows that K_2^S has at most $2m-1$ sign changes if Q^η differs from Q_m^{G} and Q_{m+1}^{Lo}, and has at most $2m-2$ sign changes otherwise. If $K_2^S(\eta)$ were negative, two sign changes would vanish. Since Lemma 6.4.1 says that K_2^S has at least $2m-2$ sign changes, we have proved the positivity of $K_2^S(\eta)$ and therefore the theorem. □

By an analogous argument, one can show that the theorem also holds with $\mathbb{P}_{2m-1}$ in place of $\mathbb{P}_{2m}$.

The situation becomes more complicated if we consider lower bounds for second Peano kernels of general positive rules or Peano kernels of order $3, \dots, 2m-1$. In these cases, we know only that the maximal and minimal values at a point in $[-1,1]$ are attained by a Krein-type rule, but not which one.

THEOREM 6.4.9 (Petras 1992a). *Denote by $\mathbf{Q}^{\mathrm{Kr},2m+1}$ the set of Krein rules with index at most $2m+1$ which are exact for $\mathbb{P}_{2m-1}$. For $2 \le s \le 2m$ and all $x \in [-1,1]$, we have*

$$\inf_{Q \in \mathbf{Q}^+(I, \mathbb{P}_{2m-1})} K_s(Q, x) = \inf_{Q \in \mathbf{Q}^{\mathrm{Kr},2m+1}} K_s(Q, x)$$

and

$$\sup_{Q \in \mathbf{Q}^+(I, \mathbb{P}_{2m-1})} K_s(Q, x) = \sup_{Q \in \mathbf{Q}^{\mathrm{Kr},2m+1}} K_s(Q, x).$$

Again, an analogous result holds for Peano kernels of rules from $\mathbf{Q}^+(I, \mathbb{P}_{2m})$. We shall prove the theorem via one-sided approximation of the functions

$$h_{z,s}(x) := \frac{(x-z)_+^{s-1}}{(s-1)!}.$$

Note that $K_s(Q, z) = R[h_{z,s}]$. The main idea is that if the best one-sided approximation $p \in \mathbb{P}_{2m-1}$ coincides with $h_{z,s}$ at the nodes of a positive quadrature rule which is exact for $\mathbb{P}_{2m-1}$, then the error is $I[h_{z,s} - p]$. As in Section 3.3, a "best" one-sided approximation will always be understood in regard to a functional I; here we choose

$$I[f] = \int_{-1}^{1} f(x)w(x)\,dx.$$

First, we consider the trivial cases.

LEMMA 6.4.3. Denote by x_ν^{G} the nodes of Q_m^{G}. Let $s \le 2m$ and $z > x_1^{\mathrm{G}}$. Then a best one-sided approximation of $h_{z,s}$ from $\mathbb{P}_{2m-1}$ may have infinitely many points in common with $h_{z,s}$ if and only if $z \ge x_m^{\mathrm{G}}$ and the approximation is from below. In this case the approximation is identically zero.

PROOF. Let $z \ge x_m^{\mathrm{G}}$ and let $p \in \mathbb{P}_{2m-1}$ be a best one-sided approximation of $h_{z,s}$ from below. Then $I[p] = Q_m^{\mathrm{G}}[p] \le 0$, and $\widetilde{p}(x) \equiv 0$ is a best one-sided approximation as well.

Now suppose $z < x_m^{\mathrm{G}}$. If x^* is chosen so that $z < x^* < x_m^{\mathrm{G}}$, we can find an $\varepsilon > 0$ such that

$$p \in \mathbb{P}_{2m-1}, \qquad p(x) = \varepsilon(x - x^*) \prod_{\nu=1}^{m-1} (x - x_\nu^{\mathrm{G}})^2, \tag{6.35}$$

is a one-sided approximation of $h_{z,s}$ from below and satisfies the condition $I[p] = Q_m^{\mathrm{G}}[p] = a_m^{\mathrm{G}} p(x_m^{\mathrm{G}}) > 0$. Hence $\widetilde{p}(x) \equiv 0$ is not a best one-sided approximation.

Thus, the best approximation can be zero only under the conditions given in the statement of the lemma. Furthermore, $q_{z,s}$ defined by

$$q_{z,s}(x) := \frac{(x - z)^{s-1}}{(s-1)!}$$

may be the best approximation from below if s is even and the best approximation from above if s is odd. We would then subtract $q_{z,s}$ from $h_{z,s}$ and consider $\widetilde{w}(x) = w(-x)$. If $q_{z,s}$ were the solution of the original approximation problem, the zero function would be the solution of the transformed problem, and this case has already been treated above. If the best approximation is not identically zero and not equal to $q_{z,s}$, then it can coincide with $h_{z,s}$ only at finitely many points. □

Obviously, corresponding results also hold when $x < x_1^{\mathrm{G}}$.

PROOF OF THEOREM 6.4.9. We have to prove the statement for $s > 2$ and the lower bound for $s = 2$. Note that for each best one-sided approximation there is a positive quadrature formula having the zeros of the error as its nodes; see Theorem 3.3.2.

Suppose that the best one-sided approximation is not any of the trivial polynomials mentioned in the proof of Lemma 6.4.3. Furthermore, let $p \in \mathbb{P}_{2m-1}$ be a best one-sided L_1-approximation of $h_{z,s}$, and suppose that $p - h_{z,s}$ has l zeros, where zeros are counted with multiplicities. Hence $(p - h_{z,s})^{(s-2)}$ has at least $l - s + 2$ zeros. Upon differentiating twice the restrictions to the intervals $[-1, z]$ and $[z, 1]$, we see that $p^{(s)} \in \mathbb{P}_{2m-s-1}$ has at least $l - s - 2$ zeros, so that $l - s - 2 \le 2m - s - 1$, i.e. $l - 1 \le 2m$. Since each interior zero of $h_{z,s} - p$ is a multiple zero, the index of the quadrature formula corresponding to the one-sided approximation p is less than or equal to l and hence less than or equal to $2m + 1$. □

6.5. Peano kernels for $w = 1$

Throughout this section, we take $w = 1$ and $[a, b] = [-1, 1]$.

Using Theorem A.12, we can specialize Theorem 6.4.4 to this case and obtain the following result.

THEOREM 6.5.1. We have

$$R_n^{\mathrm{G}}[f] = \frac{1}{2n+1} \cdot \frac{2^{2n+1}(n!)^4}{[(2n)!]^2} \cdot \frac{f^{(2n)}(\xi)}{(2n)!},$$

$$R_n^{\mathrm{Ra},\pm 1}[f] = \mp\frac{1}{n} \cdot \frac{2^{2n+1}(n!)^4}{[(2n)!]^2} \cdot \frac{f^{(2n-1)}(\xi)}{(2n-1)!},$$

$$R_n^{\mathrm{Lo}}[f] = -\frac{2n-1}{n(n-1)} \cdot \frac{2^{2n+1}(n!)^4}{[(2n)!]^2} \cdot \frac{f^{(2n-2)}(\xi)}{(2n-2)!},$$

for some $\xi \in [-1, 1]$, which may be different in each case.

To get an idea of the size of the terms, note that

$$\frac{2^{2n+1}(n!)^4}{[(2n)!]^2} = \frac{\pi}{2^{2n}} \cdot (2n + \eta) \quad \text{with } \eta \in [0, 1], \tag{6.36}$$

which can be proved by applying Wallis's formula.

EXAMPLE 6.5.1. Gauss (1816), in a fundamental article, applied his method to

$$\int_{10^5}^{2\cdot 10^5} \frac{dx}{\ln x} =: \int_{10^5}^{2\cdot 10^5} f_0(x)\, dx.$$

He determined $Q_1^{\mathrm{G}}[f_0]$, $Q_2^{\mathrm{G}}[f_0], \ldots,$ $Q_7^{\mathrm{G}}[f_0]$ to six digits of precision and observed that all the computed digits of $Q_6^{\mathrm{G}}[f_0]$ and $Q_7^{\mathrm{G}}[f_0]$ coincided. He did not perform an error estimation. Let us see what can be achieved with Theorem 6.5.1; we start with Q_1^{G}. Transforming from $[-1, 1]$ to a general interval $[a, b]$ gives

$$\int_a^b f(x)\, dx = (b-a)\, f\left(\frac{a+b}{2}\right) + \frac{(b-a)^3}{24} f''(\xi).$$

In our case, we have

$$Q_1[f_0] = \frac{10^5}{\ln 150\,000} = 8390.39\ldots.$$

The monotonicity of the second derivative

$$f_0''(x) = \frac{1}{(x \ln x)^2}\left(1 + \frac{2}{\ln x}\right)$$

implies that

$$8390.39 + 8.13 \le \int_{10^5}^{2\cdot 10^5} \frac{dx}{\ln x} \le 8390.40 + 36.90$$

This means that the true value of the integral lies in the interval $[8398.52\,,\, 8427.30]$. By taking the midpoint of this interval as an estimate, we obtain a relative error of less than $2 \cdot 10^{-3}$, even though only one function value was used. Of course, this is due to the fact that higher derivatives of the integrand vary only slightly. We will discuss a different situation in Example 6.6.1.

For $Q_2^{\mathrm{G}}, Q_3^{\mathrm{G}}$ etc., the same analysis as for the Q_1^{G} case becomes more and more difficult as the higher derivatives become more and more complicated. These difficulties can be overcome by using the methods of automatic differentiation and interval arithmetic; see Appendix C. Automatic differentiation provides a way of computing derivatives without actually writing down explicit expressions for them. By substituting an interval $[\alpha, \beta]$ for the argument of the function and using interval arithmetic, we get enclosures for the derivatives on $[\alpha, \beta]$. Usually one would then subdivide the whole interval into smaller intervals, thus obtaining tighter bounds (for the most part). In our case, the derivatives have no sign changes, so the values at the boundary enclose all derivative values. We compute the derivatives at $y_1 = 10^5$ and $y_2 = 2 \cdot 10^5$ via automatic differentiation. The first step is to initialize the vector $u = (y_i, 1, 0, 0, \ldots)$ with $d+1$ components, where d is the number of derivatives that we want to compute. Then, the automatic differentiation rule for the logarithm (Theorem C.3) gives

$$v = (v_0, v_1, v_2, \ldots, v_d) = \left(\ln y_i, \, \frac{1}{y_i}, \, -\frac{1}{2y_i^2}, \, \frac{1}{3y_i^3}, \, \ldots \right)$$

and, finally, the automatic differentiation rule for division (Theorem C.1) yields

$$\begin{aligned} w &= (w_0, w_1, w_2, \ldots, w_d) \\ &= \left(\frac{1}{v_0}, \, -w_0 \cdot v_1 w_0, \, -w_0 \cdot (v_1 w_1 + v_2 w_0), \, -w_0 \cdot (v_1 w_2 + v_2 w_1 + v_3 w_0), \, \ldots \right) \\ &= \left(f(y_i), f'(y_i), \, \frac{f''(y_i)}{2!}, \, \frac{f'''(y_i)}{3!}, \, \ldots, \, \frac{f^{(d)}(y_i)}{d!} \right). \end{aligned}$$

For more details, see Appendix C.

Now we can use Q_n^{G} along with its error representation from Theorem 6.5.1 to obtain the enclosures tabulated below:

n	Enclosure for $I[f_0]$
1	[8398.531807, 8427.290759]
2	[8406.032838, 8407.386341]
3	[8406.237686, 8406.303848]
4	[8406.242981, 8406.246375]
5	[8406.243117, 8406.243298]
6	[8406.243120, 8406.243131]
7	[8406.243120, 8406.243122]

It is also possible to obtain bounds without computing derivatives. Specifically, it is easy to prove that

$$f_0^{(2\nu)}(x) > 0 \quad \text{for } \nu = 0, 1, \ldots.$$

According to Theorem 6.5.1, this means that Gaussian rules yield lower bounds for the integral while Lobatto rules yield upper bounds. Upon computing $Q_2^{\mathrm{Lo}}[f_0]$, we obtain

$$I[f_0] \in [8390.39, \, 8439.27].$$

With the pair Q_2^{G} and Q_3^{Lo} we find that

$$I[f_0] \in [8405.95,\ 8406.67],$$

and with the pair Q_3^{G} and Q_4^{Lo} we get

$$I[f_0] \in [8406.23,\ 8406.26].$$

Theorem 6.4.5 gave lower and upper bounds for Peano kernels of arbitrary positive quadrature rules. For the $w = 1$ case, we can add the following statement.

THEOREM 6.5.2 (Brass and Schmeisser 1981). *For a positive quadrature rule Q with $R[\mathbb{P}_{2p-1}] = \{0\}$, the Peano kernel K_{2p} satisfies*

$$|K_{2p}(x)| \le |K_{2p}^{\mathrm{Lo}}(x)|,$$

where $K_{2p}^{\mathrm{Lo}}(x)$ is the $2p$th Peano kernel of Q_{p+1}^{Lo}.

PROOF. By applying the theory of reduced functionals from Theorem 4.2.8, we shall show that

$$S := R_p^{\mathrm{G}} + R_{p+1}^{\mathrm{Lo}}$$

is definite of order $2p$, i.e. its $2p$th Peano kernel has no sign changes. Now, S is the remainder functional of a positive quadrature rule for $2I$ with nodes at the boundary; therefore it can only be negative definite, so that

$$K_{2p}^{\mathrm{G}}(x) + K_{2p}^{\mathrm{Lo}}(x) \le 0.$$

Theorem 6.4.5 then gives the asserted result.

To prove definiteness, we consider the $(2p-1)$st reduction of S according to Theorem 4.2.8,

$$S^{\mathrm{red}} := S^{(x_1, x_2, \dots, x_p, y_2, \dots, y_p)},$$

where $x_1, \dots, x_p$ are the nodes of Q_p^{G} and $y_2, \dots, y_p$ are the interior nodes of Q_{p+1}^{Lo}. It remains to show that S^{red} is definite of order one. We have

$$\prod_{\nu=1}^{p} (x - x_\nu) = \vartheta_1 P_p(x)$$

and

$$\prod_{\nu=2}^{p} (x - y_\nu) = \vartheta_2 \cdot \frac{P_{p+1}(x) - P_{p-1}(x)}{x^2 - 1} = \vartheta_3 P_p'(x)$$

for some positive numbers ϑ_i (see Corollary 6.1.3 and Theorem A.14(iv)). Therefore,

$$S^{\mathrm{red}}[f] = \vartheta_1 \vartheta_3 \biggl(2 \int_{-1}^{1} f(x) P_p(x) P_p'(x)\, dx \\ - a_1^{\mathrm{Lo}} f(-1) P_p(-1) P_p'(-1) - a_{p+1}^{\mathrm{Lo}} f(1) P_p(1) P_p'(1) \biggr).$$

We now use

$$a_1^{\mathrm{Lo}} = a_{p+1}^{\mathrm{Lo}} = \frac{2}{p(p+1)}$$

(from Corollary 6.1.3) and

$$P_p(\pm 1) = (\pm 1)^p, \qquad P_p'(\pm 1) = (\pm 1)^{p-1} \cdot \frac{p(p+1)}{2}$$

(i.e. (A.7) and Theorem A.14(i), respectively) to obtain

$$S^{\mathrm{red}}[f] = \vartheta_1\vartheta_3\Big(2\int_{-1}^{1} f(x)P_p(x)P_p'(x)\,dx + f(-1) - f(1)\Big).$$

Integration by parts gives

$$S^{\mathrm{red}}[f] = \vartheta_1\vartheta_3\Big(-\int_{-1}^{1} f'(x)P_p^2(x)\,dx\Big),$$

which establishes the definiteness. □

This result raises the obvious question, namely how large $\|K_{2p}^{\mathrm{Lo}}\|$ is. The estimates from Section 5.7 are rather poor when applied to Peano kernels of high order. We shall employ a different method, which can also be used to prove more general results.

THEOREM 6.5.3. For $p = 1, 2, \ldots$, we have

$$\sup\big\{\|K_{2p}\| \,:\, Q \text{ positive}, R[\mathbb{P}_{2p-1}] = \{0\}\big\} = \frac{\sqrt{2\pi p}}{2^{2p}(2p)!}\Big(1 + \frac{\vartheta_p}{p}\Big),$$

with $\vartheta_p \in [-3, 3]$.

PROOF. According to the previous theorem, we need only consider the Lobatto rule Q_{p+1}^{Lo}. Theorem 6.4.3 shows that $|K_{2p}^{\mathrm{Lo}}|$ has exactly one local maximum. For symmetry reasons, this is $|K_{2p}^{\mathrm{Lo}}(0)|$. In order to estimate this value, we start with the expansion

$$\frac{|x|^{2p-1}}{(2p-1)!} = r_{2p-2}(x) + \frac{4}{\pi}\sum_{\mu=p}^{\infty}\frac{(-1)^{p+\mu}T_{2\mu}(x)}{(2\mu-2p+1)(2\mu-2p+3)\cdots(2\mu+2p-1)} \tag{6.37}$$

with some $r_{2p-2} \in \mathbb{P}_{2p-2}$. To prove this for $p = 1$, we make explicit use of the formula for the Chebyshev expansion,

$$f = \frac{\alpha_0}{2} + \sum_{\lambda=1}\alpha_\lambda T_\lambda \quad \text{where } \alpha_\lambda = \frac{2}{\pi}\int_{-1}^{1}\frac{f(x)T_\lambda(x)}{\sqrt{1-x^2}}\,dx.$$

For $f(x) = |x|$, we have $\alpha_{2\mu-1} = 0$ from symmetry and

$$\alpha_{2\mu} = \frac{4}{\pi}\int_0^1 \frac{xT_{2\mu}}{\sqrt{1-x^2}}\,dx = \frac{4}{\pi}\int_0^{\pi/2}\cos t\cos 2\mu t\,dt = \frac{4(-1)^{\mu+1}}{\pi(2\mu-1)(2\mu+1)},$$

which is just (6.37) in the $p = 1$ case. For induction over p, we multiply both sides of (6.37) by x^2 and use the relation

$$4x^2T_{2\nu}(x) = T_{2\nu+2}(x) + 2T_{2\nu}(x) + T_{2\nu-2}(x),$$

which follows from two applications of

$$2xT_{2\nu}(x) = T_{2\nu+1}(x) + T_{2\nu-1}(x);$$

see Theorem A.9(ii).

Next, apply R_{p+1}^{Lo} to both sides of (6.37) and obtain

$$K_{2p}^{\mathrm{Lo}}(0) = \frac{2}{\pi}\sum_{\mu=p}^{\infty}\frac{(-1)^{p+\mu}}{(2\mu-2p+1)(2\mu-2p+3)\cdots(2\mu+2p-1)}R_{p+1}^{\mathrm{Lo}}[T_{2\mu}].$$

We use only the first summand, and estimate the remaining part as follows:

$$\begin{aligned}
&\frac{2}{\pi}\sum_{\mu=p+1}^{\infty}\frac{\|R\|}{(2\mu-2p+1)(2\mu-2p+3)\cdots(2\mu+2p-1)}\\
&\quad=\frac{2\|R\|}{\pi(4p-2)}\sum_{\mu=p+1}^{\infty}\left(\frac{1}{(2\mu-2p+1)(2\mu-2p+3)\cdots(2\mu+2p-3)}\right.\\
&\qquad\qquad\qquad\left.-\frac{1}{(2\mu-2p+3)(2\mu-2p+5)\cdots(2\mu+2p-1)}\right)\\
&\quad=\frac{8}{\pi(4p-2)}\cdot\frac{1}{3\cdot 5\cdots(4p-1)}.
\end{aligned}$$

This gives

$$K_{2p}^{\mathrm{Lo}}(0)=\frac{2}{\pi}\cdot\frac{1}{3\cdot 5\cdots(4p-1)}\left(R_{p+1}^{\mathrm{Lo}}[T_{2p}]+\frac{\vartheta}{2p-1}\right),\quad\text{with }\vartheta\in[-2,2].$$

Theorem 6.5.1 tells us that

$$\frac{\pi}{2}\cdot\frac{p+1/2}{p}\le\left|R_{p+1}^{\mathrm{Lo}}[T_{2p}]\right|\le\frac{\pi}{2}\cdot\frac{p+1/2}{p}\cdot\frac{p+3/2}{p+1},$$

and Wallis's product gives

$$\sqrt{\frac{8p}{\pi}}\cdot\frac{1}{2^{2p}(2p)!}\le\frac{2}{\pi}\cdot\frac{1}{3\cdot 5\cdots(4p-1)}\le\sqrt{\frac{8(p+1/4)}{\pi}}\cdot\frac{1}{2^{2p}(2p)!}.$$

By way of these identities, the theorem can be proved without difficulty. □

To deal with convex integrands, we will need pointwise bounds for the second Peano kernels of positive quadrature rules. The derivation of these bounds is based mainly on our results for the Gaussian rule.

THEOREM 6.5.4 (Förster and Petras 1990b). *For every positive quadrature rule Q with $R[\mathbb{P}_{2p-1}]=\{0\}$, we have*

$$|K_2(x)|\le\frac{8\pi^2}{(2p+1)^2}(1-x^2).$$

PROOF. Define

$$D_2:=K_2^{\mathrm{G}}-K_2$$

where K_2^{G} is the second Peano kernel of Q_p^{G}. We shall show that

$$|D_2(x)|\le\frac{4\pi^2}{(2p+1)^2}(1-x^2).\tag{6.38}$$

This implies

$$|K_2^{\mathrm{G}}(x)|\le\frac{4\pi^2}{(2p+1)^2}(1-x^2),\tag{6.39}$$

because $\|K_2\|$ can be arbitrarily small for appropriate Q. Inequalities (6.38) and (6.39) together give the theorem. Without loss of generality, we shall prove (6.38) for $x\ge 0$ (in the $x<0$ case, the result follows from the "reflected" rule).

(i) For simplicity, we write x_ν for x_ν^{G} and a_ν for a_ν^{G}; we also set $x_0=-1$ and $x_{p+1}=1$. From Theorem 6.4.8 we have

$$D_2(x_\nu)\ge 0\quad\text{for }\nu=1,\ldots,p.\tag{6.40}$$

(ii) For the function $D_1 := K_1^{\mathrm{G}} - K_1$, Theorem 6.4.1 gives

$$D_1(x_\nu + 0) = K_1^{\mathrm{G}}(x_\nu + 0) - K_1(x_\nu + 0) \le K_1^{\mathrm{G}}(x_\nu + 0) - K_1^{\mathrm{G}}(x_\nu - 0) = a_\nu$$

and, analogously,

$$D_1(x_{\nu+1} - 0) \ge -a_{\nu+1}. \tag{6.41}$$

Since D_1 is decreasing on $]x_\nu, x_{\nu+1}[$, we have

$$-a_{\nu+1} \le D_1(x \pm 0) \le a_\nu \quad \text{for } x \in\,]x_\nu, x_{\nu+1}[. \tag{6.42}$$

(iii) Theorem 6.3.2(i) shows that for $\nu \ge (p+1)/2$,

$$x_{\nu+1} - x_\nu = \cos\theta_\nu - \cos\theta_{\nu+1} = 2\sin\frac{\theta_{\nu+1}+\theta_\nu}{2}\sin\frac{\theta_{\nu+1}-\theta_\nu}{2} \le \frac{\pi}{p+1/2}\sin\theta_\nu.$$

Theorem 6.3.4 then gives

$$a_\nu(x_{\nu+1} - x_\nu) \le \left(\frac{\pi}{p+1/2}\right)^2 (1 - x_\nu^2). \tag{6.43}$$

(iv) Using parts (i), (ii) and (iii), we prove the upper bound for D_2 as follows.

Let $x \in [x_\nu, x_{\nu+1}]$, $\nu < n$ and $x \ge 0$. The positivity of Q implies that D_2 is convex between any two consecutive nodes of Q_p^{G} and is non-negative on both $[x_0, x_1]$ and $[x_n, x_{n+1}]$. Therefore, according to Lemma 6.4.1, it is only possible for D_2 to have at least $2n-2$ sign changes if it has two sign changes on every $[x_\mu, x_{\mu+1}]$, for $\mu = 1, \ldots, n-1$.

Taking (6.42), (6.43) and $\xi \in [x_\nu, x_{\nu+1}]$, we obtain

$$D_2(x) = \int_x^\xi D_1(u)\,du \le |\xi - x| a_\nu \le a_\nu(x_{\nu+1} - x_\nu) \le \left(\frac{\pi}{p+1/2}\right)^2 (1 - x_\nu^2).$$

In particular,

$$D_2(x_\nu) \le \left(\frac{\pi}{p+1/2}\right)^2 (1 - x_\nu^2). \tag{6.44}$$

Since D_2 is convex on $x \in [x_\nu, x_{\nu+1}]$ and $(1 - x^2)$ is concave, this establishes the upper bound for arbitrary x.

(v) Again, let $x \in [x_\nu, x_{\nu+1}]$. The inequalities (6.40) and (6.41) give

$$D_2(x) = D_2(x_{\nu+1}) + \int_x^{x_{\nu+1}} D_1(u)\,du \ge -(x_{\nu+1} - x)a_{\nu+1} =: u(x).$$

We have to show that this is greater than $v(x) := -4\pi^2(1-x^2)/(2p+1)^2$. Since v is convex and $u \in \mathbb{P}_1$, we only need to prove the inequality for $x \in \{x_\nu, x_{\nu+1}\}$. For $x = x_{\nu+1}$ it is trivial, whereas (6.43) in conjunction with the monotonicity of the coefficients described in Corollary 6.3.1 gives

$$u(x_\nu) \ge -(x_{\nu+1} - x_\nu)a_{\nu+1} \ge -(x_{\nu+1} - x_\nu)a_\nu \ge -\left(\frac{\pi}{p+1/2}\right)^2 (1 - x_\nu^2) = v(x_\nu),$$

which verifies the inequality for the $x = x_\nu$ case. □

Applying the preceding results to Theorem 4.10.2 leads to the corollary below.

COROLLARY 6.5.1 (Förster and Petras 1990b). For the modified estimation framework of Section 4.10, we have

$$\rho(Q_n^{\mathrm{G}}, \mathcal{C}_M^{FP}) \le \frac{8\pi^2 M}{(2n+1)^2}. \tag{6.45}$$

For every positive quadrature rule Q with $R[\mathbb{P}_{2p-1}] = \{0\}$, we have

$$\rho(Q, \mathcal{C}_M^{FP}) \le \frac{16\pi^2 M}{(2p+1)^2}.$$

It was proved in Petras (1993c) that the constant $8\pi^2$ on the right-hand side of (6.45) can be replaced by $2\pi^2/3$ and that $\lim_{n\to\infty} (2n+1)^2 \rho(Q_n^{\mathrm{G}}, \mathcal{C}_M^{FP}) = 2\pi^2 M/3$.

The inequalities for K_2 from Theorem 6.5.4 can be extended by induction to Peano kernels of higher order.

THEOREM 6.5.5 (Petras 1992b). There exist constants c_r such that

$$|K_r(x)| \le \frac{c_r}{(2p+1)^r}\left(1-x^2\right)^{r/2}, \quad r = 2, 3, \ldots, \tag{6.46}$$

holds for Peano kernels K_r of any positive quadrature rule Q with $R[\mathbb{P}_{2p-1}] = \{0\}$.

We remark that (6.46) does not necessarily hold for $r = 1$, since there are quadrature rules with a node at -1, implying that $K_1(-1+0) \neq 0$. Theorems 5.7.1 and 5.7.2 give substitute estimates for this case.

PROOF. Let D_r be the rth Peano kernel of $S := R_p^{\mathrm{G}} - R$. By the proof of Theorem 6.5.4(ii), D_2 has exactly $2p$ zeros,

$$-1 = y_0^{(2)} < y_1^{(2)} < \cdots < y_{2p-1}^{(2)} = 1,$$

and we have

$$x_{\lceil \nu/2 \rceil}^{\mathrm{G}} \le y_\nu^{(2)} \le x_{\lceil \nu/2 \rceil + 1}^{\mathrm{G}} \qquad (\text{with } x_0^{\mathrm{G}} = -1,\ x_{p+1}^{\mathrm{G}} = 1). \tag{6.47}$$

From Rolle's theorem and Lemma 6.4.1, we obtain the existence of exactly $2p-r+2$ zeros, which we denote by $y_\nu^{(r)}$. These zeros satisfy

$$-1 = y_0^{(r+1)} < y_1^{(r)} < y_1^{(r+1)} < y_2^{(r)} < \cdots < y_{2p-r}^{(r)} < y_{2p-r}^{(r+1)} = 1.$$

Combining the above with (6.47), we get

$$x_{\lceil \nu/2 \rceil}^{\mathrm{G}} \le y_\nu^{(r)} \le x_{\lceil (\nu+r)/2 \rceil}^{\mathrm{G}}. \tag{6.48}$$

We start the inductive proof of (6.46) with the $r = 2$ case, which is just Theorem 6.5.4. Let $x \in [y_k^{(r)}, y_{k+1}^{(r)}]$ and, without loss of generality, take $x = -\cos\theta \in [-1, 0]$. Then

$$|K_{r-1}(x)| \le \frac{c_{r-1}}{(2p+1)^{r-1}}\left(1-x^2\right)^{(r-1)/2}$$

implies

$$|K_r(x)| \le \left| \int_{y_k^{(r)}}^x K_{r-1}(t)\, dt \right| \le \left(x - y_k^{(r)}\right) \cdot \frac{c_{r-1}}{(2p+1)^{r-1}} \cdot \left(1-x^2\right)^{(r-1)/2}. \tag{6.49}$$

Setting $y_k^{(r)} = -\cos\theta_k^{(r)}$ and using (6.48) together with Theorem 6.3.2, we obtain

$$\begin{aligned}
x - y_k^{(r)} &= \cos\theta_k^{(r)} - \cos\theta \\
&= 2\sin\frac{\theta+\theta_k^{(r)}}{2}\sin\frac{\theta-\theta_k^{(r)}}{2} \\
&\le (\theta - \theta_k^{(r)})\sin\theta \\
&\le (\theta_{k+1}^{(r)} - \theta_k^{(r)})\sin\theta \\
&\le \big(\theta_{\lceil (k+r+1)/2\rceil} - \theta_{\lceil k/2\rceil}\big)\sin\theta \\
&\le \frac{r+3}{2}\cdot\frac{\pi}{p+1/2}\cdot\sin\theta \\
&= \frac{(r+3)\pi}{2p+1}\cdot\sqrt{1-x^2}.
\end{aligned}$$

Plugging this into (6.49) completes the induction. □

6.6. Error bounds

In this section, we use the basic interval $[-1,1]$ and start with bounds based on the modulus of continuity. Such bounds are of interest because they are applicable to every continuous function. In practice, however, they often overestimate by a huge amount. Lebesgue's inequality, together with (5.4), shows that

$$\rho\big(Q_n^{\mathrm{G}}, C_\omega^{\mathrm{mc}}\big) \le \|R_n^{\mathrm{G}}\|\cdot\omega\left(\frac{\pi}{2n}\right) \tag{6.50}$$

for arbitrary weight functions. In the case of $w=1$, we have equality in Proinov's bound from Theorem 4.5.5:

$$\rho\big(Q_n^{\mathrm{G}}, C_\omega^{\mathrm{mc}}\big) = \int_{-1}^{1}\omega\big(|K_1(x)|\big)\,dx$$

(Petras 2005). In general, the numerical evaluation of this integral is difficult. More convenient is the result of Ligun, Theorem 4.5.6. In conjunction with Theorem 6.6.2 below, it says that

$$\rho\big(Q_n^{\mathrm{G}}, C_\omega^{\mathrm{mc}}\big) \le 2\,\omega\left(\frac{\pi^2}{16n}\right) \tag{6.51}$$

for a concave ω.

Next, we turn to error bounds involving the total variation. Combining Theorem 4.4.4 with Theorem 5.7.1 gives

$$\rho\big(Q_n^{\mathrm{G}}, C_M^{\mathrm{Var}}\big) \le \frac{\pi M}{2n}\sup_{-1\le x\le 1} w(x)\sqrt{1-x^2}. \tag{6.52}$$

This bound is optimal when $w(x) = \big(1-x^2\big)^{-1/2}$; see Example 5.5.2 (in this case the Polya rule is just the Gaussian rule). For $w=1$, a minor improvement is possible (see Förster and Petras 1991):

$$\rho\big(Q_n^{\mathrm{G}}, C_1^{\mathrm{Var}}\big) = \begin{cases} \dfrac{1}{2}a_{(n+1)/2,n}^{\mathrm{G}} & \text{for odd } n, \\[2ex] \big|x_{n/2,n}^{\mathrm{G}}\big| & \text{for even } n. \end{cases}$$

In order to compare this result with (6.52), note that we have the explicit representation

$$a^{\mathrm{G}}_{(n+1)/2,n} = 2 \prod_{\nu=1}^{(n-1)/2} \left(\frac{2\nu}{2\nu+1}\right)^2,$$

which follows from Corollary 6.1.1 and the definition of Legendre polynomials (see Definition A.4). For odd n, we can deduce that

$$\frac{M\pi}{2n+2} \le \rho\big(Q_n^{\mathrm{G}}, \mathcal{C}_M^{\mathrm{Var}}\big) \le \frac{M\pi}{2n+1},$$

and for even n a similar result can be obtained.

Let us now consider the classical co-observation

$$\mathcal{C}_M^{(2n)} = \{f : \|f^{(2n)}\| \le M\}.$$

Markoff (1885) was already able to show that

$$\rho\big(Q_n^{\mathrm{G}}, \mathcal{C}_M^{(2n)}\big) = \frac{M}{\delta_n^2 (2n)!},$$

which follows immediately from Theorem 6.4.4. For $w = 1$, this means that

$$\rho\big(Q_n^{\mathrm{G}}, \mathcal{C}_M^{(2n)}\big) = \frac{2^{2n+1}(n!)^4}{[(2n)!]^2(2n+1)} \cdot \frac{M}{(2n)!}, \tag{6.53}$$

and for $w(x) = \left(1-x^2\right)^{-1/2}$ we get

$$\rho\big(Q_n^{\mathrm{G}}, \mathcal{C}_M^{(2n)}\big) = \frac{\pi}{2^{2n-1}} \cdot \frac{M}{(2n)!}. \tag{6.54}$$

Often, though not always, these error bounds lead to useful results.

EXAMPLE 6.6.1. Let $w = 1$ and

$$f_d(x) = \frac{1}{x^2 + d^2}.$$

It is easily checked that

$$\|f_d^{(2n)}\| = |f^{(2n)}(0)| = \frac{(2n)!}{|d|^{2n+2}},$$

so (6.53) gives

$$\big|R_n^{\mathrm{G}}[f_d]\big| \le \frac{2^{2n+1}(n!)^4}{[(2n)!]^2(2n+1)d^{2n+2}} = \frac{\pi}{d^2(2d)^{2n}}\big(1 + O(n^{-1})\big).$$

For $|d| < 1/2$, this error bound tends to infinity exponentially fast; on the other hand, the error itself tends to zero exponentially fast, because f_d is analytic.

In a situation as in the example above, error estimates involving lower-order derivatives have much better properties. If we do not wish to put restrictions on the weight function, we are left with the approximation-theoretic method. The bound (5.6) gives

$$\rho\big(Q_n^{\mathrm{G}}, \mathcal{C}_M^{(r)}\big) \le \mathcal{K}_r \cdot \frac{(2n-r)!}{(2n)!} \|R_n^{\mathrm{G}}\| M, \tag{6.55}$$

where $\mathcal{K}_r$ is the Favard constant. With $w = 1$ and $r = 1$, we obtain

$$\rho(Q_n^{\mathrm{G}}, \mathcal{C}_M^{(1)}) \le \frac{\pi}{n} M.$$

Below, we will improve this result to

$$\rho(Q_n^{\mathrm{G}}, \mathcal{C}_M^{(r)}) \le \frac{\pi^2}{8n} M,$$

but that will require much greater technical effort.

The quality of these error bounds can be judged with the following theorem.

THEOREM 6.6.1. *Suppose that w satisfies Assumption P from Section 6.2 for all $[c,d] \subset]-1,1[$. Then*

$$\lim_{n\to\infty} n^r \rho(Q_n^{\mathrm{G}}, \mathcal{C}_M^{(r)}) = M \cdot \pi^r \int_0^1 |B_r(y)|\, dy \cdot \int_{-1}^1 w(x)\big(1-x^2\big)^{r/2}\, dx.$$

PROOF. This assertion is simply a consequence of Corollary 5.8.1 and Theorem 6.2.7. □

Theorem 6.6.1 is applicable, for instance, to $w = 1$ and yields

$$\rho(Q_n^{\mathrm{G}}, \mathcal{C}_M^{(1)}) = \frac{\pi^2}{8n} \cdot M \cdot \big(1 + o(1)\big),$$

$$\rho(Q_n^{\mathrm{G}}, \mathcal{C}_M^{(2)}) = \frac{2\sqrt{3}\,\pi^2}{81n^2} \cdot M \cdot \big(1 + o(1)\big),$$

$$\rho(Q_n^{\mathrm{G}}, \mathcal{C}_M^{(3)}) = \frac{\pi^4}{512n^3} \cdot M \cdot \big(1 + o(1)\big).$$

Asymptotically sharp bounds for these cases were proved by Petras (1989, 1993c, 1999c). We illustrate the techniques used by considering the simplest of the three cases.

THEOREM 6.6.2 (Petras 1989). *For $w = 1$, we have*

$$\rho(Q_n^{\mathrm{G}}, \mathcal{C}_M^{(1)}) \le \frac{\pi^2}{8n} \cdot M.$$

PROOF. Take $x_\nu = -\cos\theta_\nu$ with $\theta_\nu \in [0,\pi]$, $x_0 = -1$, $x_{n+1} = 1$ and $N = n + 1/2$. Let n be even; the case of odd n can be treated similarly.

Between a pair of nodes, K_1 is a line with slope -1. Furthermore, Theorem 6.4.3(i) for $r = 1$ implies that K_1 has a zero on each interval between two nodes (i.e. that Q_n^{G} is a Riemann sum). We obtain the representation

$$\int_{x_\nu}^{x_{\nu+1}} |K_1(x)|\, dx = \frac{1}{4}(x_{\nu+1} - x_\nu)^2 + K_1^2\left(\frac{x_\nu + x_{\nu+1}}{2}\right).$$

Hence,

$$\rho(Q_n^{\mathrm{G}}, \mathcal{C}_1^{(1)}) = \int_{-1}^1 |K_1(x)|\, dx = \frac{1}{4}\sum_{\nu=0}^{n}(x_{\nu+1} - x_\nu)^2 + \sum_{\nu=0}^{n} K_1^2\left(\frac{x_\nu + x_{\nu+1}}{2}\right).$$

We shall show that

$$S(x_0, x_1, \ldots, x_{n+1}) := \sum_{\nu=0}^{n} \frac{1}{4}(x_{\nu+1} - x_\nu)^2 \le \frac{\pi^2}{8n} - \frac{\pi^2}{16nN} \tag{6.56}$$

and

$$\left|K_1\left(\frac{x_\nu + x_{\nu+1}}{2}\right)\right| \le \frac{7\pi}{48N^2}\ln 563N \quad \text{for } \nu = 0,\dots,n. \tag{6.57}$$

These relations lead to

$$\int_{-1}^{1} |K_1(x)|\,dx \le \frac{\pi^2}{8n} - \frac{\pi^2}{16nN} + \frac{2N+1}{2N}\cdot\frac{49\pi^2}{2304N^3}\ln^2 563N.$$

Therefore, the assertion of the theorem follows if

$$\frac{2N+1}{2N}\cdot\frac{49\pi^2}{2304N^3}\ln^2 563N \le \frac{\pi^2}{16nN},$$

or

$$\frac{\ln^2 563N}{N} \le \frac{144}{49}.$$

The latter inequality holds for $n \ge 33$. For $n < 33$, we determine $\int_{-1}^{1} |K_1(x)|\,dx$ numerically.

Consider the sum $S(x_0,\dots,x_{n+1})$ from (6.56); we want to show that

$$S(x_0,\dots,x_{n+1}) \le S(y_0,\dots,y_{n+1}), \tag{6.58}$$

where

$$y_{n+1} = -y_0 = 1 \quad \text{and} \quad y_\nu = -\cos\left(\frac{\nu - 1/4}{N}\pi\right) \quad \text{for } \nu = 1,\dots,n.$$

First, note that $x_{\nu+1} - 2x_\nu + x_{\nu-1} > 0$ for $\nu \le n/2$ (Theorem 6.3.3) and that the reverse inequality holds for the other ν. Analogous inequalities can be proved explicitly for the y_ν. Setting $z_\nu(\lambda) := (1-\lambda)x_\nu + \lambda y_\nu$, we see that for $\lambda \in [0,1]$,

$$\begin{aligned}\frac{d}{d\lambda}S\big(z_0(\lambda), z_1(\lambda),\dots,z_{n+1}(\lambda)\big) &= \sum_{\nu=1}^{n}\frac{\partial S}{\partial z_\nu}\cdot\frac{\partial z_\nu}{\partial\lambda}\\ &= -\frac{1}{2}\sum_{\nu=1}^{n}(z_{\nu+1} - 2z_\nu + z_{\nu-1})(y_\nu - x_\nu) \ge 0.\end{aligned}$$

(Note that, being a convex combination, $z_{\nu+1}(\lambda) - 2z_\nu(\lambda) + z_{\nu-1}(\lambda)$ has the same sign property as $x_{\nu+1} - 2x_\nu + x_{\nu-1}$ and $y_{\nu+1} - 2y_\nu + y_{\nu-1}$; also note that $|x_\nu| \le |y_\nu|$ by Theorem 6.3.2). We have therefore proved (6.58). Hence,

$$\begin{aligned}&S(x_0,\dots,x_{n+1})\\ &\le \frac{1}{4}\sum_{\nu=1}^{n-1}\left\{\cos\left(\frac{\nu-1/4}{N}\pi\right) - \cos\left(\frac{\nu+3/4}{N}\pi\right)\right\}^2 + \frac{1}{2}\left(1 - \cos\frac{3\pi}{4N}\right)^2\\ &= \sin^2\frac{\pi}{2N}\sum_{\nu=1}^{n-1}\sin^2\left(\frac{\nu+1/4}{N}\pi\right) + 2\sin^4\frac{3\pi}{8N}\\ &= \frac{1}{2}\sin^2\frac{\pi}{2N}\cdot\left(n - 1 + \frac{\sin(3\pi/(2N))}{\sin(\pi/N)}\right) + 2\sin^4\frac{3\pi}{8N}.\end{aligned}$$

The concavity of the sine function gives $\sin\left(\frac{3}{2}\cdot\frac{\pi}{N}\right) \le \frac{3}{2}\cdot\sin\frac{\pi}{N}$, so

$$S \le \frac{N}{2}\sin^2\frac{\pi}{2N} + 2\sin^4\frac{3\pi}{8N}.$$

The first three terms in the Taylor expansion of $\sin^2 x$ form an upper bound:

$$S \le \frac{\pi^2}{8N} - \frac{\pi^4}{96N^3} + \frac{\pi^6}{2880N^5} + \frac{81\pi^4}{2048N^4}.$$

The right-hand side is smaller than $\pi^2/(8N)$ for $n \ge 4$, which is exactly the upper bound stated in (6.56). The proof of (6.57) is more difficult, however.

From Theorem 4.2.5(i), we obtain the identity

$$K_1\left(\frac{x_\nu + x_{\nu-1}}{2}\right) - K_1\left(\frac{x_{\nu+1} + x_\nu}{2}\right) = \frac{x_{\nu+1} - x_{\nu-1}}{2} - a_\nu =: D_\nu.$$

Furthermore, the symmetry $K_1(x) = -K_1(-x)$ gives $K_1(0)$. Together, these yield the identity

$$K_1\left(\frac{x_{\nu+1} + x_\nu}{2}\right) = \sum_{\mu=\nu+1}^{n/2} D_\mu. \tag{6.59}$$

First, we derive an upper bound for D_μ. Using Theorem 6.3.2 for $\mu \le n/2$, we obtain

$$\begin{aligned}
\frac{x_{\mu+1} - x_{\mu-1}}{2} &= \frac{1}{2}\big[\cos(\theta_\mu - (\theta_\mu - \theta_{\mu-1})) - \cos(\theta_\mu + (\theta_{\mu+1} - \theta_\mu))\big] \\
&\le \frac{1}{2}\big[\cos(\theta_\mu - (\theta_{\mu+1} - \theta_\mu)) - \cos(\theta_\mu + (\theta_{\mu+1} - \theta_\mu))\big] \\
&= \sin\theta_\mu \, \sin(\theta_{\mu+1} - \theta_\mu) \\
&\le \frac{\pi}{N}\sin\theta_\mu.
\end{aligned}$$

Owing to symmetry, this inequality holds for all μ. Theorem 6.3.4 then gives

$$D_\mu \le \frac{\pi}{4N^3 \sin\theta_\mu}. \tag{6.60}$$

To prove a lower bound for D_μ in the case where $2 \le \mu \le n/2$, we derive, as above,

$$\frac{x_{\mu+1} - x_{\mu-1}}{2} \ge \sin\theta_\mu \, \sin(\theta_\mu - \theta_{\mu-1}).$$

By (6.24), we have

$$\sin(\theta_\mu - \theta_{\mu-1}) \ge \sin\frac{\pi}{N}\left(1 + \frac{1}{4N^2\sin^2\theta_{\mu-1}}\right)^{-1/2}.$$

The estimate

$$\frac{d}{dx}\sin\frac{\pi}{N\sqrt{1+x}} = -\frac{\pi}{2N}(1+x)^{-3/2}\cos\frac{\pi}{N\sqrt{1+x}} \ge -\frac{\pi}{2N} \quad \text{for } x > 0$$

shows that

$$\sin(\theta_\mu - \theta_{\mu-1}) \ge \sin\frac{\pi}{N} - \frac{\pi}{8N^3\sin^2\theta_{\mu-1}}.$$

Theorem 6.3.2 implies that

$$\frac{\sin\theta_\mu}{\sin\theta_{\mu-1}} \le \frac{\theta_\mu}{\theta_{\mu-1}} \le \frac{\theta_{\mu-1} + \pi/N}{\theta_{\mu-1}} \le 1 + \frac{\pi}{N\theta_{\mu-1}} \le 1 + \frac{1}{\mu - 5/4} \le \frac{7}{3},$$

so the estimate $a_\mu \le (\pi/N)\sin\theta_\mu$ from Theorem 6.3.4 gives

$$D_\mu \ge -\frac{\pi^3}{6N^3} - \frac{7\pi}{24N^3\sin\theta_{\mu-1}}.$$

Now, using $\sin\theta_{\mu-1} \geq \frac{2}{\pi}\theta_{\mu-1} \geq \frac{2}{N}(\mu - \frac{5}{4})$, we get the final bound for $|D_\mu|$:

$$|D_\mu| \leq \frac{\pi^3}{6N^3} + \frac{7\pi}{48N^2(\mu - 5/4)}.$$

Plugging this result into (6.59) for $\nu \geq 1$ gives

$$\begin{aligned}\left|K_1\left(\frac{x_{\nu+1}+x_\nu}{2}\right)\right| &\leq \frac{\pi^3}{12N^2} + \frac{7\pi}{48N^2}\sum_{\mu=2}^{n/2}\frac{1}{\mu - 5/4} \\ &\leq \frac{\pi^3}{12N^2} + \frac{7\pi}{48N^2}\int_{1/4}^{n/2-3/4}\frac{dx}{x} \\ &\leq \frac{7\pi}{48N^2}\left(\ln 2N + \frac{4\pi^2}{7}\right) \leq \frac{7\pi}{48N^2}\ln 563N.\end{aligned}$$

The above bound has been established only for $\nu \geq 1$, so we add the following case:

$$\left|K_1\left(\frac{x_1+x_0}{2}\right)\right| = \frac{1}{2}(1+x_1) = \frac{1}{2}(1-\cos\theta_1) = \sin^2\frac{\theta_1}{2} \leq \frac{\pi^2}{4N^2} \leq \frac{7\pi}{48N^2}\ln 563N.$$

Thus (6.57), and hence the theorem, is proved. □

We remark that (6.55) is quite a poor estimate for high-order derivatives. In those cases, the approximation-theoretic method applied to (5.7) rather than (5.6) gives more reasonable bounds. We obtain

$$\rho(Q_n^{\mathrm{G}}, \mathcal{C}_M^{(r)}) \leq \frac{1}{r!\,2^{r-1}}\left(\frac{(2r)!(2n-r)!}{(2n+r-1)!\,(2r-1)}\right)^{1/2} M\|R_n^{\mathrm{G}}\| \tag{6.61}$$

and emphasize the special case where $r = n$,

$$\rho(Q_n^{\mathrm{G}}, \mathcal{C}_M^{(n)}) \leq \frac{1}{n!\,2^{n-1}}\left(\frac{(2n)!\,n!}{(3n-1)!\,(2n-1)}\right)^{1/2} M\|R_n^{\mathrm{G}}\|,$$

which implies

$$\rho(Q_n^{\mathrm{G}}, \mathcal{C}_M^{(n)}) \leq \frac{5.2\sqrt[4]{n}}{n!(3\sqrt{3})^n} M\|R_n^{\mathrm{G}}\|. \tag{6.62}$$

For $w = 1$, Förster (1993a) improved this bound, obtaining

$$\rho(Q_n^{\mathrm{G}}, \mathcal{C}_M^{(n)}) \leq \frac{2}{n!(3\sqrt{3})^{n-1}} M.$$

For further information on bounds involving high-order derivatives, see Brass and Förster (1989).

We now turn to error estimates that require holomorphy of the integrand. Most papers on this subject require the domains of holomorphy to be ellipses $\mathcal{E}_r$. The important case of circular domains then follows, since the disc centered at 0 with radius $(r + 1/r)/2$ contains $\mathcal{E}_r$. It can be shown that this gives the correct order for the error constant (see Petras 1995). Error bounds valid for arbitrary weight functions can be obtained from Lebesgue's inequality and (5.9):

$$\rho(Q_n^{\mathrm{G}}, \mathcal{C}_M(\mathcal{E}_r)) \leq \frac{2M\|R_n^{\mathrm{G}}\|}{r^{2n}(1-r^{-1})}. \tag{6.63}$$

We already proved this inequality in Corollary 4.9.1. The method used there shows that the denominator in (6.63) can be replaced by $r^{2n}(1 - r^{-2})$ for symmetric weight functions. Using a different method, von Sydow (1977) showed that this improvement holds for arbitrary w.

On combining Lebesgue's inequality with (5.10), we obtain

$$\rho\big(Q_n^{\mathrm{G}}, \mathcal{C}_M(\mathcal{E}_r)\big) \le \frac{8}{\pi} \cdot \frac{M\|R_n^{\mathrm{G}}\|}{r^{2n}+r^{-2n}} \tag{6.64}$$

(Stenger 1966), which is better than (6.63) at least for $r \le 1/(1-\pi/4) = 4.659\ldots$.

Numerous bounds are known for specific weight functions; see, for example, Chawla and Jain (1968a), Chawla (1969, 1970), Kumar (1974), Akrivis (1984, 1985), Notaris (1990a), Peherstorfer (1993b) or Hunter (1995).

EXAMPLE 6.6.2. Let $w(x) = \big(1-x^2\big)^{-1/2}$. The corresponding Q_n^{G} is the Gauss–Chebyshev rule

$$Q_n^{\mathrm{G}}[f] = \frac{\pi}{n}\sum_{\nu=1}^{n} f\left(-\cos\left(\frac{2\nu-1}{2n}\pi\right)\right);$$

see Examples 5.1.1 or 6.1.2. The simplicity of this case is due to the fact that

$$R_n^{\mathrm{G}}[T_\lambda] = \begin{cases} (-1)^{\sigma+1}\pi & \text{for } \lambda = 2\sigma n \text{ where } \sigma = 1,2,\ldots, \\ 0 & \text{otherwise.} \end{cases} \tag{6.65}$$

To prove this relation, note that

$$T_{2\sigma n \pm l}[x_{\nu,n}] = (-1)^\sigma T_l[x_{\nu,n}]$$

from an explicit calculation; therefore

$$Q_n^{\mathrm{G}}[T_{2\sigma n\pm l}] = (-1)^\sigma Q_n^{\mathrm{G}}[T_l] = (-1)^\sigma I[T_l] = \begin{cases} (-1)^\sigma \pi & \text{for } l = 0, \\ 0 & \text{for } 0 < l \le n, \end{cases}$$

which is (6.65). Plugging (6.65) into Theorem 4.9.4 gives

$$\begin{aligned}
\rho\big(Q_n^{\mathrm{G}}, \mathcal{C}_M(\mathcal{E}_r)\big) &\le \frac{M}{\pi}\int_0^{2\pi}\left|\sum_{\nu=1}^{\infty} r^{-2\nu n}(-1)^{\nu+1}\pi e^{-i2\nu n t}\right| dt \\
&= M\int_0^{2\pi}\left|\frac{r^{-2n}e^{-i2nt}}{1+r^{-2n}e^{-i2nt}}\right| = M\int_0^{2\pi}\frac{dt}{|r^{2n}e^{i2nt}+1|} \\
&= M\int_0^{2\pi}\frac{dt}{\left[\left(r^{2n}+1\right)^2 - 4r^{2n}\sin^2 nt\right]^{1/2}} \\
&= M\int_0^{2\pi}\frac{dt}{\left[\left(r^{2n}+1\right)^2 - 4r^{2n}\sin^2 t\right]^{1/2}} \\
&= \frac{4M}{r^{2n}+1}\int_0^{\pi/2}\left[1-\left(\frac{2}{r^n+r^{-n}}\right)^2\sin^2 t\right]^{-1/2} dt. \qquad (6.66)
\end{aligned}$$

This integral is a complete elliptic integral of the first kind. Properties and tables of values of the integral can be found in Abramowitz and Stegun (1964: Chapter 17). The equality (6.66) comes from Hunter (1995). The integral-free bound

$$\rho\big(Q_n^{\mathrm{G}}, \mathcal{C}_M(\mathcal{E}_r)\big) \le \frac{2\pi M r^{-2n}}{\sqrt{1-r^{-4n}}} \tag{6.67}$$

found by Chawla (1969) is a special case of (4.50). From its derivation, one can see that it has to be worse than Hunter's bound; however, the difference is small. This is evident from the lower bound

$$\rho\big(Q_n^{\mathrm{G}}, \mathcal{C}_M(\mathcal{E}_r)\big) \geq \frac{2\pi M r^{-2n}}{1+r^{-4n}},$$

which comes from an application of Theorem 4.9.5.

The more important case of $w = 1$ is also more complicated, since we have no simple representation of $R_n^{\mathrm{G}}[T_\nu]$ for all ν. Theorem 6.5.1 gives

$$R_n^{\mathrm{G}}[T_{2n}] = \frac{1}{2n+1} \cdot \frac{2^{2n+1}(n!)^4}{[(2n)!]^2} \cdot \frac{T_{2n}^{(2n)}(\xi)}{(2n)!} = \frac{2^{4n}}{2n+1} \cdot \frac{(n!)^4}{[(2n)!]^2},$$

and Stirling's formula shows that

$$\big|R_n^{\mathrm{G}}[T_{2n}]\big| \geq \frac{\pi}{2}\left(1 - \frac{1}{4n}\right),$$

so we have

$$\rho\big(Q_n^{\mathrm{G}}, \mathcal{C}_M(\mathcal{E}_r)\big) \geq R_n^{\mathrm{G}}\left[\frac{2MT_{2n}}{r^{2n}+r^{-2n}}\right] \geq \frac{\pi\big(1-(4n)^{-1}\big)}{r^{2n}+r^{-2n}} M.$$

A simple upper bound can be obtained by using

$$\big|R_n^{\mathrm{G}}[T_{2k}]\big| \leq \big|I[T_{2k}]\big| + \big|Q_n^{\mathrm{G}}[T_{2k}]\big| \leq \frac{2}{4k^2-1} + 2 \leq \frac{2}{4n^2-1} + 2$$

for $k \geq n$ and $R_n^{\mathrm{G}}[T_{2k}] = 0$ for $k < n$. Then, Theorem 4.9.4 gives

$$\rho\big(Q_n^{\mathrm{G}}, \mathcal{C}_M(\mathcal{E}_r)\big) \leq \frac{4Mr^{-2n}}{1-r^{-2}}\left(1 + \frac{1}{4n^2-1}\right). \tag{6.68}$$

Rabinowitz (1969), who proved this bound, compared it with bounds derived by other methods and did not observe any major differences. Bounds significantly better than (6.63) and (6.68) seem to require a great deal more technical effort to derive. From a very thorough examination of $R_n^{\mathrm{G}}[T_\nu]$, Petras (1995) was able to prove that

$$\rho\big(Q_n^{\mathrm{G}}, \mathcal{C}_M(\mathcal{E}_r)\big) \leq \frac{\pi M}{r^{2n}}\left(1+r^{-2}\right)\left(1 + \frac{3}{2nr^2} + \frac{9}{2r^{n+1}}\right). \tag{6.69}$$

We now have a collection of error bounds. In the following simple but non-trivial example, we shall demonstrate that even for an analytic integrand, it is necessary to use not only the error bound involving analyticity. Error bounds based on lower-order derivatives are sometimes much better. Furthermore, this example shows that error bounds can be applied in a systematic way.

EXAMPLE 6.6.3. Consider

$$I[f_a] = \int_{-1}^{1} f_a(x)\,dx \quad \text{where } f_a(x) = \frac{e^{-x^2}}{a^2+x^2} \quad \text{with } a > 0.$$

Analytical expressions for the derivatives become more and more complicated as the order of the derivative increases. Moreover, the derivatives are not monotone. For higher-order derivatives, we do not have explicit expressions for the extremal values. We can, however, estimate these higher derivatives on intervals. The algorithmic computation of derivatives for real arguments has already been demonstrated in Example 6.5.1. The only difference here is that, rather than computing with numbers, we use intervals as described in Appendix C.2. Applying the techniques of automatic differentiation and interval arithmetic directly to the whole interval in a single computation hugely overestimates the derivatives; but upon subdividing the interval of integration into, for example, ten pieces of equal size, reasonable bounds can be obtained. Using the bounds (6.53), (6.55) (note that (6.53) is better than (6.55) if $n \geq 6$) and (6.61), as well as the one from Theorem 6.6.2, we obtain the following error bounds for $\left|R_{20}^{\mathrm{G}}[f_a]\right|$:

r	$a = 0.25$	$a = 0.5$	$a = 1$	$a = 2$	$a = 4$
1	2.75	0.4	0.07	0.015	$3.5 \cdot 10^{-3}$
5	$3.0 \cdot 10^{-2}$	$2.8 \cdot 10^{-4}$	$4.4 \cdot 10^{-6}$	$2.6 \cdot 10^{-7}$	$4.0 \cdot 10^{-8}$
10	$8.3 \cdot 10^{-3}$	$2.5 \cdot 10^{-6}$	$1.3 \cdot 10^{-9}$	$4.9 \cdot 10^{-12}$	$3.4 \cdot 10^{-13}$
20	4.4	$5.8 \cdot 10^{-7}$	$8.9 \cdot 10^{-13}$	$4.8 \cdot 10^{-19}$	$1.5 \cdot 10^{-21}$
30	$1.4 \cdot 10^{7}$	$1.8 \cdot 10^{-3}$	$2.6 \cdot 10^{-13}$	$7.3 \cdot 10^{-23}$	$7.3 \cdot 10^{-28}$
40	$4.1 \cdot 10^{17}$	$5.0 \cdot 10^{4}$	$7.1 \cdot 10^{-9}$	$1.6 \cdot 10^{-21}$	$1.2 \cdot 10^{-30}$

Again, we can see that it is unpredictable which of the derivative orders will give the best result.

For each Gaussian rule, there are a finite number of error constants (which can be computed numerically) in classes of type $\mathcal{C}_M^{(r)}$. For a given sufficiently differentiable integrand and a given number of nodes, we can compute all error bounds and choose the smallest one. With this purpose in mind, note that automatic differentiation gives all derivatives up to a certain degree simultaneously.

If we seek error bounds using bounds in the complex plane, then we cannot precompute all of them; therefore we are lucky to have analytic expressions, such as (6.68), for estimating error bounds. Suppose we want to use (6.68) to do an error estimation for f_a and $n = 20$. First, we need to be able to estimate the integrand on, say, ellipses $\mathcal{E}_r$. To do this, we can cover $\mathcal{E}_r$ with a finite number of (in our case, five) axial parallel rectangles and apply rectangular complex interval arithmetic (see Appendix C.3). The choice of r can then be optimized heuristically. For this example we chose a simple algorithm because the numerically computed values indicated that the error bounds obtained from the procedure, call them $e(r)$, could form a convex function. Take equidistant r_1, r_2, r_3, r_4, r_5 from the interval $[1.01,\, 0.99(a + \sqrt{a^2 - 1})]$, and compute error bounds $e(r_i)$ as described above. Pick out the three r_i values—r_1, r_2 and r_3, say—that correspond to the smallest error bound; then take also the midpoint between r_1 and r_2 as well as the midpoint between r_2 and r_3. Thus, we again have five r-values and can repeat the step. After performing this procedure five times, we came up with the following results:

	$a = 0.25$	$a = 0.5$	$a = 1$	$a = 2$	$a = 4$
r	1.2515	1.5787	2.3518	4.1072	7.5127
Error bound	$7.9 \cdot 10^{-2}$	$3.2 \cdot 10^{-6}$	$2.8 \cdot 10^{-13}$	$3.2 \cdot 10^{-22}$	$2.2 \cdot 10^{-29}$

Asymptotically we get the best behaviour with this kind of estimate for our integrands. On the other hand, for $n = 20$, there is always an error bound involving derivatives that is superior.

6.7. Asymptotics of the error

Our goal in this section is to study the asymptotic behaviour of $R_n^{\mathrm{G}}[f]$ for some very special, but in a sense typical, integrands. We shall look at the following functions, all defined on $[-1,1]$, listed in increasing order of singularity:

$$\begin{aligned}
f_1(x) &= e^x, && \\
f_2(x) &= \frac{1}{c-x} && \text{with } c>1, \\
f_3(x) &= (c-x)^\beta && \text{with } c>1,\ \beta \notin \mathbb{N}, \\
f_4(x) &= (1-x)^\beta && \text{with } \beta>0,\ \beta \notin \mathbb{N}, \\
f_5(x) &= |x-c|^\beta && \text{with } c \in \,]-1,1[,\ \beta>0,\ \frac{\beta}{2} \notin \mathbb{N}.
\end{aligned}$$

Using the results on $R_n^{\mathrm{G}}[f_i]$ that we develop subsequently, we can discuss $R_n^{\mathrm{G}}[f]$ for more general f. The simplest idea is to split f as follows:

$$f = \text{const} \cdot f_i + g \quad \text{where const} \neq 0,\ i \in \{1,\dots,5\}.$$

If we can prove that

$$R_n^{\mathrm{G}}[g] = o(R_n^{\mathrm{G}}[f_i]),$$

for example by applying the results from the previous section, then we evidently have

$$R_n^{\mathrm{G}}[f] = \text{const} \cdot R_n^{\mathrm{G}}[f_i]\big(1+o(1)\big).$$

Moreover, direct generalizations are possible in all the cases we consider by using the techniques below.

Our investigation has two aims. First, we want to be able to judge the quality of our upper bounds on $\left|R_n^{\mathrm{G}}[f]\right|$, which have been a central topic up to now, by also establishing lower bounds on the error for specific integrands. Second, we wish to compare the Gaussian method against other methods in regard to quality, with the goal of demonstrating its superiority (or at least non-inferiority) for some sample functions and their modifications.

We restrict our comparisons to $R_n^{\mathrm{CC}}[f_i]$ and $R_n^{\mathrm{Tr}}[f_i]$, since these are particularly important methods. Note that R_n^{Lo} and R_n^{Ra} are similar to R_n^{G}, whereas R_n^{Fi} behaves similarly to R_n^{CC}.

In all cases, we focus on $w = 1$. An overview of our observations is given in the following table, where $q := c + \sqrt{c^2-1}$ and $a_n = \Theta(b_n)$ means that both of the relations $a_n = O(b_n)$ and $b_n = O(a_n)$ hold.

i	$\lvert R_n^{\mathrm{G}}[f_i]\rvert$	$\lvert R_n^{\mathrm{CC}}[f_i]\rvert$	$\lvert R_n^{\mathrm{Tr}}[f_i]\rvert$
1	$\Theta\left(\frac{1}{2^{2n}(2n)!}\right)$	$\Theta\left(\frac{1}{2^n(n+4)!}\right)$	$\Theta\left(n^{-2}\right)$
2	$\Theta\left(q^{-2n}\right)$	$\Theta\left(n^{-3}q^{-n}\right)$	$\Theta\left(n^{-2}\right)$
3	$\Theta\left(n^{-1-\beta}q^{-2n}\right)$	$\Theta\left(n^{-4-\beta}q^{-n}\right)$	$\Theta\left(n^{-2}\right)$
4	$\Theta\left(n^{-2-2\beta}\right)$	$\Theta\left(n^{-2-2\beta}\right)$	$\Theta\left(n^{-\min\{2,1+\beta\}}\right)$
5	$O\left(n^{-1-\beta}\right)$	$O\left(n^{-1-\beta}\right)$	$O\left(n^{-\min\{2,1+\beta\}}\right)$

The simplest way of deriving error asymptotics is to use Theorem 4.2.7.

THEOREM 6.7.1. *Let $f \in C^{2n+2}[-1,1]$. Then there is a $\xi \in [-1,1]$ such that*

$$R_n^{\mathrm{G}}[f] = \frac{2^{2n+1}}{2n+1} \cdot \frac{(n!)^4}{[(2n)!]^2} \cdot \left(\frac{f^{(2n)}(0)}{(2n)!} + \frac{(2n+1)(n^2+n-1)}{(2n-1)(2n+3)} \cdot \frac{f^{(2n+2)}(\xi)}{(2n+2)!} \right).$$

PROOF. According to Theorem 4.2.7, we have to determine $R_n^{\mathrm{G}}[p_{2n}]$, $R_n^{\mathrm{G}}[p_{2n+1}]$ and $R_n^{\mathrm{G}}[p_{2n+2}]$. Theorem 6.5.1 yields

$$R_n^{\mathrm{G}}[p_{2n}] = \frac{2^{2n+1}}{2n+1} \cdot \frac{(n!)^4}{[(2n)!]^2},$$

and $R_n^{\mathrm{G}}[p_{2n+1}] = 0$ follows from symmetry. It remains to prove that

$$R_n^{\mathrm{G}}[p_{2n+2}] = \frac{(2n+1)(n^2+n-1)}{(2n-1)(2n+3)} R_n^{\mathrm{G}}[p_{2n}]. \tag{6.70}$$

We will use $I[P_n P_{n+2}] = 0 = Q_n^{\mathrm{G}}[P_n P_{n+2}]$. Note that

$$P_k(x) = \delta_k x^k + \eta_k x^{k-2} + \cdots$$
$$\text{where } \delta_k = \frac{1}{2^k}\binom{2k}{k}, \quad \eta_k = -\frac{1}{2^{k+1}}\binom{2k}{k} \cdot \frac{k(k-1)}{2k-1}$$

(which follows immediately from the definition of Legendre polynomials). Upon applying R_n^{G} to both sides of

$$P_n(x)P_{n+2}(x) = \delta_n \delta_{n+2} x^{2n+2} + (\delta_n \eta_{n+2} + \delta_{n+2}\eta_n) x^{2n} + \cdots,$$

we obtain

$$0 = \delta_n \delta_{n+2} R_n^{\mathrm{G}}[p_{2n+2}] + (\delta_n \eta_{n+2} + \delta_{n+2}\eta_n) R_n^{\mathrm{G}}[p_{2n}],$$

that is,

$$R_n^{\mathrm{G}}[p_{2n+2}] = -\left(\frac{\eta_{n+2}}{\delta_{n+2}} + \frac{\eta_n}{\delta_n} \right) R_n^{\mathrm{G}}[p_{2n}],$$

and this is precisely (6.70). □

From Theorem 6.7.1 and formula (6.36), we immediately obtain that

$$R_n^{\mathrm{G}}[f_1] = \frac{\pi}{2^{2n}(2n)!} \cdot \big(1 + O(n^{-1})\big). \tag{6.71}$$

Chawla and Jain (1968b) studied $R_n^{\mathrm{G}}[f_1]$ using another method, but did not find the simple expression (6.71).

The same method can be applied to $R_n^{\mathrm{Fi}}[f_1]$ and $R_n^{\mathrm{Lo}}[f_1]$; however, it does not work for $R_n^{\mathrm{CC}}[f_1]$, owing to lack of definiteness. For $R_n^{\mathrm{CC}}[f_1]$, we use the trick from Lemma 5.4.1 to obtain

$$R_n^{\mathrm{CC}}[f_1] = \frac{1}{2^{n-4}(n+4)!} \cdot \big(1 + o(1)\big)$$

for odd n (Brass 1984). This shows that the Gaussian method is significantly better. For the trapezoidal rule, the relation

$$R_n^{\mathrm{Tr}}[f_1] = \frac{1}{3n^2}(e - e^{-1}) + O(n^{-3})$$

will be proved in the next chapter; so we can see that this method is not competitive for the integrand f_1.

Next, let us look at $f_2(x) = (c-x)^{-1}$. Denote by intpol$[f]$ the polynomial that interpolates f at the points $x_1, x_2, \ldots, x_n$. Evidently, we have

$$f_2(x) - \mathrm{intpol}[f_2](x) = \frac{q(x)}{c - x} \tag{6.72}$$

with $q \in \mathbb{P}_n$ and $q(x_\nu) = 0$. This determines q up to a constant factor. Multiplying both sides by $c - x$ and then letting $x \to c$ gives

$$f_2(x) - \mathrm{intpol}[f_2](x) = \frac{\prod_{\nu=1}^{n}(x - x_\nu)}{(c - x)\prod_{\nu=1}^{n}(c - x_\nu)} \tag{6.73}$$

for an arbitrary set of points $x_1, x_2, \ldots, x_n$. For the Gaussian rule, an integration gives

$$R_n^{\mathrm{G}}[f_2] = \frac{1}{P_n(c)} \int_{-1}^{1} \frac{P_n(x)}{c - x}\, dx. \tag{6.74}$$

Following Freud (1973), we transform the above into

$$\begin{aligned} R_n^{\mathrm{G}}[f_2] &= \frac{1}{P_n(c)P_{n+1}(c)} \int_{-1}^{1} \frac{[P_n(x)P_{n+1}(c) - P_n(c)P_{n+1}(x)] + P_n(c)P_{n+1}(x)}{c - x}\, dx \\ &= \frac{1}{P_n(c)P_{n+1}(c)} \cdot \frac{2}{n+1} + \frac{1}{P_{n+1}(c)} \int_{-1}^{1} \frac{P_{n+1}(x)}{c - x}\, dx, \end{aligned}$$

where we have made use of Theorem A.5. Thus we have shown that

$$R_n^{\mathrm{G}}[f_2] = \frac{2}{(n+1)P_n(c)P_{n+1}(c)} + R_{n+1}^{\mathrm{G}}[f_2].$$

By repeating this step and using the convergence of the Gaussian method, we obtain

$$R_n^{\mathrm{G}}[f_2] = \sum_{\nu=n}^{\infty} \frac{2}{(\nu+1)P_\nu(c)P_{\nu+1}(c)}.$$

We may now use the asymptotics of P_n as given in Theorem A.17 and obtain

$$R_n^{\mathrm{G}}[f_2] = \frac{2\pi}{\left(c+\sqrt{c^2-1}\right)^{2n+1}} \cdot \left(1+o(1)\right). \tag{6.75}$$

Again, we wish to compare this with R_n^{CC}. We also start with (6.73) and obtain, as a counterpart of (6.74),

$$R_n^{\mathrm{CC}}[f_2] = \frac{1}{(c^2-1)U_{n-2}(c)} \int_{-1}^{1} \frac{(x^2-1)U_{n-2}(x)}{c-x}\,dx.$$

For even n, three integrations by parts lead to

$$R_n^{\mathrm{CC}}[f_2] = \frac{1}{(c^2-1)U_{n-2}(c)} \left(\frac{4c}{c^2-1}n^{-3} + O(n^{-4})\right).$$

Now

$$U_{n-2}(c) = \frac{1}{2\sqrt{c^2-1}} \left(\left(c+\sqrt{c^2-1}\right)^{n-1} - \left(c-\sqrt{c^2-1}\right)^{n-1}\right)$$

(see Theorem A.9(vii)) gives

$$R_n^{\mathrm{CC}}[f_2] = \frac{8c\left(c^2-1\right)^{-3/2}}{n^3\left(c+\sqrt{c^2-1}\right)^{n-1}} \cdot \left(1+o(1)\right)$$

for even n. The Gaussian method again yields much better results. Later on we will encounter results which show immediately that the trapezoidal rule is not competitive (see Section 7.1). Let us reiterate that our comparison statements concerning $R_n^{\mathrm{CC}}[f_2]$ and $R_n^{\mathrm{G}}[f_2]$ are of an asymptotic character. For small n, the superiority of the Gaussian rule melts away—see Weidemann and Trefethen (2007)—but in this regime it seems difficult to obtain precise and general results.

We start our discussion of $R_n^{\mathrm{G}}[f_3]$ with a very simple method, which gives the correct order of the error but not the exact asymptotic behaviour.

According to Theorem 4.2.6(ii), there exists $\xi \in [-1,1]$ such that

$$R_n^{\mathrm{G}}[f] = \frac{f^{(2n)}(\xi)}{g^{(2n)}(\xi)} R_n^{\mathrm{G}}[g].$$

We set $g(x) = f_2(x) = (c-x)^{-1}$ and $f(x) = f_3(x) = (c-x)^{\beta}$, and obtain

$$R_n^{\mathrm{G}}[f_3] = \binom{\beta}{2n}(c-\xi)^{\beta+1} R_n^{\mathrm{G}}[f_2]. \tag{6.76}$$

The preceding result for f_2, together with

$$\lim_{n\to\infty} (2n)^{\beta+1}\binom{\beta}{2n} = \frac{1}{\Gamma(-\beta)},$$

gives

$$R_n^{\mathrm{G}}[f_3] = \frac{\Lambda_n}{\Gamma(-\beta)\, n^{\beta+1}\left(c+\sqrt{c^2-1}\right)^{2n+1}},$$

where all the Λ_n lie between two fixed positive numbers that depend on c and β.

A technique that allows us to obtain more precise results is expansion in Chebyshev polynomials.

THEOREM 6.7.2 (Brass 1979b). Let

$$f = \sum_{\nu=0}^{\infty} \delta_\nu q^{-\nu} T_\nu,$$

where $q > 1$ and the δ_ν are such that

$$\lim_{\nu\to\infty} \frac{\delta_{\nu+1}}{\delta_\nu} = 1. \tag{6.77}$$

Then

$$R_n^{\mathrm{G}}[f] = \frac{\pi}{2} \cdot (1-q^{-2})q^{-2n} \cdot \delta_{2n} \cdot \big(1+o(1)\big).$$

PROOF. Given $\varepsilon > 0$, (6.77) implies

$$1-\varepsilon < \frac{\delta_{\nu+1}}{\delta_\nu} < 1+\varepsilon$$

for all $\nu \geq n_0(\varepsilon)$ and hence

$$(1-\varepsilon)^\sigma < \frac{\delta_{\nu+\sigma}}{\delta_\nu} < (1+\varepsilon)^\sigma,$$

that is,

$$\left|\frac{\delta_{\nu+\sigma}}{\delta_\nu} - 1\right| < \max\big\{(1+\varepsilon)^\sigma - 1,\, 1-(1-\varepsilon)^\sigma\big\} = (1+\varepsilon)^\sigma - 1.$$

For $2n \geq n_0(\varepsilon)$, this yields

$$\sum_{\sigma=0}^{\infty} \left|\frac{\delta_{2n+\sigma}}{\delta_{2n}} - 1\right| q^{-\sigma} < \sum_{\sigma=0}^{\infty} \frac{(1+\varepsilon)^\sigma - 1}{q^\sigma} = \frac{q}{q-1-\varepsilon} - \frac{q}{q-1}.$$

So we obtain

$$\sum_{\sigma=0}^{\infty} \left|\frac{\delta_{2n+\sigma}}{\delta_{2n}} - 1\right| q^{-\sigma} = o(1),$$

or equivalently

$$\sum_{\sigma=0}^{\infty} |\delta_{2n+\sigma} - \delta_{2n}|\, q^{-2n-\sigma} = o(\delta_{2n} q^{-2n}),$$

as $n \to \infty$. Since $R_n^{\mathrm{G}}[T_\nu] = O(1)$, it follows that

$$\sum_{\sigma=0}^{\infty} (\delta_{2n+\sigma} - \delta_{2n})\, q^{-2n-\sigma} R_n^{\mathrm{G}}[T_{2n+\sigma}] = o(\delta_{2n} q^{-2n}),$$

and so

$$R_n^{\mathrm{G}}[f] = \delta_{2n} R_n^{\mathrm{G}}\left[\sum_{\sigma=0}^{\infty} q^{-\nu} T_\nu\right] + o(\delta_{2n} q^{-2n}).$$

Combining this with (4.49) and (6.75) completes the proof. □

We have

$$f_3 = \frac{\alpha_0}{2} + \sum_{\nu=1}^{\infty} \alpha_\nu T_\nu \quad \text{with } \alpha_\nu = \frac{2\big(c^2-1\big)^{\beta/2}}{\Gamma(-\beta)\,\nu^{\beta+1}} \left(c + \sqrt{c^2-1}\right)^{-\nu} \cdot \big(1+o(1)\big)$$

(see Elliot 1964), so Theorem 6.7.2 can be applied. This gives

$$R_n^{\mathrm{G}}[f_3] = \frac{2\pi\big(c^2-1\big)^{(\beta+1)/2}}{\Gamma(-\beta)\,(2n)^{\beta+1}} \left(c + \sqrt{c^2-1}\right)^{-2n-1} \cdot \big(1+o(1)\big).$$

A similar method shows that

$$R_n^{\mathrm{CC}}[f_3] = \frac{8c\big(c^2-1\big)^{(\beta-2)/2}}{\Gamma(-\beta)\, n^{\beta+4}} \Big(c + \sqrt{c^2-1}\Big)^{1-n} \cdot (1+o(1))$$

for even n. We shall see in the next chapter that $R_n^{\mathrm{Tr}}[f_3]$ is approximately a nonzero constant multiple of n^{-2}. Comparison of the three methods for f_3 therefore yields a similar conclusion as in the previous cases.

The situation is different for our model function f_4, owing to a lack of smoothness. It will be shown below that we have convergence at the rate of $n^{-2\beta-2}$ for all positive quadrature methods of interpolatory type, and the Gaussian method affords no improvement. For the trapezoidal method, we shall prove (cf. Theorem 7.1.3) that

$$R_n^{\mathrm{Tr}}[f_4] = c(\beta) n^{-\min\{2,1+\beta\}} \cdot \big(1+o(1)\big), \quad \text{where } c(\beta) \neq 0.$$

Hence, the Gaussian and Clenshaw–Curtis methods still have a certain advantage.

To prove our assertion in the case of positive quadrature methods of interpolatory type, we use Theorem 6.5.5, which says that

$$|K_r(x)| \leq \frac{\mathrm{const}_r}{n^r} \big(1-x^2\big)^{r/2}. \tag{6.78}$$

After performing integration by parts, we obtain

$$\int_{-1}^{1-\varepsilon} f_4'(x) K_1(x)\, dx = -\sum_{\nu=1}^{\lambda-1} f_4^{(\nu)}(1-\varepsilon) K_{\nu+1}(1-\varepsilon) + \int_{-1}^{1-\varepsilon} f_4^{(\lambda)}(x) K_\lambda(x)\, dx.$$

We choose $\varepsilon = n^{-2}$ and $\lambda = \lfloor 2\beta \rfloor + 3$. Then, it follows from (6.78) that all summands are of the order $O(n^{-2\beta-2})$. Now let $0 < \delta < n^{-2}$. An integration by parts gives

$$\begin{aligned} &\int_{1-n^{-2}}^{1-\delta} f_4'(x) K_1(x)\, dx \\ &= -f_4'(1-\delta) K_2(1-\delta) + f_4'(1-n^{-2}) K_2(1-n^{-2}) + \int_{1-n^{-2}}^{1-\delta} f_4''(x) K_2(x)\, dx. \end{aligned}$$

We again apply (6.78), obtaining the order $O(\delta^\beta n^{-2})$ or $O(n^{-2\beta-2})$ for the summands. Letting $\delta \to 0$ then gives the stated result.

In particular, we have

$$R_n^{\mathrm{G}}[f_4] = O(n^{-2\beta-2}) \tag{6.79}$$

(see DeVore and Scott 1984), which gives the correct order, as can be seen from the following result of Barrett (1960):

$$R_n^{\mathrm{G}}[f_4] = \frac{2^{1-\beta} \sin \pi\beta}{\pi n^{2\beta+2}} \int_0^\infty \frac{t^{2\beta-1} K_0(t)}{I_0(t)}\, dt \cdot \big(1+o(1)\big), \tag{6.80}$$

where K_0 and I_0 are the modified Bessel functions. The proof of (6.80), which is rather long, will be omitted here. The correct order of $R_n^{\mathrm{CC}}[f_4]$ is also $O(n^{-2\beta-2})$; see Kütz (1984a).

Finally, we turn to f_5. For guidance, let us first look at the simple example of

$$f_5^*(x) = |x|.$$

Because $|x| = 2(x-0)_+ - x$, the equation $R_n[f_5^*] = 2K_2(0)$ holds for all rules for which K_2 exists. Therefore, we have

$$R_n[f_5^*] = O(n^{-2})$$

for all methods of interest. Theorem 6.4.7 tells us that

$$\left|R_n^{\mathrm{G}}[f_5^*]\right| \geq \mathrm{const}\cdot n^{-2}, \quad \text{where const} > 0.$$

Thus, the Gaussian method offers no real advantage over other methods of interest; for the most part, however, there is also no disadvantage to using the Gaussian method, since for any $Q \in \mathbf{Q}_n$ with second Peano kernel we have that

$$\sup_c R_n\big[\,|\cdot - c|\,\big] = 2\sup_c |K_2(c)| \geq \frac{(b-a)^2}{8\big(n-1+2^{-1/2}\big)^2}$$

(see the remark after Example 4.6.1).

Let us now proceed with general f_5. We have the simple result

$$R_n[f_5] = O(n^{-\beta-1}) \tag{6.81}$$

for any positive quadrature method of interpolatory type. To prove this, we use an "interval splitting" method similar to the one used for f_4:

$$R_n[f_5] = \int_{-1}^{1} f_5'(x)K_1(x)\,dx = \int_{I_n} f_5'(x)K_1(x)\,dx + \int_{[-1,1]\setminus I_n} f_5'(x)K_1(x)\,dx,$$

where I_n is an interval centered at c of length $2/n$. Now, the integration by parts technique (which was already applied to f_4), along with the estimate $K_r(x) = O(n^{-r})$ (see Theorem 5.7.1), gives (6.81). More precise results were obtained by Petras (1993a). In particular, it was shown in that paper that (6.81) cannot be improved upon.

Using the same proof as above, we find that

$$R_n^{\mathrm{Tr}}[f_5] = O(n^{-\beta-1}) \quad \text{for } \beta \leq 1$$

but that

$$R_n^{\mathrm{Tr}}[f_5] \neq o(n^{-2})$$

for each $\beta \geq 1$; see Theorem 7.2.7.

6.8. Extremal properties of Gaussian rules

Again we focus on the weight function $w = 1$, but most of the results can be extended to more general weight functions. Let φ denote a real-valued function on $\mathbf{Q}^+(I,\mathbb{P}_m)$. Then $\hat{Q} \in \mathbf{Q}^+(I,\mathbb{P}_m)$ is said to be *extremal* with respect to φ if

$$\varphi(\hat{Q}) = \sup\{\varphi(Q) \,:\, Q \in \mathbf{Q}^+(I,\mathbb{P}_m)\}.$$

THEOREM 6.8.1. Suppose $m = 2r-1$. Then Q_r^{G} is extremal with respect to:

(i) $\varphi(Q) = x_i$ for $i \in \{1,\ldots,r\}$;
(ii) $\varphi(Q) = \inf x_i - \sup x_i$, where x_i are the nodes of Q;
(iii) $\varphi(Q) = a_1$;
(iv) $\varphi(Q) = \max_i\{a_i\}$ for odd r;
(v) $\varphi(Q) = K_1(x_i^{\mathrm{G}}+0)$ and $\varphi(Q) = -K_1(x_i^{\mathrm{G}}-0)$;
(vi) $\varphi(Q) = K_2(x_i^{\mathrm{G}})$;
(vii) $\varphi(Q) = K_{2r}(x)$ for $x \in [-1,1]$;
(viii) $\varphi(Q) = R[f]$ if $f^{(2r)} \geq 0$.

PROOF. (i) and (ii) are direct consequences of Theorem 3.4.1. Since each Q has a node smaller than x_1^{G}, (6.29) gives (iii). (iv) is one of the cases in Corollary 6.2.1. (v) and (vi) are Theorem 6.4.1 (Markoff–Stieltjes) and Theorem 6.4.8. Finally, (viii) follows from (vii), which is Theorem 6.4.5 (Markoff's theorem). □

Sometimes, it is not the Gaussian formula but rather the Lobatto formula that is extremal. One example is given in Theorem 6.5.2, where $\varphi(Q) = |K_{2r}(x)|$.

Such extremal characteristics are often used to obtain information about properties of arbitrary $Q \in \mathbf{Q}^+(I, \mathbb{P}_m)$. Further information can be extracted if we consider Gaussian rules together with their relatives, the Krein rules. We therefore define

$$\mathbf{Q}_m^{\mathrm{Kr}} = \{Q : Q \in \mathbf{Q}^+(I, \mathbb{P}_m) \text{ is a Krein rule}\}.$$

Note that $\mathbf{Q}_m^{\mathrm{Kr}}$ is a one-parameter family, and we can deal much more easily with it than with the whole set $Q \in \mathbf{Q}^+(I, \mathbb{P}_m)$.

THEOREM 6.8.2. $\sup\{\varphi(Q) : Q \in \mathbf{Q}_m^{\mathrm{Kr}}\} = \sup\{\varphi(Q) : Q \in \mathbf{Q}^+(I, \mathbb{P}_m)\}$ for:

(i) $\varphi(Q) = \{\sup_\nu (x_{\nu+1} - x_\nu) : x_\nu \text{ and } x_{\nu+1} \text{ are nodes of } Q\}$;
(ii) $\varphi(Q) = A(x, Q)$;
(iii) $\varphi(Q) = K_s(x)$ for $s \le m + 1$;
(iv) $\varphi(Q) = -K_s(x)$ for $s \le m + 1$;
(v) $\varphi(Q) = \|K_s\|$ for $s \le m + 1$.

PROOF. (i) follows from Theorem 3.4.1, (ii) is Theorem 3.4.2, and (iii)–(v) are consequences of Theorem 6.4.9. □

Besides the explicit extremal properties that hold in some cases, we also know in certain other cases that Gaussian rules at least come close to being extremal as the number of nodes increases. In these cases, it could be that Gaussian rules (or sometimes Lobatto rules) are indeed extremal but that the extremality just has not been proved (or disproved) yet. We shall say that a quadrature method $Q^{(1)}, Q^{(2)}, \ldots$, where $Q^{(m)} \in \mathbf{Q}^+(I, \mathbb{P}_m)$, is *asymptotically extremal* with respect to the functional φ if

$$\lim_{m\to\infty} \frac{\varphi(Q^{(m)})}{\sup\{\varphi(Q) : Q \in \mathbf{Q}^+(I, \mathbb{P}_m)\}} = 1.$$

THEOREM 6.8.3. The method $Q^{(0)}, Q^{(1)}, Q^{(2)}, \ldots$, where $Q^{(2r-2)} = Q^{(2r-1)} = Q_r^{\mathrm{G}}$, is asymptotically extremal with respect to:

(i) $\varphi(Q) = \|K_s\|$ for $s \in \{1, 2, \ldots\}$;
(ii) $\varphi(Q) = \rho\big(Q, C_M^{(s)}\big)$ for $s \in \{1, 2, \ldots\}$;
(iii) $\varphi(Q) = \sum\limits_{a_i \text{ is coefficient of } Q} a_i^2$.

Note that (i) follows from Theorems 6.2.7, 6.4.7 and 6.8.2(v); see Petras (1992a). The proofs of (ii) and (iii) require a lot more effort and are therefore omitted here; the interested reader can find these proofs in Brass (1993b) and Förster (1987a), respectively.

Finally, we present an extremal property of a different kind. A *generalized* quadrature rule is a functional of the form

$$Q_n^{\mathrm{gen}}[f] = \sum_{\nu=1}^{n} \big(a_\nu f(x_\nu) + b_\nu f'(x_\nu)\big).$$

For given $x_1, \ldots, x_n$, there is a unique choice of $a_1, \ldots, a_n$ and $b_1, \ldots, b_n$ such that Q^{gen} is exact for all polynomials of degree $2n-1$ (the proof is easy with Theorem 2.2.2). Note that Q_n^{G} is a generalized quadrature rule with the special property that $b_1 = \cdots = b_n = 0$.

THEOREM 6.8.4 (Filippi 1966).

$$\inf \Big\{ \rho\big(Q_n^{\text{gen}}, \mathcal{C}_M^{(2n)}\big) : \\ Q_n^{\text{gen}} \text{ is a generalized quadrature rule with } R_n^{\text{gen}}[\mathbb{P}_{2n-1}] = \{0\} \Big\} \\ = \rho\big(Q_n^{\text{G}}, \mathcal{C}_M^{(2n)}\big),$$

and the infimum is attained only by the Gaussian rule.

PROOF. Let

$$p(x) = \prod_{\nu=1}^{n} (x - x_\nu).$$

Then

$$\rho\big(Q_n^{\text{gen}}, \mathcal{C}_M^{(2n)}\big) \geq \frac{M}{(2n)!} I[p^2]$$

because $M \cdot p^2/(2n)! \in \mathcal{C}_M^{(2n)}$. Parseval's identity (Theorem A.8) gives

$$I[p^2] \geq I^2[pq_n] = \delta_n^{-2},$$

and this implies

$$\rho\big(Q_n^{\text{gen}}, \mathcal{C}_M^{(2n)}\big) \geq \frac{M}{(2n)!\,\delta_n^2}.$$

Theorem 6.4.4 now yields the assertion. □

6.9. Why Gaussian quadrature?

When there is no restriction on the choice of observation, the Gaussian method is recommended. Some exceptions exist to this rule of thumb; a main one, for instance, is the case of a $(b-a)$-periodic integrand (see Chapter 8), where there are good reasons for using midpoint-type rules.

The first argument for employing the Gaussian method is that it is simply more convenient to use easily available and well-studied rules. Selecting and constructing a rule specially adapted to the problem at hand can be very time-consuming and rarely yields a worthwhile advantage.

However, the main reason for our endorsement of the Gaussian method is its "universality". Saying a rule is universal means that, for many different co-observations, no other rules are available that can improve the error bounds to a significant extent. This somewhat vague description can be made precise in various ways; see Babuska (1966, 1968), Lötzbeyer (1971), Brass (1974b) and Petras (1996b) as well as Section 7.4. Let us start by considering the "loss"

$$\text{loss}^*(Q_n^{\text{G}}, \mathcal{C}) = \frac{\rho(Q_n^{\text{G}}, \mathcal{C})}{\rho_n^{\text{best}}(\mathcal{C})}$$

of the Gaussian rule with respect to the best rule in class $\mathcal{C}$. In the following we restrict ourselves to the case where $w(x) = 1$ and $[a, b] = [-1, 1]$.

THEOREM 6.9.1. We have

$$\sup\left\{\operatorname{loss}^*(Q_n^{\mathrm{G}},\mathcal{C}) \,:\, n \text{ such that } \rho_n^{\mathrm{best}}(\mathcal{C})<\infty\right\}<\infty \tag{6.82}$$

for

$$\mathcal{C}\in\left\{\mathcal{C}_\omega^{\mathrm{mc}},\ \mathcal{C}_M^{(r)}\ (r=1,2,\ldots),\ \mathcal{C}_M^{\mathrm{Var},r}\ (r=0,1,\ldots),\ \mathcal{C}_M^{\mathrm{Sa},r}\ (r=1,2,\ldots),\ \mathcal{C}_M(\mathcal{E}_r)\right\}.$$

Although condition (6.82) appears quite weak, it actually does not hold for some very natural co-observations. Indeed, Petras (1998a) proved that among all co-observations $\mathcal{C}_M(\mathcal{G})$ with $\mathcal{G}$ being symmetric with respect to the real axis, (6.82) holds only for $\mathcal{G}=\mathcal{E}_r$. So, for many regions, substantially better rules exist. Nevertheless, one also has to take into account the fast convergence for holomorphic functions; thus, often, use of the Gaussian method does not lead to a considerable increase in the number of nodes needed. Further discussion relating to this issue can be found in Petras (1998a).

PROOF OF THEOREM 6.9.1. We start with $\mathcal{C}_\omega^{\mathrm{mc}}$. Using Theorem 4.5.3 and (6.50), we obtain

$$\operatorname{loss}^*(Q_n^{\mathrm{G}},\rho_\omega^{\mathrm{mc}})\le\frac{4\,\omega\left(\frac{\pi}{2n}\right)}{2n\int_0^{1/n}\omega(x)\,dx}\le\frac{4\,\omega\left(\frac{\pi}{2n}\right)}{2n\int_{1/(2n)}^{1/n}\omega(x)\,dx}\le\frac{4\,\omega\left(\frac{4}{2n}\right)}{\omega\left(\frac{1}{2n}\right)}\le 16.$$

To prove our assertion for $\mathcal{C}_M^{(r)}$, it is sufficient to compare Theorem 4.3.3 with Theorem 4.3.4.

For $\mathcal{C}_M^{\mathrm{Var},r}$, we combine Theorem 4.6.1 with Theorem 5.7.1 to get

$$\left|R_n^{\mathrm{G}}[f]\right|\le\frac{\mathrm{const}}{n^{r+1}}\cdot\operatorname{Var}f^{(r)}.$$

It is impossible to find a quadrature method with error bounds of lower order, since from

$$\left|R_n[f]\right|\le c_n\cdot\operatorname{Var}f^{(r)}$$

it follows that

$$\left|R_n[f]\right|\le c_n(b-a)\|f^{(r+1)}\|,$$

and Theorem 4.3.3 gives

$$c_n\ge\frac{\widetilde{\mathrm{const}}}{n^{r+1}}.$$

Similar arguments apply to the $\mathcal{C}_M^{\mathrm{Sa},r}$ case.

Finally, for $\mathcal{C}_M(\mathcal{E}_r)$ the assertion follows from Corollaries 4.9.1 and 4.9.2. □

Theorem 6.9.1 does not seem very convincing, since the supremum involved, while finite, may be a very large number. It should also be emphasized that Theorem 6.9.1 fails to hold if the Gaussian method is replaced by any other method (for example, $\rho_1^{\mathrm{best}}(\mathcal{C}_M^{(2)})<\infty$ but $\rho_1(Q_1,\mathcal{C}_M^{(2)})=\infty$ if $Q_1\ne Q_1^{\mathrm{G}}$). Even if we modify the assumption to refer to "sufficiently large n", a result like Theorem 6.9.1 cannot be established for nearly all of the standard methods; the only exceptions are the Lobatto and Radau rules.

For special co-observations it is possible to say more; $\mathcal{C}_M^{(r)}$ and $\mathcal{C}(\mathcal{E}_r)$ are the most interesting cases. As n increases, we have

$$\rho_n^{\mathrm{best}}(\mathcal{C}_M^{(r)})=\frac{2M\mathcal{K}_r}{\pi^r n^r}\cdot\left(1+o(1)\right)$$

(see Section 7.2) and

$$\rho(Q_n^{\mathrm{G}}, \mathcal{C}_M^{(r)}) = \frac{M\pi^r}{n^r} \int_0^1 |B_r(y)|\, dy \cdot \int_{-1}^1 (1-x^2)^{r/2}\, dx \cdot (1+o(1))$$

(see Theorem 6.6.1). For odd r, we have

$$\mathcal{K}_r = (2\pi)^r \int_0^1 |B_r(y)|\, dy$$

(Theorem B.8) and hence

$$\lim_{n\to\infty} \operatorname{loss}^*(Q_n^{\mathrm{G}}, \mathcal{C}_M^{(r)}) = \left(\frac{\pi}{2}\right)^{r+1} \frac{1\cdot 3\cdot 5\cdots r}{2\cdot 4\cdot 6\cdots (r+1)}.$$

Therefore, as $n \to \infty$ the asymptotic loss with respect to the best rule is an increasing, unbounded function when r is odd. Similar results can be derived for even r.

For $\mathcal{C}_M(\mathcal{E}_r)$, we have lower bounds (see Corollary 4.9.2) and upper bounds (see (6.69)) for the error. Taken together, these say that Gaussian rules are "almost best" for moderate to large r and large n.

Bounds of the type

$$\left|R_n^{\mathrm{G}}[f]\right| \le c_n \|f^{(n)}\|$$

occupy a middle position between those for $\mathcal{C}_M^{(r)}$ with fixed r and those for holomorphic functions. The advantage of the Gaussian rule over many other rules becomes quite clear if we compare

$$\lim_{n\to\infty} \left(n!\, c_n\right)^{1/n} = \frac{1}{3\sqrt{3}} = 0.192\ldots$$

for Q_n^{G} (see (6.62)) with

$$\lim_{n\to\infty} \left(n!\, c_n\right)^{1/n} = \frac{1}{2}$$

for Q_n^{CC} (see Theorem 5.4.2) and its relatives.

EXAMPLE 6.9.1. We examine $\operatorname{loss}^*(Q_n^{\mathrm{G}}, \mathcal{C}_\omega^{\mathrm{mc}})$ when ω is concave (which is true for all cases of interest). From Theorem 4.5.3, we deduce that

$$\rho_n^{\mathrm{best}}(\mathcal{C}_\omega^{\mathrm{mc}}) = 2n \int_0^{1/n} \omega(x)\, dx \ge 2n \cdot \frac{1}{2n}\, \omega\!\left(\frac{1}{n}\right) \ge \omega\!\left(\frac{\pi^2}{16n}\right).$$

Comparison with (6.51) gives a loss of at most 2. This does not seem serious, but we should also take into account the slow convergence. If we have, for instance,

$$\omega(\delta) = \frac{1}{|\ln \delta|},$$

then to halve the error bound means to replace n by n^2. We could, instead, get this halving for free by using Q_n^{Mi} ($\equiv Q_n^{\mathrm{opt}}$) rather than Q_n^{G}.

The above example shows that one should not judge the usefulness of the Gaussian method based only on its loss. We have shown that the loss in $\mathcal{C}_M^{(r)}$ is large, but the significance of this fact is reduced by the following theorem.

THEOREM 6.9.2. We have

$$\rho(Q_{2n}^{\mathrm{G}}, \mathcal{C}_M^{(r)}) \le \rho_n^{\mathrm{best}}(\mathcal{C}_M^{(r)})$$

for $r = 1, 2 \ldots$ and $n \ge r$.

PROOF. From the results in Section 8.2, we obtain

$$\rho_n^{\mathrm{best}}(\mathcal{C}_M^{(r)}) \ge \frac{2\mathcal{K}_r}{(\pi n)^r} M, \tag{6.83}$$

and (6.55) gives

$$\rho(Q_n^{\mathrm{G}}, \mathcal{C}_M^{(r)}) \le \mathcal{K}_r \frac{(2n-r)!}{(2n)!} \cdot 4M.$$

It is therefore sufficient to prove that

$$\frac{4\mathcal{K}_r}{4n(4n-1)\cdots(4n-r+1)} \le \frac{2\mathcal{K}_r}{(\pi n)^r}$$

or, equivalently,

$$\prod_{\nu=1}^{r-1} \left(1 - \frac{\nu}{4n}\right) \ge 2\left(\frac{\pi}{4}\right)^r. \tag{6.84}$$

Since the left-hand side is an increasing function of n, we just need to show that

$$\prod_{\nu=1}^{r-1} \left(1 - \frac{\nu}{4r}\right) \ge 2\left(\frac{\pi}{4}\right)^r. \tag{6.85}$$

This statement can be verified for $r = 6$ by direct computation and for $r \ge 7$ by using Stirling's formula. For $r = 5$, (6.85) does not hold; however, (6.84) does hold for $n \ge 6$ and, also, for $r = 4$ and $n \ge 6$ as well as $r = 3$ and $n \ge 24$. In the remaining cases with $r = 3, 4, 5$, we can determine $\rho(Q_{2n}^{\mathrm{G}}, \mathcal{C}_M^{(r)})$ and verify the asserted inequality numerically or by explicit calculation.

When $r = 1$, we can apply Theorem 6.6.2; and when $r = 2$, we can use a result of Petras (1993c):

$$\rho(Q_n^{\mathrm{G}}, \mathcal{C}_M^{(2)}) \le \frac{2\sqrt{3}\,\pi^2}{81 n^2} M.$$

This finishes the proof of the theorem. □

The results in this section admit many refinements and generalizations, but for the purpose of justifying the "rule of thumb" stated at the beginning of this section, they should be quite sufficient.

Before moving on, let us mention that our recommendation of the Gaussian rule is not quite undisputed. Trefethen (2008) writes that "In practice Clenshaw–Curtis does not turn out to be half as efficient as Gauss. In fact, for most integrands, the two formulas are about equally accurate." Our results for holomorphic integrands in Section 6.7 do not agree with this judgement. On the other hand, Trefethen's numerical experiments do support his thesis. The explanation of this apparent contradiction is that he uses the exact value of $R_n[f]$, whereas we use only the asymptotic value. In most practical problems, however, $R_n[f]$ is unknown, and the decision on whether to use Q_n^{G} or Q_n^{CC} has to be based on one's knowledge of $\rho(Q_n, \mathcal{C})$ for some co-observation; in almost all such cases, Q_n^{G} comes out better.

6.10. The Kronrod method

To get a hint as to the error of a quadrature value $Q[f]$, attempts were made early on to obtain a better estimate by employing another quadrature rule $\widetilde{Q}$. The argument is that

$$R[f] = I[f] - Q[f] \approx \widetilde{Q}[f] - Q[f],$$

which would be plausible only if $\widetilde{Q}$ uses many more nodes. However, the main effort in obtaining a quadrature value lies in the computation of function values. Therefore, for this idea to be useful in practice, we shall require that $\widetilde{Q}$ reuse all the function values that have already been computed for the application of Q. In other words, the nodal set of $\widetilde{Q}$ should contain that of Q. This consideration leads to the following definition.

DEFINITION 6.10.1. Let

$$Q[f] = \sum_{\nu=1}^{n} a_\nu f(x_\nu).$$

The quadrature rule $\widetilde{Q}$ is said to be an *extension* of Q if it has the form

$$\widetilde{Q}[f] = \sum_{\nu=1}^{n} b_\nu f(x_\nu) + \sum_{\mu=1}^{m} c_\mu f(y_\mu), \quad \text{where } y_\mu \notin \{x_1, \dots, x_m\},\ b_\nu \neq 0.$$

Without the requirement that $b_\nu \neq 0$, all quadrature rules would be extensions of each other; usually, however, we will not be so strict as to insist on all the b_ν being non-zero, and will allow rare accidental zero coefficients among the b_ν.

We need to find a compromise between m being not too large and $\widetilde{Q}$ being significantly better than Q. One attempt at achieving this balance is to choose m to be the minimal possible number such that $\deg \widetilde{Q} > \deg Q$. Applying this principle to $Q = Q_n^{\mathrm{G}}$, we see that only $m > n$ makes sense, because any quadrature rule with no more than $2n$ nodes given and degree greater than or equal to $2n-1$ is uniquely determined and must therefore be the Gaussian rule. Taking any $n+1$ new nodes and using the quadrature rule of interpolatory type on all $2n+1$ nodes, we obtain an extension of degree $2n$ or greater. We want to use our freedom in the choice of the $n+1$ new nodes to make this degree as large as possible. The following is what Kronrod (1964) proposed.

DEFINITION 6.10.2. Let q_n be the nth orthonormal polynomial with respect to the weight function w. The monic polynomial $E_{n+1} \in \mathbb{P}_{n+1}$ is called the *nth Stieltjes polynomial with respect to I* if

$$I[q_n E_{n+1} p] = 0 \quad \text{for all } p \in \mathbb{P}_n. \tag{6.86}$$

According to (6.86), we can regard E_{n+1} as something like an orthogonal polynomial, where the "weight function" $w \cdot q_n$ has sign changes.

THEOREM 6.10.1. *For given n and I, there exists exactly one Stieltjes polynomial.*

PROOF. Set

$$E_{n+1}(x) = x^{n+1} + \sum_{\nu=0}^{n} \alpha_\nu x^\nu.$$

Then the orthogonality of q_n, together with condition (6.86) in which we subsequently take $p(x) = x^0, x^1, \ldots, x^n$, gives a triangular linear system of equations for the coefficients α_ν. $\square$

A consequence of the uniqueness is that

$$E_{n+1}(x) = (-1)^{n+1} E_{n+1}(b + a - x) \tag{6.87}$$

if w is symmetric.

Theorem 5.1.3 leads to the following result.

THEOREM 6.10.2. If E_{n+1} has $n+1$ pairwise distinct zeros y_μ on $[a,b]$ that are also different from $x_1^{\mathrm{G}}, \ldots, x_n^{\mathrm{G}}$, then the quadrature rule of interpolatory type with nodes $x_1^{\mathrm{G}}, \ldots, x_n^{\mathrm{G}}, y_1, \ldots, y_{n+1}$ has degree at least $3n+1$.

DEFINITION 6.10.3. A quadrature rule as in Theorem 6.10.2 is referred to as a $(2n+1)$st Kronrod rule and is denoted by Q_{2n+1}^{Kro}.

The conditions in Theorem 6.10.2 are not always satisfied—verifying them is in fact one of the major problems in the theory of Kronrod rules. For $w = 1$, the proof will be presented in the next section. If the weight function has the form $w(x) = \left(1 - x^2\right)^\beta$, existence of Kronrod rules is known for $\beta \in [-1/2, 3/2]$ (Szegö 1935) and non-existence for $\beta \in \left]5/2, \infty\right[$ and sufficiently large n (Peherstorfer and Petras 2000). Peherstorfer (1992, 1995) proved existence for $w(x) = v(x)\sqrt{1-x^2}$ where $v \in C^2[-1,1]$ and $v > 0$ on $[-1,1]$.

Gautschi and Rivlin (1988) demonstrated that Kronrod rules for weight functions on $[-1,1]$ given by

$$w(x) = \frac{\sqrt{1-x^2}}{(1+\gamma)^2 - 4\gamma t^2}, \quad \gamma \in \left]-1, 1\right[,$$

can be treated almost explicitly. Even more special is the case where

$$w(x) = \sqrt{1-x^2}$$

with

$$q_n = \sqrt{\frac{2}{\pi}}\, U_n \quad \text{and} \quad E_{n+1} = 2^{-n} T_{n+1},$$

which follow from $U_n \cdot T_{n+1} = U_{2n+1}$. The latter also implies $Q_{2n+1}^{\mathrm{Kro}} = Q_{2n+1}^{\mathrm{G}}$, i.e. $\deg Q_{2n+1}^{\mathrm{Kro}} = 4n+1$, which is an exception. A similarly simple case is where

$$w(x) = \frac{1}{\sqrt{1-x^2}} \quad \text{with} \quad q_n = \sqrt{\frac{2}{\pi}}\, T_n \quad \text{and} \quad E_{n+1}(x) = (1-x^2) U_{n-1}(x).$$

Gautschi and Notaris (1989) and Notaris (1990b) found further weight functions for which Kronrod rules exist.

Once existence is established, the next important question concerns positivity (we have already mentioned many nice properties that follow from positivity and a high degree). The main tool used in such investigations is the following relation involving divided differences:

$$Q_{2n+1}^{\mathrm{Kro}}[f] - Q_n^{\mathrm{G}}[f] = R_n^{\mathrm{G}}[p_{2n}] \cdot \operatorname{dvd}(x_1^{\mathrm{G}}, \ldots, x_n^{\mathrm{G}}, y_1, \ldots, y_{n+1}) \tag{6.88}$$

where $p_{2n}(x) = x^{2n}$; this equation follows easily from the definition of divided differences (Definition 5.5.2).

Equation (5.42) for the coefficients of divided differences immediately implies the following fact:

For

$$\operatorname{dvd}(z_1, \dots, z_m) = \sum_{\nu=1}^{m} \beta_\nu f(z_\nu) \qquad \textit{with } z_1 < \dots < z_m,$$

we have

$$\operatorname{sgn}(\beta_\nu) = (-1)^{m-\nu}.$$

A simple consequence of this fact and the positivity of $R_n^{\mathrm{G}}[p_{2n}]$ (Theorem 6.4.4) is the next theorem.

THEOREM 6.10.3 (Monegato 1976). If there exists a Kronrod rule

$$Q_{2n+1}^{\mathrm{Kro}}[f] = \sum_{\nu=1}^{n} b_\nu f(x_\nu^{\mathrm{G}}) + \sum_{\mu=1}^{n+1} c_\mu f(y_\mu) \tag{6.89}$$

with

$$y_1 < x_1^{\mathrm{G}} < y_2 < x_2^{\mathrm{G}} < \cdots < x_n^{\mathrm{G}} < y_{n+1},$$

then we have

$$c_\mu > 0 \quad \text{for } \mu = 1, \dots, n+1.$$

Investigation of the positivity (or non-positivity) of the remaining coefficients is often much more difficult. For $w = 1$, the proof of positivity in the next section will be based on the relation

$$b_\nu = a_\nu^{\mathrm{G}} + \frac{R_n^{\mathrm{G}}[p_{2n}]}{E_{n+1}(x_\nu) \prod\limits_{\substack{\mu=1 \\ \mu \neq \nu}}^{n} (x_\nu^{\mathrm{G}} - x_\mu^{\mathrm{G}})}, \tag{6.90}$$

which follows from (6.88) and (5.42).

The numerical determination of Kronrod rules is discussed in numerous articles. Monegato (2001) provides a survey, and tables of nodes and coefficients can be found in Kronrod (1964) and Piessens *et al.* (1983).

Up to now we have been taking Q_n^{G} as our starting point, but we can also extend Q_n^{Lo} in the same way. In this case, $n-1$ additional nodes are needed to achieve a higher degree. These nodes can be chosen so that the extended rule has degree at least $3n-3$ (Patterson 1968).

The concept of extension given in Definition 6.10.1 encompasses many special cases of interest; see Gautschi (1988). Among them, iterated extensions have attracted particular attention.

Finally, we emphasize that aiming to identify a "better rule" that has "higher degree" is, of course, not the only justifiable approach. This idea is essentially based on an overall intention to be independent of co-observation. Nonetheless, the goal of achieving "high degree" has led to a very rich theory.

6.11. Kronrod rules for $w = 1$

In this section we again set $[a,b] = [-1,1]$ and $w = 1$. Even in this standard case, many difficulties arise in proving existence and positivity. First, we need some knowledge about solutions of the Legendre differential equation

$$[(1-x^2)y']' + n(n+1)y = 0 \tag{6.91}$$

that are linearly independent of the Legendre polynomials.

LEMMA 6.11.1. Denote by $x_1, \ldots, x_n$ the zeros of the Legendre polynomial P_n. Then

$$S_n(x) := \frac{1}{2}P_n(x)\ln\frac{1+x}{1-x} - \sum_{\nu=1}^{n}\frac{1}{\left[P_n'(x_\nu)\right]^2(1-x_\nu^2)}\cdot\frac{P_n(x)}{(x-x_\nu)}$$

is a solution of (6.91) on $]-1,1[$ which is linearly independent of P_n.

PROOF. The usual technique for obtaining a second solution is to plug $y = u{\cdot}P_n$ into (6.91). This results in the differential equation

$$u'' + 2\left(\frac{P_n'}{P_n} - \frac{x}{1-x^2}\right)u' = 0,$$

whose first integral is

$$u'(x) = \frac{\text{const}}{(1-x^2)P_n^2(x)}.$$

To integrate this expression, we use the partial fraction decomposition

$$\frac{1}{(1-x^2)P_n^2(x)} = \frac{c_0}{1-x} + \frac{c_1}{1+x} + \sum_{\nu=1}^{n}\left(\frac{d_\nu}{(x-x_\nu)^2} + \frac{e_\nu}{x-x_\nu}\right).$$

All of the e_ν vanish, or else the solution uP_n would have singularities in $]-1,1[$. We find the remaining coefficients to be

$$c_1 = c_0 = \frac{1}{2P_n^2(1)} = \frac{1}{2} \quad \text{and} \quad d_\nu = \frac{1}{(1-x_\nu^2)\left[P_n'(x_\nu)\right]^2},$$

and this proves the lemma. □

LEMMA 6.11.2.

$$S_n(x) = \frac{1}{2}P_n(x)\ln\frac{1+x}{1-x} - \frac{1}{2}\int_{-1}^{1}\frac{P_n(y)-P_n(x)}{y-x}\,dy.$$

PROOF. By Lemma 6.11.1, it remains to show that

$$\frac{1}{2}\int_{-1}^{1}\frac{P_n(y)-P_n(x)}{y-x}\,dy = \sum_{\mu=1}^{n}\frac{1}{\left[P_n'(x_\mu)\right]^2(1-x_\mu^2)}\cdot\frac{P_n(x)}{(x-x_\mu)}.$$

We have polynomials of degree $n-1$ or less on both sides, so they are identical if they coincide at the n points $x_1, \ldots, x_n$. Indeed, the left-hand side evaluated at x_ν gives

$$\frac{1}{2}\int_{-1}^{1}\frac{P_n(y)}{y-x_\nu}\,dy,$$

while the right-hand side gives

$$\frac{1}{P_n'(x_\nu)(1-x_\nu^2)}.$$

Both values are equal to $\frac{1}{2}P_n'(x_\nu)a_\nu^{\mathrm{G}}$ (see (5.2) and (6.4)), so we are done. □

The proof of Lemma 6.11.2 also shows the following:

LEMMA 6.11.3.

$$a_\nu^{\mathrm{G}} = -\frac{2S_n(x_\nu)}{P_n'(x_\nu)}.$$

Another consequence of Lemma 6.11.2 is that the recurrence relation for Legendre polynomials carries over to the S_n.

LEMMA 6.11.4. For $n = 2, 3, \ldots$, we have

$$nS_n(x) = (2n-1)xS_{n-1}(x) - (n-1)S_{n-2}(x).$$

From this, we can derive an expansion for S_n as follows:

LEMMA 6.11.5. For $t \in \,]0, \pi[$, we have

$$S_n(\cos t) + i\,\frac{\pi}{2}P_n(\cos t) = 2 \cdot \frac{2 \cdot 4 \cdots (2n)}{3 \cdot 5 \cdots (2n+1)} \sum_{\nu=0}^{\infty} f_{\nu,n} e^{i(n+2\nu+1)t},$$

where

$$f_{\nu,n} = \prod_{\mu=1}^{\nu} \frac{2\mu-1}{2\mu} \cdot \frac{n+\mu}{n+\mu+\frac{1}{2}} \quad \text{for } \nu = 1, 2, 3, \ldots \text{ (and } f_{0,n} = 1).$$

PROOF. We use induction over n. For $n = 0$, Fourier series expansion shows that

$$\frac{1}{2}\ln\frac{1+\cos t}{1-\cos t} = \sum_{\nu=0}^{\infty} \frac{2}{2\nu+1}\cos\bigl((2\nu+1)t\bigr)$$

and

$$\frac{\pi}{2} = \sum_{\nu=0}^{\infty} \frac{2}{2\nu+1}\sin\bigl((2\nu+1)t\bigr).$$

For $n = 1$, we apply the identity

$$S_1(\cos t) + i\,\frac{\pi}{2}\cos t = \Bigl(S_0(\cos t) + i\frac{\pi}{2}\Bigr)\cos t - 1.$$

The induction step is standard if we take into account the fact that S_n and P_n satisfy the same recurrence relation. □

Now we can prove the main theorem.

THEOREM 6.11.1 (Szegö 1935). The zeros of E_{n+1} lie in $]-1, 1[$, are simple and interlace with those of P_n.

PROOF. We show details for the case where n is even; for odd n, only slight modifications are necessary. Let us fix n so that we can omit the second index in $f_{\nu,n}$. Since E_{n+1} is an odd function and its leading coefficient is 1, we have

$$E_{n+1} = \sum_{\mu=0}^{n/2} \gamma_\mu T_{n-2\mu+1}, \qquad \gamma_0 = 2^{-n}. \tag{6.92}$$

The orthogonality of $E_{n+1}P_n$ to all polynomials of degree $n-1$ gives

$$\sum_{\mu=0}^{n/2} \gamma_\mu \int_{-1}^{1} P_n(x)T_{n-2\mu+1}(x)U_{2\lambda-1}(x)\,dx = 0 \quad \text{for } \lambda = 1, 2, \ldots, n/2.$$

Setting $U_{-i}(x) = -U_{i-2}(x)$ for $i \geq 0$, trigonometric identities show that

$$T_i U_j = \frac{1}{2}(U_{j+i} + U_{j-i}),$$

and we get the conditions

$$\sum_{\mu=0}^{n/2} \gamma_\mu \int_{-1}^{1} P_n(x)\big(U_{n-2\mu+2\lambda}(x) + U_{2\lambda+2\mu-n-2}(x)\big)\, dx = 0.$$

The orthogonality of P_n simplifies the above equation to

$$\sum_{\mu=0}^{\lambda} \gamma_\mu \int_{-1}^{1} P_n(x) U_{n-2\mu+2\lambda}(x)\, dx = 0.$$

Owing to the fact that

$$\int_{-1}^{1} P_n(x) U_{n+2k}(x)\, dx = \int_0^\pi P_n(\cos t) \sin\big((n+2k+1)t\big)\, dt,$$

the integral is given in Lemma 6.11.5 and so our conditions finally become

$$\sum_{\mu=0}^{\lambda} \gamma_\mu f_{\lambda-\mu} = 0 \quad \text{for } \lambda = 1, 2, \ldots, n/2. \tag{6.93}$$

We now define numbers $\gamma_\mu = \gamma_{\mu,n}$ recursively by

$$\gamma_0 f_0 = 2^{-n}, \qquad \sum_{\mu=0}^{\lambda} \gamma_\mu f_{\lambda-\mu} = 0 \quad \text{for } \lambda = 1, 2, 3, \ldots. \tag{6.94}$$

It can then be seen that $\gamma_0, \gamma_1 \ldots, \gamma_{n/2}$ are identical to the numbers having the same name and which are defined by (6.92). As a consequence, we obtain

$$f_0 + f_1 z + f_2 z^2 + \cdots = \frac{2^{-n}}{\gamma_0 + \gamma_1 z + \gamma_2 z^2 + \cdots} \tag{6.95}$$

for $|z| < 1$. Next, we want to show that

$$\gamma_0 > 0 \quad \text{while} \quad \gamma_1 < 0,\ \gamma_2 < 0,\ \gamma_3 < 0, \ldots \tag{6.96}$$

and that

$$\gamma_0 + \gamma_1 + \gamma_2 + \cdots = 0. \tag{6.97}$$

The sign properties of γ_0 and γ_1 follow directly from (6.94). To prove the remaining statements, we combine two consecutive equations in (6.94) to eliminate γ_0. This gives

$$\gamma_{\lambda+1} f_0 = \sum_{\mu=1}^{\lambda} \gamma_\mu f_{\lambda-\mu} \cdot \left(\frac{f_{\lambda+1}}{f_\lambda} - \frac{f_{\lambda+1-\mu}}{f_{\lambda-\mu}} \right) \quad \text{for } \lambda = 1, 2, \ldots$$

Since the sequence of $f_{\nu+1}/f_\nu$ values is increasing, the coefficients of this recurrence relation for the γ_μ must be positive, hence (6.96) follows. Equation (6.97) is a consequence of the divergence of $\sum_{\nu=0}^\infty f_\nu$.

For the next step of the proof, we define

$$e_{n+1}(t) := \sum_{\mu=0}^{n/2} \gamma_\mu \sin\big((n - 2\mu + 1)t\big),$$

so that

$$E_{n+1}(\cos t) - ie_{n+1}(t) = \sum_{\mu=0}^{n/2} \gamma_\mu e^{-i(n-2\mu+1)t}.$$

By Lemma 6.11.5 we have that for all $t \in]0, \pi[$,

$$\begin{aligned}
&\left(S_n(\cos t) + i\frac{\pi}{2}P_n(\cos t)\right)\left(E_{n+1}(\cos t) - ie_{n+1}(t)\right) \\
&= 2 \cdot \frac{2 \cdot 4 \cdots 2n}{3 \cdot 5 \cdots (2n+1)} \sum_{\nu=0}^{\infty} f_\nu e^{i(n+2\nu+1)t} \sum_{\mu=0}^{n/2} \gamma_\mu \cdot e^{-i(n-2\mu+1)t} \\
&= 2 \cdot \frac{2 \cdot 4 \cdots 2n}{3 \cdot 5 \cdots (2n+1)} \sum_{\nu=0}^{\infty} f_\nu e^{i \cdot 2\nu t} \sum_{\mu=0}^{n/2} \gamma_\mu \cdot e^{i \cdot 2\mu t} \\
&= 2 \cdot \frac{2 \cdot 4 \cdots 2n}{3 \cdot 5 \cdots (2n+1)} \cdot 2^{-n} \left(\sum_{\mu=0}^{n/2} \gamma_\mu \cdot e^{i \cdot 2\mu t}\right) \cdot \left(\sum_{\mu=0}^{\infty} \gamma_\mu \cdot e^{i \cdot 2\mu t}\right)^{-1} \\
&= 2 \cdot \frac{n!}{3 \cdot 5 \cdots (2n+1)} \cdot \frac{1}{1+z}
\end{aligned}$$

with

$$z = \left(\sum_{\mu=1+n/2}^{\infty} \gamma_\mu \cdot e^{i \cdot 2\mu t}\right) \cdot \left(\sum_{\mu=0}^{n/2} \gamma_\mu \cdot e^{i \cdot 2\mu t}\right)^{-1}.$$

We also have

$$|z| \le \left(\sum_{\mu=1+n/2}^{\infty} |\gamma_\mu|\right)\left(|\gamma_0| - \sum_{\mu=1}^{n/2} |\gamma_\mu|\right)^{-1} = -\left(\sum_{\mu=1+n/2}^{\infty} \gamma_\mu\right)\left(\sum_{\mu=0}^{n/2} \gamma_\mu\right)^{-1} = 1.$$

Using the fact that $\inf_{|z| \le 1} \Re(1+z)^{-1} \ge 1/2$, we obtain the key result

$$S_n(\cos t)E_{n+1}(\cos t) + \frac{\pi}{2}P_n(\cos t)e_{n+1}(t) \ge \frac{n!}{3 \cdot 5 \cdots (2n+1)} \quad \text{for } 0 < t < \pi. \tag{6.98}$$

We conclude that, at the zeros of P_n and for $x \to \pm 1$, E_{n+1} has the same sign as S_n. Now, recall that S_n is a solution of the Legendre differential equation and is linearly independent of P_n. A well-known result, which is a variant of Sturm's comparison theorem (see Theorem 6.3.1), says that the zeros of two linearly independent solutions interlace. Thus the proof of Theorem 6.11.1 is complete. □

THEOREM 6.11.2 (Monegato 1978a). *The Kronrod rules for $w = 1$ are positive.*

PROOF. For nodes that are at the zeros of Stieltjes polynomials, the assertion follows immediately on combining Theorems 6.10.3 and 6.11.1.

For coefficients corresponding to the other nodes, we have the representation (6.90):

$$a_\nu^{\mathrm{G}} + \frac{R_n^{\mathrm{G}}[p_{2n}]}{E_{n+1}(x_\nu) \prod_{\substack{\mu=1 \\ \mu \ne \nu}}^{n} (x_\nu^{\mathrm{G}} - x_\mu^{\mathrm{G}})}.$$

We plug in the formulas

$$a_\nu^{\mathrm{G}} = -\frac{2S_n(x_\nu^{\mathrm{G}})}{P_n'(x_\nu^{\mathrm{G}})}$$

from Lemma 6.11.3,

$$R_n^{\mathrm{G}}[p_{2n}] = \frac{1}{2n+1} \cdot \frac{2^{2n+1}(n!)^4}{[(2n)!]^2}$$

from Theorem 6.5.1, and

$$P_n(x) = \frac{(2n)!}{2^n(n!)^2}x^n + \cdots$$

from Definition A.4 to get that the coefficient under consideration is

$$\frac{-2}{P_n'(x_\nu^{\mathrm{G}})E_{n+1}(x_\nu^{\mathrm{G}})} \cdot \left(S_n(x_\nu^{\mathrm{G}})E_{n+1}(x_\nu^{\mathrm{G}}) - \frac{n!}{3\cdot 5 \cdots (2n+1)} \right).$$

Inequality (6.98) tells us that the term in brackets is non-negative, so it remains to show that

$$P_n'(x_\nu^{\mathrm{G}})E_{n+1}(x_\nu^{\mathrm{G}}) < 0,$$

and this follows directly from the interlacing of zeros of P_n and E_{n+1}. □

The use of Kronrod quadrature is justified only if $Q_{2n+1}^{\mathrm{Kro}}[f]$ is a better approximation for $I[f]$ than is $Q_n^{\mathrm{G}}[f]$. One result in this direction (which can be generalized) is the following:

THEOREM 6.11.3 (Brass and Förster 1987).

$$\frac{\rho\big(Q_{2n+1}^{\mathrm{Kro}}, \mathcal{C}_M^{(2n)}\big)}{\rho\big(Q_n^{\mathrm{G}}, \mathcal{C}_M^{(2n)}\big)} \le \mathrm{const}\cdot\sqrt[4]{n}\left(\frac{16}{25\sqrt{5}}\right)^n = \mathrm{const}\cdot\sqrt[4]{n}\left(\frac{1}{3.493\ldots}\right)^n.$$

PROOF. Theorem 6.5.1 says that

$$\rho\big(Q_n^{\mathrm{G}}, \mathcal{C}_M^{(2n)}\big) = \frac{1}{2n+1}\cdot\frac{2^{2n+1}(n!)^4}{[(2n)!]^3}\cdot M,$$

while equation (5.7) and Lebesgue's inequality tell us that

$$\begin{aligned}\rho\big(Q_{2n+1}^{\mathrm{Kro}}, \mathcal{C}_M^{(2n)}\big) &\le 4 \sup_{\|f^{(2n)}\|\le M} \mathrm{dist}(f, \mathbb{P}_{3n+1}) \\ &\le \frac{1}{(2n)!\,2^{2n-3}}\left(\frac{(4n)!(n+2)!}{(5n+1)!(4n-1)}\right)^{1/2} M.\end{aligned}$$

Stirling's formula then gives the assertion. □

Another way of testing the quality of Kronrod rules is to compare them with Q_{2n+1}^{G}: is $\rho\big(Q_{2n+1}^{\mathrm{G}}, \mathcal{C}\big)$ much smaller than $\rho\big(Q_{2n+1}^{\mathrm{Kro}}, \mathcal{C}\big)$? If so, Q_{2n+1}^{Kro} is justified only as an error indicator for Q_n^{G} but not as a good formula on its own. Ehrich (1994a) has shown that this is the case for estimates involving derivatives of highest possible order. His result says that for $n \ge 15$,

$$\frac{\rho\big(Q_{2n+1}^{\mathrm{G}}, \mathcal{C}_M^{(3n+2+\lambda)}\big)}{\rho\big(Q_{2n+1}^{\mathrm{Kro}}, \mathcal{C}_M^{(3n+2+\lambda)}\big)} < 3^{1-n} \quad \text{with } \lambda = \begin{cases} 0 & \text{for even } n, \\ 1 & \text{for odd } n. \end{cases}$$

Analogous results for other co-observations can be found in Ehrich (1993).

Kronrod rules are covered in many references. For instance, see Rabinowitz (1980) for further information on degree, Rabinowitz (1986b) for non-definiteness,

Monegato (1978b), Notaris (1993) and Ehrich (1994a, 1994b, 1996a) for error bounds, Ehrich (1995) for asymptotics of E_{n+1} and a_ν^{Kro}, Ehrich (1996b) for precise asymptotics of error bounds, Ehrich (1998) for analogues of the theorems of Sloan and Smith, Ehrich (1999) for a general survey, and Monegato (2001) for computational aspects.

CHAPTER 7

Quadrature Rules with Equidistant Nodes

For practical reasons, quadrature rules with equidistant nodes are sometimes employed. Since the principle of constructing quadrature formulas by integrating interpolating polynomials does not yield useful methods, we shall integrate piecewise linear functions instead. The trapezoidal method results from this approach and, while converging slowly itself (except in the case of $(b-a)$-periodic integrands—see Chapter 8), forms the basis of the more powerful methods of Gregory and Romberg. These have much better error asymptotics, and the Romberg methods, in particular, are often used in computing practice.

7.1. The trapezoidal method and the Euler–Maclaurin formula

Throughout our discussion of the trapezoidal rule and its refinements, we shall restrict our attention to the weight function $w = 1$. The nodes

$$x_\nu = a + (\nu - 1)h \quad \text{for } \nu = 1, 2, \ldots, n+1, \quad \text{where } h := \frac{b-a}{n},$$

will be used in Sections 7.1–7.7. In Section 7.8 we shall consider another set of equidistant nodes.

The trapezoidal method was introduced in Example 2.1.1, and we repeat its definition here.

DEFINITION 7.1.1. The quadrature rule defined by

$$Q_{n+1}^{\mathrm{Tr}}[f] := h \cdot \left(\frac{1}{2} f(a) + \sum_{\nu=2}^{n} f(x_\nu) + \frac{1}{2} f(b) \right)$$

is the $(n+1)$st trapezoidal rule. The sequence $Q_2^{\mathrm{Tr}}, Q_3^{\mathrm{Tr}}, \ldots$ is called the trapezoidal method.

The trapezoidal method is perhaps the best-known and, after the Gaussian method, most-studied quadrature method. Note that Q_{n+1}^{Tr} is a projection rule (see Definition 2.2.2) corresponding to a simple spline space. A short calculation shows that

$$Q_{n+1}^{\mathrm{Tr}}[f] = \int_a^b s(x)\, dx,$$

where s denotes the spline function of degree one with knots $x_2, \ldots, x_n$ that interpolates f at the nodes $x_1, \ldots, x_{n+1}$. We begin our investigation with a famous formula:

THEOREM 7.1.1 (Euler–Maclaurin formula). Suppose that $f^{(r-1)}$ is absolutely continuous. Then

$$R_{n+1}^{\mathrm{Tr}}[f] = -\sum_{\mu=2}^{r} h^{\mu} B_{\mu}(0) \int_a^b f^{(\mu)}(x)\,dx + (-1)^r h^r \int_a^b B_r^*\left(\frac{x-a}{h}\right) f^{(r)}(x)\,dx.$$

For the definition of the Bernoulli polynomials B_μ and B_μ^*, see Appendix B.

PROOF. We start with the expansion of f in Bernoulli polynomials given in equation (B.5). The special case obtained by setting $z = c = x_\nu$ and $d = x_{\nu+1}$ yields

$$f(x_\nu) = \sum_{\mu=0}^{r} h^{\mu-1} B_\mu(0) \int_{x_\nu}^{x_{\nu+1}} f^{(\mu)}(x)\,dx + (-h)^{r-1} \int_{x_\nu}^{x_{\nu+1}} B_r^*\left(\frac{x-x_\nu}{h}\right) f^{(r)}(x)\,dx$$

for $\nu = 1, 2, \ldots, n$. Using

$$B_r^*\left(\frac{x-x_\nu}{h}\right) = B_r^*\left(\frac{x-a}{h} - (\nu-1)\right) = B_r^*\left(\frac{x-a}{h}\right),$$

summation over ν gives

$$\sum_{\nu=1}^{n} f(x_\nu) = \sum_{\mu=0}^{r} h^{\mu-1} B_\mu(0) \int_a^b f^{(\mu)}(x)\,dx + (-h)^{r-1} \int_a^b B_r^*\left(\frac{x-a}{h}\right) f^{(r)}(x)\,dx.$$

Now we use $B_0(0) = 1$ and $B_1(0) = -1/2$ to obtain the assertion after a simple rearrangement. □

The Euler–Maclaurin formula decomposes the error of the trapezoidal rule into summands of order n^{-2}, n^{-4}, and so on (since $B_{2\nu+1}(0) = 0$), plus the remainder

$$\rho_r := (-1)^r h^r \int_a^b B_r^*\left(\frac{x-a}{h}\right) f^{(r)}(x)\,dx.$$

For a thorough discussion of this remainder term, see Ostrowski (1969). All conclusions drawn from Theorem 7.1.1 are based on assuming that ρ_r is small. Our next theorem makes this assumption more precise.

THEOREM 7.1.2.

(i) $\rho_r = o(n^{-r})$ if $f^{(r-1)}$ is absolutely continuous.

(ii) $|\rho_r| \le n^{-r}(b-a)^{r+1} \int_0^1 |B_r(y)|\,dy \cdot \|f^{(r)}\|$ if $f^{(r)}$ is bounded.

(iii) $|\rho_r| \le n^{-r-1}(b-a)^{r+2}\,|B_{r+1}(0)| \cdot \|f^{(r+1)}\|$ if $f^{(r+1)}$ is bounded and r is odd.

(iv) $\rho_r = -n^{-r-1}(b-a)^{r+2} B_{r+1}(0) \cdot f^{(r+1)}(\xi)$ for some $\xi \in [a,b]$ if $f^{(r+1)}$ is continuous and r is odd.

Before proving these assertions, let us mention some important special cases:

$$R_{n+1}^{\mathrm{Tr}}[f] = o(n^{-1}) \tag{7.1}$$

if f is absolutely continuous;

$$R_{n+1}^{\mathrm{Tr}}[f] = -\frac{(b-a)^2}{12n^2}\left[f'(b) - f'(a)\right] + o(n^{-2}) \tag{7.2}$$

if f' is absolutely continuous;

$$|R_{n+1}^{\mathrm{Tr}}[f]| \leq \frac{(b-a)^2}{4n}\|f'\| \tag{7.3}$$

if f' is bounded;

$$|R_{n+1}^{\mathrm{Tr}}[f]| \leq \frac{(b-a)^3}{12n^2}\|f''\| \tag{7.4}$$

if f'' is bounded;

$$R_{n+1}^{\mathrm{Tr}}[f] = -\frac{(b-a)^3}{12n^2}f''(\xi) \quad \text{for some } \xi \in [a,b] \tag{7.5}$$

if f'' is continuous; and

$$R_{n+1}^{\mathrm{Tr}}[f] = -\frac{(b-a)^2}{12n^2}\left[f'(b) - f'(a)\right] + \frac{(b-a)^5}{720n^4}f^{(4)}(\xi) \quad \text{for some } \xi \in [a,b] \tag{7.6}$$

if $f^{(4)}$ is continuous.

The bounds (7.3), (7.4) and (7.5) are also easy consequences of the Peano kernel theorem. For the proof of Theorem 7.1.2 we need the following lemma.

LEMMA 7.1.1 (Polya 1933). Let φ be a bounded integrable function with period $b-a$. Then

$$\lim_{m\to\infty} \int_a^b \varphi(mx)f(x)\,dx = \frac{1}{b-a}\int_a^b \varphi(x)\,dx \int_a^b f(x)\,dx$$

for any integrable f.

PROOF. We prove the lemma under the additional assumption that

$$\int_a^b \varphi(x)\,dx = 0. \tag{7.7}$$

For more general φ, the assertion of the lemma follows upon replacing φ by

$$\varphi^* := \varphi - \frac{1}{b-a}\int_a^b \varphi(x)\,dx.$$

For each $\varepsilon > 0$, there is an absolutely continuous f_1 such that

$$\int_a^b |f(x) - f_1(x)|\,dx < \varepsilon.$$

Hence

$$\begin{aligned}\left|\int_a^b \varphi(mx)f(x)\,dx\right| &\leq \left|\int_a^b \varphi(mx)\left[f(x) - f_1(x)\right]dx\right| + \left|\int_a^b \varphi(mx)f_1(x)\,dx\right| \\ &\leq \varepsilon \sup_x |\varphi(x)| + \left|\int_a^b \varphi(mx)f_1(x)\,dx\right|.\end{aligned}$$

To estimate the last integral, we use integration by parts with the notation

$$\Phi(x) := \int_a^x \varphi(u)\,du.$$

From the periodicity and (7.7), the boundedness of Φ follows and we have

$$\left|\int_a^b \varphi(mx)f_1(x)\,dx\right| = \left|\left[\frac{1}{m}f_1(x)\Phi(mx)\right]_a^b - \frac{1}{m}\int_a^b f_1'(x)\Phi(mx)\,dx\right| \leq \frac{\mathrm{const}}{m},$$

which verifies the claim. □

PROOF OF THEOREM 7.1.2. (i) follows immediately from the definition of ρ_r and the lemma.

(ii) is proved by the simplest estimate for the integral:

$$|\rho_r| \le h^r \|f^{(r)}\| \int_a^b \left| B_r^*\left(n\frac{x-a}{b-a}\right)\right| dx = h^{r+1}\|f^{(r)}\| \int_0^n |B_r^*(y)|\, dy$$
$$= h^{r+1}\|f^{(r)}\| \cdot n \int_0^1 |B_r(y)|\, dy.$$

To prove (iv), we replace r in the Euler–Maclaurin formula by $r+1$, obtaining

$$\rho_r = -h^{r+1}B_{r+1}(0)\int_a^b f^{(r+1)}(x)\,dx + (-1)^{r+1}h^{r+1}\int_a^b B_{r+1}^*\left(\frac{x-a}{h}\right) f^{(r+1)}(x)\,dx$$
$$= h^{r+1}\int_a^b f^{(r+1)}(x)\left(-B_{r+1}(0) + B_{r+1}^*\left(\frac{x-a}{h}\right)\right) dx.$$

From the trigonometric expansion of B_{r+1}^* (see Theorem B.7), we conclude that $B_{r+1}^* - B_{r+1}(0)$ has no sign changes. Thus the generalized mean value theorem gives

$$\rho_r = h^{r+1}f^{(r+1)}(\xi)\int_a^b \left(B_{r+1}^*\left(\frac{x-a}{h}\right) - B_{r+1}(0)\right) dx$$
$$= h^{r+2}f^{(r+1)}(\xi)\int_0^n \left(B_{r+1}^*(y) - B_{r+1}(0)\right)\, dy$$
$$= -h^{r+2}f^{(r+1)}(\xi)\cdot nB_{r+1}(0),$$

and this is statement (iv). The proof of (iii) is similar. □

The usefulness of the Euler–Maclaurin formula is limited by the requirement that $f^{(r)}$ exist. For problems of the form

$$\int_0^1 x^\alpha g(x)\,dx, \quad \alpha \in\,]0,1[, \tag{7.8}$$

with a well-behaved g, we do not get much information about the asymptotics of R_n^{Tr}. However, there is a modification suitable for integrals like (7.8). For simplicity, we consider the fundamental interval $[0,1]$.

THEOREM 7.1.3 (Navot 1961). *Suppose that $f(x) = x^\alpha g(x)$ with $\alpha \in\,]0,1[$ and that $g \in C^{(m+1)}[0,1]$. Then*

$$R_{n+1}^{\mathrm{Tr}}[f] = -\sum_{\mu=2}^{m} n^{-\mu}B_\mu(0)f^{(\mu-1)}(1) - \sum_{\mu=0}^{m-1} n^{-\mu-1-\alpha}c_{\mu,\alpha}\frac{g^{(\mu)}(0)}{\mu!} + O(n^{-m-1})$$

where

$$c_{\mu,\alpha} = (-1)^\mu \alpha(\alpha+1)(\alpha+2)\cdots(\alpha+\mu)\int_0^\infty B_{\mu+1}^*(y)y^{\alpha-1}\,dy. \tag{7.9}$$

It is not difficult to replace the O-term by a (rather complicated) bound.

Concerning the numbers $c_{\mu,\alpha}$, we first have to show the existence of the improper integral in (7.9). This can be done with the help of the second mean value theorem:

$$\begin{aligned}\left|\int_c^d B^*_{\mu+1}(y)y^{\alpha-1}\,dy\right| &= \left|c^{\alpha-1}\int_c^\xi B^*_{\mu+1}(y)\,dy + d^{\alpha-1}\int_\xi^d B^*_{\mu+1}(y)\,dy\right| \\ &\le c^{\alpha-1}\left|B^*_{\mu+2}(\xi)-B^*_{\mu+2}(c)\right| + d^{\alpha-1}\left|B^*_{\mu+2}(d)-B^*_{\mu+2}(\xi)\right| \\ &\le 2\,\|B_{\mu+2}\|\,(c^{\alpha-1}+d^{\alpha-1}).\end{aligned}$$

Next, we give two other representations of $c_{\mu,\alpha}$. If we substitute the trigonometric expansion of $B^*_{\mu+1}$ (see Theorem B.7) into (7.9) and use

$$\int_0^\infty y^{\alpha-1}\sin y\,dy = \Gamma(\alpha)\sin\frac{\alpha\pi}{2} \quad\text{and}\quad \int_0^\infty y^{\alpha-1}\cos y\,dy = \Gamma(\alpha)\cos\frac{\alpha\pi}{2}$$

(see, e.g., Erdelyi 1953: p. 13), we obtain

$$c_{\mu,\alpha} = \frac{2\Gamma(\alpha+\mu+1)\cos\big((\alpha+\mu+1)\pi/2\big)}{(2\pi)^{\alpha+\mu+1}}\sum_{\nu=1}^\infty \frac{1}{\nu^{\alpha+\mu+1}}.$$

A known identity for Riemann's zeta function (see Erdelyi 1955: p. 190) gives

$$c_{\mu,\alpha} = \zeta(-\mu-\alpha).$$

This is a very simple, but perhaps not too useful, formula.

The following special case of Theorem 7.1.3 is the main tool used in its proof.

LEMMA 7.1.2. *Let $p(x) = x^{\alpha+\mu}$ with $\alpha\in\,]0,1[$ and $\mu\in\{0,1,2,\dots\}$. Then, for any $m>\mu$,*

$$R^{\mathrm{Tr}}_{n+1}[p] = -\sum_{\lambda=2}^m n^{-\lambda}B_\lambda(0)p^{(\lambda-1)}(1) - \frac{c_{\mu,\alpha}}{n^{\mu+\alpha+1}} + \rho,$$

where $c_{\mu,\alpha}$ is as given in (7.9) and

$$|\rho| \le \frac{4\,\|B_{m+1}\|\,|p^{(m)}(1)|}{n^{m+1}}.$$

PROOF. We start with the Euler–Maclaurin formula:

(7.10)

$$\begin{aligned}&R^{\mathrm{Tr}}_{n+1}[p] \\ &= -\sum_{\lambda=2}^{\mu+1} n^{-\lambda}B_\lambda(0)p^{(\lambda-1)}(1) - (-1)^\mu n^{-\mu-1}\int_0^1 B^*_{\mu+1}(nx)p^{(\mu+1)}(x)\,dx \\ &= -\sum_{\lambda=2}^{\mu+1} n^{-\lambda}B_\lambda(0)p^{(\lambda-1)}(1) - (-1)^\mu n^{-\mu-1-\alpha}\int_0^n B^*_{\mu+1}(y)p^{(\mu+1)}(y)\,dy \\ &= -\sum_{\lambda=2}^{\mu+1} n^{-\lambda}B_\lambda(0)p^{(\lambda-1)}(1) \\ &\quad - (-1)^\mu n^{-\mu-1-\alpha}\left(\int_0^\infty B^*_{\mu+1}(y)p^{(\mu+1)}(y)\,dy - \int_n^\infty B^*_{\mu+1}(y)p^{(\mu+1)}(y)\,dy\right).\end{aligned}$$

Using repeated integration by parts together with $B_{2\sigma+1}(0) = 0$, we obtain

$$\begin{aligned}
&\int_n^\infty B_{\mu+1}^*(y) p^{(\mu+1)}(y)\,dy \\
&= \sum_{\nu=0}^{m-\mu-2} (-1)^{\nu-1} B_{\mu+\nu+2}(0) p^{(\mu+1+\nu)}(n) + (-1)^{m-\mu-1} \int_n^\infty B_m^*(y) p^{(m)}(y)\,dy \\
&= (-1)^{\mu+1} \sum_{\nu=0}^{m-\mu-2} n^{\alpha-1-\nu} B_{\mu+\nu+2}(0) p^{(\mu+1+\nu)}(1) \\
&\quad + (-1)^{m-\mu-1} \int_n^\infty B_m^*(y) p^{(m)}(y)\,dy.
\end{aligned}$$

We plug the above expression into (7.10) to get

$$\begin{aligned}
R_{n+1}^{\mathrm{Tr}}[p] &= -\sum_{\lambda=2}^{\mu+1} n^{-\lambda} B_\lambda(0) p^{(\lambda-1)}(1) - n^{-\mu-1-\alpha} c_{\mu,\alpha} \\
&\quad - \sum_{\nu=0}^{m-\mu-2} n^{-\mu-\nu-2} B_{\mu+\nu+2}(0) p^{(\mu+1+\nu)}(1) \\
&\quad - (-1)^m n^{-\mu-1-\alpha} \int_n^\infty B_m^*(y) p^{(m)}(y)\,dy \\
&= -\sum_{\lambda=2}^{m} n^{-\lambda} B_\lambda(0) p^{(\lambda-1)}(1) - n^{-\mu-1-\alpha} c_{\mu,\alpha} \\
&\quad - (-1)^m n^{-\mu-1-\alpha} \int_n^\infty B_m^*(y) p^{(m)}(y)\,dy.
\end{aligned}$$

The last integral can be estimated in a manner quite similar to the proof of existence of $\int_0^\infty B_m^*(y) y^{\alpha-1}\,dy$ given above. □

PROOF OF THEOREM 7.1.3. Define

$$h(x) := \sum_{i=0}^{m-1} \frac{g^{(i)}(0)}{i!}\, x^{i+\alpha}.$$

Then $(f-h)^{(\lambda)}$ exists and is absolutely continuous for $\lambda = 0, 1, \ldots, m$. We have

$$(f-h)^{(\lambda)}(0) = 0 \quad \text{for } \lambda = 0, 1, \ldots, m,$$

so Theorem 7.1.1 gives

$$R_{n+1}^{\mathrm{Tr}}[f-h] = -\sum_{\mu=2}^{m} n^{-\mu} B_\mu(0) \Big(f^{(\mu-1)}(1) - h^{(\mu-1)}(1) \Big) + O(n^{-m-1}).$$

We then write

$$R_{n+1}^{\mathrm{Tr}}[f] = R_{n+1}^{\mathrm{Tr}}[f-h] + R_{n+1}^{\mathrm{Tr}}[h]$$

and apply Lemma 7.1.2 to $R_{n+1}^{\mathrm{Tr}}[h]$. □

7.2. More on the trapezoidal method

Why would one use the trapezoidal method? If f' is absolutely continuous and $f'(a) \neq f'(b)$, then (7.2) shows that

$$\lim_{n\to\infty} n^2 R_{n+1}^{\mathrm{Tr}}[f] \neq 0.$$

This implies rather slow convergence, and existence of derivatives of higher order does not improve matters. On the other hand, for any given r there exists a method with equidistant nodes such that $R_n[f] = O(n^{-r})$ if $f \in C^{(r)}$. The Euler–Maclaurin formula demonstrates that Q_n^{Tr} is able to compete with all such methods if

$$f^{(r)}(a) = f^{(r)}(b) \quad \text{for } r = 0, 1, 2, \ldots.$$

Under these conditions, f can be extended to a $(b-a)$-periodic function. We shall discuss periodic integrands in the next chapter.

The trapezoidal method can also be recommended in situations where the integrand is of low smoothness. The trapezoidal rule Q_n^{Tr} is strongly optimal for the co-observations $\mathcal{C}_M^{(1)}$ (Example 2.1.1), $\mathcal{C}_M^{\mathrm{Var}}$ (Theorem 4.4.2) and $\mathcal{C}_M^{\mathrm{Sa},1}$ (Theorem 4.7.1), and it is optimal for $\mathcal{C}_\omega^{\mathrm{mc}}$ (Theorem 4.5.2).

Let us now examine more closely the bounds that can be obtained for the co-observations discussed in Chapter 4. Sometimes we will use the Peano kernels of the trapezoidal rules; their representations

$$K_1(x)\big|_{]x_\nu, x_{\nu+1}[}(x) = \frac{x_{\nu+1} + x_\nu}{2} - x$$

and

$$K_2(x)\big|_{]x_\nu, x_{\nu+1}[}(x) = \frac{1}{2}(x - x_\nu)(x - x_{\nu+1})$$

are easily checked (note that the one for K_1 is none other than the Euler–Maclaurin formula for $r = 1$). In particular, Q_{n+1}^{Tr} is negative definite.

We begin with $\mathcal{C}^{(1)} \equiv \mathrm{Lip}_M\, 1$. The inequality

$$\left|R_{n+1}^{\mathrm{Tr}}[f]\right| \leq \frac{(b-a)^2}{4n}\|f'\| \tag{7.11}$$

that follows from $\rho\big(Q_{n+1}^{\mathrm{Tr}}, \mathcal{C}_M^{(1)}\big) = M(b-a)^2/(4n)$ cannot be improved upon: for any n, there exists a function $f \in \mathcal{C}_M^{(1)}$ for which equality holds in (7.11). Nevertheless, it is almost sure that (7.11) leads to an overestimation, since for any fixed $f \in \mathcal{C}_M^{(1)}$ we have

$$R_n^{\mathrm{Tr}}[f] = o(n^{-1}) \qquad \text{and} \qquad \sum_{\nu=2}^{\infty} \left|R_n^{\mathrm{Tr}}[f]\right| \nu^\beta < \infty \quad \text{if } \beta < 1/2$$

(see (7.1) and Theorem 7.2.3 below). On average, therefore, $\left|R_n^{\mathrm{Tr}}[f]\right|$ is substantially smaller than the bound in (7.11). Knowing this does not help us to find a better bound, however. As we have shown in Example 2.1.1, a better bound is possible, namely

$$\left|R_{n+1}^{\mathrm{Tr}}[f]\right| \leq \frac{(b-a)^2}{4n}\|f'\| - \frac{1}{4\|f'\|}\sum_{\nu=1}^{n}\left[f(x_{\nu+1}) - f(x_\nu)\right]^2.$$

This improvement on (7.11) was achievable because we used the full information on f, not just $f \in \mathcal{C}_M^{(1)}$, to determine the bound.

For $f \in \mathcal{C}_M^{(2)}$ we have, from $\rho(Q_{n+1}^{\mathrm{Tr}}, \mathcal{C}_M^{(2)}) = M(b-a)^3/(12n^2)$,

$$|R_{n+1}^{\mathrm{Tr}}[f]| \leq \frac{(b-a)^3}{12n^2} \|f''\|. \tag{7.12}$$

Formula (7.2) shows that this bound gives the correct order of $R_{n+1}^{\mathrm{Tr}}[f]$ if $f'(a) \neq f'(b)$. Compared with (7.11) this is an advantage, but the non-optimality of Q_n^{Tr} in $\mathcal{C}_M^{(2)}$ is a disadvantage. Optimal rules in this case (see Köhler 1988a) have a rather complicated structure. This is surprising, since the best rule (Theorem 4.3.2) is quite simple. Not much can be gained from using optimal rules, as can be seen from

$$\rho_{n+1}^{\mathrm{best}}(\mathcal{C}_M^{(2)}) = \frac{M(b-a)^3}{32(n+\sqrt{3/4})^2}.$$

Moreover, for the simple rule

$$Q_{n+1}^A[f] := h\left(\frac{13}{32}f(x_1) + \frac{35}{32}f(x_2) + \sum_{\nu=3}^{n-1} f(x_\nu) + \frac{35}{32}f(x_n) + \frac{13}{32}f(x_{n+1})\right),$$

(see Brass 1974b), the Peano kernel method gives

$$|R_{n+1}^A[f]| \leq \frac{(b-a)^3}{32n^2}\left(1 + \frac{533}{384n}\right)\|f''\|, \tag{7.13}$$

and Köhler (1988a) proved that

$$\rho(Q_{n+1}^A, \mathcal{C}_M^{(2)}) \leq \rho_n^{\mathrm{opt}}(\mathcal{C}_M^{(2)}).$$

The next co-observation we consider is $\mathcal{C}_M^{\mathrm{Var}}$. According to Theorem 4.4.4, we have

$$|R_{n+1}^{\mathrm{Tr}}[f]| \leq \frac{b-a}{2n} \operatorname{Var} f \tag{7.14}$$

and, in contrast to the situation in $\mathcal{C}_M^{(1)}$, changing from ρ^{opt} to ρ^{intr} does not yield an improvement.

Dubuc and Todor (1983) constructed an increasing function g such that

$$R_{2^k+1}^{\mathrm{Tr}}[g] = 2^{-k} \quad \text{for } k = 1, 2, \ldots.$$

Hence the sequence $\left(nR_{n+1}^{\mathrm{Tr}}[f]\right)_{n\in\mathbb{N}}$ does not necessarily tend to zero. Large terms are rare, however, because it can be proved (see Petras 1994b) that this sequence is Césaro-summable to zero, which means

$$\lim_{n\to\infty} \frac{1}{n} \sum_{\nu=1}^{n} \nu R_{\nu+1}^{\mathrm{Tr}}[f] = 0.$$

We can therefore say that (7.14) leads to overestimation in most cases. However, we should also emphasize the wide applicability of (7.14).

Now we look at the co-observation $\mathcal{C}_M^{\mathrm{Var},1}$. Theorem 4.6.1 gives

$$|R_{n+1}^{\mathrm{Tr}}[f]| \leq \frac{(b-a)^2}{8n^2} \operatorname{Var} f'. \tag{7.15}$$

Although $Q_{n+1}^{\mathrm{Tr}}[f]$ is not optimal in $\mathcal{C}_M^{\mathrm{Var},1}$ for $n > 1$, significant improvement is not possible since Arnold (1977) showed that

$$\rho_{n+1}^{\mathrm{opt}}(\mathcal{C}_M^{\mathrm{Var},1}) = \frac{3-\sqrt{8}}{2n^2}(b-a)^2 M.$$

In the class $\mathcal{C}_\omega^{\mathrm{mc}}$, specializing Theorem 4.5.2 gives

$$\left|R_{n+1}^{\mathrm{Tr}}[f]\right| \leq 2n \int_0^{h/2} \omega(t)\, dt. \tag{7.16}$$

As already remarked in Section 4.1, this bound gives an unrealistic overestimation in most cases of practical interest, even though it is the best possible in $\mathcal{C}_\omega^{\mathrm{mc}}$.

EXAMPLE 7.2.1. The computation of

$$\int_0^{e^{-2}} f_0(x)\, dx \quad \text{where } f_0(x) = \frac{1}{|\ln x|}$$

to a moderate precision using the trapezoidal method seems impossible if we bound the error by (7.16). Specifically, the modulus of continuity of f_0 is $\omega(t) = f_0(t)$ so that the right-hand side of (7.16) is greater than

$$\frac{1}{2e^2(2+\ln 2n)}$$

(the concavity of the integrand implies that the two-point trapezoidal rule gives a lower bound for the integral). This does not guarantee an error of less than $\frac{1}{2}\cdot 10^{-4}$ even for $n = 10^{500}$.

We could do better by using (7.14) instead of (7.16). In this case we obtain

$$\left|R_{n+1}^{\mathrm{Tr}}[f_0]\right| = O(n^{-1}),$$

and Theorem 7.2.5 below even gives

$$\left|R_{n+1}^{\mathrm{Tr}}[f_0]\right| = O\big((n \ln n)^{-1}\big).$$

Is further improvement possible? The answer is provided by the asymptotic behaviour

$$R_{n+1}^{\mathrm{Tr}}[f_0] = \frac{1}{2e^2 n \ln n}\big(1+o(1)\big),$$

which we shall prove by a method that can be used for many similar integrands. We start with

$$R_{n+1}^{\mathrm{Tr}}[f_0] = \int_0^h f_0(x)\, dx - \frac{h}{2} f_0(h) + \widetilde{R}_n^{\mathrm{Tr}}[f_0],$$

where $\widetilde{R}_n^{\mathrm{Tr}}$ denotes the error of the trapezoidal rule on the interval $[h, e^{-2}]$. For this term, we use (7.15) to obtain

$$\left|\widetilde{R}_n^{\mathrm{Tr}}[f_0]\right| \leq \frac{(e^{-2}-h)^2}{8(n-1)^2} \cdot \frac{1}{h \ln^2 h} = o\big((n \ln n)^{-1}\big).$$

We still have to show that

$$\int_0^h f_0(x)\, dx - \frac{h}{2} f_0(h) = \frac{h}{2} f_0(h)\big(1+o(1)\big),$$

that is,

$$\lim_{h\to 0+0} \frac{1}{h f_0(h)} \left(\int_0^h f_0(x)\, dx - h f_0(h) \right) = 0.$$

This can be achieved by applying the Bernoulli–l'Hospital rule, which tells us that the left-hand side is

$$\lim_{h\to 0+0} \frac{f_0(h) - [f_0(h) + h f_0'(h)]}{f_0(h) + h f_0'(h)} = \lim_{h\to 0+0} \frac{1}{\ln h - 1} = 0.$$

To examine the problem of overestimation more closely, we shall concentrate on the co-observation $\mathrm{Lip}_M\,\alpha$, which is a special case of $\mathcal{C}_\omega^{\mathrm{mc}}$. Let us recall the definition:

$$f \in \mathrm{Lip}_M\,\alpha \iff \omega(f;\delta) \le M\delta^\alpha.$$

The bound (7.16) yields

$$R_n^{\mathrm{Tr}}[f] = O(n^{-\alpha}) \tag{7.17}$$

for any $f \in \mathrm{Lip}_M\,\alpha$, and we obtain

$$R_n^{\mathrm{Tr}}[f] = o(n^{-1})$$

from (7.1) for an absolutely continuous function f, and certainly for $f \in \mathrm{Lip}_M\,1$.

We shall show that this kind of improvement of (7.17) is not possible for $\alpha < 1$ but that (7.17) fails to provide sufficiently concrete information about error.

We shall use the theory of Fourier series as a tool in our investigations. Let us first recall the definitions of the Fourier coefficients of f:

$$\alpha_\nu = \alpha_\nu[f] := \frac{1}{\pi}\int_0^{2\pi} f(x)\cos\nu x\,dx \quad \text{for } \nu = 0,1,2,\ldots,$$
$$\beta_\nu = \beta_\nu[f] := \frac{1}{\pi}\int_0^{2\pi} f(x)\sin\nu x\,dx \quad \text{for } \nu = 0,1,2,\ldots.$$

The Fourier series of f is given by

$$\frac{1}{2}\alpha_0[f] + \sum_{\nu=1}^{\infty}\bigl(\alpha_\nu[f]\cos\nu x + \beta_\nu[f]\sin\nu x\bigr).$$

A major result is the following theorem.

Theorem. *The Fourier series of f converges uniformly to f under one of the following sets of assumptions:*

(i) *f is continuous, 2π-periodic and of bounded variation*[1].

(ii) *f is continuous, 2π-periodic and such that*

$$\lim_{\delta\to 0+0}\omega(f;\delta)\ln\delta = 0.$$

There are many interrelations between Fourier coefficients and errors of the trapezoidal method. One of the simplest is the following:

THEOREM 7.2.1. Suppose that f satisfies assumption (i) or assumption (ii) of the previous theorem. The error of the trapezoidal method applied over the fundamental interval $[0, 2\pi]$ is

$$R_{n+1}^{\mathrm{Tr}}[f] = -2\pi\sum_{k=1}^{\infty}\alpha_{k\cdot n}[f].$$

This theorem is a rather immediate consequence of the lemma below.

[1]Here "bounded variation" means that f restricted to $[0, 2\pi]$ is of bounded variation.

LEMMA 7.2.1. On the fundamental interval $[0, 2\pi]$, we have

$$R_{n+1}^{\mathrm{Tr}}[\cos(\lambda\,\cdot\,)] = \begin{cases} -2\pi & \text{for } \lambda = n, 2n, 3n, \ldots, \\ 0 & \text{otherwise,} \end{cases}$$

$$R_{n+1}^{\mathrm{Tr}}[\sin(\lambda\,\cdot\,)] = 0.$$

PROOF. From the geometric series, we deduce that

$$\sum_{\nu=1}^{n} e^{i\lambda(\nu-1)h} = \begin{cases} n & \text{for } \lambda = 0, n, 2n, \ldots, \\ 0 & \text{otherwise.} \end{cases}$$

Taking real and imaginary parts of the above relation gives $h^{-1}Q_{n+1}^{\mathrm{Tr}}[\cos(\lambda\,\cdot\,)]$ and $h^{-1}Q_{n+1}^{\mathrm{Tr}}[\sin(\lambda\,\cdot\,)]$, respectively. The rest of the proof is trivial. □

Using Lemma 7.2.1, we can prove the existence of continuous functions that have $R_{n+1}^{\mathrm{Tr}}[f]$ tending to zero very slowly. Theorem 3.1.3 tells us that this result holds for any quadrature method, but the following theorem is more concrete.

THEOREM 7.2.2. Suppose that $(\sigma_n)_{n\in\mathbb{N}}$ decreases to zero, and let

$$f(x) = \frac{1}{2\pi}\sum_{\nu=1}^{\infty}(\sigma_{\nu+1} - \sigma_\nu)\cos(2^\nu x). \tag{7.18}$$

Then

$$R_{2^n+1}^{\mathrm{Tr}}[f] = \sigma_n.$$

PROOF. Evidently, the series (7.18) converges uniformly. So

$$\begin{aligned} R_{2^n+1}^{\mathrm{Tr}}[f] &= \frac{1}{2\pi}\sum_{\nu=1}^{\infty}(\sigma_{\nu+1} - \sigma_\nu)R_{2^n+1}^{\mathrm{Tr}}[\cos(2^\nu\cdot\,)] \\ &= \frac{1}{2\pi}\sum_{\nu=n}^{\infty}(\sigma_{\nu+1} - \sigma_\nu)\cdot(-2\pi) = \sigma_n, \end{aligned}$$

as claimed. □

Our next goal is to prove a certain degree of smoothness for the function (7.18). To this end, we use a classical theorem from approximation theory. With the notation

$$\mathbb{T}_n := \left\{\sum_{\nu=0}^{n}\left(\alpha_\nu\cos(\nu\,\cdot\,) + \beta_\nu\sin(\nu\,\cdot\,)\right) \,:\, \alpha_\nu, \beta_\nu \in \mathbb{R}\right\},$$

$$\operatorname{dist}(f, \mathbb{T}_n) := \inf\{\|f - t\| \,:\, t \in \mathbb{T}_n\},$$

this theorem can be formulated as follows:

Theorem. *Let f be 2π-periodic and let $\alpha \in \,]0,1[$. Then the following conditions are equivalent:*

(i) $f^{(k)} \in \operatorname{Lip}\alpha := \bigcup_M \operatorname{Lip}_M \alpha$.
(ii) $\operatorname{dist}(f, \mathbb{T}_n) = O(n^{-k-\alpha})$.

This theorem is due to Jackson and Bernstein; see DeVore and Lorentz (1993: p. 201).

If we choose the partial sums in (7.18) as approximating functions t, we obtain

$$\operatorname{dist}(f, \mathbb{T}_\lambda) \le \frac{1}{2\pi} \cdot \sigma_n \qquad \text{for } 2^{n-1} \le \lambda < 2^n.$$

For the special case where $\sigma_n := 2^{-n\alpha}$, this yields

$$\operatorname{dist}(f, \mathbb{T}_\lambda) \le \frac{1}{2\pi\,(2^n)^\alpha} \le \frac{1}{2\pi\lambda^\alpha}$$

and the Jackson–Bernstein theorem shows that $f \in \operatorname{Lip}\alpha$. Thus, we have constructed a function $f \in \operatorname{Lip}\alpha$ such that

$$R^{\mathrm{Tr}}_{2^n+1}[f] = \frac{1}{2\pi\,(2^n)^\alpha} \qquad \text{for } n = 1, 2, \dots.$$

Therefore (7.17) cannot be improved upon, as asserted. Nevertheless, large values of $R^{\mathrm{Tr}}_n[f]$ are exceptional. This is shown by the next theorem.

THEOREM 7.2.3 (Brass 1979a). *Let $f \in \operatorname{Lip}\gamma$ and $0 \le \delta < \gamma - 1/2$. Then*

$$\sum_{\nu=1}^{\infty} \nu^\delta \left|R^{\mathrm{Tr}}_{\nu+1}[f]\right| < \infty.$$

PROOF. Obviously, the location of the fundamental interval is not significant, so let us choose $[0, 2\pi]$. We then replace f by its even (relative to π) part, and obtain after periodic continuation a function g with the properties that

$$g \in \operatorname{Lip}\gamma \qquad \text{and} \qquad R^{\mathrm{Tr}}_{\nu+1}[g] = R^{\mathrm{Tr}}_{\nu+1}[f].$$

According to Theorem 7.2.1, we have

$$\begin{aligned}
\sum_{\nu=1}^{p} \nu^\delta \left|R^{\mathrm{Tr}}_{\nu+1}[f]\right| &= \sum_{\nu=1}^{p} \nu^\delta \left|R^{\mathrm{Tr}}_{\nu+1}[g]\right| = 2\pi \sum_{\nu=1}^{p} \nu^\delta \sum_{\mu=1}^{\infty} \left|\alpha_{\nu\mu}[g]\right| \\
&\le 2\pi \sum_{\sigma=1}^{\infty} \left|\alpha_\sigma[g]\right| \sum_{\nu|\sigma} \nu^\delta \le 2\pi \sum_{\sigma=1}^{\infty} \left|\alpha_\sigma[g]\right| \sigma^\delta d(\sigma),
\end{aligned}$$

where $d(\sigma)$ denotes the number of divisors of σ. It can be proved that

$$d(\sigma) = O(\sigma^\varepsilon)$$

for any $\varepsilon > 0$ (see Hardy and Wright 1960: p. 260). Using this result together with the next lemma, the proof of the theorem will be complete. □

LEMMA 7.2.2 (Hardy). *For $h \in \operatorname{Lip}\gamma$, we have*

$$\sum_{\sigma=1}^{\infty} \left|\alpha_\sigma[h]\right| \sigma^\tau < \infty$$

for $\tau \in [0, \gamma - 1/2[$.

PROOF. We need two well-known and elementary facts from the theory of Fourier series. Writing $S_n[h]$ for the nth partial sum of the Fourier series of h, we have

$$\int_0^{2\pi} \left(h(x) - S_n[h](x)\right)^2 dx = \pi \sum_{\sigma>n} \left(\alpha_\sigma^2[h] + \beta_\sigma^2[h]\right)$$

and

$$\int_0^{2\pi} \left(h(x) - S_n[h](x)\right)^2 dx \le \int_0^{2\pi} \left(h(x) - t(x)\right)^2 dx$$

for any $t \in \mathbb{T}_n$. Combining these gives

$$\pi \sum_{\sigma > n} \left(\alpha_\sigma^2[h] + \beta_\sigma^2[h]\right) \le \int_0^{2\pi} \left(h(x) - t(x)\right)^2 dx.$$

We use the t from the Jackson–Bernstein theorem and obtain

$$\sum_{\sigma > n} \left(\alpha_\sigma^2[h] + \beta_\sigma^2[h]\right) = O(n^{-2\gamma})$$

for any $h \in \operatorname{Lip}\gamma$. Now, with $\alpha_\nu := \alpha_\nu[h]$,

$$\begin{aligned}
\sum_{\sigma=2}^{2^m} |\alpha_\sigma| \sigma^\tau = \sum_{\mu=0}^{m-1} \sum_{\nu=2^\mu+1}^{2^{\mu+1}} |\alpha_\nu| \nu^\tau &\le \sum_{\mu=0}^{m-1} \left(\sum_{\nu=2^\mu+1}^{2^{\mu+1}} \alpha_\nu^2 \sum_{\nu=2^\mu+1}^{2^{\mu+1}} \nu^{2\tau} \right)^{1/2} \\
&\le \sum_{\mu=0}^{m-1} \left(\sum_{\nu > 2^\mu} \alpha_\nu^2 \cdot 2^\mu \cdot 2^{(\mu+1)2\tau} \right)^{1/2} \\
&\le \text{const} \cdot \sum_{\mu=0}^{m-1} 2^{-\mu\gamma + \mu/2 + (\mu+1)\tau} \\
&= \text{const} \cdot 2^\tau \sum_{\mu=0}^{m-1} 2^{(\tau - \gamma + 1/2)\mu}.
\end{aligned}$$

The last series converges, since $\tau - \gamma + 1/2 < 0$. □

Turning to the next co-observation $\mathcal{C}_M^{\mathrm{Sa},1}$, with Peano kernel methods we obtain

$$|R_{n+1}^{\mathrm{Tr}}[f]| \le \left(\frac{(b-a)^3}{12n^2} \int_a^b \left[f'(x)\right]^2 dx \right)^{1/2}. \tag{7.19}$$

Improvement is possible if we use ρ^{intr} instead of ρ^{opt}. We gave ρ^{intr} for any system of nodes in Example 2.3.1. Specializing to equidistant nodes gives

$$|R_{n+1}^{\mathrm{Tr}}[f]| \le \left(\frac{(b-a)^3}{12n^2} \int_a^b \left[f'(x)\right]^2 dx - \frac{(b-a)^2}{12n} \sum_{\nu=1}^{n} \left[f(x_{\nu+1}) - f(x_\nu)\right]^2 \right)^{1/2}.$$

This formula appeared in the pioneering paper of Golomb and Weinberger (1959).

For the co-observation $\mathcal{C}_M^{\mathrm{Sa},2}$, we obtain

$$|R_{n+1}^{\mathrm{Tr}}[f]| \le \left(\frac{(b-a)^5}{120n^4} \int_a^b \left[f''(x)\right]^2 dx \right)^{1/2}.$$

The trapezoidal rule Q_{n+1}^{Tr} is not optimal in $\mathcal{C}_M^{\mathrm{Sa},2}$. In fact,

$$\rho_n^{\mathrm{opt}} = \left(\frac{(b-a)^5}{720n^4} \right)^{1/2} \cdot \left(1 + O(n^{-1})\right) \cdot M$$

(see Kershaw 1972; Meyers and Sard 1950). However, as we argued in Section 4.7, the small gain in the error constant is in most cases not enough to compensate for the loss in simplicity of the rule.

In order to apply Sard's bound (7.19), we have to compute $\int_a^b [f'(x)]^2\,dx$. In most cases this is more difficult than our original problem, the determination of $\int_a^b f(x)dx$. The bound

$$\int_a^b [f'(x)]^2\,dx \le (b-a)\|f'\|^2$$

does not help, because this calls for using the co-observation $\mathcal{C}_M^{(1)}$, and then (7.11) would be better. So (7.19) is only to be recommended if $\int_a^b [f'(x)]^2\,dx$ can easily be found exactly or almost exactly. The following example illustrates such a situation.

EXAMPLE 7.2.2. Compute

$$\int_0^{\pi/2} f_0(x)\,dx \quad \text{where } f_0(x) = \sin^{3/2} x.$$

We have

$$\int_0^{\pi/2} [f_0'(x)]^2\,dx = \frac{9}{4}\int_0^{\pi/2} \sin x \cos^2 x\,dx = -\frac{9\cos^3 x}{12}\bigg|_0^{\pi/2} = \frac{3}{4},$$

so (7.19) gives

$$|R_{n+1}^{\mathrm{Tr}}[f_0]| \le \frac{0.492}{n}. \tag{7.20}$$

Now, (7.11) would give

$$|R_{n+1}^{\mathrm{Tr}}[f_0]| \le \frac{0.575}{n}$$

and (7.14)

$$|R_{n+1}^{\mathrm{Tr}}[f_0]| \le \frac{0.786}{n}.$$

The Golomb–Weinberger bound leads, in any case, to an improvement of (7.20), for example in the $n=7$ case from

$$|R_8^{\mathrm{Tr}}[f_0]| \le 0.0703$$

to

$$|R_8^{\mathrm{Tr}}[f_0]| \le 0.0086.$$

However, (7.15) gives an asymptotically much better bound, namely

$$|R_{n+1}^{\mathrm{Tr}}[f_0]| \le \frac{0.575}{n^2},$$

and, finally, Theorem 7.1.3 gives the correct asymptotic:

$$R_{n+1}^{\mathrm{Tr}}[f_0] = \frac{0.079\ldots}{n^{5/2}} + O(n^{-3}).$$

For the application of Davis's method to the trapezoidal rule (which means considering the co-observation $\mathcal{C}_M^{\mathrm{Da}}$), we refer to Theorem 4.8.1.

The convexity of f is a useful property here, since it allows us to obtain good and simple bounds as well as other results concerning R_n^{Tr}. The simplest of these results is the following:

THEOREM 7.2.4. Let f be convex. Then

$$R_n^{\mathrm{Tr}}[f] \le 0. \tag{7.21}$$

PROOF. This is an immediate consequence of Lemma 4.10.1.

The classical proof, which is quite illustrative, uses the definition of $Q_n^{\mathrm{Tr}}[f]$ as the integral of the interpolating broken line function and the fact that f lies below or on any chord.

An alternative proof is based on the fact that $Q_{2^\nu n+1}^{\mathrm{Tr}}[f]$, $\nu = 0, 1, 2, \ldots$, is a decreasing sequence with limit $I[f]$. The monotonicity then follows from the identity

$$Q_{n+1}^{\mathrm{Tr}}[f] - Q_{2n+1}^{\mathrm{Tr}}[f] = \frac{h}{4} \sum_{\nu=1}^{n} \Big(f\big(a + (\nu-1)h\big) - 2f\big(a + (\nu - 1/2)h\big) + f\big(a + \nu h\big) \Big),$$

where $h = \dfrac{b-a}{n}$. □

The next theorem gives a quite elementary, but very useful, error bound.

THEOREM 7.2.5 (Wolfe 1959). Let f be convex. Then

$$|R_{n+1}^{\mathrm{Tr}}[f]| \le \frac{h}{2}\Big(f(a) - f\Big(a + \frac{h}{4}\Big)\Big) + \frac{h}{2}\Big(f(b) - f\Big(b - \frac{h}{4}\Big)\Big).$$

PROOF. By virtue of (7.21), we only have to prove that

$$R_{n+1}^{\mathrm{Tr}}[f] \ge \frac{h}{2}\Big(f\Big(a + \frac{h}{4}\Big) - f(a)\Big) + \frac{h}{2}\Big(f\Big(b - \frac{h}{4}\Big) - f(b)\Big) =: L[f]. \tag{7.22}$$

The quadrature rule $Q := Q_{n+1}^{\mathrm{Tr}} + L$ is composed of Q_1^{Mi} applied to $[a, a+h/2]$ and $[b-h/2, b]$ and Q_{n-1}^{Mi} applied to $[a+h/2, b-h/2]$. All of these rules are positive definite of order two and therefore give a lower bound for the respective integrals (see Lemma 4.10.1). □

Our next aim is to prove that the bound given in Theorem 7.2.5 is of the order of the error, at least in the classes $\mathrm{Lip}\,\alpha$. To this end, we have to first prove an "inverse theorem":

THEOREM 7.2.6 (Brass 1978a). Let f be continuous and convex and let $\alpha \in]0, 1]$. Then

$$R_n^{\mathrm{Tr}}[f] = O(n^{-1-\alpha})$$

implies

$$f \in \mathrm{Lip}\,\alpha.$$

PROOF. We take $[0, 2\pi]$ as the fundamental interval. Denote by g the even part of f; then g is convex and hence of bounded variation. The periodic extension of g has a uniformly convergent Fourier series

$$g(x) = \frac{\alpha_0}{2} + \sum_{\nu=1}^{\infty} \alpha_\nu \cos \nu x.$$

Theorem 7.2.1 gives

$$R_{n+1}^{\mathrm{Tr}}[f] = R_{n+1}^{\mathrm{Tr}}[g] = -2\pi \sum_{k=1}^{\infty} \alpha_{k\cdot n}. \tag{7.23}$$

Then, we apply the following lemma, whose proof will be given afterwards:

LEMMA 7.2.3. Suppose that the restriction of f to $[0, 2\pi]$ is convex. Then $\alpha_\nu[f] \ge 0$ for $\nu = 1, 2, \ldots$.

Using Lemma 7.2.3 and (7.23), we deduce that

$$|\alpha_\nu| \le \frac{1}{2\pi}\left|R^{\mathrm{Tr}}_{\nu+1}[f]\right|.$$

Hence, for any $h > 0$ we have

$$\begin{aligned} |g(h)-g(0)| &= \left|\sum_{\nu=1}^{\infty}\alpha_\nu(\cos\nu h - 1)\right| = 2\sum_{\nu=1}^{\infty}\alpha_\nu \sin^2\frac{\nu h}{2} \\ &\le \frac{h^2}{2}\sum_{\nu=1}^{n}\alpha_\nu\nu^2 + \sum_{\nu>n}\alpha_\nu \\ &\le \frac{h^2}{4\pi}\sum_{\nu=1}^{n}\left|R^{\mathrm{Tr}}_{\nu+1}[f]\right|\nu^2 + \sum_{\nu>n}\left|R^{\mathrm{Tr}}_{\nu+1}[f]\right|. \end{aligned} \tag{7.24}$$

Using the assumption that $R^{\mathrm{Tr}}_\nu[f] = O(\nu^{-1-\alpha})$, we choose $n = \lfloor 1/h \rfloor$ and obtain

$$|g(h)-g(0)| = O(h^\alpha).$$

This relation means that

$$|f(h)+f(2\pi-h)-f(0)-f(2\pi)| = O(h^\alpha),$$

or

$$\frac{f(h)-f(0)}{h}\,h^{1-\alpha} - \frac{f(2\pi)-f(2\pi-h)}{h}\,h^{1-\alpha} = O(1). \tag{7.25}$$

The convexity of f implies that the first quotient is bounded from above while the second one is bounded from below. Therefore, both summands on the left-hand side of (7.25) are uniformly bounded, which means that

$$|f(h)-f(0)| = O(h^\alpha) \qquad \text{and} \qquad |f(2\pi)-f(2\pi-h)| = O(h^\alpha).$$

Now, $f \in \operatorname{Lip}\alpha$ follows from the monotonicity of $f(\,\cdot+h)-f(\cdot)$. □

PROOF OF LEMMA 7.2.3. We have

$$\pi\alpha_n[f] = \int_0^{2\pi} f(x)\cos nx\,dx = \sum_{\nu=0}^{4n-1}\int_{\nu h/2}^{(\nu+1)h/2} f(x)\cos nx\,dx.$$

Set $h = \pi/n$, and define a new variable y via the relation

$$x = (-1)^\nu\left(y - \frac{h}{4}\right) + \frac{(2\nu+1)h}{4}.$$

Thus,

$$\begin{aligned} \pi\alpha_n[f] &= \sum_{\mu=0}^{n-1}\int_0^{h/2}\Big[f(y+2\mu h) - f\big(-y+(2\mu+1)h\big) \\ &\qquad\qquad - f\big(y+(2\mu+1)h\big) + f\big(-y+(2\mu+2)h\big)\Big]\cos ny\,dy \\ &= \sum_{\mu=0}^{n-1}\int_0^{h/2}\Big(\big[f\big((2\mu+1)h-y+h\big) - f\big((2\mu+1)h-y\big)\big] \\ &\qquad\qquad - \big[f(2\mu h+y+h) - f(2\mu h+y)\big]\Big)\cos ny\,dy. \end{aligned}$$

Since $f(\,\cdot+h)-f(\cdot)$ is increasing, the assertion follows. □

A slight modification of the proof leads to another interesting result:

THEOREM 7.2.7 (Gaier 1970/71). Let f be convex. Then

$$R_n^{\mathrm{Tr}}[f] = o(n^{-2})$$

implies

$$f \in \mathbb{P}_1.$$

PROOF. Set $\varepsilon_\nu = \nu^2 \big| R_{\nu+1}^{\mathrm{Tr}}[f] \big|$ (which tends to zero). Then

$$\sum_{\nu=1}^{n} \nu^2 \big| R_{\nu+1}^{\mathrm{Tr}}[f] \big| = n \, \frac{\varepsilon_1 + \varepsilon_2 + \cdots + \varepsilon_n}{n} = o(n)$$

and

$$\sum_{\nu > n} \big| R_{\nu+1}^{\mathrm{Tr}}[f] \big| = O\big(n^{-1} \max_{\nu > n} \varepsilon_\nu\big) = o(n^{-1}).$$

Next, we use (7.24). Proceeding as in the previous proof, we obtain, in place of (7.25),

$$\frac{f(h) - f(0)}{h} - \frac{f(2\pi) - f(2\pi - h)}{h} = o(1),$$

i.e. $f'_+(0) = f'_-(2\pi)$. For any convex f, the one-sided derivatives are increasing and $f'_+(x) \le f'_-(y)$ holds for $x < y$. So we obtain

$$f'_+(0) \le f'_+(x) \le f'_-(y) \le f'_-(2\pi) = f'_+(0),$$

and this implies the existence and constancy of f'. □

Finally, in our discussion of R_n^{Tr} for convex integrands, we present a fairly simple bound:

THEOREM 7.2.8 (Stopping inequality). Let f be convex. Then

$$-R_{2n+1}^{\mathrm{Tr}}[f] \le Q_{n+1}^{\mathrm{Tr}}[f] - Q_{2n+1}^{\mathrm{Tr}}[f].$$

This formula suggests the following algorithm of "automatic integration": Compute successively $Q_{n+1}^{\mathrm{Tr}}[f], Q_{2n+1}^{\mathrm{Tr}}[f], Q_{4n+1}^{\mathrm{Tr}}[f], \ldots$. Note that, given $Q_{n+1}^{\mathrm{Tr}}[f]$, the evaluation of $Q_{2n+1}^{\mathrm{Tr}}[f]$ does not require more work than the evaluation of $Q_{n+2}^{\mathrm{Tr}}[f]$. Any step will give an improvement. Stop when the improvement is smaller than the desired precision.

PROOF OF THEOREM 7.2.8. The assertion is equivalent to

$$I[f] - \big(2Q_{2n+1}^{\mathrm{Tr}} - Q_{n+1}^{\mathrm{Tr}}\big) \ge 0.$$

By Lemma 4.10.1, this holds because $2Q_{2n+1}^{\mathrm{Tr}} - Q_{n+1}^{\mathrm{Tr}} = Q_n^{\mathrm{Mi}}$ and Q_n^{Mi} is positive definite. □

We have proved that the sequence $\big(Q_{2^\nu n+1}^{\mathrm{Tr}}[f]\big)_{\nu \in \mathbb{N}_0}$ is monotone if f is convex. For the monotonicity of the full sequence of trapezoidal rules, convexity is not sufficient, as simple examples show. If we additionally require f''' to be of fixed sign, we can prove even a little more:

THEOREM 7.2.9 (Newman 1974). If f''' is continuous and f'' and f''' have no sign changes, then $nR_{n+1}^{\mathrm{Tr}}[f]$ tends monotonically to zero as $n \to \infty$.

PROOF. Without loss of generality, we may assume $f'' \geq 0$ and $f''' \leq 0$; otherwise, replace f by $-f$, $f(b-\cdot\,)$ or $-f(b-\cdot\,)$. For simplicity, we work with the fundamental interval $[0,1]$.

By the Euler–Maclaurin summation formula, we have

$$\begin{aligned} R_{n+1}^{\mathrm{Tr}}[f] &= -\frac{B_2(0)}{n^2}\int_0^1 f''(x)\,dx + \frac{1}{n^2}\int_0^1 B_2^*(nx) f''(x)\,dx \\ &= \frac{1}{n^2}\int_0^1 \left[B_2^*(nx) - B_2(0)\right] f''(x)\,dx. \end{aligned}$$

An integration by parts leads to

$$R_{n+1}^{\mathrm{Tr}}[f] = -\frac{B_2(0)}{n^2} f''(1) - \frac{1}{n^3}\int_0^1 \left[B_3^*(nx) - nxB_2(0)\right] f'''(x)\,dx,$$

and therefore

$$-nR_{n+1}^{\mathrm{Tr}}[f] = \frac{B_2(0)}{n} f''(1) + \int_0^1 x^2 f'''(x) C(nx)\,dx,$$

where

$$C(y) := \frac{B_3^*(y) - yB_2(0)}{y^2}.$$

We have to prove that C is increasing; together with $B_2(0) > 0$ and the assumptions on the signs of f'' and f''', this would yield that $\left(-nR_{n+1}^{\mathrm{Tr}}[f]\right)_{n\in\mathbb{N}_0}$ is decreasing. We define

$$C_\nu(z) := C(\nu + z) = \frac{B_3(z) - (z+\nu)B_2(0)}{(z+\nu)^2} \quad \text{for } z \in [0,1],\ \nu = 0,1,2,\ldots.$$

It suffices to show that C_ν is increasing on $[0,1]$, but this follows immediately from

$$C_\nu(z) = \frac{2z^3 - 3z^2 - \nu}{12(z+\nu)^2} = \frac{2z-3}{12}\cdot\left(\frac{z}{z+\nu}\right)^2 - \frac{\nu}{12(z+\nu)^2},$$

and so the proof is complete. □

We know from the Euler–Maclaurin summation formula that the trapezoidal method converges rapidly if f is $(b-a)$-periodic with a high order of differentiability. We wish to prove a sort of converse of this fact. Trivially, we have $R_n^{\mathrm{Tr}}[h] = 0$ for every odd function h, so the odd part of f has no influence on the convergence properties of the trapezoidal method. For even functions g, the converse statement is almost correct; we only have to assume absolute continuity to deduce a high degree of smoothness from a high speed of convergence (and then the possibility of periodic extension follows from the Euler–Maclaurin formula).

THEOREM 7.2.10 (Brass 1978a). *Let g be an even and absolutely continuous function such that*

$$R_n^{\mathrm{Tr}}[g] = O(n^{-k-\beta}) \quad \text{for } k \geq 1,\ \beta \in\ ]0,1[.$$

Then $g^{(k-1)}$ exists and satisfies a Lipschitz condition of order β. Furthermore, $g^{(\nu)}(a) = g^{(\nu)}(b)$ for $\nu = 0,1,2,\ldots,k-1$.

It is not possible to remove the assumption of absolute continuity—Ching has provided a counterexample; see Blakely *et al.* (1973).

PROOF. We work with the fundamental interval $[0, 2\pi]$. Our main tool is a lemma that gives expressions for the Fourier coefficients $\alpha_\nu[g]$ in terms of the errors $R_n^{\mathrm{Tr}}[g]$ (and can therefore be thought of as Theorem 7.2.1 in reverse). First, we have to define the Möbius function μ.

DEFINITION 7.2.1.

$$\mu(m) := \begin{cases} 1 & \text{if } m = 1, \\ (-1)^r & \text{if } m = p_1 p_2 \cdots p_r \text{ where } p_i \text{ are distinct primes,} \\ 0 & \text{otherwise.} \end{cases}$$

LEMMA 7.2.4. Let g be an even and absolutely continuous function with Fourier cosine coefficients $\alpha_\nu[g]$. Then

$$\alpha_\nu[g] = -\frac{1}{2\pi} \sum_{\lambda=1}^{\infty} \mu(\lambda) R_{\lambda\nu+1}^{\mathrm{Tr}}[g].$$

Owing to the assumption $R_n^{\mathrm{Tr}}[g] = O(n^{-k-\beta})$, the lemma tells us that

$$\alpha_\nu := \alpha_\nu[g] = O(\nu^{-k-\beta}).$$

The Fourier series of g converges, and hence

$$\sup_x \left| g(x) - \frac{\alpha_0}{2} - \sum_{\nu=1}^{m} \alpha_\nu \cos \nu x \right| \leq \sum_{\nu > m} |\alpha_\nu| = O(m^{-k-\beta+1}).$$

This implies that

$$\operatorname{dist}(g, \mathbb{T}_m) = O(m^{-k-\beta+1}),$$

and the theorem of Jackson and Bernstein then gives our claim.

It remains to prove the lemma. A formal derivation from Theorem 7.2.1 is elementary; the difficulty lies with the convergence problem. We use a (rather deep) theorem of Davenport (1937), which says that

$$\sum_{\lambda=1}^{\infty} \frac{\mu(\lambda)}{\lambda} B_1^* \left(\frac{\lambda x}{2\pi} \right) = -\frac{\sin x}{\pi}$$

holds for almost all x and that the partial sums are bounded.

In Davenport's formula, we replace x by νx and multiply by $-2\pi g'(x)/\nu$. An integration gives

$$\sum_{\lambda=1}^{\infty} \mu(\lambda) \int_0^{2\pi} \frac{-2\pi}{\lambda\nu} B_1^* \left(\frac{\lambda\nu x}{2\pi} \right) g'(x)\, dx = \frac{2}{\nu} \int_0^{2\pi} g'(x) \sin \nu x\, dx.$$

By the Peano kernel theorem, this relation is equivalent to

$$\sum_{\lambda=1}^{\infty} \mu(\lambda) R_{\lambda\nu+1}^{\mathrm{Tr}}[g] = \frac{2}{\nu} \int_0^{2\pi} g'(x) \sin \nu x\, dx,$$

and an integration by parts completes the proof. □

Rahman and Schmeisser (1990) obtained some elegant results which can be viewed as conceptual analogues of Theorem 7.2.10. Starting with stronger assumptions about the speed of convergence, they determine the corresponding function class. For further results related to these ideas, see Rahman and Schmeisser (1991).

7.3. Simpson's method

Simpson's rules are defined only when the number of nodes is odd. Therefore, in the present section and the next one, we take

$$h := \frac{b-a}{2m} \quad \text{and} \quad x_\nu := a + (\nu - 1)h \text{ with } \nu = 1, 2, \ldots, 2m+1.$$

DEFINITION 7.3.1. The mth Simpson's rule Q_{2m+1}^{Si} is given by

$$Q_{2m+1}^{\mathrm{Si}}[f] := \sum_{\nu=1}^{2m+1} a_\nu f(x_\nu)$$

where

$$\begin{aligned} a_1 = a_{2m+1} &= \frac{1}{3}h, \\ a_{2\nu} &= \frac{4}{3}h \quad \text{for } \nu = 1, 2, \ldots, m, \\ a_{2\nu+1} &= \frac{2}{3}h \quad \text{for } \nu = 1, 2, \ldots, m-1. \end{aligned}$$

The sequence $Q_3^{\mathrm{Si}}, Q_5^{\mathrm{Si}}, \ldots$ is called Simpson's method.

For a long time, Simpson's method was the most popular quadrature method, because it combines reasonable exactness with low computational cost (as it involves almost no multiplications). The importance of the latter advantage has diminished with the wide availability of fast computers, but Simpson's method retains its position as one of the classic and most frequently applied methods for numerical integration.

The simplest Simpson's rule is

$$Q_3^{\mathrm{Si}}[f] := \frac{b-a}{6}\left[f(a) + 4f\left(\frac{a+b}{2}\right) + f(b)\right]. \tag{7.26}$$

In Example 2.2.1, we characterized this rule as the projection rule corresponding to $\mathbb{P}_2$ (i.e. the rule of interpolatory type) with nodes a, $(a+b)/2$ and b, and we proved that $R_3^{\mathrm{Si}}[\mathbb{P}_3] = \{0\}$.

Simpson's method is the so-called "compound form" of (7.26). The compound form of any given rule Q is obtained by dividing $[a, b]$ into m subintervals of equal length and applying Q (transformed to the respective subinterval) to each of them. In our special case, this construction principle gives

$$\begin{aligned} \int_a^b f(x)\,dx &= \sum_{\nu=1}^{m} \int_{x_{2\nu-1}}^{x_{2\nu+1}} f(x)\,dx \\ &= \frac{h}{3}\sum_{\nu=1}^{m}\left[f(x_{2\nu-1}) + 4f(x_{2\nu}) + f(x_{2\nu+1})\right] + \sum_{\nu=1}^{m} R_3^{\mathrm{Si},\,[x_{2\nu-1},x_{2\nu+1}]}[f] \qquad (7.27) \\ &\equiv Q_{2m+1}^{\mathrm{Si}}[f] + R_{2m+1}^{\mathrm{Si}}[f]. \end{aligned}$$

This identity shows how we can generalize the results of Examples 4.3.1 and 4.6.1 concerning R_3^{Si}. We obtain

$$\rho(Q_{2m+1}^{\mathrm{Si}}, \mathcal{C}_M^{(1)}) = \frac{5}{36}\frac{(b-a)^2}{m}M, \qquad \rho(Q_{2m+1}^{\mathrm{Si}}, \mathcal{C}_M^{\mathrm{Var}}) = \frac{1}{3}\left(\frac{b-a}{m}\right)M,$$

$$\rho(Q_{2m+1}^{\text{Si}}, \mathcal{C}_M^{(2)}) = \frac{1}{81}\,\frac{(b-a)^3}{m^2}\,M, \qquad \rho(Q_{2m+1}^{\text{Si}}, \mathcal{C}_M^{\text{Var},1}) = \frac{1}{24}\left(\frac{b-a}{m}\right)^2 M,$$

$$\rho(Q_{2m+1}^{\text{Si}}, \mathcal{C}_M^{(3)}) = \frac{1}{576}\,\frac{(b-a)^4}{m^3}\,M, \qquad \rho(Q_{2m+1}^{\text{Si}}, \mathcal{C}_M^{\text{Var},2}) = \frac{1}{324}\left(\frac{b-a}{m}\right)^3 M,$$

$$\rho(Q_{2m+1}^{\text{Si}}, \mathcal{C}_M^{(4)}) = \frac{1}{2880}\,\frac{(b-a)^5}{m^4}\,M, \qquad \rho(Q_{2m+1}^{\text{Si}}, \mathcal{C}_M^{\text{Var},3}) = \frac{1}{1152}\left(\frac{b-a}{m}\right)^4 M.$$

We note that Q_{2m+1}^{Si} is optimal in $\mathcal{C}_M^{\text{Var},2}$ (Arnold 1977) but not in the other classes.

From the explicit representation for the Peano kernel K_4 given in Example 4.2.1, we see that $K_4 \le 0$ and, using (7.27), obtain the same inequality for all Simpson's rules. We deduce that

$$\begin{aligned} R_{2m+1}^{\text{Si}}[f] &= f^{(4)}(\xi) \int_a^b K_4(x)\,dx \\ &= -f^{(4)}(\xi) \int_a^b |K_4(x)|\,dx = -f^{(4)}(\xi)\rho\big(Q_{2m+1}^{\text{Si}}, \mathcal{C}_1^{(4)}\big), \end{aligned}$$

which gives the famous formula

$$R_{2m+1}^{\text{Si}}[f] = -\frac{(b-a)^5}{2880m^4}\,f^{(4)}(\xi) \quad \text{for some } \xi \in [a,b]. \tag{7.28}$$

Another approach to the theory of Q_{2m+1}^{Si} starts with the identity

$$Q_{2m+1}^{\text{Si}} = \frac{4}{3}\,Q_{2m+1}^{\text{Tr}} - \frac{1}{3}\,Q_{m+1}^{\text{Tr}},$$

which is easy to check and implies that

$$R_{2m+1}^{\text{Si}} = \frac{4}{3}\,R_{2m+1}^{\text{Tr}} - \frac{1}{3}\,R_{m+1}^{\text{Tr}}. \tag{7.29}$$

Combining this with (7.2) gives

$$R_{2m+1}^{\text{Si}}[f] = o(m^{-2}),$$

provided f' is absolutely continuous. Hence, under the assumption that f' is absolutely continuous and $f'(a) \ne f'(b)$, Simpson's method is asymptotically better than the trapezoidal method. Unless f is well behaved, there is no particular advantage to using Simpson's method. For $f(x) = \sqrt{1-x^4}$ on $[0,1]$, for instance, Theorem 7.1.3 gives

$$R_{m+1}^{\text{Tr}}[f] = \frac{\text{const}}{m^{3/2}} + O(m^{-2}),$$

and (7.29) yields the same order for $R_{2m+1}^{\text{Si}}[f]$.

Combining (7.29) with the Euler–Maclaurin formula yields more information about the error. Here we mention two such results:

$$R_{2m+1}^{\text{Si}}[f] = -\frac{(b-a)^4}{2880m^4}\,[f'''(b) - f'''(a)] + o(m^{-4})$$

if f''' is absolutely continuous; and

$$R_{2m+1}^{\text{Si}}[f] = -\frac{(b-a)^4}{2880m^4}\,[f'''(b) - f'''(a)] + \frac{5(b-a)^7}{672m^6} \cdot \frac{f^{(6)}(\xi)}{6!} \quad \text{for some } \xi \in [a,b]$$

if $f^{(6)}$ is continuous.

If $f^{(4)}$ has no sign changes, we can use a bound of a different kind:

THEOREM 7.3.1 (Rowland and Varol 1972). Suppose that $f^{(4)} \in C[a,b]$ has no sign changes. Then

$$\left|R_{4m+1}^{\mathrm{Si}}[f]\right| \leq \left|Q_{2m+1}^{\mathrm{Si}}[f] - Q_{4m+1}^{\mathrm{Si}}[f]\right|.$$

This is a kind of "stopping inequality"; see the remarks following Theorem 7.2.8. We shall derive it as a special case of the corresponding result for Romberg's method; see (7.67).

To finish this section, we present two further characterizations of Simpson's method. Recall that Q_{n+1}^{Tr} is exact for spline functions of degree one with equidistant knots (of the splines) which coincide with the interior nodes (of the quadrature rule). If we replace "degree one" by "degree two" here, we obtain the Simpson's rules.

THEOREM 7.3.2. Q_{2m+1}^{Si} can be characterized as the quadrature rule with nodes $x_1, x_2, \ldots, x_{2m+1}$ such that

$$R[\mathcal{S}_2(x_2, x_3, \ldots, x_{2m})] = \{0\}. \tag{7.30}$$

(For the definition of the spline space, see Definition 3.3.1.)

PROOF. First, we have to check that Q_{2m+1}^{Si} gives the correct value for a basis of the space of quadratic splines; that is,

$$R_{2m+1}^{\mathrm{Si}}[p_i] = 0 \quad \text{for } p_i(x) = x^i, \;\; i = 0, 1, 2$$

(which is trivial) and

$$R_{2m+1}^{\mathrm{Si}}[f_i] = 0 \quad \text{for } f_i(x) = (x - x_i)_+^2, \;\; i = 2, 3, \ldots, 2m.$$

According to Theorem 4.2.4, the latter is equivalent to

$$K_3^{\mathrm{Si}}(x_i) = 0$$

and, indeed, this last equation holds, as can be seen from the explicit expression for K_3^{Si} in Example 4.2.1 and its compound form.

Second, we have to prove uniqueness. Suppose there exist two rules Q and Q^* with nodes $x_1, \ldots, x_{2m+1}$, both being exact on $\mathcal{S}_2(x_2, x_3, \ldots, x_{2m})$. The difference $D := Q - Q^*$ has the form

$$D[f] = \sum_{\nu=1}^{2m+1} c_\nu f(x_\nu)$$

and vanishes on $\mathcal{S}_2(x_2, x_3, \ldots, x_{2m})$. From $D[f_{2m}] = 0$ we deduce that $c_{2m+1} = 0$. We then apply this argument with f_{2m-1} in place of f_{2m} and obtain $c_{2m} = 0$. Continuing in this manner, using p_0 in the last step, we end up with $D \equiv 0$, i.e. $Q = Q^*$. □

For our final characterization, we define a space of piecewise quadratic polynomials:

$$\mathcal{U}_{2m+1}^{\mathrm{Si}} := \left\{ f \in C[a,b] \;:\; f\big|_{[x_{2\nu-1}, x_{2\nu+1}]} \in \mathbb{P}_2 \text{ for } \nu = 1, 2, \ldots, m \right\}$$

(note that this is not a spline space). Moreover, we define a projection $\mathrm{proj}_{2m+1}^{\mathrm{Si}}$ from $C[a,b]$ onto $\mathcal{U}_{2m+1}^{\mathrm{Si}}$ by

$$\mathrm{proj}_{2m+1}^{\mathrm{Si}}[f]\big|_{[x_{2\nu-1}, x_{2\nu+1}]} := \mathrm{intpol}(x_{2\nu-1}, x_{2\nu}, x_{2\nu+1})[f], \tag{7.31}$$

where $\mathrm{intpol}(\ldots)[f]$ denotes the interpolation polynomial with the given nodes.

Considering the compound form of Q^{Si}_{2m+1}, we see immediately that

$$Q^{\mathrm{Si}}_{2m+1}[f] = \int_a^b \mathrm{proj}^{\mathrm{Si}}_{2m+1}[f](x)\,dx.$$

Thus, Q^{Si}_{2m+1} is a projection rule.

7.4. The Filon method

The rapidly oscillating integrand in problems such as

$$I[f] = \int_a^b f(x)\cos kx\,dx \qquad \text{or} \qquad I[f] = \int_a^b f(x)\sin kx\,dx \tag{7.32}$$

where k is large gives rise to difficulties in their numerical solution. For such problems, Filon (1928/29) proposed applying the projection $\mathrm{proj}^{\mathrm{Si}}_{2m+1}$ from (7.31).

DEFINITION 7.4.1. The mth Filon rule Q^{Fil}_{2m+1} is given by

$$Q^{\mathrm{Fil}}_{2m+1} := \int_a^b \mathrm{proj}^{\mathrm{Si}}_{2m+1}[f](x)e^{ikx}\,dx.$$

To deal simultaneously with both integrals in (7.32), we take $\int_a^b f(x)e^{ikx}\,dx$ to be the underlying integral. We assume that f is real, so that the separation of real and imaginary parts is easy.

Before beginning the discussion of Filon rules, let us emphasize that the important case of a $(b-a)$-periodic integrand has many peculiarities; we shall deal with this special case in the next chapter. First, we give a more explicit form of the Filon rule.

THEOREM 7.4.1.

$$Q^{\mathrm{Fil}}_{2m+1}[f] = \alpha Q^{\mathrm{Tr}}_{m+1}[g] + \beta Q^{\mathrm{Mi}}_{m}[g] + \gamma ih\big[g(a) - g(b)\big]$$

with $g(x) := f(x)e^{ikx}$ and

$$\alpha = \alpha(\theta) := \frac{\theta + \theta\cos^2\theta - \sin 2\theta}{\theta^3},$$

$$\beta = \beta(\theta) := \frac{2(\sin\theta - \theta\cos\theta)}{\theta^3},$$

$$\gamma = \gamma(\theta) := \frac{\theta^2 + \theta\sin\theta\cos\theta - 2\sin^2\theta}{\theta^3},$$

where $\theta = k\cdot h$.

As $\theta \to 0$, we have

$$\alpha(\theta) = \frac{1}{3} + O(\theta^2),$$

$$\beta(\theta) = \frac{2}{3} + O(\theta^2),$$

$$\gamma(\theta) = \frac{2}{45}\theta^3 + O(\theta^5).$$

In particular, for $\theta = 0$ we obtain

$$Q^{\mathrm{Fil}}_{2m+1} = \frac{1}{3}Q^{\mathrm{Tr}}_{m+1} + \frac{2}{3}Q^{\mathrm{Mi}}_{m} = Q^{\mathrm{Si}}_{2m+1},$$

as it should be.

PROOF OF THEOREM 7.4.1.

$$
\begin{aligned}
&Q_{2m+1}^{\mathrm{Fil}}[f] \\
&= \int_a^b \mathrm{proj}_{2m+1}^{\mathrm{Si}}[f](x) e^{ikx}\, dx \\
&= \sum_{\nu=1}^{m} \int_{x_{2\nu-1}}^{x_{2\nu+1}} \mathrm{proj}_{2m+1}^{\mathrm{Si}}[f](x) e^{ikx}\, dx \\
&= \sum_{\nu=1}^{m} \int_{x_{2\nu-1}}^{x_{2\nu+1}} \left[f(x_{2\nu-1}) \frac{(x-x_{2\nu})(x-x_{2\nu+1})}{2h^2} - f(x_{2\nu}) \frac{(x-x_{2\nu-1})(x-x_{2\nu+1})}{h^2} \right. \\
&\qquad\qquad \left. + f(x_{2\nu+1}) \frac{(x-x_{2\nu-1})(x-x_{2\nu})}{2h^2} \right] e^{ikx}\, dx \\
&= \sum_{\nu=1}^{m} f(x_{2\nu-1}) e^{ikx_{2\nu-1}} \int_{-h}^{h} \frac{y(y-h)}{2h^2} e^{ik(y+h)}\, dy \\
&\quad - \sum_{\nu=1}^{m} f(x_{2\nu}) e^{ikx_{2\nu}} \int_{-h}^{h} \frac{y^2-h^2}{h^2} e^{iky}\, dy \\
&\quad + \sum_{\nu=1}^{m} f(x_{2\nu+1}) e^{ikx_{2\nu+1}} \int_{-h}^{h} \frac{y(y+h)}{2h^2} e^{ik(y-h)}\, dy.
\end{aligned}
$$

The claim follows upon calculation of the integrals. □

There are many variations and generalizations of the Filon rules; see, for instance, Luke (1954), van de Vooren and van Linde (1966), Tuck (1967), Marsden and Taylor (1972) and Belward (1986).

Next, we turn to the issue of deriving an error bound. Since the method of proof allows for considerable generalization of the underlying functional, we define

$$
Q_{2m+1}^{\mathrm{gen\,Fil}}[f] := \int_a^b \mathrm{proj}_{2m+1}^{\mathrm{Si}}[f](x) w(x)\, dx
$$

as an estimate for

$$
I[f] = \int_a^b f(x) w(x)\, dx.
$$

THEOREM 7.4.2 (Köhler 1993). *Suppose $f \in C^4[a,b]$ and $w \in C^1[a,b]$. Then*

$$
\left| R_{2m+1}^{\mathrm{gen\,Fil}}[f] \right| \le \frac{(b-a)^5}{2880 m^4} \left(\|f^{(4)}\| \, \|w\| + 4 \|f'''\| \, \|w'\| \right).
$$

The factor $(b-a)^5/(2880m^4)$ is the best possible, as a comparison with (7.28) shows.

PROOF. We have

$$
\begin{aligned}
R_{2m+1}^{\mathrm{gen\,Fil}}[f] &= \int_a^b \Big(f(x) - \mathrm{proj}_{2m+1}^{\mathrm{Si}}[f](x) \Big) w(x)\, dx \\
&= \sum_{\nu=1}^{m} w(x_{2\nu}) \int_{x_{2\nu-1}}^{x_{2\nu+1}} \Big(f(x) - \mathrm{proj}_{2m+1}^{\mathrm{Si}}[f](x) \Big)\, dx \\
&\quad + \sum_{\nu=1}^{m} \int_{x_{2\nu-1}}^{x_{2\nu+1}} \Big(f(x) - \mathrm{proj}_{2m+1}^{\mathrm{Si}}[f](x) \Big) \big(w(x) - w(x_{2\nu}) \big)\, dx.
\end{aligned}
$$

For the first sum, we use (7.28) and obtain

$$-\sum_{\nu=1}^{m} w(x_{2\nu})\frac{(b-a)^5}{2880m^5} f^{(4)}(x_{2\nu}+\vartheta_\nu) \quad \text{for some } |\vartheta_\nu| \le h. \tag{7.33}$$

In the second sum, we use

$$w(x) - w(x_{2\nu}) = (x - x_{2\nu})w'(x_{2\nu}+\eta_\nu) \quad \text{for some } |\eta_\nu| \le h$$

and

$$\begin{aligned} f(x) &- \mathrm{proj}^{\mathrm{Si}}_{2m+1}[f](x) \\ &= f(x) - \mathrm{intpol}(x_{2\nu-1}, x_{2\nu}, x_{2\nu+1})[f](x) \\ &= (x - x_{2\nu-1})(x - x_{2\nu})(x - x_{2\nu+1})\,\frac{f'''(x_{2\nu}+\xi_\nu)}{6} \quad \text{for some } |\xi_\nu| \le h. \end{aligned}$$

Setting $y = x - x_{2\nu}$, we obtain

$$\begin{aligned} &\int_{x_{2\nu-1}}^{x_{2\nu+1}} \Big(f(x) - \mathrm{proj}^{\mathrm{Si}}_{2m+1}[f](x)\Big)\big(w(x) - w(x_{2\nu})\big)\,dx \\ &\quad = \int_{-h}^{h} \frac{(y-h)y^2(y+h)}{6} w'(x_{2\nu}+\eta_\nu) f'''(x_{2\nu}+\xi_\nu)\,dx \\ &\quad = -\frac{2}{45}\,h^5 w'(\rho_\nu) f'''(\sigma_\nu) \quad \text{for some } \rho_\nu, \sigma_\nu \in [x_{2\nu-1}, x_{2\nu+1}]. \end{aligned} \tag{7.34}$$

Upon replacing all function values by their suprema, the result follows. □

An asymptotic result can be derived using almost the same method:

THEOREM 7.4.3 (Petras 1990). *Under the assumptions of Theorem 7.4.2, we have*

$$R^{\mathrm{gen\,Fil}}_{2m+1}[f] = -\frac{(b-a)^4}{2880m^4}\int_a^b \Big(f^{(4)}(x)w(x) + 4f'''(x)w'(x)\Big)\,dx + o(m^{-4}).$$

PROOF. Expressions (7.33) and (7.34) together with

$$\begin{aligned} \lim_{m\to\infty} 2h\sum_{\nu=1}^{m} f^{(4)}(x_{2\nu}+\vartheta_\nu)w(x_{2\nu}) &= \lim_{m\to\infty} 2h\sum_{\nu=1}^{m} f^{(4)}(x_{2\nu})w(x_{2\nu}) \\ &= \int_a^b f^{(4)}(x)w(x)\,dx \end{aligned}$$

and

$$\lim_{m\to\infty} 2h\sum_{\nu=1}^{m} f'''(x_{2\nu}+\xi_\nu)w'(x_{2\nu}+\eta_\nu) = \int_a^b f'''(x)w'(x)\,dx$$

yield the assertion. □

For more information about the error term, see Ehrenmark (1987).

7.5. Gregory methods

The Euler–Maclaurin formula gives rather precise information about $R_n^{\mathrm{Tr}}[f]$. Gregory methods use this information to improve the trapezoidal rule. The simplest case is that of

$$R_{n+1}^{\mathrm{Tr}}[f] = -\frac{h^2}{12}[f'(b) - f'(a)] + O(h^3),$$

or

$$\int_a^b f(x)\,dx = Q_{n+1}^{\mathrm{Tr}}[f] - \frac{h^2}{12}\big[f'(b) - f'(a)\big] + O(h^3).$$

We plug in

$$f'(b) = \frac{f(b) - f(b-h)}{h} + O(h)$$

and

$$f'(a) = \frac{f(a+h) - f(a)}{h} + O(h)$$

to obtain

$$\int_a^b f(x)\,dx = Q_{n+1}^{\mathrm{Tr}}[f] - \frac{h}{12}\big[f(b) - f(b-h) - f(a+h) + f(a)\big] + O(h^3).$$

This gives rise to the "Durand method"

$$Q_{n+1}^{\mathrm{Du}}[f] := h\bigg(\frac{5}{12}f(x_1) + \frac{13}{12}f(x_2) + \sum_{\nu=3}^{n-1} f(x_\nu) + \frac{13}{12}f(x_n) + \frac{5}{12}f(x_{n+1})\bigg),$$

which yields better results than the trapezoidal method, at least asymptotically, when $f'(a) \neq f'(b)$.

We can continue improving Q_{n+1}^{Tr} in this way by using more terms in the Euler–Maclaurin formula and better approximations for the derivatives. The next case (under the assumption that f''' is absolutely continuous) is

$$R_{n+1}^{\mathrm{Tr}}[f] = -\frac{h^2}{12}\big[f'(b) - f'(a)\big] + O(h^4)$$

with

$$f'(b) = \frac{3f(b) - 4f(b-h) + f(b-2h)}{2h} + O(h^2)$$

and

$$f'(a) = \frac{-3f(a) + 4f(a+h) - f(a+2h)}{2h} + O(h^2).$$

This results in the "Lacroix method"

$$Q_{n+1}^{\mathrm{La}}[f] := h\bigg(\frac{3}{8}f(x_1) + \frac{7}{6}f(x_2) + \frac{23}{24}f(x_3) + \sum_{\nu=4}^{n-2} f(x_\nu) + \frac{23}{24}f(x_{n-1}) + \frac{7}{6}f(x_n) + \frac{3}{8}f(x_{n+1})\bigg).$$

All further methods that can be obtained in this manner have the form

$$Q_{n+1}[f] = Q_{n+1}^{\mathrm{Tr}}[f] + h\sum_{\nu=1}^{r+1} b_\nu\big[f(x_\nu) + f(x_{n+2-\nu})\big], \qquad n \geq r, \tag{7.35}$$

where r is fixed and the b_ν are independent of n. In all situations of practical interest, r is a small number ($r < 8$, approximately) so that (7.35) is computationally very simple. All methods of the form (7.35) are convergent, since we have

$$Q_{n+1}[f] = Q_{n+1}^{\mathrm{Tr}}[f] + O(h).$$

The starting point for investigating methods of type (7.35) is the following theorem.

THEOREM 7.5.1 (Martensen 1964). For each fixed $r \in \mathbb{N}$, there is exactly one method of the form (7.35) such that

$$R_{n+1}[f] = O(n^{-r-2}) \tag{7.36}$$

for all f with absolutely continuous $f^{(r+1)}$.

DEFINITION 7.5.1. The method characterized by Theorem 7.5.1 is called the Gregory method of rank r. We denote it by $\left(Q_n^{\mathrm{Gr},r}\right)_{(n-r)\in\mathbb{N}}$.

Thus,

$$Q_n^{\mathrm{Gr},1} = Q_n^{\mathrm{Du}} \qquad \text{and} \qquad Q_n^{\mathrm{Gr},2} = Q_n^{\mathrm{La}}$$

PROOF OF THEOREM 7.5.1. (i) We start with the Euler–Maclaurin formula

$$R_{n+1}^{\mathrm{Tr}}[f] = -\sum_{\mu=2}^{r+1} h^\mu B_\mu(0)\left[f^{(\mu-1)}(b) - f^{(\mu-1)}(a)\right] + O(h^{r+2})$$

and use the following lemma, whose proof will be given later.

LEMMA 7.5.1. For $\mu \le r+1$, there exist numbers $c_{\nu,\mu}$ which are independent of n and f such that

$$h^\mu f^{(\mu-1)}(a) = h\sum_{\nu=1}^{r+1} c_{\nu,\mu} f(x_\nu) + O(h^{r+2}) \tag{7.37}$$

for any f with absolutely continuous $f^{(r+1)}$.

We apply this lemma to $f := g(b + a - \cdot\,)$ and obtain

$$h^\mu g^{(\mu-1)}(b) = (-1)^{\mu-1} h\sum_{\nu=1}^{r+1} c_{\nu,\mu} g(x_{n+2-\nu}) + O(h^{r+2}). \tag{7.38}$$

If we substitute (7.37) and (7.38) in the Euler–Maclaurin formula and take into account the fact that $B_{2\mu+1}(0) = 0$, then the existence of the method is proved.

(ii) To prove uniqueness, suppose that there are two methods $\left(Q_n\right)_{n\in\mathbb{N}}$ and $\left(Q_n^*\right)_{n\in\mathbb{N}}$ which satisfy (7.35) and (7.36). This would imply that

$$R_{n+1}[f] - R_{n+1}^*[f] = h\sum_{\nu=1}^{r+1}(b_\nu^* - b_\nu)\left[f(x_\nu) + f(x_{n+2-\nu})\right] = O(h^{r+2}).$$

We apply this statement to

$$f_\mu(x) := \prod_{\substack{\nu=1\\ \nu\ne\mu}}^{r+1}(x - x_\nu)\cdot\prod_{\nu=1}^{r+1}(x - x_{n+2-\nu})$$

and obtain

$$(b_\mu^* - b_\mu) \prod_{\substack{\nu=1 \\ \nu\neq\mu}}^{r+1} (x_\mu - x_\nu) = O(h^{r+1}).$$

This yields $b_\mu - b_\mu^* = O(h)$ and hence $b_\mu^* = b_\mu$. □

PROOF OF LEMMA 7.5.1. Define $(c_{\nu,\mu})_{\nu=1,\ldots,r+1}$ to be the solution of the linear system

$$\sum_{\nu=0}^{r} c_{\nu+1,\mu}\nu^k = \begin{cases} 0 & \text{for } k = 0, 1, \ldots, r, \quad k \neq \mu - 1, \\ (\mu-1)! & \text{for } k = \mu - 1. \end{cases}$$

It is well known that this system is non-singular, since it has a Vandermonde matrix. Applying Taylor's theorem gives

$$\begin{aligned} h\sum_{\nu=1}^{r+1} c_{\nu,\mu} f(x_\nu) &= h\sum_{\nu=0}^{r} c_{\nu+1,\mu} f(a+\nu h) \\ &= h\sum_{\nu=0}^{r} c_{\nu+1,\mu}\left[\sum_{k=0}^{r} \nu^k h^k \frac{f^{(k)}(a)}{k!} + O(h^{r+1})\right] \\ &= h\sum_{k=0}^{r} h^k \frac{f^{(k)}(a)}{k!} \sum_{\nu=0}^{r} c_{\nu+1,\mu}\nu^k + O(h^{r+2}) \\ &= h^\mu f^{(\mu-1)}(a) + O(h^{r+2}), \end{aligned}$$

which rearranges to (7.37). □

EXAMPLE 7.5.1. The method above allows us to determine explicit formulas for $Q_n^{\mathrm{Gr},r}$. An example for $n \geq 11$ is

$$\begin{aligned} Q_{n+1}^{\mathrm{Gr},5}[f] = \frac{h}{60480}\Big[& 19087\big(f(x_1) + f(x_{n+1})\big) + 84199\big(f(x_2) + f(x_n)\big) \\ & + 37738\big(f(x_3) + f(x_{n-1})\big) + 75242\big(f(x_4) + f(x_{n-2})\big) \\ & + 55031\big(f(x_5) + f(x_{n-3})\big) + 61343\big(f(x_6) + f(x_{n-4})\big)\Big] \\ & + h\sum_{\nu=7}^{n-5} f(x_\nu). \end{aligned}$$

This rule is positive. In fact, by explicit calculation it can be shown that all Gregory rules of rank less than or equal to 6 are positive, but $Q_9^{\mathrm{Gr},7}$ is not. Also, $Q_n^{\mathrm{Gr},8}$ is non-positive for all $n \neq 10$.

Our next theorem gives two expressions for the Gregory rules which are very informative. Recall the definition of the Laplace coefficient from Section 5.2,

$$L_\lambda := (-1)^{\lambda+1} \int_0^1 \binom{x}{\lambda}\, dx,$$

and denote by $\mathrm{intpol}(y_1, \ldots, y_s)[f]$ the polynomial interpolating f at $y_1, \ldots, y_s$.

THEOREM 7.5.2. We have

$$Q_{n+1}^{\mathrm{Gr},r}[f] = \sum_{\nu=1}^{n-r} \int_{x_\nu}^{x_{\nu+1}} \mathrm{intpol}(x_\nu, \dots, x_{\nu+r+1})[f](x)\,dx + \int_{x_{n-r+1}}^{x_{n+1}} \mathrm{intpol}(x_{n-r+1}, \dots, x_{n+1})[f](x)\,dx \tag{7.39}$$

and

$$Q_{n+1}^{\mathrm{Gr},r}[f] = Q_{n+1}^{\mathrm{Tr}}[f] - h\sum_{\nu=1}^{r} L_{\nu+1}\big(\Delta^\nu f(x_{n+1-\nu}) + (-1)^\nu \Delta^\nu f(x_1)\big). \tag{7.40}$$

PROOF. (i) First, we prove that the right-hand sides of (7.39) and (7.40) are equal. We start with Newton's form of the interpolation polynomial,

$$\mathrm{intpol}(x_\nu, \dots, x_{\nu+r+1})[f](x) = f(x_\nu) + \sum_{\mu=1}^{r+1} \frac{\Delta^\mu f(x_\nu)}{h^\mu \mu!} \prod_{\lambda=\nu}^{\nu+\mu-1} (x - x_\lambda).$$

Since

$$\int_{x_\nu}^{x_{\nu+1}} \prod_{\lambda=\nu}^{\nu+\mu-1} (x - x_\lambda)\,dx = h^{\mu+1} \int_0^1 \prod_{\lambda=0}^{\mu-1} (y-\lambda)\,dy = h^{\mu+1}\mu!(-1)^{\mu+1} L_\mu, \tag{7.41}$$

we have

$$\int_{x_\nu}^{x_{\nu+1}} \mathrm{intpol}(x_\nu, \dots, x_{\nu+r+1})[f](x)\,dx = h\Big[f(x_\nu) + \sum_{\mu=1}^{r+1} (-1)^{\mu+1} L_\mu \Delta^\mu f(x_\nu)\Big]$$

and therefore

$$\begin{aligned} &\sum_{\nu=1}^{n-r} \int_{x_\nu}^{x_{\nu+1}} \mathrm{intpol}(x_\nu, \dots, x_{\nu+r+1})[f](x)\,dx \\ &= h\sum_{\nu=1}^{n-r} f(x_\nu) + h\sum_{\mu=1}^{r+1} (-1)^{\mu+1} L_\mu \sum_{\nu=1}^{n-r} \Delta^\mu f(x_\nu) \\ &= h\sum_{\nu=1}^{n-r} f(x_\nu) + h\sum_{\mu=1}^{r+1} (-1)^{\mu+1} L_\mu \big(\Delta^{\mu-1} f(x_{n-r+1}) - \Delta^{\mu-1} f(x_1)\big). \end{aligned} \tag{7.42}$$

We now deal with the last term in (7.39). Transforming Lemma 5.2.1 to the interval $[x_{n-r+1}, x_{n+1}]$ gives

$$\begin{aligned} &\int_{x_{n-r+1}}^{x_{n+1}} \mathrm{intpol}(x_{n-r+1}, \dots, x_{n+1})[f](x)\,dx \\ &= h \sum_{\nu=n-r+1}^{n+1} f(x_\nu) - h\sum_{\mu=1}^{r+1} L_\mu \big(\Delta^{\mu-1} f(x_{n-\mu+2}) + (-1)^{\mu-1} \Delta^{\mu-1} f(x_{n+1-r})\big). \end{aligned}$$

Combining the latter equation and (7.42) yields, for the right-hand side of (7.39),

$$\begin{aligned} &h\sum_{\nu=1}^{n+1} f(x_\nu) - h\sum_{\mu=1}^{r+1} L_\mu \big(\Delta^{\mu-1} f(x_{n-\mu+2}) + (-1)^{\mu-1}\Delta^{\mu-1} f(x_1)\big) \\ &\qquad = Q_{n+1}^{\mathrm{Tr}}[f] - h\sum_{\nu=1}^{r} L_{\nu+1}\big(\Delta^\nu f(x_{n+1-\nu}) + (-1)^\nu \Delta^\nu f(x_1)\big), \end{aligned}$$

and this is the same as the right-hand side of (7.40).

(ii) By applying Theorem 7.5.1, we shall identify the rule just studied as $Q_{n+1}^{\mathrm{Gr},r}$. From (7.40) we see that the rule has the form (7.35), and from (7.39) we obtain

$$R_{n+1}[f] = \sum_{\nu=1}^{n-r} \int_{x_\nu}^{x_{\nu+1}} \Big(f(x) - \mathrm{intpol}(x_\nu, \ldots, x_{\nu+r+1})[f](x)\Big)\,dx \tag{7.43}$$
$$+ \int_{x_{n-r+1}}^{x_{n+1}} \Big(f(x) - \mathrm{intpol}(x_{n-r+1}, \ldots, x_{n+1})[f](x)\Big)\,dx.$$

Next, we use the well-known formula

$$f(x) - \mathrm{intpol}(y_1, \ldots, y_m)[f](x) = \frac{f^{(m)}(\xi)}{m!} \prod_{\nu=1}^{m} (x - y_\nu) \tag{7.44}$$

together with simple bounds for the integrals to conclude that

$$R_{n+1}[f] = \sum_{\nu=1}^{n-r} O(h^{r+3}) + O(h^{r+2}) = O(h^{r+2}),$$

which gives (7.36). The proof is thus complete. □

A first important consequence of (7.39) is that

$$R_{n+1}^{\mathrm{Gr},r}[\mathbb{P}_r] = \{0\}. \tag{7.45}$$

We remark that (7.45) may replace (7.36) in the characterization of Gregory methods if and only if r is even.

A further consequence of (7.39) is the following simple expression for the error:

THEOREM 7.5.3 (Barret 1952; Martensen 1964). *For $f \in C^{(r+2)}[a, b]$ where r is even, there exists $\eta \in [a, b]$ such that*

$$R_{n+1}^{\mathrm{Gr},r}[f] = -h^{r+3}\big((n - r - 1)L_{r+2} + 2L_{r+3}\big) f^{(r+2)}(\eta).$$

PROOF. The last summand in (7.43) is $R_{r+1}^{\mathrm{NC}}[f]$ for the nodes $x_{n-r+1}, \ldots, x_{n+1}$. According to Theorem 5.2.4, we have

$$R_{r+1}^{\mathrm{NC}}[f] = -h^{r+3}(2L_{r+3} - L_{r+2}) f^{(r+2)}(\xi). \tag{7.46}$$

For the other summands in (7.43) we use (7.44), the generalized mean value theorem and (7.41) to get

$$\int_{x_\nu}^{x_{\nu+1}} \Big(f(x) - \mathrm{intpol}(x_\nu, \ldots, x_{\nu+r+1})[f](x)\Big)\,dx$$
$$= \int_{x_\nu}^{x_{\nu+1}} \frac{f^{(r+2)}(\xi_\nu)}{(r+2)!} \prod_{\mu=\nu}^{\nu+r+1} (x - x_\mu)\,dx = \frac{f^{(r+2)}(\eta_\nu)}{(r+2)!} h^{r+3} (r+2)!\, L_{r+2} (-1)^{r+1}.$$

Since $f^{(r+2)}(\xi)$ in (7.46) and $f^{(r+2)}(\eta_\nu)$ in the last equation have factors of the same sign, we may combine all contributions to $R_{n+1}^{\mathrm{Gr},r}[f]$ and the assertion follows. □

$R_{n+1}^{\mathrm{Gr},r}[f]$ is not definite if r is odd. Rather than giving more complicated expressions for the error in this case (see, e.g., Martensen 1964, 1970; Brass 1974a), we shall just describe the asymptotics:

THEOREM 7.5.4. Let $f^{(r+2)}$ be absolutely continuous. Then

$$R_{n+1}^{\mathrm{Gr},r}[f] = -L_{r+2}h^{r+2}\big[f^{(r+1)}(b) - (-1)^r f^{(r+1)}(a)\big] + O(h^{r+3}).$$

PROOF. From (7.40) we immediately obtain

$$Q_{n+1}^{\mathrm{Gr},r+1}[f] = Q_{n+1}^{\mathrm{Gr},r}[f] - L_{r+2}h\big[\Delta^{r+1}f(x_{n-r}) - (-1)^r\Delta^{r+1}f(x_1)\big],$$

that is,

$$R_{n+1}^{\mathrm{Gr},r}[f] = R_{n+1}^{\mathrm{Gr},r+1}[f] - L_{r+2}h\big[\Delta^{r+1}f(x_{n-r}) - (-1)^r\Delta^{r+1}f(x_1)\big].$$

Applying

$$\begin{aligned}\Delta^{r+1}f(x_{n-r}) &= h^{r+1}f^{(r+1)}(\xi) = h^{r+1}f^{(r+1)}(b) + O(h^{r+2}),\\ \Delta^{r+1}f(x_1) &= h^{r+1}f^{(r+1)}(\eta) = h^{r+1}f^{(r+1)}(a) + O(h^{r+2})\end{aligned}$$

and

$$R_{n+1}^{\mathrm{Gr},r+1}[f] = O(h^{r+3})$$

then gives the theorem. □

The computational simplicity of the Gregory rules would be worthless if not accompanied by a small error. We shall prove the asymptotic optimality of $Q_n^{\mathrm{Gr},r}[f]$ in $\mathcal{C}_M^{(2s+1)}$ for $s = 0, 1, \ldots, \lfloor r/2 \rfloor$. With slight modifications to the proof, it can be shown that the same holds in $\mathcal{C}_M^{\mathrm{Sa},2s+1}$. For more results on the near-optimality of the Gregory rules and their relatives, see Brass (1974b), Weinberger (1978) and Köhler (1995).

THEOREM 7.5.5. For $s = 1, 2, \ldots, 2\lfloor r/2 \rfloor + 1$, we have

$$\rho\big(Q_{n+1}^{\mathrm{Gr},r}[f], \mathcal{C}_M^{(s)}\big) = \frac{(b-a)^{s+1}}{n^s}\int_0^1 |B_s(y)|\,dy \cdot M + O(n^{-s-1}).$$

This result is especially interesting as in the next chapter we will prove that

$$\rho_n^{\mathrm{opt}}(\mathcal{C}_M^{(s)}) = \frac{(b-a)^{s+1}}{n^s}\int_0^1 |B_s(y)|\,dy \cdot M + O(n^{-s-1})$$

if s is odd.

PROOF. We shall investigate the structure of the Peano kernel K_s of $Q_{n+1}^{\mathrm{Gr},r}$. From the explicit expression for Peano kernels, we easily obtain that

$$\sup\big\{|K_s(x)| \,:\, x \in [x_1, x_{r+1}] \cup [x_{n-r+1}, x_{n+1}]\big\} = O(h^s).$$

The theorem then follows from

$$K_s\big|_{[x_{r+1},x_{n-r+1}]}(x) = (-h)^s B_s^*\left(\frac{x-a}{h}\right)$$

for $s \le 2\lfloor r/2 \rfloor + 1$ (in the $s = 1$ case excluding the nodes), which we shall prove next. To this end, let us recall that

$$K_1\big|_{]x_\nu,x_{\nu+1}[}(x) = -x + a + \sum_{\mu=1}^{\nu} a_\mu. \tag{7.47}$$

Since $a_\mu = h$ for $\mu = r+2, \ldots, n-r$ and $a_\lambda = a_{n+2-\lambda}$, we have

$$\sum_{\nu=1}^{r+1} a_\nu = \sum_{\nu=1}^{n+1} a_\nu - \sum_{\nu=r+2}^{n-r} a_\nu - \sum_{\nu=n-r+1}^{n+1} a_\nu = b - a - (n-2r-1)h - \sum_{\nu=1}^{r+1} a_\nu$$

so that

$$\sum_{\nu=1}^{r+1} a_\nu = \frac{2r+1}{2}\, h.$$

From (7.47) we obtain

$$K_1\big|_{]x_{r+1},x_{r+2}[}(x) = -x + a + \left(r+\tfrac{1}{2}\right)h$$

and hence

$$K_1\big|_{]x_\nu,x_{\nu+1}[}(x) = -x + a + \left(\nu-\tfrac{1}{2}\right)h = -h\cdot\frac{x-a}{h} + \nu h - \frac{h}{2}$$

for $r+1 \le \nu \le n-r$. This means that

$$K_1(x) = -hB_1^*\left(\frac{x-a}{h}\right) \qquad \text{for } x \in \left]x_{r+1}, x_{n-r+1}\right[,$$

which is the assertion in the $s = 1$ case.

For the remaining s we shall prove the statement by induction. We therefore assume it has already been established that

$$K_s(x) = (-h)^s B_s^*\left(\frac{x-a}{h}\right) \qquad \text{for } x \in \left]x_{r+1}, x_{n-r+1}\right[$$

for some $s \le 2\lfloor r/2 \rfloor$. Then

$$\begin{aligned}
K_{s+1}(x) &= \int_x^b K_s(u)\,du = \int_x^{x_{n-r+1}} K_s(u)\,du + \int_{x_{n-r+1}}^b K_s(u)\,du \\
&= (-h)^s \int_x^{x_{n-r+1}} B_s^*\left(\frac{u-a}{h}\right) du + \int_{x_{n-r+1}}^b K_s(u)\,du \\
&= (-h)^s \left[hB_{s+1}^*(n-r) - B_{s+1}^*\left(\frac{x-a}{h}\right)\right] + K_{s+1}(x_{n-r+1}) \\
&= (-h)^{s+1} B_{s+1}^*\left(\frac{x-a}{h}\right) + \left[K_{s+1}(x_{n-r+1}) - (-h)^{s+1} B_{s+1}(0)\right] \\
&\equiv (-h)^{s+1} B_{s+1}^*\left(\frac{x-a}{h}\right) + C
\end{aligned}$$

for $x \in \left]x_{r+1}, x_{n-r+1}\right[$. It suffices to show that $C = 0$. Using the symmetry of K_{s+1}, we obtain that

$$\begin{aligned}
0 &= \int_a^b K_{s+1}(u)\,du \\
&= \int_a^{x_{r+1}} K_{s+1}(u)\,du + \int_{x_{r+1}}^{x_{n-r+1}} K_{s+1}(u)\,du + \int_{x_{n-r+1}}^b K_{s+1}(u)\,du \\
&= (-1)^{s+1} \int_{x_{n-r+1}}^b K_{s+1}(u)\,du \; + (x_{n-r+1} - x_{r+1})C + \int_{x_{n-r+1}}^b K_{s+1}(u)\,du \\
&= \left(1 + (-1)^{s+1}\right) K_{s+2}(x_{n-r+1}) + (n-2r)hC.
\end{aligned}$$

A straightforward calculation yields

$$K_{s+2}(x_{n-r+1}) = \delta \cdot h^{s+2} \qquad \text{and} \qquad C = \gamma \cdot h^{s+1}$$

for some δ and γ that are independent of n. Our previous equation therefore becomes

$$\left(1+(-1)^{s+1}\right)\delta \cdot h^{s+2} + (n-2r)\gamma h^{s+2} = 0,$$

and this is possible only if $\gamma = 0$. □

EXAMPLE 7.5.2. Let us take a closer look at the Lacroix method. On specializing Theorem 7.5.3, we get

$$R_{n+1}^{\mathrm{La}}[f] = -\frac{(b-a)^5}{n^4}\left(\frac{19}{720} - \frac{1}{24n}\right) f^{(4)}(\xi).$$

For even n, a comparison with

$$R_{n+1}^{\mathrm{Si}}[f] = -\frac{(b-a)^5}{n^4} \cdot \frac{4}{720} \cdot f^{(4)}(\xi)$$

shows that the Simpson rule has an advantage, and the asymptotics

$$R_{n+1}^{\mathrm{La}}[f] = -\left(\frac{b-a}{n}\right)^4 \cdot \frac{19}{720} \cdot \left[f'''(b) - f'''(a)\right] + O(n^{-5}),$$

$$R_{n+1}^{\mathrm{Si}}[f] = -\left(\frac{b-a}{n}\right)^4 \cdot \frac{4}{720} \cdot \left[f'''(b) - f'''(a)\right] + O(n^{-5})$$

bear this out as well. On the other hand, Q_{n+1}^{La} is better for the co-observations $\mathcal{C}_M^{(1)}$ and $\mathcal{C}_M^{(3)}$. Overall, there is no considerable difference in the errors and the usefulness of the two methods.

It is quite interesting that $Q_{n+1}^{\mathrm{La}}[f]$, like $Q_{2n+1}^{\mathrm{Si}}[f]$, can be obtained by integrating certain piecewise quadratic polynomials. We have the representation

$$\begin{aligned} Q_{n+1}^{\mathrm{La}}[f] = &\int_{x_1}^{x_1+h/2} \operatorname{intpol}(x_1, x_2, x_3)[f](x)\,dx \\ &+ \sum_{\nu=2}^{n} \int_{x_\nu - h/2}^{x_\nu + h/2} \operatorname{intpol}(x_{\nu-1}, x_\nu, x_{\nu+1})[f](x)\,dx \\ &+ \int_{x_n+h/2}^{x_{n+1}} \operatorname{intpol}(x_{n-1}, x_n, x_{n+1})[f](x)\,dx, \end{aligned}$$

which can be verified by direct calculation of the integrals. Another representation is

$$\begin{aligned} Q_{n+1}^{\mathrm{La}}[f] = \frac{1}{2}\Bigg[&\int_{x_1}^{x_2} \operatorname{intpol}(x_1, x_2, x_3)[f](x)\,dx \\ &+ \sum_{\nu=2}^{n} \int_{x_{\nu-1}}^{x_{\nu+1}} \operatorname{intpol}(x_{\nu-1}, x_\nu, x_{\nu+1})[f](x)\,dx \\ &+ \int_{x_n}^{x_{n+1}} \operatorname{intpol}(x_{n-1}, x_n, x_{n+1})[f](x)\,dx \Bigg]. \end{aligned}$$

For generalizations and further variants, see Martensen (1964) and Köhler (1995).

Rather than using individual Gregory methods, we could use a sequence of Gregory rules of increasing rank. One would begin with the computation of $Q_{n+1}^{\mathrm{Tr}}[f]$ and add correction terms according to

$$\begin{aligned} Q_{n+1}^{\mathrm{Tr}}[f] &- \frac{h}{12}[\Delta f(x_n) - \Delta f(x_1)] \\ &- \frac{h}{24}[\Delta^2 f(x_{n-1}) + \Delta^2 f(x_1)] \\ &- \frac{19h}{720}[\Delta^3 f(x_{n-2}) - \Delta^3 f(x_1)] \\ &\dots . \end{aligned}$$

The highest possible number of correction terms is n, which means that we have exactness for $f \in \mathbb{P}_n$ with $n+1$ nodes, i.e. $Q_{n+1}^{\mathrm{NC}}[f]$. For larger n, the procedure does not always lead to a permanent improvement, though this is the case if all derivatives of f are positive: then, $Q_{n+1}^{\mathrm{Tr}}[f]$ is too large and all correction terms reduce this value while $Q_{n+1}^{\mathrm{NC}}[f]$ is still too large (see Theorem 5.2.4). In particular, this implies convergence of the Newton–Cotes method for this class of functions.

For more information on Gregory rules and their generalizations, see Hamming and Pinkham (1966), Lötzbeyer (1971, 1972), Martensen (1973), Brass (1976), Förster (1987c), de Villiers (1993), Köhler (1995), Martensen (2005) and Velleman (2005).

7.6. Romberg methods

Romberg quadrature methods are obtained by applying Richardson's convergence acceleration procedure to the sequence $Q_{2^m+1}^{\mathrm{Tr}}[f]$, $m = 0, 1, \dots$, so first we discuss this procedure. Its underlying principle can be formulated as follows:

LEMMA 7.6.1. Suppose that the sequence $\left(y_m^{(0)}\right)_{m=0,1,\dots}$ has the asymptotic expansion

$$y_m^{(0)} = y + \alpha_1 q_1^m + \alpha_2 q_2^m + \cdots + \alpha_s q_s^m + o(q_s^m),$$

where $1 > q_1 > q_2 > \cdots > q_s > 0$ are fixed numbers. Then the sequence

$$y_m^{(1)} := \frac{y_{m+1}^{(0)} - q_1 y_m^{(0)}}{1 - q_1}, \quad m = 0, 1, \dots,$$

has the asymptotic expansion

$$y_m^{(1)} = y + \alpha_2 \frac{q_2 - q_1}{1 - q_1} q_2^m + \alpha_3 \frac{q_3 - q_1}{1 - q_1} q_3^m + \cdots + \alpha_s \frac{q_s - q_1}{1 - q_1} q_s^m + o(q_s^m).$$

The proof is trivial. The main consequence is that

$$\lim_{m \to \infty} \frac{y_m^{(1)} - y}{y_m^{(0)} - y} = 0 \quad \text{if } \alpha_1 \neq 0,$$

which means that changing from the first sequence to the second leads to faster convergence. Since the structure of the asymptotic expansion is not altered, we can repeat the process by defining

$$y_m^{(2)} := \frac{y_{m+1}^{(1)} - q_2 y_m^{(1)}}{1 - q_2}.$$

Again, this sequence converges faster than the $y_m^{(1)}$ sequence if $\alpha_2 \neq 0$. Generalizing this idea, the Richardson procedure is defined iteratively by

$$y_m^{(\sigma)} := \frac{y_{m+1}^{(\sigma-1)} - q_\sigma y_m^{(\sigma-1)}}{1-q_\sigma} \quad \text{for } m = 0,1,\dots \quad \text{and} \quad \sigma = 1,2,\dots,s. \tag{7.48}$$

We arrange the computed values in a triangular scheme

$$\begin{array}{ccccccccc}
y_0^{(0)} & & & & & & & & \\
& \searrow & & & & & & & \\
y_1^{(0)} & \rightarrow & y_0^{(1)} & & & & & & \\
& \searrow & & \searrow & & & & & \\
y_2^{(0)} & \rightarrow & y_1^{(1)} & \rightarrow & y_0^{(2)} & & & & \\
& \searrow & & \searrow & & \searrow & & & \\
y_3^{(0)} & \rightarrow & y_2^{(1)} & \rightarrow & y_1^{(2)} & \rightarrow & y_0^{(3)} & & \\
& \searrow & & \searrow & & \searrow & & \searrow & \\
y_4^{(0)} & \rightarrow & y_3^{(1)} & \rightarrow & y_2^{(2)} & \rightarrow & y_1^{(3)} & \rightarrow & y_0^{(4)} \\
& \ddots & & \ddots & & \ddots & & \ddots &
\end{array} \tag{7.49}$$

All the columns are convergent with limit y, and the further to the right the faster the convergence. It can easily be proved that

$$\begin{aligned} y_m^{(\sigma)} &= y + \alpha_{\sigma+1} \frac{(q_{\sigma+1}-q_1)(q_{\sigma+1}-q_2)\cdots(q_{\sigma+1}-q_\sigma)}{(1-q_1)(1-q_2)\cdots(1-q_\sigma)} q_{\sigma+1}^m \\ &\quad + \alpha_{\sigma+2} \frac{(q_{\sigma+2}-q_1)(q_{\sigma+2}-q_2)\cdots(q_{\sigma+2}-q_\sigma)}{(1-q_1)(1-q_2)\cdots(1-q_\sigma)} q_{\sigma+2}^m \\ &\quad + \cdots + o(q_s^m). \end{aligned} \tag{7.50}$$

The improvement in convergence rate does not necessarily mean that the last-computed (i.e. the southeast) value is the best. In some cases, however, there is a simple sufficient condition:

If, for $k = 0, 1, \dots, r$, the columns $\left(y_\nu^{(k)}\right)_{\nu=0,1,\dots}$ are all decreasing, then $y_0^{(r)}$ is the best value in the scheme (up to the $(r+1)$st row).

This follows from the fact that we would have decreasing sequences in the rows as well, since

$$y_m^{(\sigma)} - y_{m+1}^{(\sigma-1)} = \frac{q_\sigma}{1-q_\sigma}\left(y_{m+1}^{(\sigma-1)} - y_m^{(\sigma-1)}\right).$$

Any element of the scheme (7.49) is a linear combination of elements of the first column. While the actual computation of (7.49) is always done recursively via the relation (7.48), any deeper investigation of its properties requires that we find an explicit expression for these linear combinations.

With this purpose in mind, we define an operator T by

$$T[(z_0, z_1, z_2, \dots)] = (z_1, z_2, \dots),$$

where $(z_0, z_1, z_2, \dots)$ is any sequence. If we denote $(y_0^{(\sigma)}, y_1^{(\sigma)}, y_2^{(\sigma)}, \dots)$ by $y^{(\sigma)}$, we can rewrite (7.48) as

$$y^{(\sigma)} = \frac{1}{1-q_\sigma}\,(T - q_\sigma\,\mathrm{id})y^{(\sigma-1)},$$

where id stands for the identity. By repeated application of this recursive formula, we obtain

$$y^{(\sigma)} = \frac{1}{\prod_{\nu=1}^{\sigma}(1-q_\nu)} (T - q_\sigma \,\mathrm{id})(T - q_{\sigma-1}\,\mathrm{id}) \cdots (T - q_1 \,\mathrm{id})\, y^{(0)}.$$

We have thus proved the following fact.

LEMMA 7.6.2. Let p_σ be given by

$$p_\sigma(x) := \prod_{\nu=1}^{\sigma}(x - q_\nu),$$

and let the numbers $\beta_\mu^{(\sigma)}$ be such that

$$p_\sigma(x) = \sum_{\mu=0}^{\sigma} \beta_\mu^{(\sigma)} x^\mu.$$

Then

$$y_m^{(\sigma)} = \frac{1}{p_\sigma(1)} \sum_{\mu=0}^{\sigma} \beta_\mu^{(\sigma)} y_{m+\mu}^{(0)}. \tag{7.51}$$

Now we are ready to get back to quadrature theory. According to what Romberg (1955) proposed, we choose

$$y_m^{(0)} = Q_{2^m+1}^{\mathrm{Tr}}[f]. \tag{7.52}$$

In this special case, (7.49) is called the Romberg scheme. This scheme is not a quadrature method in the sense that we have defined the notion; any column is a quadrature method, but the most interesting one consists of the rules along the upper diagonal: $y_0^{(0)}, y_0^{(1)}, y_0^{(2)}, \ldots$.

First, we have to prove that for (7.52), the assumptions of Lemma 7.6.1 are satisfied. This follows from the Euler–Maclaurin formula (Theorems 7.1.1 and 7.1.2):

$$Q_{2^m+1}^{\mathrm{Tr}}[f] = \int_a^b f(x)\,dx + \sum_{\mu=1}^{s} \frac{(b-a)^{2\mu}}{\left(2^{2\mu}\right)^m} B_{2\mu}(0)\left[f^{(2\mu-1)}(b) - f^{(2\mu-1)}(a)\right] + o(2^{-2sm})$$

for functions with absolutely continuous $(2s-1)$st derivative. We then deduce that

$$q_\nu = 4^{-\nu} \quad \text{and} \quad \alpha_\nu = (b-a)^{2\nu} B_{2\nu}(0)\left[f^{(2\nu-1)}(b) - f^{(2\nu-1)}(a)\right].$$

Based on (7.48) and (7.52), we make the following definition.

DEFINITION 7.6.1. The Romberg method of rank σ is defined recursively by

$$Q_{2^{m+\sigma}+1}^{\mathrm{Ro}(\sigma)} := \frac{Q_{2^{m+\sigma}+1}^{\mathrm{Ro}(\sigma-1)} - 4^{-\sigma} Q_{2^{m+\sigma-1}+1}^{\mathrm{Ro}(\sigma-1)}}{1 - 4^{-\sigma}}, \quad m = 0, 1, 2, \ldots,$$

and

$$Q_{2^m+1}^{\mathrm{Ro}(0)} := Q_{2^m+1}^{\mathrm{Tr}}.$$

The connection with the notation in the general scheme is given by

$$y_m^{(\sigma)} = Q_{2^{m+\sigma}+1}^{\mathrm{Ro}(\sigma)}[f].$$

EXAMPLE 7.6.1. On the fundamental interval $[0, 1]$, we have

$$\begin{aligned} Q_9^{\mathrm{Ro}(3)}[f] &= \frac{1}{2835}\left(-Q_2^{\mathrm{Tr}}[f] + 84\, Q_3^{\mathrm{Tr}}[f] - 1344\, Q_5^{\mathrm{Tr}}[f] + 4096\, Q_9^{\mathrm{Tr}}[f]\right) \\ &= \frac{1}{5670}\Bigg(217\, f(0) + 1024\, f\left(\frac{1}{8}\right) + 352\, f\left(\frac{2}{8}\right) + 1024\, f\left(\frac{3}{8}\right) + 436\, f\left(\frac{4}{8}\right) \\ &\qquad + 1024\, f\left(\frac{5}{8}\right) + 352\, f\left(\frac{6}{8}\right) + 1024\, f\left(\frac{7}{8}\right) + 217\, f(1)\Bigg) \end{aligned}$$

We emphasize that these expressions are meant to serve as an illustration, rather than an instruction, for the computation.

We begin our study of Romberg methods with four simple statements:

(i) The Romberg method of rank one is a partial sequence of Simpson's method.
(ii) The Romberg method of rank σ is exact for $f \in \mathbb{P}_{2\sigma+1}$.
(iii) All Romberg formulas are symmetric.
(iv) The application of Romberg rules to $(b-a)$-periodic functions is not justified (at least not by our derivation, since we have $\alpha_\nu = 0$).

As a special case of (7.50), we obtain the following asymptotic result after a short calculation.

THEOREM 7.6.1. *Let $f^{(2\sigma+1)}$ be absolutely continuous. Then, as $m \to \infty$,*

$$\begin{aligned} &R_{2^{m+\sigma}+1}^{\mathrm{Ro}(\sigma)}[f] \\ &= (-1)^{\sigma+1}\left(\frac{b-a}{2^{m+\sigma/2}}\right)^{2\sigma+2} \cdot B_{2\sigma+2}(0)\left[f^{(2\sigma+1)}(b) - f^{(2\sigma+1)}(a)\right] + o(2^{-(2\sigma+2)m}). \end{aligned}$$

From this result we deduce that for $\sigma \geq 4$, the asymptotic behaviour of the Gregory method of rank 2σ (see Theorem 7.5.4) is better, for $\sigma = 8$ by a factor of around 10^{-10}. This is not an argument against using the "diagonal" Romberg method $Q_{2^\sigma+1}^{\mathrm{Ro}(\sigma)}$, $\sigma = 0, 1, \ldots$, although it turns out that a similar non-asymptotic result holds too; see Theorem 7.6.5.

Our next task is to examine the Peano kernels. In our special case, (7.51) yields

$$Q_{2^{m+\sigma}+1}^{\mathrm{Ro}(\sigma)} = \frac{1}{p_\sigma(1)} \sum_{\mu=0}^{\sigma} \beta_\mu^{(\sigma)} Q_{2^{m+\mu}+1}^{\mathrm{Tr}} \tag{7.53}$$

where

$$p_\sigma(x) = \prod_{\nu=1}^{\sigma} (x - 4^{-\nu}) \equiv \sum_{\mu=0}^{\sigma} \beta_\mu^{(\sigma)} x^\mu. \tag{7.54}$$

Since

$$p_\sigma(1) = \sum_{\mu=0}^{\sigma} \beta_\mu^{(\sigma)},$$

we may write

$$R_{2^{m+\sigma}+1}^{\mathrm{Ro}(\sigma)} = \frac{1}{p_\sigma(1)} \sum_{\mu=0}^{\sigma} \beta_\mu^{(\sigma)} R_{2^{m+\mu}+1}^{\mathrm{Tr}}$$

instead of (7.53). The Peano kernels of the trapezoidal rules are known (see Theorem 7.1.1), so we obtain the following expression for the first Peano kernel K_1 of $R^{\mathrm{Ro}(\sigma)}_{2^{m+\sigma}+1}$:

$$K_1(x) = -\frac{1}{p_\sigma(1)} \sum_{\mu=0}^{\sigma} \beta_\mu^{(\sigma)} \frac{b-a}{2^{m+\mu}} B_1^* \left(2^{m+\mu} \frac{x-a}{b-a}\right).$$

From this formula we can obtain the higher Peano kernels by forming negative primitives with mean value zero, as described in the following theorem.

THEOREM 7.6.2. The rth Peano kernel K_r of $R^{\mathrm{Ro}(\sigma)}_{2^{m+\sigma}+1}$ has the form

$$K_r(x) = \frac{(-1)^r}{p_\sigma(1)} \sum_{\mu=0}^{\sigma} \beta_\mu^{(\sigma)} \left(\frac{b-a}{2^{m+\mu}}\right)^r B_r^* \left(2^{m+\mu} \frac{x-a}{b-a}\right), \quad r = 1, 2, \ldots, 2\sigma+1,$$

where p_σ and $\beta_\mu^{(\sigma)}$ are given by (7.54).

As a first corollary, we shall prove the positivity of all Romberg rules (see Bauer *et al.* 1963), which implies the convergence of all "diagonal" methods. The positivity follows from this next result.

THEOREM 7.6.3 (Baker 1968). All Romberg rules are Riemann sums.

Recall that Q_n is a Riemann sum if there are numbers

$$a = y_0 \le x_1 \le y_1 \le x_2 \le \cdots \le y_n = b$$

such that

$$Q_n[f] = \sum_{\nu=1}^{n} (y_\nu - y_{\nu-1}) f(x_\nu).$$

From the explicit expression for K_1, we can see that

$$K_1(x_\nu - 0) \le 0 \le K_1(x_\nu + 0) \quad \text{for } \nu = 1, \ldots, n \tag{7.55}$$

is a necessary and sufficient condition for Q_n to be a Riemann sum.

PROOF OF THEOREM 7.6.3. From Theorem 7.6.2 we have

$$\begin{aligned} K_1(x_\nu \pm 0) &= -\frac{1}{p_\sigma(1)} \sum_{\mu=0}^{\sigma} \beta_\mu^{(\sigma)} \frac{b-a}{2^{m+\mu}} B_1^* \left(\frac{\nu-1}{2^{\sigma-\mu}} \pm 0\right) \\ &= -\frac{1}{p_\sigma(1)} \cdot \frac{b-a}{2^m} \left(\sum_{\mu=0}^{\sigma-1} \frac{\beta_\mu^{(\sigma)}}{2^\mu} B_1^* \left(\frac{\nu-1}{2^{\sigma-\mu}} \pm 0\right) \mp \frac{1}{2^{\sigma+1}} \right). \end{aligned}$$

Therefore, (7.55) would follow if we could show that

$$\frac{1}{2^{\sigma+1}} \ge \left| \sum_{\mu=0}^{\sigma-1} \frac{\beta_\mu^{(\sigma)}}{2^\mu} B_1^* \left(\frac{\nu-1}{2^{\sigma-\mu}} \pm 0\right) \right|.$$

Since $\|B_1^*\| = 1/2$, the above is a consequence of

$$\frac{1}{2^\sigma} \ge \sum_{\mu=0}^{\sigma-1} \frac{|\beta_\mu^{(\sigma)}|}{2^\mu} = \sum_{\mu=0}^{\sigma} \frac{(-1)^{\sigma-\mu}}{2^\mu} \beta_\mu^{(\sigma)} - \frac{\beta_\sigma^{(\sigma)}}{2^\sigma} = (-1)^\sigma p_\sigma\left(-\tfrac{1}{2}\right) - 2^{-\sigma}.$$

The first equality in the previous statement comes from the fact that the sign of $\beta_\mu^{(\sigma)}$ can be determined from (7.54), so it remains to prove that

$$2^{1-\sigma} \geq (-1)^\sigma p_\sigma\Big(-\frac{1}{2}\Big) = \prod_{\nu=1}^{\sigma}\Big(\frac{1}{2}+4^{-\nu}\Big) = 2^{-\sigma}\prod_{\nu=1}^{\sigma}\Big(1+\frac{2}{4^\nu}\Big).$$

But this is true because

$$\prod_{\nu=1}^{\sigma}\Big(1+\frac{2}{4^\nu}\Big) = \exp\Big(\sum_{\nu=1}^{\sigma}\ln\Big(1+\frac{2}{4^\nu}\Big)\Big) \leq \exp\sum_{\nu=1}^{\sigma}\frac{2}{4^\nu} \leq \exp\frac{2}{3} < 1.95.$$

Hence the proof is complete. □

We consider error bounds next.

THEOREM 7.6.4 (Brass and Fischer 1999a).

$$\rho\big(Q_{2^{m+\sigma}+1}^{\mathrm{Ro}(\sigma)}, \mathcal{C}_1^{(r)}\big) \leq 6\,\frac{(b-a)^{r+1}}{2^{r(m+\sigma)}}\cdot\frac{2^{(r-1)^2/4}}{(2\pi)^r}. \tag{7.56}$$

It is possible to reduce the constant 6 in the right-hand side; however, replacing it by 1.4 (see Brass and Fischer 1999a) would not give a correct bound for every σ and every r. Using the method employed in the proof of Theorem 7.6.4 and working through the details more carefully, we can prove that

$$\rho\big(Q_{2^{m+\sigma}+1}^{\mathrm{Ro}(\sigma)}, \mathcal{C}_1^{(1)}\big) \leq 0.64\,\frac{(b-a)^2}{2^{m+\sigma}}.$$

Comparing this bound with

$$\rho_{n+1}^{\mathrm{opt}}(\mathcal{C}_1^{(1)}) = 0.25\,\frac{(b-a)^2}{n}$$

shows that, in $\mathcal{C}_M^{(1)}$, all Romberg rules are able to compete with the optimal rules. However, as r gets larger, the loss incurred from applying a Romberg rule instead of the optimal rule in $\mathcal{C}_M^{(r)}$ increases fast.

EXAMPLE 7.6.2. Let us apply Theorem 7.6.4 to the problem of computing

$$\int_a^b f(x)\,dx \equiv \int_{-1}^1 \frac{1}{x^2+c^2}\,dx.$$

This simple integrand is useful for illustrating the method, as $\|f^{(2r)}\|$ can be found easily by expanding in partial fractions. We have

$$\sup_{-1\leq x\leq 1} |f^{(2r)}(x)| = |f^{(2r)}(0)| = \frac{(2r)!}{c^{2r+2}}.$$

From Theorem 7.6.4 we obtain

$$\big|R_{2^\sigma+1}^{\mathrm{Ro}(\sigma)}[f]\big| \leq 6\,\frac{2^{r^2+r+5/4}(2r)!}{2^{2\sigma r}(2\pi)^{2r}c^{2r+2}} \quad \text{for } r = 1, 2, \ldots, \sigma.$$

Initially we get an improvement of the bound as r increases, but thereafter the quality deteriorates. A typical case is that where $\sigma = 5$ and $c^2 = 0.1$:

r	1	2	3	4	5
$\lvert R[f]\rvert \leq$	$2.9\cdot 10^{-1}$	$1.4\cdot 10^{-2}$	$6.4\cdot 10^{-3}$	$2.3\cdot 10^{-2}$	$5.1\cdot 10^{-1}$

The classical bound uses the co-observation $\mathcal{C}_M^{(12)}$ (see Theorem 7.6.5 below) and yields 19.3. This is useless, given the fact that $|I[f]| \le 20$. Bounds that use higher derivatives become satisfactory and, eventually, good if we move the singularity ic away from the fundamental interval $[-1,1]$, that is, if we increase $|c|$.

PROOF OF THEOREM 7.6.4. By Theorem 7.6.2, we have

$$\begin{aligned} \int_a^b |K_r(x)|\,dx &\le \frac{1}{p_\sigma(1)} \sum_{\mu=0}^{\sigma} |\beta_\mu^{(\sigma)}| \left(\frac{b-a}{2^{m+\mu}}\right)^r \int_a^b \left| B_r^*\left(2^{m+\mu}\frac{x-a}{b-a}\right)\right| dx \\ &= \frac{1}{p_\sigma(1)} \sum_{\mu=0}^{\sigma} |\beta_\mu^{(\sigma)}| \frac{(b-a)^{r+1}}{2^{r(m+\mu)}} \int_0^1 |B_r(y)|\,dy \\ &= \frac{1}{p_\sigma(1)} \frac{(b-a)^{r+1}}{2^{rm}} \int_0^1 |B_r(y)|\,dy \cdot |p_\sigma(-2^{-r})|. \end{aligned} \tag{7.57}$$

Now,

$$\begin{aligned} |p_\sigma(-2^{-r})| &= \prod_{\nu=1}^{\sigma} (2^{-r} + 2^{-2\nu}) \\ &= 2^{-r\sigma} \prod_{\nu=1}^{\lfloor r/2 \rfloor} (1 + 2^{r-2\nu}) \prod_{\nu=\lfloor r/2\rfloor+1}^{\sigma} (1 + 2^{-(2\nu-r)}) \\ &= \begin{cases} 2^{-r\sigma}\cdot 2^{(r-1)^2/4}\cdot 2^{-1/4} \prod\limits_{\mu=0}^{(r-2)/2} (1+2^{-2\mu}) \prod\limits_{\mu=1}^{\sigma-r/2} (1+2^{-2\mu}) & \text{for even } r \\ 2^{-r\sigma}\cdot 2^{(r-1)^2/4} \prod\limits_{\mu=1}^{(r-1)/2} (1+2^{1-2\mu}) \prod\limits_{\mu=1}^{\sigma-(r+1)/2} (1+2^{1-2\mu}) & \text{for odd } r \end{cases} \\ &\le \begin{cases} 2^{-r\sigma}\cdot 2^{(r-1)^2/4}\cdot 2^{3/4} \prod\limits_{\mu=1}^{\infty} (1+2^{-2\mu})^2 & \text{for even } r \\ 2^{-r\sigma}\cdot 2^{(r-1)^2/4} \prod\limits_{\mu=1}^{\infty} (1+2^{-2\mu+1})^2 & \text{for odd } r \end{cases} \\ &\le 2^{-r\sigma}\cdot 2^{(r-1)^2/4}\cdot 3.092. \end{aligned} \tag{7.58}$$

Using the fact that

$$p_\sigma(1) \ge \prod_{\mu=1}^{\infty} (1 - 4^{-\mu}) > 0.688,$$

the estimate (7.57) gives

$$\int_a^b |K_r(x)|\,dx \le 4.5 \cdot \frac{(b-a)^{r+1}}{2^{r(m+\sigma)}} \cdot 2^{(r-1)^2/4} \int_0^1 |B_r(y)|\,dy.$$

The last step in the proof is to apply the bound

$$\int_0^1 |B_r(y)|\,dy \le \frac{1.3}{(2\pi)^r}. \tag{7.59}$$

Inequality (7.59) can be verified for $r \leq 5$ by direct calculation using the explicit representation of B_r. For $r \geq 6$, we use the Fourier series of B_r (see Theorem B.7) and the relations

$$\int_0^1 \left|\cos(2\nu\pi x)\right| dx = \int_0^1 \left|\sin(2\nu\pi x)\right| dx = \frac{2}{\pi}$$

to obtain

$$\int_0^1 |B_r(y)|\, dy \leq \frac{2}{\pi}\sum_{\nu=1}^{\infty} \frac{2}{(2\nu\pi)^r} \leq \frac{4}{\pi(2\pi)^r}\sum_{\nu=1}^{\infty}\nu^{-6} = \frac{4}{\pi(2\pi)^r}\cdot\frac{(2\pi)^6 B_6(0)}{2} < \frac{1.296}{(2\pi)^r}.$$

This completes the proof of the theorem. □

Theorem 7.6.4 does not apply to bounds using the highest possible derivative (i.e. $f^{(2\sigma+2)}$ for Romberg rules of rank σ), but a variant of its proof would work to give bounds in that case. With a different method, one can obtain a fairly explicit expression for the corresponding error constant. The key point is the definiteness of $R^{\mathrm{Ro}(\sigma)}_{2^{m+\sigma}+1}$ of order $2\sigma + 2$.

THEOREM 7.6.5 (Bauer 1961). For $f \in C^{(2\sigma+2)}[a, b]$, there exists $\xi \in [a, b]$ such that

$$R^{\mathrm{Ro}(\sigma)}_{2^{m+\sigma}+1}[f] = -\frac{(b-a)^{2\sigma+3}|B_{2\sigma+2}(0)|}{2^{(2m+\sigma)(\sigma+1)}} f^{(2\sigma+2)}(\xi). \tag{7.60}$$

PROOF. First, we establish an explicit expression for the Peano kernel $K_{2\sigma+2}$ by using Theorem 7.6.2 and the relation

$$K_{2\sigma+2}(x) = \int_x^b K_{2\sigma+1}(y)\, dy.$$

We obtain

(7.61)

$$K_{2\sigma+2}(x) = \frac{1}{p_\sigma(1)}\sum_{\mu=0}^{\sigma}\beta_\mu^{(\sigma)}\left(\frac{b-a}{2^{m+\mu}}\right)^{2\sigma+2}\left[B^*_{2\sigma+2}\left(2^{m+\mu}\frac{x-a}{b-a}\right) - B_{2\sigma+2}(0)\right].$$

The next step is to prove that

$$K_{2\sigma+2}(x) \leq 0. \tag{7.62}$$

Since

$$K_{2\sigma+2}\left(x + \frac{b-a}{2^m}\right) = K_{2\sigma+2}(x),$$

we may restrict our attention to

$$x = a + y\cdot\frac{b-a}{2^m} \quad \text{where } y \in [0,1].$$

Therefore, we take $[a, b] = [0, 1]$ and $m = 0$ from now on.

We shall prove by induction that $K_{2\sigma+2}\left[R^{\mathrm{Ro}(\sigma)}_{2^\sigma+1}\right]$ is strictly decreasing on $[0, \frac{1}{2}]$. From this, (7.62) would follow immediately by virtue of symmetry and the fact that $K_{2\sigma+2}(0) = 0$. We begin the induction with

$$K_2\left[Q_2^{\mathrm{Ro}(0)}\right](x) = \frac{x(x-1)}{2}.$$

The induction step is based on the relation

$$K_{2\sigma}\left[R^{\mathrm{Ro}(\sigma)}_{2^\sigma+1}\right](x) = \frac{1}{4^\sigma-1}\left(K_{2\sigma}\left[R^{\mathrm{Ro}(\sigma-1)}_{2^{\sigma-1}+1}\right](2x) - K_{2\sigma}\left[R^{\mathrm{Ro}(\sigma-1)}_{2^{\sigma-1}+1}\right](x)\right) \tag{7.63}$$

for $x \in [0,1/2]$. To prove (7.63), we combine

$$K_{2\sigma}\left[R^{\mathrm{Ro}(\sigma)}_{2^\sigma+1}\right] = \frac{1}{4^\sigma-1}\left(4^\sigma K_{2\sigma}\left[R^{\mathrm{Ro}(\sigma-1)}_{2^\sigma+1}\right] - K_{2\sigma}\left[R^{\mathrm{Ro}(\sigma-1)}_{2^{\sigma-1}+1}\right]\right),$$

which comes from the recursion formula, with

$$K_{2\sigma}\left[R^{\mathrm{Ro}(\sigma-1)}_{2^{\sigma-1}+1}\right](2x) = 4^\sigma K_{2\sigma}\left[R^{\mathrm{Ro}(\sigma-1)}_{2^\sigma+1}\right](x),$$

which follows from (7.61).

As a consequence of the induction hypothesis, we now have that on $]0,1/4[$, the first summand in (7.63) is smaller than the second one and hence

$$K_{2\sigma}\left[R^{\mathrm{Ro}(\sigma)}_{2^\sigma+1}\right](x) < 0 \quad \text{for } x \in]0,1/4[.$$

On $[1/4,1/2]$, the first summand in (7.63) is strictly increasing while the second one is decreasing, so there is at most one zero in $]0,1/2[$. The relations $K'_{2\sigma+1} = -K_{2\sigma}$ and $K_{2\sigma+1}(0) = K_{2\sigma+1}(1/2) = 0$ imply that

$$K_{2\sigma+1}\left[R^{\mathrm{Ro}(\sigma)}_{2^\sigma+1}\right](x) > 0 \quad \text{for } x \in]0,1/2[.$$

Therefore $K_{2\sigma+2}\left[R^{\mathrm{Ro}(\sigma)}_{2^\sigma+1}\right]$ is decreasing and the induction step is done.

We have now proved the definiteness of order $2\sigma+2$ of $R^{\mathrm{Ro}(\sigma)}_{2^\sigma+1}$ and hence of $R^{\mathrm{Ro}(\sigma)}_{2^{m+\sigma}+1}$, so

$$R^{\mathrm{Ro}(\sigma)}_{2^{m+\sigma}+1}[f] = f^{(2\sigma+2)}(\xi)\int_a^b K_{2\sigma+2}\left[R^{\mathrm{Ro}(\sigma)}_{2^{m+\sigma}+1}\right](x)\,dx.$$

Applying (7.61) gives

$$\begin{aligned}\int_a^b K_{2\sigma+2}(x)\,dx &= -\frac{1}{p_\sigma(1)}\sum_{\mu=0}^{\sigma}\beta^{(\sigma)}_\mu\left(\frac{b-a}{2^{m+\mu}}\right)^{2\sigma+2}(b-a)B_{2\sigma+2}(0)\\ &= -\frac{(b-a)^{2\sigma+3}}{2^{(2\sigma+2)m}}B_{2\sigma+2}(0)\cdot\frac{p_\sigma(2^{-2\sigma-2})}{p_\sigma(1)}.\end{aligned}$$

So finally,

$$\begin{aligned}p_\sigma(2^{-2\sigma-2}) &= \prod_{\nu=1}^{\sigma}(2^{-2\sigma-2}-2^{-2\nu})\\ &= 2^{-\sigma(\sigma+1)}\prod_{\nu=1}^{\sigma}(2^{-2\sigma-2+2\nu}-1) = 2^{-\sigma(\sigma+1)}(-1)^\sigma p_\sigma(1),\end{aligned}$$

and the theorem is proved. □

A comparison of Theorem 7.6.5 with Theorem 7.5.3 shows that for the Gregory rules of rank 2σ and the Romberg rules of rank σ, we have error representations of the form

$$R_{n+1}[f] = c(\sigma)\frac{(b-a)^{2\sigma+3}}{n^{2\sigma+2}}f^{(2\sigma+2)}(\xi)$$

(in the case of Romberg rules only for $n = 2^{m+\sigma}$, $m = 0, 1, \ldots$). It is easy to prove that

$$\lim_{\sigma\to\infty} \frac{c_{Ro}(\sigma)}{c_{Gr}(\sigma)} = \infty,$$

which shows that the Romberg rules are inferior for large σ. For the $\sigma = 1$ case see Example 7.5.2.

The Romberg scheme is often applied without any consideration of the error and, clearly, this may lead to false conclusions. It is important to note that these can be avoided if we assume

$$f^{(2\sigma+2)} \geq 0 \quad \text{for } \sigma = 0, 1, 2, \ldots, \tag{7.64}$$

because then we can prove monotonicity and stopping inequalities (see the remarks after Theorem 7.2.8 for explanation of this concept).

It is practical to simplify the notation by using the abbreviation

$$y_m^{(\sigma)} := Q_{2^{m+\sigma}+1}^{\mathrm{Ro}(\sigma)}[f].$$

THEOREM 7.6.6 (Ström 1967). $y_m^{(\sigma)}$ is decreasing for increasing m if $f^{(2\sigma+2)} \geq 0$.

PROOF. We have to show that

$$R_{2^{m+\sigma}+1}^{\mathrm{Ro}(\sigma)} - R_{2^{m+\sigma+1}+1}^{\mathrm{Ro}(\sigma)} \tag{7.65}$$

is a negative definite functional of order $2\sigma + 2$. This can be done by replacing the $R_{2^{m+\sigma}+1}^{\mathrm{Ro}(\sigma)}$ in the proof of Theorem 7.6.5 by (7.65); the arguments are almost identical. □

Here is a consequence we would like to emphasize: the element $y_0^{(\sigma)}$ computed last in the Romberg scheme is at least as good as all the other values in the scheme if $f^{(2\nu)} \geq 0$ for $\nu = 1, 2, \ldots, \sigma + 1$. This follows from Theorem 7.6.6 and an earlier statement concerning the general Richardson scheme.

We now define

$$\widetilde{y}_m^{(\sigma)} := 2y_{m+1}^{(\sigma)} - y_m^{(\sigma)}.$$

As we shall see in Section 7.8, these numbers are rather natural counterparts of the Romberg values. A first example of their usefulness is provided by the following theorem.

THEOREM 7.6.7 (Albrecht 1972; Ström 1972). $\widetilde{y}_m^{(\sigma)}$ is increasing for increasing m if $f^{(2\sigma+2)} \geq 0$.

PROOF. We again use the same method as in the proof of Theorem 7.6.5, but this time with

$$2R_{2^{m+\sigma+2}+1}^{\mathrm{Ro}(\sigma)} - 3R_{2^{m+\sigma+1}+1}^{\mathrm{Ro}(\sigma)} + R_{2^{m+\sigma}+1}^{\mathrm{Ro}(\sigma)}$$

in place of $R_{2^{m+\sigma}+1}^{\mathrm{Ro}(\sigma)}$. □

Under the assumption that $f^{(2\sigma+2)} \geq 0$, we have shown that

$$2y_{m+1}^{(\sigma)} - y_m^{(\sigma)} \leq I[f] \tag{7.66}$$

or, equivalently,

$$y_{m+1}^{(\sigma)} - I[f] \leq y_m^{(\sigma)} - y_{m+1}^{(\sigma)}.$$

According to Theorem 7.6.6, both sides of the preceding inequality are positive, so we have a stopping inequality for the method of rank σ as follows:

$$|y_{m+1}^{(\sigma)} - I[f]| \leq |y_m^{(\sigma)} - y_{m+1}^{(\sigma)}| \quad \text{if } f^{(2\sigma+2)} \geq 0. \tag{7.67}$$

More important is the following stopping inequality for the "diagonal" Romberg method:

THEOREM 7.6.8 (Förster 1982a).

$$|y_m^{(\sigma+1)} - I[f]| \leq \left(1 - \frac{2}{4^{\sigma+1}}\right) |y_m^{(\sigma)} - y_m^{(\sigma+1)}| \quad \text{if } f^{(2\sigma+2)} \geq 0.$$

PROOF. The recursion relation implies that

$$y_{m+1}^{(\sigma)} = (1 - 4^{-\sigma-1}) y_m^{(\sigma+1)} + 4^{-\sigma-1} y_m^{(\sigma)}. \tag{7.68}$$

Combining this identity with (7.66) gives

$$I[f] \geq 2(1 - 4^{-\sigma-1}) y_m^{(\sigma+1)} + (2 \cdot 4^{-\sigma-1} - 1) y_m^{(\sigma)},$$

or

$$\begin{aligned} y_m^{(\sigma+1)} - I[f] &\leq \left(1 - \frac{2}{4^{\sigma+1}}\right) (y_m^{(\sigma)} - y_m^{(\sigma+1)}) \\ &\leq \left(1 - \frac{2}{4^{\sigma+1}}\right) |y_m^{(\sigma)} - y_m^{(\sigma+1)}|. \end{aligned} \tag{7.69}$$

Using Theorem 7.6.6 and (7.68), we obtain

$$\begin{aligned} I[f] - y_m^{(\sigma+1)} &\leq y_{m+1}^{(\sigma)} - y_m^{(\sigma+1)} = 4^{-\sigma-1}(y_m^{(\sigma)} - y_m^{(\sigma+1)}) \\ &\leq 4^{-\sigma-1} |y_m^{(\sigma)} - y_m^{(\sigma+1)}| \leq \left(1 - \frac{2}{4^{\sigma+1}}\right) |y_m^{(\sigma)} - y_m^{(\sigma+1)}|. \end{aligned}$$

This, together with (7.69), yields the assertion. □

One disadvantage of the classical Romberg method, as discussed here, is the rapidly increasing number of nodes. Any new row doubles the computational effort. We could do better by applying Richardson's acceleration procedure to $Q_{n_i}^{\mathrm{Tr}}[f]$ with a slowly increasing sequence (n_i). Bulirsch (1964) proposed using the following sequence:

$$n_i = h_i 2^{i/2} \quad \text{where} \quad h_i := \begin{cases} 1 & \text{for } i = 0, \\ \sqrt{2} & \text{for odd } i, \\ 3/2 & \text{for even } i > 0. \end{cases}$$

Detailed examination of this method (see Fischer 2002) reveals that it yields similar results as in the classical case. However, the Bulirsch method is not positive and is not based on equidistant nodes.

7.7. Equidistant nodes and the degree of polynomial exactness

Our investigation of the Newton–Cotes methods has shown that rules using equidistant nodes and the highest possible degree of exactness are not very practical. But if the degree requirement is lowered, we can obtain good methods with equidistant nodes, as described in the following theorem.

THEOREM 7.7.1 (Locher 1973a). There is a quadrature method $(Q_n)_{n\in\mathbb{N}}$, with Q_{n+1} having equidistant nodes and degree $\lfloor\sqrt{n}\rfloor$, such that

$$\lim_{n\to\infty} \|Q_n\| = b - a.$$

PROOF. We need the following lemma, whose proof will be given afterwards.

LEMMA 7.7.1 (Ehlich and Zeller 1964). Suppose that $x_1, \ldots, x_{n+1}$ are equidistant points in $[a, b]$. For $p \in \mathbb{P}_{\lfloor\sqrt{n}\rfloor}$ with $|p(x_\nu)| \le 1$ for $\nu = 1, 2, \ldots, n+1$, we have $\|p\| \le 2$.

Under the assumptions of this lemma, we obtain

$$\left|\int_a^b p(x)\,dx\right| \le 2\big(\lfloor\sqrt{n}\rfloor - 1\big)\frac{b-a}{n} + b - a, \tag{7.70}$$

since $|p|$ can exceed 1 only in (parts of) the $\lfloor\sqrt{n}\rfloor - 1$ or fewer intervals $[x_\nu, x_{\nu+1}]$ where p has a relative extremum.

Now, we apply Theorem 3.2.1 to

$$X := \{x_\nu \,:\, \nu = 1, 2, \ldots, n+1\}, \quad U := \mathbb{P}_{\lfloor\sqrt{n}\rfloor}\big|_X \quad \text{and} \quad I[f] := \int_a^b f(x)\,dx.$$

The integral I is defined on U, but evidently we may regard I as being defined on $\mathbb{P}_{\lfloor\sqrt{n}\rfloor}$. According to (7.70), we have $M = b - a + o(1)$, and this gives the assertion. □

PROOF OF LEMMA 7.7.1. A theorem of Markoff (see, e.g., DeVore and Lorentz 1993: p. 98) yields

$$\|q'\| \le \frac{2m^2\|q\|}{b-a}$$

for any $q \in \mathbb{P}_m$. In the situation of the lemma, we therefore have

$$\|p''\| \le \frac{4n^2\|p\|}{(b-a)^2}.$$

Let ξ be such that $|p(\xi)| = \|p\|$. The assertion is trivial if $\xi \in \{a, b\}$, so we may assume that $\xi \in\,]a, b[$. Then, Taylor's theorem gives

$$p(x_\nu) = p(\xi) + \frac{(x_\nu - \xi)^2}{2} p''(\eta).$$

With a suitable x_ν, we obtain

$$|p(x_\nu)| \ge \|p\| - \frac{(b-a)^2}{8n^2} \cdot \frac{4n^2}{(b-a)^2} \cdot \|p\| = \frac{1}{2}\|p\|,$$

or

$$\|p\| \le 2 \sup_\nu |p(x_\nu)| \le 2$$

as claimed. □

The degree $\lfloor\sqrt{n}\rfloor$ in Theorem 7.7.1 is quite disappointing, but unfortunately it is not possible to obtain any significant improvement, because by slightly modifying the proof of Theorem 5.6.2 we get the following result.

THEOREM 7.7.2. Any quadrature method that has equidistant nodes and $\deg Q_{n+1} > 2.5\sqrt{n}\ln n$ is divergent for some continuous f.

For related results, see Bernstein (1937a, 1937b), Wilson (1969, 1970a, 1970b), Brakhage and Brombeer (1969) and Brombeer (1975).

7.8. The midpoint method

The midpoint rule was defined in Example 2.1.1. Let us recall the definition here.

DEFINITION 7.8.1. The quadrature rule given by

$$Q_n^{\mathrm{Mi}}[f] := h \sum_{\nu=1}^{n} f(a + (\nu - \tfrac{1}{2})h), \qquad h = \frac{b-a}{n}$$

is called the nth midpoint rule and the sequence $Q_1^{\mathrm{Mi}}, Q_2^{\mathrm{Mi}}, \ldots$ the midpoint method.

The quadrature rule Q_n^{Mi} is best in $\mathcal{C}_M^{\mathrm{Var}}$ (by Theorem 4.4.3), in $\mathcal{C}_\omega^{\mathrm{mc}}$ (by Theorem 4.5.3), in $\mathcal{C}_M^{(1)}$ in particular, and—it is easy to prove—in $\mathcal{C}_M^{\mathrm{Sa},1}$. Interest in Q_n^{Mi} stems from these properties as well as from the exceptional simplicity of the rule.

A simple approach to the theory of the midpoint method is by way of the identity

$$Q_n^{\mathrm{Mi}}[f] = 2Q_{2n+1}^{\mathrm{Tr}}[f] - Q_{n+1}^{\mathrm{Tr}}[f],$$

or equivalently

$$R_n^{\mathrm{Mi}}[f] = 2R_{2n+1}^{\mathrm{Tr}}[f] - R_{n+1}^{\mathrm{Tr}}[f]. \tag{7.71}$$

Some immediate consequences are that

$$R_n^{\mathrm{Mi}}[f] = o(n^{-1}) \quad \text{if } f \text{ is absolutely continuous}$$

and

$$R_n^{\mathrm{Mi}}[f] = \frac{(b-a)^2}{24n^2}[f'(b) - f'(a)] + o(n^{-2}) \quad \text{if } f' \text{ is absolutely continuous.} \tag{7.72}$$

A comparison of this last relation with (7.2) shows that

$$R_n^{\mathrm{Mi}}[f] = -\frac{1}{2} R_{n+1}^{\mathrm{Tr}}[f](1 + o(1))$$

if f' is absolutely continuous and $f'(b) \neq f'(a)$ holds. This is a modest advantage of Q_n^{Mi} over Q_{n+1}^{Tr}, which can be observed in many other classes of functions as well.

By combining (7.71) with Theorem 7.1.1, one obtains a cunterpart of the Euler–Maclaurin formula. From this or the Peano kernel theorem, we derive the following error bounds:

$$\begin{aligned}
|R_n^{\mathrm{Mi}}[f]| &\le \frac{(b-a)^2}{4n}\,\|f'\| && \text{if } f \text{ has a bounded derivative,}\\
|R_n^{\mathrm{Mi}}[f]| &\le \frac{b-a}{2n}\,\operatorname{Var} f && \text{if } f \text{ is of bounded variation,}\\
|R_n^{\mathrm{Mi}}[f]| &\le \frac{(b-a)^2}{8n^2}\,\operatorname{Var} f' && \text{if } f' \text{ is of bounded variation,}\\
R_n^{\mathrm{Mi}}[f] &= \frac{(b-a)^3}{24n^2}\,f''(\xi) && \text{if } f'' \text{ is continuous,}
\end{aligned}$$

and

$$R_n^{\mathrm{Mi}}[f] = \frac{(b-a)^2}{24n^2}[f'(b) - f'(a)] - \frac{7(b-a)^5}{5760n^4} f^{(4)}(\xi) \quad \text{if } f^{(4)} \text{ is continuous,}$$

where the ξ in the latter two equations are (unknown) numbers from $[a, b]$.

Just as in the case of the trapezoidal rule, the property of convexity is of importance. The positive definiteness of R_n^{Mi} gives the following:

THEOREM 7.8.1. For convex f,

$$R_n^{\mathrm{Mi}}[f] \geq 0.$$

This result, combined with Theorem 7.2.4, leads to the following bracketing property.

COROLLARY 7.8.1. For convex f,

$$Q_n^{\mathrm{Mi}}[f] \leq I[f] \leq Q_{n+1}^{\mathrm{Tr}}[f]. \tag{7.73}$$

Here, Q_n^{Mi} and $Q_{n+1}^{\mathrm{Tr}}[f]$ use distinct evaluation points. We could save time if the upper and lower bounds used the same nodes; this can be achieved by replacing Q_{n+1}^{Tr} by the generalized trapezoidal rule with nodes

$$a,\ a+\frac{1}{2}h,\ a+\frac{3}{2}h,\ a+\frac{5}{2}h, \ldots,\ a+\frac{2n-1}{2}h,\ b.$$

The above rule has the explicit representation

$$Q_{n+2}^*[f] := Q_n^{\mathrm{Mi}}[f] + \frac{h}{4}\left[f(b) - f\left(b-\frac{h}{2}\right) - f\left(a+\frac{h}{2}\right) + f(a)\right].$$

As remarked in Example 4.2.2, $Q_n^{\mathrm{gen\,Tr}}[f] \geq I[f]$ holds for convex functions if $x_1 = a$ and $x_n = b$. Instead of (7.73), we have

$$Q_n^{\mathrm{Mi}}[f] \leq I[f] \leq Q_{n+2}^*[f]. \tag{7.74}$$

The main point is that the interval of uncertainty in (7.74) is asymptotically of the same size as that in (7.73), at least when f' is absolutely continuous with $f'(b) \neq f'(a)$. To see this, we use (7.2) and (7.72) to get

$$Q_{n+1}^{\mathrm{Tr}}[f] - Q_n^{\mathrm{Mi}}[f] = R_n^{\mathrm{Mi}}[f] - R_{n+1}^{\mathrm{Tr}}[f] = \frac{h^2}{8}\left[f'(b) - f'(a)\right] + o(h^2).$$

On the other hand,

$$Q_{n+2}^*[f] - Q_n^{\mathrm{Mi}}[f] = \frac{h}{4}\left[f(b) - f\left(b-\frac{h}{2}\right) - f\left(a+\frac{h}{2}\right) + f(a)\right] = \frac{h^2}{8}\left[f'(b) - f'(a)\right] + o(h^2).$$

The midpoint of the interval given by (7.73) is $Q_{2n+1}^{\mathrm{Tr}}[f]$ and, since $Q_{2n+1}^{\mathrm{Tr}}[f] \geq I[f]$, the integral lies in the left half of this interval, which means that

$$\left|R_n^{\mathrm{Mi}}[f]\right| \leq \left|R_{n+1}^{\mathrm{Tr}}[f]\right|$$

for convex functions. Again, the midpoint rule is better than the trapezoidal rule. Another result (Brass 1995a) which illustrates the same phenomenon is that

$$\sum_{\nu=1}^{\infty} R_\nu^{\mathrm{Mi}}[f]$$

converges for any convex function, whereas the same is not true of the trapezoidal method (see Example 7.2.1).

We can view the superiority of the midpoint method as a special case of the principle of "avoiding the singularity" (Rabinowitz 1979), which recommends that rules not use a node where the integrand has a singularity. For convex functions, singularities can exist (namely, with $f'(a+0) = -\infty$ or $f'(b-0) = \infty$) only at the boundary points a and b, so Q_n^{Mi} is preferable from that standpoint.

In Chapter 1 we mentioned the following "inverse theorem".

THEOREM 7.8.2. Let f be convex and suppose that $R_n^{\mathrm{Mi}}[f] = O(n^{-1-\alpha})$ with $\alpha \in]0, 1]$. Then we have $f \in \operatorname{Lip} \alpha$.

This is an easy consequence of the corresponding theorem for the trapezoidal method (Theorem 7.2.6). The link between the two methods is given by the following lemma.

LEMMA 7.8.1. Let f be convex and suppose that $R_n^{\mathrm{Mi}}[f] = O(n^{-s})$ with $s > 1$. Then $R_n^{\mathrm{Tr}}[f] = O(n^{-s})$.

PROOF. The assertion follows from the relation

$$R_{n+1}^{\mathrm{Tr}}[f] = -\sum_{\nu=0}^{\infty} 2^\nu R_{n\cdot 2^\nu}^{\mathrm{Mi}}[f],$$

which can be deduced from

$$R_{n+1}^{\mathrm{Tr}}[f] = -\sum_{\nu=0}^{k-1} 2^\nu R_{n\cdot 2^\nu}^{\mathrm{Mi}}[f] + 2^k R_{n\cdot 2^k+1}^{\mathrm{Tr}}[f] \tag{7.75}$$

and the fact that

$$\lim_{m\to\infty} m R_m^{\mathrm{Tr}}[f] = 0$$

(see Theorem 7.2.5). A k-fold application of

$$R_{n+1}^{\mathrm{Tr}}[f] = -R_n^{\mathrm{Mi}}[f] + 2R_{2n+1}^{\mathrm{Tr}}[f]$$

establishes the identity (7.75). □

The improvements made to the trapezoidal method that led to the Gregory and Romberg methods can be done on the midpoint method in an analogous fashion.

EXAMPLE 7.8.1. Proceeding as in the derivation of the Durand method, but now starting with (7.72), we obtain the rule

$$Q_n^{**}[f] := Q_n^{\mathrm{Mi}}[f] + \frac{h}{24}\left[f\left(a+\frac{h}{2}\right) - f\left(a+\frac{3h}{2}\right) - f\left(b-\frac{3h}{2}\right) + f\left(b-\frac{h}{2}\right)\right].$$

This rule is an improvement because

$$R_n^{**}[f] = o(n^{-2}) \quad \text{if } f' \text{ is absolutely continuous}$$

and

$$R_n^{**}[f] = O(n^{-3}) \quad \text{if } f'' \text{ is absolutely continuous,}$$

whereas

$$R_n^{\mathrm{Mi}}[f] = O(n^{-2})$$

is best possible in both classes. In a variant, we approximate $f'(a)$ by

$$\frac{2}{h}\cdot\left[f\left(a+\frac{h}{2}\right) - f(a)\right]$$

(that is, we use nodes that are not among the nodes of Q_n^{Mi}), and similarly for $f'(b)$. This yields the method of Piobert and Parmentier:

$$Q_{n+2}^{\mathrm{PP}}[f] := Q_n^{\mathrm{Mi}}[f] + \frac{h}{12}\left[f(a) - f\left(a+\frac{h}{2}\right) - f\left(b-\frac{h}{2}\right) + f(b)\right].$$

Many methods can be constructed in this manner—for a general scheme see Brass (1976)—but they seem to offer no significant advantage over the Gregory methods. Such schemes are therefore of little practical value.

To derive the analogue of the Romberg scheme (which is contained in Romberg's original paper of 1955), we start with Lemma 7.6.1 in the special case where

$$y_m^{(0)} = Q_{2^m}^{\mathrm{Mi}}, \qquad y = \int_a^b f(x)\,dx, \qquad q_\nu = 4^{-\nu}$$

and

$$\alpha_\nu = (b-a)^{2\nu}\left(2^{1-2\nu}-1\right)B_{2\nu}(0)\left[f^{(2\nu-1)}(b) - f^{(2\nu-1)}(a)\right].$$

Let us denote by $\overline{y}_m^{(\sigma)}$ the element in the resulting Romberg midpoint scheme that corresponds to the $y_m^{(\sigma)}$ in the general Richardson scheme. Romberg midpoint rules are not necessarily positive, as

$$\overline{y}_0^{(1)} = \frac{b-a}{3}\left[2f\left(\frac{3a+b}{4}\right) - f\left(\frac{a+b}{2}\right) + 2f\left(\frac{a+3b}{4}\right)\right]$$

shows.

The Romberg midpoint scheme uses the same recursion relation as does the classical Romberg scheme, and a similar theory could be developed. This theory can be based on the following simple connection between the two schemes:

$$\overline{y}_m^{(\sigma)} = 2Q_{2^{m+\sigma+1}+1}^{\mathrm{Ro}(\sigma)}[f] - Q_{2^{m+\sigma}+1}^{\mathrm{Ro}(\sigma)}[f]. \tag{7.76}$$

When discussing the classical Romberg scheme, we introduced the numbers $\widetilde{y}_m^{(\sigma)}$. Equation (7.76) tells us that

$$\overline{y}_m^{(\sigma)} = \widetilde{y}_m^{(\sigma)},$$

so the $\widetilde{y}_m^{(\sigma)}$ are indeed natural counterparts of the Romberg values. The proof of (7.76) is by induction. For $\sigma = 0$, the relation reads

$$Q_{2^m}^{\mathrm{Mi}}[f] = 2Q_{2^{m+1}+1}^{\mathrm{Tr}}[f] - Q_{2^m+1}^{\mathrm{Tr}}[f],$$

which is correct. The induction step is trivial, since we apply the same recursion formula to both sides.

Owing to (7.76), the simultaneous construction of both schemes does not require much additional effort but is very useful when $f^{(2\sigma+2)} > 0$, because then the bracketing property

$$\widetilde{y}_m^{(\sigma)} \le I[f] \le Q_{2^{m+\sigma}+1}^{\mathrm{Ro}(\sigma)}[f]$$

holds, according to Theorems 7.6.5 and 7.6.6.

CHAPTER 8

Periodic Integrands

We examine the computation of an integral of a periodic function over one complete period. For the most important special cases $w(x) \equiv 1$ and $w(x) = \cos px$, the trapezoidal method is of paramount significance in numerical practice. We discuss reasons for this exceptional position and give error bounds specially adapted to periodicity.

8.1. The special role of the trapezoidal rule for $w = 1$

We denote by $\widetilde{\mathcal{C}}$ the space of all 2π-periodic real-valued continuous functions. All integrands in this chapter will belong to $\widetilde{\mathcal{C}}$. The case of an arbitrary period can be treated by applying an appropriate linear transformation.

The first problem we will consider is that of computing

$$I[f] = \int_0^{2\pi} f(x)\,dx \ \left(= \int_\alpha^{2\pi+\alpha} f(x)\,dx \quad \text{for arbitrary } \alpha \right).$$

In the interval of integration there is no special point (in contrast to the case of non-periodic functions, where the endpoints a and b occupy an exceptional position). Therefore, in the present section and the next one, we use the notation

$$x_\nu = x_\nu(\alpha) := \nu h - \alpha \quad \text{for } \nu = 1, 2, \ldots, n, \ \ \alpha \in \,]0, h], \ \ h := \frac{2\pi}{n}. \tag{8.1}$$

Because $f(0) = f(2\pi)$, we have

$$Q_{n+1}^{\mathrm{Tr}}[f] = h \sum_{\nu=1}^{n} f(\nu h)$$

so that Q_{n+1}^{Tr} can be regarded as a rule with n nodes.

In what follows, we shall only be interested in co-observations that share with the set of periodic functions the fundamental property of being translation invariant.

DEFINITION 8.1.1. A set $X \subseteq \widetilde{\mathcal{C}}$ is said to be *translation invariant* if for any $f \in X$ and any $c \in \mathbb{R}$, the function $g = f(\,\cdot + c)$ also belongs to X.

For translation-invariant co-observations $\mathcal{C}$, the choice of α in (8.1) is irrelevant, since we evidently have

$$\sup_{f \in \mathcal{C}} \left| R[f] \right| = \sup_{f \in \mathcal{C}} \left| R^*[f] \right| \tag{8.2}$$

for the rules

$$Q[f] = \sum_{\nu=1}^{n} a_\nu f(\nu h) \quad \text{and} \quad Q^*[f] = \sum_{\nu=1}^{n} a_\nu f(\nu h - \alpha).$$

In particular, we have

$$\rho(Q_{n+1}^{\mathrm{Tr}}, \mathcal{C}) = \rho(Q_n^{\mathrm{Mi}}, \mathcal{C}) \tag{8.3}$$

for any translation-invariant $\mathcal{C}$.

The first main result that reveals the exceptional position of the trapezoidal rule (or, by (8.3), the midpoint rule) is the following theorem.

THEOREM 8.1.1. For the observation (8.1) and a co-observation $\mathcal{C}$ that is translation invariant (and satisfies the General Assumption from page 8), there exists a constant c such that

$$\rho_n^{\mathrm{opt}}(\mathcal{C}) = \rho(c \cdot Q_{n+1}^{\mathrm{Tr}},\, \mathcal{C}).$$

If all constant functions belong to $\mathcal{C}$, then $c = 1$.

So Q_{n+1}^{Tr} is optimal for many co-observations, such as for

$$\widetilde{\mathcal{C}}_M^{(r)} = \{f \in \widetilde{\mathcal{C}} \,:\, \sup_x |f^{(r)}(x)| \le M\}. \tag{8.4}$$

In fact, even more can be said: Motornyi (1974) proved that Q_{n+1}^{Tr} is best in $\widetilde{\mathcal{C}}_M^{(r)}$, and the same is true of many other translation-invariant co-observations; see Žensykbaev (1977a, 1977b) and Brombeer (1978). For still more general results, see Žensykbaev (1981), Ligun (1978) and Bojanov (1981).

PROOF OF THEOREM 8.1.1. Starting with any rule

$$Q[f] = \sum_{\nu=1}^{n} a_\nu f(x_\nu),$$

we define further rules by

$$\begin{aligned} Q^{(\lambda)}[f] &:= Q[f(\,\cdot + \lambda h)] \quad \text{for } \lambda = 1, 2, \ldots, n, \\ \overline{Q}[f] &:= \frac{1}{n}\sum_{\lambda=1}^{n} Q^{(\lambda)}[f]. \end{aligned}$$

Evidently, we have

$$\rho(Q^{(\lambda)},\, \mathcal{C}) = \rho(Q,\, \mathcal{C})$$

and hence

$$\rho(\overline{Q},\, \mathcal{C}) \le \rho(Q,\, \mathcal{C}).$$

The theorem follows from the easily verified identity

$$\overline{Q}[f] = \frac{1}{n}\sum_{\nu=1}^{n} a_\nu \sum_{\mu=1}^{n} f(x_\mu) = \frac{1}{2\pi}\left(\sum_{\nu=1}^{n} a_\nu\right) Q_{n+1}^{\mathrm{Tr}}[f] = c \cdot Q_{n+1}^{\mathrm{Tr}}[f],$$

which means that $c \cdot Q_{n+1}^{\mathrm{Tr}}$ is at least as good as any other quadrature rule with equidistant nodes. If arbitrary constants all belong to $\mathcal{C}$, then a Q with finite ρ has to satisfy

$$Q[1] = \sum_{\nu=1}^{n} a_\nu = I[1] = 2\pi,$$

since otherwise $R[\mathrm{const}] = \mathrm{const} \cdot R[1]$ could attain any value. This implies that $c = 1$. $\square$

The construction of rules according to the principle $R[u] = \{0\}$ for a suitable subspace U requires that $U \subset \widetilde{\mathcal{C}}$ for the problems of this chapter. The most popular choice of U is the space of trigonometric polynomials,

$$\mathbb{T}_n := \big\{t \, : \, t(x) = \frac{\alpha_0}{2} + \sum_{\nu=1}^{n} \alpha_\nu \cos \nu x + \beta_\nu \sin \nu x \;\text{ with } \alpha_\nu, \beta_\nu \in \mathbb{R}\big\}.$$

Another reason for the trapezoidal rule's exceptional position is given by the next theorem.

THEOREM 8.1.2 (Schmidt 1949). For any rule $Q \in \mathbf{Q}_n$, we have $R[\mathbb{T}_{n-1}] = \{0\}$ if and only if the nodes are equidistant as in (8.1) and

$$a_\nu = \frac{2\pi}{n} \quad \text{for } \nu = 1, 2, \ldots, n.$$

There is no $Q \in \mathbf{Q}_n$ with $R[\mathbb{T}_n] = \{0\}$.

For $U = \mathbb{T}_{n-1}$, the trapezoidal rule is an "almost double precision" rule, i.e. equivalent to the Gaussian rule in some sense, because $\dim \mathbb{T}_{n-1} = 2n - 1$. For other spaces U that have this property, see Babenko (1984).

To prove Theorem 8.1.2 we need the following lemma.

LEMMA 8.1.1 (Szasz 1960). Let $0 \le \xi_1 < \xi_2 < \cdots < \xi_n < 2\pi$. If

$$\int_0^{2\pi} t(x)\, dx = 0$$

for all $t \in \mathbb{T}_{n-1}$ with $t(\xi_1) = t(\xi_2) = \cdots = t(\xi_n) = 0$, then there is an α such that

$$\xi_\nu = \nu \frac{2\pi}{n} - \alpha \quad \text{for } \nu = 1, 2, \ldots, n.$$

PROOF. We begin by defining

$$h(x) := \prod_{\nu=1}^{n} (e^{ix} - e^{i\xi_\nu}) \equiv \sum_{\mu=0}^{n-1} c_\mu e^{i\mu x} + e^{inx},$$

$$h_\lambda(x) := e^{-i\lambda x} h(x) \quad \text{for } \lambda = 1, 2, \ldots, n-1.$$

Obviously, we have $h_\lambda(\xi_\nu) = 0$, and, since $\Re h_\lambda \in \mathbb{T}_{n-1}$ and $\Im h_\lambda \in \mathbb{T}_{n-1}$ (where $\Re$ and $\Im$ denote the real and imaginary parts, respectively), our assumption gives

$$\int_0^{2\pi} h_\lambda(x)\, dx = 0.$$

On the other hand, direct integration yields

$$\int_0^{2\pi} h_\lambda(x)\, dx = c_\lambda \cdot 2\pi.$$

Hence

$$h(x) = c_0 + e^{inx} = \prod_{\nu=1}^{n} \left(e^{ix} - e^{i(\nu \frac{2\pi}{n} - \alpha)}\right),$$

where α is determined by

$$e^{-in\alpha} = -c_0 = (-1)^{n+1} e^{i(\xi_1 + \ldots + \xi_n)}.$$

Thus the lemma is proved. □

PROOF OF THEOREM 8.1.2. According to Theorem 2.2.1, the lemma implies that equidistant nodes are necessary for rules satisfying $R[\mathbb{T}_{n-1}] = \{0\}$. The rules

$$Q_n[f] = \frac{2\pi}{n}\sum_{\nu=1}^{n} f(x_\nu), \qquad x_\nu = \nu\frac{2\pi}{n} - \alpha,$$

prove the sufficiency, for we have

$$\sum_{\nu=1}^{n} e^{i\lambda x_\nu} = \begin{cases} n & \text{for } \lambda = 0, \\ 0 & \text{for } \lambda = 1, 2, \ldots, n-1. \end{cases}$$

We have shown that there are rules $Q \in \mathbf{Q}_n$ with $R[\mathbb{T}_{n-1}] = \{0\}$ and that all these rules are of the form

$$Q_n[f] = \sum_{\nu=1}^{n} a_\nu f(x_\nu), \qquad x_\nu = \nu\frac{2\pi}{n} - \alpha.$$

The proof that $a_\nu = 2\pi/n$ necessarily holds relies on the functions

$$t_1(x) = \prod_{\nu=2}^{n} \frac{1-\cos(x - x_\nu)}{1 - \cos(x_1 - x_\nu)},$$

$$t_\mu(x) = t_1\Big(x - (\mu - 1)\frac{2\pi}{n}\Big) \quad \text{for } \mu = 2, 3, \ldots, n.$$

Clearly, $t_\mu \in \mathbb{T}_{n-1}$ and

$$t_\mu(x_\nu) = \begin{cases} 1 & \text{for } \mu = \nu, \\ 0 & \text{otherwise.} \end{cases}$$

Therefore

$$a_\mu = \int_0^{2\pi} t_\mu(x)\,dx = \int_0^{2\pi} t_1\Big(x - (\mu-1)\frac{2\pi}{n}\Big)\,dx = \int_0^{2\pi} t_1(x)\,dx = a_1.$$

The claim now follows from the fact that

$$2\pi = \sum_{\mu=1}^{n} a_\mu = na_1.$$

Finally, observe that for all of these rules we have

$$R_n[f_0] = -2\pi \quad \text{for } f_0(x) = \cos n(x + \alpha).$$

Thus they cannot be exact for all $f \in \mathbb{T}_n$. □

8.2. Error bounds for the trapezoidal rule

We start with the co-observation $\widetilde{C}_M^{(r)}$; see (8.4). We shall need the Favard constants $\mathcal{K}_r$ defined by

$$\mathcal{K}_r := \frac{4}{\pi}\sum_{\nu=0}^{\infty} \frac{(-1)^{\nu(r+1)}}{(2\nu+1)^{r+1}} \quad \text{for } r = 1, 2, \ldots.$$

THEOREM 8.2.1 (Krylov and Sul'gina 1964).

$$\rho(Q_{n+1}^{\mathrm{Tr}}, \widetilde{C}_M^{(r)}) = \frac{2\pi\mathcal{K}_r}{n^r} M.$$

Combining this with Theorem 8.1.1 yields the explicit value of $\rho_n^{\text{opt}}(\widetilde{\mathcal{C}}_M^{(r)})$. Since the behaviour of the error in class $\mathcal{C}_M^{(r)}$ is surely no more favourable than in the subclass $\widetilde{\mathcal{C}}_M^{(r)}$, we deduce that

$$\rho_n^{\text{opt}}(\mathcal{C}_M^{(r)}) \geq \frac{2\pi\mathcal{K}_r}{n^r} M$$

for equidistant nodes and $[a, b] = [0, 2\pi]$. By the result of Motornyi (1974) mentioned in the preceding section, we could even get

$$\rho_n^{\text{best}}(\mathcal{C}_M^{(r)}) \geq \frac{2\pi\mathcal{K}_r}{n^r} M.$$

Upon transforming to an arbitrary interval $[a, b]$, this means that

$$\rho_n^{\text{best}}(\mathcal{C}_M^{(r)}) \geq \frac{(b-a)^{r+1}}{(2\pi)^r} \cdot \frac{\mathcal{K}_r}{n^r} M.$$

For odd r, Theorem B.8 reads

$$\mathcal{K}_r = (2\pi)^r \int_0^1 |B_r(x)|\, dx.$$

Therefore

$$\rho_n^{\text{best}}(\mathcal{C}_M^{(r)}) \geq \frac{(b-a)^{r+1}}{n^r} M \cdot \int_0^1 |B_r(x)|\, dx.$$

A comparison with Theorem 7.5.5 shows the asymptotic optimality of $Q_n^{\text{Gr},s}$ in $\mathcal{C}_M^{(r)}$ for $r = 1, 3, \ldots, 2\lfloor s/2 \rfloor + 1$. So, for odd r, we have that

$$\rho_n^{\text{best}}(\mathcal{C}_M^{(r)}) = \frac{(b-a)^{r+1}}{n^r} M \cdot \int_0^1 |B_r(x)|\, dx + O(n^{-r-1}). \tag{8.5}$$

The same method does not apply to even r, but it is possible to obtain asymptotically optimal methods by slightly modifying the Gregory rules (Brass 1974b).

PROOF OF THEOREM 8.2.1. By the Euler–Maclaurin formula (Theorem 7.1.1), we obtain that for $f \in \widetilde{\mathcal{C}}_M^{(r)}$,

$$R_{n+1}^{\text{Tr}}[f] = -(-1)^r h^r \int_0^{2\pi} B_r^*\left(\frac{x}{h}\right) f^{(r)}(x)\, dx.$$

The periodicity gives $f^{(r-1)}(0) = f^{(r-1)}(2\pi)$, and therefore

$$R_{n+1}^{\text{Tr}}[f] = -(-1)^r h^r \int_0^{2\pi} \left[B_r^*\left(\frac{x}{h}\right) - c\right] f^{(r)}(x)\, dx$$

for any constant c. Thus,

$$\rho(Q_{n+1}^{\text{Tr}}, \widetilde{\mathcal{C}}_M^{(r)}) \leq M \cdot h^r \int_0^{2\pi} \left|B_r^*\left(\frac{x}{h}\right) - c\right| dx.$$

Equality would hold if there were to exist $f \in \widetilde{\mathcal{C}}_M^{(r)}$ such that

$$f^{(r)}(x) = M \operatorname{sgn}\left[B_r^*\left(\frac{x}{h}\right) - c\right].$$

It is easy to prove that g is a derivative of a periodic function if $\int_0^{2\pi} g(x)\, dx = 0$; that is, we have to determine c from the condition

$$\int_0^{2\pi} \operatorname{sgn}\left[B_r^*\left(\frac{x}{h}\right) - c\right] dx = 2\pi \int_0^1 \operatorname{sgn}\left[B_r(y) - c\right] dy = 0.$$

Taking into account the positions of the sign changes of B_r (see Theorems B.3 and B.5), we obtain

$$c = \begin{cases} 0 & \text{for odd } r, \\ B_r(\frac{1}{4}) & \text{for even } r. \end{cases}$$

We have therefore proved that

$$\rho(Q_{n+1}^{\mathrm{Tr}}, \widetilde{\mathcal{C}}_M^{(r)}) = \frac{(2\pi)^{r+1}}{n^r} M \cdot \begin{cases} \displaystyle\int_0^1 |B_r(y)|\, dy & \text{for odd } r, \\ \displaystyle\int_0^1 \left|B_r(y) - B_r\left(\tfrac{1}{4}\right)\right| dy & \text{for even } r. \end{cases}$$

For the last step, we refer to Theorem B.8. □

For $f^{(r)} \in \widetilde{\mathcal{C}}$, the bounds given in Theorem 8.2.1 can be improved by using the modulus of continuity. Malozemov (1967a, 1967b) established best bounds for these co-observations.

An important subset of $\widetilde{\mathcal{C}}$ consists of those functions which admit an analytic continuation. Any such function is holomorphic in a strip

$$S_r := \{z \,:\, |\Im z| < r\} \quad \text{for some } 0 < r \leq \infty.$$

We will write

$$H_r := \{f \in \widetilde{\mathcal{C}} \,:\, f \text{ is holomorphic in } S_r\}.$$

Recall that our functions are real-valued on the real axis.

The relevant co-observation of most use is

$$\mathcal{C}(S_r) := \{f \in H_r \,:\, \sup_{z \in S_r} |f(z)| \leq 1\}. \tag{8.6}$$

The following classes are similar and occasionally useful:

$$\mathcal{C}_r^{\mathrm{Sch}} := \{f \in H_r \,:\, \sup_{z \in S_r} |\Re f(z)| \leq 1\} \tag{8.7}$$

(Schönhage 1975) and

$$\mathcal{C}_r^{\mathrm{Kr}} := \{f \in H_r \,:\, \sup_{z \in S_r} |\Im f(z)| \leq 1\} \tag{8.8}$$

(Kress 1978). Further co-observations are discussed in Kress (1972, 1978), Forst (1976) and Knauff (1977a, 1977b).

In preparation for the derivation of error bounds for the classes (8.6), (8.7) and (8.8), we prove some representations of the Fourier coefficients

$$\alpha_\nu = \frac{1}{\pi} \int_0^{2\pi} f(x) \cos(\nu x)\, dx =: \Re \gamma_\nu, \quad \text{where } \gamma_\nu = \frac{1}{\pi} \int_0^{2\pi} f(x) e^{i\nu x}\, dx.$$

By Cauchy's theorem, we have

$$\int_R f(z) e^{i\nu z}\, dz = 0$$

if the integration is over the rectangle R with vertices 0, 2π, $2\pi + ip$ and ip, where $p < r$. Owing to periodicity, the contributions from the vertical segments cancel out. So we obtain

$$\gamma_\nu = \frac{1}{\pi} \int_0^{2\pi} f(x + ip) e^{i\nu(x+ip)}\, dx, \tag{8.9}$$

or

$$e^{\nu p}\gamma_\nu = \frac{1}{\pi}\int_0^{2\pi} f(x+ip)e^{i\nu x}\,dx,$$

and analogously

$$e^{-\nu p}\gamma_\nu = \frac{1}{\pi}\int_0^{2\pi} f(x-ip)e^{i\nu x}\,dx.$$

Combining these equations gives

$$\left(e^{\nu p} \pm e^{-\nu p}\right)\gamma_\nu = \frac{1}{\pi}\int_0^{2\pi}\left[f(x+ip) \pm f(x-ip)\right]e^{i\nu x}\,dx,$$

that is,

$$\gamma_\nu = \frac{1}{\pi\cosh\nu p}\int_0^{2\pi}\Re\left[f(x+ip)\right]e^{i\nu x}\,dx \tag{8.10}$$

$$= \frac{i}{\pi\sinh\nu p}\int_0^{2\pi}\Im\left[f(x+ip)\right]e^{i\nu x}\,dx. \tag{8.11}$$

The connection with the error of the trapezoidal rule is given by Theorem 7.2.1:

$$R_{n+1}^{\mathrm{Tr}}[f] = -2\pi(\alpha_n + \alpha_{2n} + \alpha_{3n} + \cdots). \tag{8.12}$$

From (8.9) we immediately obtain

$$|\alpha_\nu| \le |\gamma_\nu| \le 2e^{-\nu p}\sup_x |f(x+ip)|,$$

and plugging this into the sum in (8.12) gives the following result.

THEOREM 8.2.2 (Davis 1959). *For every $f \in H_r$, we have*

$$\left|R_{n+1}^{\mathrm{Tr}}[f]\right| \le \frac{4\pi}{e^{np}-1}\sup_x |f(x+ip)| \quad \text{for all } 0 < p < r.$$

Another proof of this result can be found in Kress (1971).

From Theorem 8.2.2 we infer that for $f \in H_r$,

$$R_{n+1}^{\mathrm{Tr}}[f] = O(e^{-np}) \quad \text{for all } 0 < p < r. \tag{8.13}$$

This statement can almost be inverted: If (8.13) holds and f is absolutely continuous, then we have $g \in H_r$, where g is the even part of f. We recall that $R_{n+1}^{\mathrm{Tr}}[f] = R_{n+1}^{\mathrm{Tr}}[g]$, so nothing can be said about the odd part.

For the proof of this result we use Lemma 7.2.4, which gives

$$\left|\alpha_\nu[g]\right| \le \frac{1}{2\pi}\sum_{\lambda=1}^{\infty}\left|R_{\lambda\nu+1}^{\mathrm{Tr}}[g]\right|.$$

Thus, using (8.13), we get

$$\left|\alpha_\nu[g]\right| = O(e^{-\nu p}).$$

Since $\sup_{z\in S_q}|\cos\nu z| \le e^{\nu q}$, the function g given by

$$g(z) = \frac{\alpha_0}{2} + \sum_{\nu=1}^{\infty}\alpha_\nu\cos\nu z$$

is analytic in S_p, and hence $g \in H_r$.

More precise results of this kind have been proved by Rahman and Schmeisser (1990). In Rahman and Schmeisser (1991), the assumption on the asymptotics of $R_{n+1}^{\mathrm{Tr}}[f]$ is replaced by one for $R_{n+1}^{\mathrm{Tr}}[f_h]$, where $f_h(x) = f(x+h)$ for all small h.

This modified assumption leads to perfect inverse theorems, in particular without the difficulties concerning odd and even parts.

EXAMPLE 8.2.1. In Chapter 1, we claimed that

$$\left|R_5^{\mathrm{Mi}}[f_0]\right| \leq 0.5 \cdot 10^{-8} \quad \text{for } f_0(x) = e^{\sin x}.$$

Here we try to prove this result with Theorem 8.2.2; recall that

$$\sup_{f\in\mathcal{C}}\left|R_n^{\mathrm{Mi}}[f]\right| = \sup_{f\in\mathcal{C}}\left|R_{n+1}^{\mathrm{Tr}}[f]\right|$$

for any translation invariant $\mathcal{C}$, so Theorem 8.2.2 is applicable. We have

$$\sup_{x\in S_p} |f_0(z)| = e^{\cosh p},$$

and hence

$$\left|R_{n+1}^{\mathrm{Tr}}[f_0]\right| \leq \frac{4\pi e^{-np+\cosh p}}{1-e^{-np}}$$

where p can be chosen arbitrarily. The minimum of the numerator is attained when $\sinh p = n$ and then the denominator is close to one. With this choice of p, we obtain

$$\left|R_{11}^{\mathrm{Tr}}[f_0]\right| \leq 2.78 \cdot 10^{-8} \tag{8.14}$$

and

$$\left|R_{12}^{\mathrm{Tr}}[f_0]\right| \leq 1.32 \cdot 10^{-9}.$$

For most practical applications this is entirely sufficient; nevertheless, the assertion has not yet been proved.

In a first step to improve the bounds, we take advantage of the symmetry of f_0. We have the remarkable result that if

$$f(x+\pi) = f(x) \tag{8.15}$$

for all x, then

$$R_{2n+1}^{\mathrm{Tr}}[f] = R_{n+1}^{\mathrm{Tr}}[f] \quad \text{for odd } n.$$

The proof of this fact is simple.

The relation $R_n^{\mathrm{Mi}} = 2R_{2n+1}^{\mathrm{Tr}} - R_{n+1}^{\mathrm{Tr}}$ implies that

$$R_n^{\mathrm{Mi}}[f] = R_{2n+1}^{\mathrm{Tr}}[f]. \tag{8.16}$$

Unfortunately, f_0 does not have the property (8.15); but its even part satisfies (8.15), therefore we may apply (8.16) and (8.14) to obtain

$$\left|R_5^{\mathrm{Mi}}[f_0]\right| \leq 2.78 \cdot 10^{-8}.$$

We are now much closer to proving the assertion, but for a definitive argument, another quite specific method is necessary.

We start with the Fourier series expansion of f_0 (see Abramowitz and Stegun 1964: p. 376):

$$e^{\sin x} = I_0(1) + 2\sum_{k=0}^{\infty}(-1)^k I_{2k+1}(1)\sin\big((2k+1)x\big) + 2\sum_{k=1}^{\infty}(-1)^k I_{2k}(1)\cos(2kx)$$

where I_k denotes the modified Bessel function, which is defined by

$$I_k(x) = \sum_{\mu=0}^{\infty} \frac{x^{k+2\mu}}{2^{k+2\mu}\mu!(k+\mu)!}.$$

We use (8.16) and Theorem 7.2.1 to obtain

$$R_n^{\mathrm{Mi}}[f_0] = 4\pi(I_{2n}(1) - I_{4n}(1) + I_{6n}(1)\cdots)$$

for odd n. This is an alternating series with decreasing terms, so

$$4\pi\big(I_{2n}(1) - I_{4n}(1)\big) \le R_n^{\mathrm{Mi}}[f_0] \le 4\pi I_{2n}(1).$$

Referring again to Abramowitz and Stegun (1964: p. 428), we deduce our final result:

$$R_5^{\mathrm{Mi}}[f_0] = (0.3459\ldots)\cdot 10^{-8}.$$

More generally, the same method yields

$$R_n^{\mathrm{Mi}}[f_0] = \begin{cases} \dfrac{\pi}{2^{2n-2}(2n)!}\big(1+o(1)\big) & \text{for odd } n,\\[2ex] \dfrac{(-1)^{n/2}\pi}{2^{n-2}n!}\big(1+o(1)\big) & \text{for even } n. \end{cases}$$

Note that the error is not monotonically decreasing; compare this example with Weideman (2002).

Theorem 8.2.2 provides an upper bound on $\rho(Q_{n+1}^{\mathrm{Tr}}, \mathcal{C}(S_r)) = \rho_n^{\mathrm{opt}}(\mathcal{C}(S_r))$, but the exact value of this quantity remains unknown. For the co-observations (8.7) and (8.8), we are in better shape. Let us choose $\mathcal{C}_r^{\mathrm{Kr}}$ to investigate, since in that class, in contrast to $\mathcal{C}_r^{\mathrm{Sch}}$, the trapezoidal rule is optimal.

THEOREM 8.2.3 (Kress 1978).

$$\rho(R_{n+1}^{\mathrm{Tr}}, \mathcal{C}_r^{\mathrm{Kr}}) = \rho_n^{\mathrm{opt}}(\mathcal{C}_r^{\mathrm{Kr}}) = 8\sum_{\nu=1}^{\infty}\frac{1}{(2\nu+1)\sinh\big((2\nu+1)nr\big)}.$$

Some simple estimations show that

$$8\ln\frac{1+e^{-nr}}{1-e^{-nr}} \le \rho_n^{\mathrm{opt}}(\mathcal{C}_r^{\mathrm{Kr}}) \le \frac{8}{1-e^{-nr}}\ln\frac{1+e^{-nr}}{1-e^{-nr}},$$

giving

$$\rho_n^{\mathrm{opt}}(\mathcal{C}_r^{\mathrm{Kr}}) = 16e^{-nr}\big(1+O(e^{-nr})\big) \quad \text{as } n\to\infty.$$

With a similar proof (Schönhage 1975), it is possible to obtain the asymptotics for $\rho(R_{n+1}^{\mathrm{Tr}},\ \mathcal{C}_r^{\mathrm{Sch}})$. Whether Theorem 8.2.2 or Theorem 8.2.3 gives the better bound depends on the particular f. In many but not all cases, the difference is small.

PROOF OF THEOREM 8.2.3. The function

$$f_v(y) = \sum_{\nu=1}^{\infty}\frac{\sin\nu y}{\sinh\nu v}$$

will play an important role in the proof. Our first result concerning f_v is that there exists a sign change at $y=\pi$ and no other sign changes in $]0,2\pi[$. This follows immediately from the identity

$$f_v(y) = \sum_{\nu=0}^{\infty}\frac{\sin y}{\cosh(2\nu+1)v - \cos y}.$$

The latter can be proved by using

$$\frac{\sin y}{\cosh u - \cos y} = 2\sum_{\mu=1}^{\infty} e^{-\mu u}\sin \mu y \quad \text{for } u > 0,$$

which is the equation relating the imaginary parts of the two sides of

$$\frac{1}{1-e^{iy-u}} = \sum_{\mu=0}^{\infty} e^{\mu(iy-u)}.$$

We use our knowledge about the sign changes to calculate that

$$\begin{aligned}\int_0^{2\pi} |f_v(y)|\,dy &= \int_0^{\pi} f_v(y)\,dy - \int_{\pi}^{2\pi} f_v(y)\,dy \\ &= 4\sum_{\nu=0}^{\infty} \frac{1}{(2\nu+1)\sinh\big((2\nu+1)v\big)}.\end{aligned} \tag{8.17}$$

After this preliminary work, we look at quadrature. Combining (8.11) and (8.12) gives

$$R_{n+1}^{\mathrm{Tr}}[f] = 2\int_0^{2\pi} \Im[f(x+ip)]f_{np}(nx)\,dx \tag{8.18}$$

and hence

$$\rho(Q_{n+1}^{\mathrm{Tr}},\, \mathcal{C}_r^{\mathrm{Kr}}) \le 2\int_0^{2\pi} |f_{np}(nx)|\,dx = 2\int_0^{2\pi} |f_{np}(y)|\,dy.$$

This holds for any $p < r$, so

$$\rho(Q_{n+1}^{\mathrm{Tr}},\, \mathcal{C}_r^{\mathrm{Kr}}) \le 2\int_0^{2\pi} |f_{nr}(y)|\,dy.$$

Equation (8.17) tells us that

$$\rho(Q_{n+1}^{\mathrm{Tr}}, \mathcal{C}_r^{\mathrm{Kr}}) \le 8\sum_{\nu=1}^{\infty} \frac{1}{(2\nu+1)\sinh\big((2\nu+1)nr\big)}. \tag{8.19}$$

The proof is completed by constructing a function $f \in \mathcal{C}_r^{\mathrm{Kr}}$ which makes $R_{n+1}^{\mathrm{Tr}}[f]$ equal to the right-hand side of (8.19); a possible choice is

$$f(z) = \frac{4}{\pi}\sum_{\nu=0}^{\infty} \frac{\cos\big((2\nu+1)nz\big)}{(2\nu+1)\sinh\big((2\nu+1)nr\big)}.$$

We have

$$\Im[f(x+ir)] = -\frac{4}{\pi}\sum_{\nu=0}^{\infty} \frac{\sin\big((2\nu+1)nx\big)}{2\nu+1},$$

and

$$\frac{4}{\pi}\sum_{\nu=0}^{\infty} \frac{\sin\big((2\nu+1)y\big)}{2\nu+1} = \begin{cases} 1 & \text{for } x \in\,]0,\pi[, \\ -1 & \text{for } x \in\,]\pi, 2\pi[\end{cases}$$

is a well-known result from the theory of Fourier series. Therefore $f \in \mathcal{C}_r^{\mathrm{Kr}}$, and from (8.18) we obtain

$$\big|R_{n+1}^{\mathrm{Tr}}[f]\big| = 2\int_0^{2\pi} |f_{np}(nx)|\,dx = 2\int_0^{2\pi} |f_{np}(y)|\,dy,$$

which finishes the proof. □

Finally, we give an example of the Hilbert space theory of Section 2.3 being applied to periodic functions. For more general results, see Babuska (1968); for improved error estimates using ρ^{intr} according to Theorem 2.3.1, see Knauff and Paulik (1978).

Consider the space of functions $f \in H_r$ such that

$$\lim_{p\to r} \int_0^{2\pi} \int_{-p}^{p} |f(x+iy)|^2 \, dy \, dx < \infty,$$

and define an inner product by

$$\varphi(f,g) = \lim_{p\to r} \int_0^{2\pi} \int_{-p}^{p} f(x+iy)\,\overline{g(x+iy)}\, dy\, dx.$$

It can be shown—see Meschkowski (1962) or Davis (1963)—that this gives a Hilbert space and that the functions

$$u_\nu(z) := \left(\frac{\nu}{2\pi \sinh(2\nu r)}\right)^{1/2} e^{i\nu z}, \quad \nu \in \mathbb{Z},$$

form an orthonormal basis for this space. As co-observation we take the unit ball

$$\mathcal{C}_r^{L_2} := \{f \, : \varphi(f,f) \le 1\},$$

and we apply Theorem 2.3.4 to $c \cdot Q_{n+1}^{\text{Tr}}$. This gives

$$\rho^2(c\cdot Q_{n+1}^{\text{Tr}},\, \mathcal{C}_r^{L_2}) = (1-c)^2\frac{\pi}{r} + c^2 S \quad \text{where } S := 4\pi n \sum_{\mu=1}^{\infty} \frac{\mu}{\sinh(2\mu n r)}. \tag{8.20}$$

The trapezoidal rule is a special case:

$$\rho(Q_{n+1}^{\text{Tr}},\, \mathcal{C}_r^{L_2}) = \sqrt{S}$$

(see Kress 1972). The expression in (8.20) is minimal when $c = c_0 := \pi/(Sr+\pi)$, and Theorem 8.1.1 gives

$$Q_n^{\text{opt}} = c_0 Q_{n+1}^{\text{Tr}}, \quad \text{i.e.} \quad \rho_n^{\text{opt}}(\mathcal{C}_r^{L_2}) = \left(\frac{\pi S}{Sr+\pi}\right)^{1/2} \rho(Q_{n+1}^{\text{Tr}},\, \mathcal{C}_r^{L_2})$$

(Knauff and Kress 1974). Unless $n \cdot r$ is very small, there is only a slight advantage to using Q_n^{opt} rather than Q_{n+1}^{Tr}. In any case, the difficulty of computing $\varphi(f,f)$ prevents $\mathcal{C}_r^{L_2}$ from being applied except in quite special circumstances.

8.3. Trigonometric interpolation

Let $(0 \le)\, \xi_1 < \xi_2 < \cdots < \xi_n \,(< 2\pi)$, and suppose that we are given a subspace $U \subset \widetilde{\mathcal{C}}$ of dimension n for which the following property holds:

$$u \in U \quad \text{with} \quad u(\xi_\nu) = 0 \ \text{ for } \nu = 1,\dots,n \qquad \text{implies} \qquad u = 0. \tag{8.21}$$

Then there exists a unique projection intpol from $\widetilde{\mathcal{C}}$ onto U such that

$$\text{intpol}[f](\xi_\nu) = f(\xi_\nu) \quad \text{for } \nu = 1,\dots,n.$$

Equivalent to property (8.21) is the existence of functions $u_1,\dots,u_n \in U$ such that

$$u_\nu(\xi_\mu) = \begin{cases} 0 & \text{if } \nu \ne \mu, \\ 1 & \text{if } \nu = \mu. \end{cases} \tag{8.22}$$

Using these functions, intpol can be expressed in the form

$$\text{intpol}[f] = \sum_{\nu=1}^{n} f(\xi_\nu) u_\nu. \tag{8.23}$$

All of the above comes from simple linear algebra.

Trigonometric interpolation is defined by choosing U to be a space of trigonometric polynomials. Recall that $\dim \mathbb{T}_q = 2q+1$, so we have the natural choice

$$U = \mathbb{T}_m$$

only for odd $n = 2m+1$. This choice of U satisfies (8.21), because any $t \in \mathbb{T}_m$ can be written in the form $e^{-imx}p(e^{ix})$ with $p \in \mathbb{P}_{2m}$, and if t had $2m+1$ zeros, then p would also have $2m+1$ zeros.

An alternative proof involves explicitly presenting functions u_ν that satisfy (8.22):

$$u_\nu(x) = \prod_{\substack{\mu=1 \\ \mu\neq\nu}}^{n} \frac{\sin\dfrac{x-\xi_\mu}{2}}{\sin\dfrac{\xi_\nu-\xi_\mu}{2}}. \tag{8.24}$$

Verifying (8.22) is trivial, and the claim that $u_\nu \in \mathbb{T}_m$ can be proved by means of trigonometric identities such as

$$2\sin\frac{x-\xi_\mu}{2}\sin\frac{x-\xi_\lambda}{2} = \cos\frac{\xi_\mu-\xi_\lambda}{2} - \cos\left(x - \frac{\xi_\mu+\xi_\lambda}{2}\right).$$

In the case of even $n = 2m$, there is no standard choice of U, and U depends on $\xi_1, \ldots, \xi_n$. We shall choose, according to a proposal of Schoenberg (1972),

$$U = \left\{ t : t(x) = \frac{\alpha_0}{2} + \sum_{\nu=1}^{m-1} (\alpha_\nu \cos \nu x + \beta_\nu \sin \nu x) + \alpha_m \sin\left(mx - \frac{\xi_1+\cdots+\xi_n}{2}\right),\ \alpha_\nu, \beta_\nu \in \mathbb{R} \right\}.$$

In this case, (8.22) is satisfied by taking

$$u_\nu(x) = \cos\frac{x-\xi_\nu}{2} \cdot \prod_{\substack{\mu=1 \\ \mu\neq\nu}}^{n} \frac{\sin\dfrac{x-\xi_\mu}{2}}{\sin\dfrac{\xi_\nu-\xi_\mu}{2}}. \tag{8.25}$$

Again, simple applications of trigonometric identities are sufficient to prove $u_\nu \in U$.

From now on, we will denote by $\text{intpol}_n^{\text{tr}}$ the operator defined by (8.23) and (8.24) for odd n and by (8.23) and (8.25) for even n.

Let us look at the important case of equidistant ξ_ν:

$$\xi_\nu = \nu\frac{2\pi}{n} - \alpha. \tag{8.26}$$

The expressions for u_ν simplify to

$$u_\nu(x) = \frac{\sin\left(\frac{n}{2}(x-\xi_\nu)\right)}{n\sin\left(\frac{1}{2}(x-\xi_\nu)\right)} \quad \text{for odd } n \tag{8.27}$$

and

$$u_\nu(x) = \frac{\sin\left(\frac{n}{2}(x-\xi_\nu)\right)}{n\sin\left(\frac{1}{2}(x-\xi_\nu)\right)} \cdot \cos\frac{x-\xi_\nu}{2} \quad \text{for even } n. \tag{8.28}$$

Obviously, (8.22) holds. To prove $u_\nu \in U$, we use the identities

$$\frac{\sin\left(\frac{2m+1}{2}z\right)}{2\sin\frac{z}{2}} = \frac{1}{2} + \cos z + \cos 2z + \cdots + \cos mz \tag{8.29}$$

and

$$\frac{\sin(mz)}{2\sin\frac{z}{2}} \cdot \cos\frac{z}{2} = \frac{1}{2} + \cos z + \cos 2z + \cdots + \cos(m-1)z + \frac{1}{2}\cos mz. \tag{8.30}$$

These identities are proved via multiplication by $\sin(z/2)$.

Finally, some calculations using (8.23) and (8.27)–(8.30) yield the following result.

LEMMA 8.3.1. Let the ξ_ν be chosen according to (8.26), i.e. equidistant, and set

$$\alpha_\mu := \frac{2}{n}\sum_{\nu=1}^{n} f(\xi_\nu)\cos(\mu\xi_\nu), \qquad \beta_\mu := \frac{2}{n}\sum_{\nu=1}^{n} f(\xi_\nu)\sin(\mu\xi_\nu).$$

Then

$$\operatorname{intpol}_n^{\mathrm{tr}}[f](x) = \frac{\alpha_0}{2} + \sum_{\mu=1}^{\lfloor n/2\rfloor}{}^{*} \left(\alpha_\mu \cos\mu x + \beta_\mu \sin\mu x\right),$$

where $\sum^*$ indicates that the last summand has to be halved if n is even.

8.4. Universality

From a practical point of view, one main objection to using optimal rules is that each is tailored to one special class of functions. There is an infinite range of possible co-observations, and any given integrand belongs to infinitely many of these. Therefore, it seems reasonable to ask whether a "universal" rule can be found. Such a rule should be able to compete with the optimal rules for many co-observations. To make this idea more concrete, we make the following definition.

DEFINITION 8.4.1. A rule Q is said to be "universal" if

$$\sup_s \frac{\rho(Q, \widetilde{C}_1^{(s)})}{\rho^{\mathrm{opt}}(\widetilde{C}_1^{(s)})} < \infty,$$

where ρ^{opt} is based on the same observation as is Q.

We remark that the same definition does not work for non-periodic functions, since $\rho(Q_n, C^{(2n+1)}) = \infty$ would hold for any Q_n. It is not straightforward to come up with a useful modification of the definition in that case. A different notion of universality which can be used with non-periodic functions is defined in Petras (1996b).

The main point here is that there are only a few universal rules for a given observation and they are easy to characterize. In the following, we shall say that Q has trigonometric degree l if $R[\mathbb{T}_l] = \{0\}$ and $R[\mathbb{T}_{l+1}] \neq \{0\}$.

THEOREM 8.4.1 (Brass 1988a). A rule Q is universal if and only if there is no rule with higher trigonometric degree (based on the same observation).

The proof relies on the well-known lemmas below (see, e.g., DeVore and Lorentz 1993: pp. 97, 213).

Lemma (Bernstein). *We have* $\|t^{(s)}\| \le l^s \|t\|$ *for all* $t \in \mathbb{T}_l$.

Lemma (Favard, Achieser and Krein). *For every* $f \in \widetilde{\mathcal{C}}_1^{(s)}$, *there exists* $t \in \mathbb{T}_l$ *with*

$$\|f - t\| \le \frac{\mathcal{K}_r}{(l+1)^s},$$

where $\mathcal{K}_r$ *is the Favard constant*

$$\mathcal{K}_r = \frac{4}{\pi} \sum_{\nu=0}^{\infty} \frac{(-1)^{\nu(r+1)}}{(2\nu+1)^{r+1}}.$$

PROOF OF THEOREM 8.4.1. (i) Sufficiency: Let the trigonometric degree of Q be l. Then there is a $t_0 \in \mathbb{T}_{l+1}$ such that $I[t_0] = 1$ and $O[t_0] = (0, 0, \dots, 0)$; otherwise, $O[t] = (0, 0, \dots, 0)$ and $t \in \mathbb{T}_{l+1}$ would imply that $I[t] = 0$, and so there would exist a rule of trigonometric degree greater than l (see Theorem 2.2.1).

We have $\|t_0^{(s)}\| \le (l+1)^s \|t_0\|$ and hence

$$\rho^{\text{opt}}(\widetilde{\mathcal{C}}_1^{(s)}) \ge I\left[\frac{t_0}{(l+1)^s \|t_0\|}\right] = \frac{1}{(l+1)^s \|t_0\|}. \tag{8.31}$$

On the other hand, according to Lebesgue's inequality (Theorem 3.1.1) and the Favard–Achieser–Krein lemma above, we have

$$\rho(Q, \widetilde{\mathcal{C}}_1^{(s)}) \le \|I - Q\| \cdot \frac{\mathcal{K}_s}{(l+1)^s} \le \|I - Q\| \cdot \frac{\pi}{2(l+1)^s} \tag{8.32}$$

for any rule of trigonometric degree l.

Inequalities (8.31) and (8.32) now yield

$$\sup_s \frac{\rho(Q, \widetilde{\mathcal{C}}_1^{(s)})}{\rho^{\text{opt}}(\widetilde{\mathcal{C}}_1^{(s)})} \le \frac{\|I - Q\|\pi}{2} \cdot \|t_0\|,$$

which means that Q is universal.

(ii) Necessity: Let Q be any rule of trigonometric degree l. Then there exists $t \in \mathbb{T}_{l+1}$ with $R[t] \ne 0$, and we obtain from

$$\frac{t}{\|t\|(l+1)^s} \in \widetilde{\mathcal{C}}_1^{(s)}$$

that

$$\rho(Q, \widetilde{\mathcal{C}}_1^{(s)}) \ge \frac{|R[t]|}{\|t\|(l+1)^s}.$$

Suppose that there exists a rule Q^* of degree greater than l. Then (8.32) shows that

$$\frac{\rho(Q, \widetilde{\mathcal{C}}_1^{(s)})}{\rho^{\text{opt}}(\widetilde{\mathcal{C}}_1^{(s)})} \ge \frac{\rho(Q, \widetilde{\mathcal{C}}_1^{(s)})}{\rho(Q^*, \widetilde{\mathcal{C}}_1^{(s)})} \ge \frac{|R[t]|}{\|t\|(l+1)^s} \cdot \frac{2(l+2)^s}{\pi\|I - Q^*\|} = \text{const} \cdot \left(\frac{l+2}{l+1}\right)^s.$$

The expression on the right-hand side tends to infinity as $s \to \infty$, so Q would not be universal. □

Theorem 8.4.1 tells us that we have to look for rules with highest possible trigonometric degree. An obvious candidate is the "standard rule"

$$Q^{\mathrm{st}} := I \circ \mathrm{intpol}^{\mathrm{tr}}.$$

Indeed, we have the following theorem.

THEOREM 8.4.2. The standard rule is universal. If n is odd, there is no other universal rule.

PROOF. We have to show that $\deg Q \leq \deg Q^{\mathrm{st}}$ for any rule Q.

(i) Let $n = 2m+1$. We have $\deg Q^{\mathrm{st}} \geq m$. For any rule Q with $\deg Q \geq m$, the coefficient of $f(x_\nu)$ is given by

$$a_\nu = Q[u_\nu] = I[u_\nu]$$

where u_ν is defined by (8.24), so the universal rule is unique.

(ii) Let $n = 2m$. Up to a constant factor,

$$t_0(x) := \prod_{\nu=1}^{n} \sin \frac{x - x_\nu}{2}$$

is the uniquely determined function in $\mathbb{T}_m$ with $t_0(x_\nu) = 0$ for $\nu = 1, \ldots, n$.

Case 1: $I[t_0] \neq 0$. The existence of Q with $\deg Q \geq m$ leads to the contradiction

$$0 = Q[t_0] = I[t_0] \neq 0.$$

Therefore Q^{st}, which is of degree $m-1$, must be universal.

Case 2: $I[t_0] = 0$. Obviously, we have

$$t - \mathrm{intpol}^{\mathrm{tr}}[t] = \mathrm{const} \cdot t_0$$

for any $t \in \mathbb{T}_m$, and hence

$$I[t] - Q^{\mathrm{st}}[t] = \mathrm{const} \cdot I[t_0] = 0.$$

Thus, $\deg Q^{\mathrm{st}} \geq m$ and, using the same argument as in (i), we conclude that there can be no $Q \neq Q^{\mathrm{st}}$ with $\deg Q \geq m$. □

There are many variations on the basic idea of universality; see, for example, Babuska (1966, 1968), Práger (1979) and Brass (1990). In most of these variations, the rules with highest trigonometric degree retain their exceptional position.

Finally, we give another expression for Q_n^{st}, which holds in the most important special case of equidistant nodes. For this purpose, we shall denote by A_ν and B_ν the Fourier coefficients of the weight function w.

THEOREM 8.4.3. Let

$$x_\nu := \nu \frac{2\pi}{n} - \alpha.$$

Then

$$Q_n^{\mathrm{st}}[f] = \frac{2\pi}{n} \sum_{\nu=1}^{n} S(x_\nu) f(x_\nu),$$

where S is the following (modified) partial sum of the Fourier series of w:

$$S(x) = \frac{A_0}{2} + \sum_{\mu=1}^{\lfloor n/2 \rfloor}{}^{*} (A_\mu \cos \mu x + B_\mu \sin \mu x).$$

As in the previous section, $\sum^*$ means that the last summand has to be halved if n is even.

PROOF. Integrating the expression for $\text{intpol}^{\text{st}}$ given in Lemma 8.3.1, we get

$$\begin{aligned}
Q_n^{\text{st}}[f] &= I\big[\text{intpol}^{\text{tr}}[f]\big] = \pi\Big(\frac{\alpha_0}{2}A_0 + \sum_{\mu=1}^{\lfloor n/2\rfloor}{}^{*}\,(\alpha_\mu A_\mu + \beta_\mu B_\mu)\Big) \\
&= \frac{2\pi}{n}\left[\frac{A_0}{2}\sum_{\nu=1}^{n} f(x_\nu) + \sum_{\mu=1}^{\lfloor n/2\rfloor}{}^{*}\Big(A_\mu \sum_{\nu=1}^{n} f(x_\nu)\cos\mu x_\nu + B_\mu \sum_{\nu=1}^{n} f(x_\nu)\sin\mu x_\nu\Big)\right] \\
&= \frac{2\pi}{n}\sum_{\nu=1}^{n}\Big(\frac{A_0}{2} + \sum_{\mu=1}^{\lfloor n/2\rfloor}{}^{*}\,(A_\mu\cos\mu x_\nu + B_\mu \sin\mu x_\nu)\Big)\cdot f(x_\nu),
\end{aligned}$$

as claimed. □

For $w \in \widetilde{\mathcal{C}}$, the outstanding quality of the trapezoidal rule suggests that we look at the value $Q_n^{\text{Tr}}[fw]$. Theorem 8.4.3 tells us that this is identical to $Q_n^{\text{st}}[f]$ if $w \in \mathbb{T}_{\lfloor (n-1)/2\rfloor}$.

Another consequence derived from Theorem 8.4.3 is a new proof of Lemma 5.3.1. The connection between the Clenshaw–Curtis rule for the weight w and Q^{st} for the weight $w^* := w(\cos\cdot)|\sin\cdot|$ and nodes $x_\nu = \nu\pi/n$ is given by

$$Q_{n+1}^{\text{CC}}[f] = \frac{1}{2}Q_{2n}^{\text{st}}[f(\cos\cdot)].$$

Error bounds for the standard rules were proved by Fischer (1997). In particular, he showed that

$$\sup_s \frac{\rho(Q, \widetilde{\mathcal{C}}_M^{(s)})}{\rho^{\text{opt}}(\widetilde{\mathcal{C}}_M^{(s)})} \le 9$$

for all standard rules with equidistant nodes and $n \le 1000$.

8.5. Standard rules for Fourier coefficients

The functionals to be approximated in this section and the next are Fourier coefficients

$$I_p^{\text{c}}[f] := \frac{1}{\pi}\int_0^{2\pi} f(x)\cos px\,dx \tag{8.33}$$

and

$$I_p^{\text{s}}[f] := \frac{1}{\pi}\int_0^{2\pi} f(x)\sin px\,dx \tag{8.34}$$

for $p = 0, 1, 2, \ldots$. We restrict our attention to the equidistant nodes

$$x_\nu := (\nu - 1)h \quad \text{where } h := \frac{2\pi}{n}.$$

The more general definition (8.26) of equidistance would lead to similar results. The standard rules for the functionals (8.33) and (8.34) are

$$Q_{p,n}^{\mathrm{st,c}}[f] = \begin{cases} \dfrac{2}{n}\displaystyle\sum_{\nu=1}^{n} f(x_\nu)\cos px_\nu & \text{if } p < \dfrac{n}{2}, \\[2ex] \dfrac{1}{n}\displaystyle\sum_{\nu=1}^{n} f(x_\nu)\cos px_\nu & \text{if } p = \dfrac{n}{2}, \\[2ex] 0 & \text{if } p > \dfrac{n}{2}; \end{cases}$$

and

$$Q_{p,n}^{\mathrm{st,s}}[f] = \begin{cases} \dfrac{2}{n}\displaystyle\sum_{\nu=1}^{n} f(x_\nu)\sin px_\nu & \text{if } p < \dfrac{n}{2}, \\[2ex] 0 & \text{if } p \geq \dfrac{n}{2}. \end{cases}$$

In practice, these rules are used almost exclusively, even though many other rules have been proposed for the computation of Fourier coefficients.

For the most part, estimates for $I_p^{\mathrm{c}}[f]$ and $I_p^{\mathrm{s}}[f]$ for $p \leq p_0$ based on the same observation are needed simultaneously. For simplicity, we assume $p_0 < n/2$ and start our investigation with the following special cases of Theorem 7.2.1:

$$R_{p,n}^{\mathrm{st,c}}[\cos q\,\cdot\,] = \begin{cases} -1 & \text{if } q = \lambda n \pm p \text{ for } \lambda = 1, 2, \dots, \\ 0 & \text{otherwise}, \end{cases} \tag{8.35}$$

$$R_{p,n}^{\mathrm{st,s}}[\sin q\,\cdot\,] = \begin{cases} \mp 1 & \text{if } q = \lambda n \pm p \text{ for } \lambda = 1, 2, \dots, \\ 0 & \text{otherwise}, \end{cases} \tag{8.36}$$

and

$$R_{p,n}^{\mathrm{st,c}}[\sin q\,\cdot\,] = R_{p,n}^{\mathrm{st,s}}[\cos q\,\cdot\,] = 0 \quad \text{for } p < \frac{n}{2} \text{ and } q = 0, 1, 2, \dots.$$

So the trigonometric degree of our rules is $n - p - 1$.

To simplify the notation, we shall omit the lower indices if the relation being considered holds for all pairs (p, n), and we shall omit the upper labels "c" and "s" if the statement holds for both cosine and sine coefficients.

All standard rules are universal, and we know from Theorem 8.4.2 that there are no other universal rules when n is odd. For the Fourier coefficients and $p < n/2$, this uniqueness extends to even n as well. The reason for this is that there is no rule $Q^* \neq Q^{\mathrm{st}}$ with n equidistant nodes and trigonometric degree greater than or equal to $n - p - 1$, as can be checked easily.

The simplest way to obtain error bounds in $\widetilde{\mathcal{C}}_1^{(s)}$ is to apply Lebesgue's inequality (Theorem 3.1.1). To this end, we have to compute $\|R^{\mathrm{st,c}}\|$ and $\|R^{\mathrm{st,s}}\|$. These numbers depend on p and n, but there is a simple and fairly sharp bound:

LEMMA 8.5.1. $\|Q^{\mathrm{st}}\| \leq \dfrac{4}{3}$ for $0 < p < n/2$.

PROOF. Define

$$C(n,p) := \sum_{\nu=1}^{n} |\cos px_\nu|,$$

and let d be the greatest common divisor of n and p. Then we immediately obtain that

$$C(n,p) = d \cdot C(n \cdot d^{-1},\ p \cdot d^{-1}).$$

If n and p are relatively prime, we have

$$C(n,p) = C(n,1),$$

since the sets $\{p, 2p, \dots, np\}$ and $\{1, 2, \dots, n\}$ are equal modulo n. By using the symmetries and applying well-known trigonometric identities, we obtain

$$\begin{aligned} C(n,1) &= 2\cot\frac{\pi}{n} && \text{if } n \equiv 0 \bmod 4, \\ C(n,1) &= 2\left(\sin\frac{\pi}{n}\right)^{-1} && \text{if } n \equiv 2 \bmod 4, \\ C(n,1) &= \left(\sin\frac{\pi}{2n}\right)^{-1} && \text{if } n \equiv 1 \bmod 2. \end{aligned}$$

For example, let

$$n \cdot d^{-1} \equiv 3 \bmod 4. \tag{8.37}$$

Then $n \cdot d^{-1} \geq 3$ and, since $x^{-1}C(x,1)$ is decreasing for increasing $x \equiv 3 \bmod 4$, we have

$$\frac{2}{n}\, C(n,p) = 2\,\frac{C(n \cdot d^{-1},\ 1)}{n \cdot d^{-1}} \leq 2\,\frac{C(3,1)}{3} = \frac{4}{3}.$$

This is the assertion of the theorem in the special case (8.37). The arguments for the other cases, as well as the proof for sine coefficients, are very similar. □

It is now easy to prove the following:

THEOREM 8.5.1. Let $s \geq 2$ and $0 < p < n/2$. Then

$$\rho\big(Q^{\mathrm{st}}_{p,n},\ \widetilde{C}^{(s)}_1\big) \leq \frac{3.37}{(n-p)^s}.$$

PROOF. The Lebesgue inequality states that

$$\big|R[f]\big| \leq \|R\| \cdot \operatorname{dist}(f, \mathbb{T}_{n-p-1}).$$

We also know from Lemma 8.5.1 that

$$\|R\| = \|I\| + \|Q^{\mathrm{st}}\| \leq \frac{4}{\pi} + \frac{4}{3} < 2.607$$

and from the Favard–Achieser–Krein lemma of Section 8.4 that

$$\operatorname{dist}(f, \mathbb{T}_{n-p-1}) \leq \frac{\mathcal{K}_s}{(n-p)^s}\, \|f^{(s)}\|.$$

Together, these yield

$$\big|R[f]\big| \leq \frac{2.607 \cdot \mathcal{K}_s}{(n-p)^s}\, \|f^{(s)}\|. \tag{8.38}$$

From the definition of the Favard constants $\mathcal{K}_s$, we obtain

$$\mathcal{K}_s \leq \mathcal{K}_3 = \frac{\pi^3}{24} \quad \text{for } s \geq 2,$$

which combined with (8.38) gives the theorem. □

Theorem 8.5.1 (or formula (8.38), which holds for $s = 1$ too) is sufficient for most applications. Nevertheless, other methods that provide further information about the error deserve some attention as well. An obvious step is to try applying the Peano kernel method. For the functional $R^{\mathrm{st,c}}$, the definition of the Peano kernel from Section 4.2 gives

$$K_1^{\mathrm{c}}\big|_{]x_\nu,x_{\nu+1}[}(x) = \frac{1}{p\pi}\left[-\sin px + \frac{ph}{2} + \frac{ph}{2}\cdot\sin\Big(\nu-\frac{1}{2}\Big)ph\cdot\Big(\sin\frac{ph}{2}\Big)^{-1}\right].$$

Since f is 2π-periodic, we have

$$R^{\mathrm{st,c}}[f] = \int_0^{2\pi} f'(x)K_1^{\mathrm{c}}(x)\,dx = \int_0^{2\pi} f'(x)\big(K_1^{\mathrm{c}}(x)-\gamma\big)\,dx$$

for any γ. We choose $\gamma = 1/n$, so that

$$\text{(8.39)}\quad \widetilde{K}_1^{\mathrm{c}}(x) := K_1^{\mathrm{c}}(x) - \frac{1}{n} = \frac{1}{p\pi}\left[-\sin px + \frac{ph}{2}\cdot\sin\Big(\nu-\frac{1}{2}\Big)ph\cdot\Big(\sin\frac{ph}{2}\Big)^{-1}\right]$$

for $x \in\,]x_\nu, x_{\nu+1}[$. This modification is important, as we have

$$\widetilde{K}_1^{\mathrm{c}}(2\pi - x) = -\widetilde{K}_1^{\mathrm{c}}(x) \quad \text{for } x \neq x_\nu$$

and therefore

$$\text{(8.40)}\qquad \int_0^{2\pi} \operatorname{sgn}\widetilde{K}_1^{\mathrm{c}}(x)\,dx = 0.$$

This implies that

$$\text{(8.41)}\qquad \rho(Q^{\mathrm{st,c}},\, \widetilde{\mathcal{C}}_1^{(1)}) = \int_0^{2\pi} |\widetilde{K}_1^{\mathrm{c}}(x)|\,dx.$$

As an exemplary case, we mention

$$2001\,\rho(Q^{\mathrm{st,c}}_{1000,2001},\, \widetilde{\mathcal{C}}_1^{(1)}) = 2.90\ldots.$$

This is a surprising result, since the "trivial rule"

$$Q^{\mathrm{triv}} := 0$$

yields

$$2001\,\rho(Q^{\mathrm{triv}}, \widetilde{\mathcal{C}}_1^{(1)}) = \frac{2001}{\pi}\int_0^{2\pi}\frac{|\sin 1000x|}{1000}\,dx = \frac{2001}{1000}\cdot\frac{4}{\pi} = 2.54\ldots$$

for the same functional, which is a better result in $\widetilde{\mathcal{C}}_1^{(1)}$.

EXAMPLE 8.5.1. As an example of what can be done with the help of the explicit expression (8.39), we derive the asymptotics of $\rho(Q^{\mathrm{st,c}}_{p,n},\, \widetilde{\mathcal{C}}_1^{(1)})$ for fixed p. We start with

$$\frac{z}{\sin z} = 1 + O(z^2) \quad \text{as } z \to 0,$$

so that (8.39) gives

$$\widetilde{K}_1^{\mathrm{c}}(x) = \frac{1}{p\pi}\left[-\sin px + \sin\Big(\nu-\frac{1}{2}\Big)ph\right] + O(n^{-2}) \quad \text{for } x \in\,]x_{\nu-1}, x_\nu[.$$

Using Taylor's formula on the bracketed terms gives

$$\begin{aligned}\widetilde{K}_1^c(x) &= \frac{1}{\pi}\left[\left(\nu-\frac{1}{2}\right)h-x\right]\cos px + O(n^{-2})\\ &= -\frac{h}{\pi}\left[\frac{x-(\nu-1)h}{h}-\frac{1}{2}\right]\cos px + O(n^{-2})\\ &= -\frac{h}{\pi}\,B_1^*\left(\frac{x}{h}\right)\cos px + O(n^{-2}),\end{aligned}$$

where B_1^* is a Bernoulli function (see Appendix B). Now, we apply Polya's lemma (Lemma 7.1.1) to get

$$\begin{aligned}\lim_{n\to\infty} n\cdot\rho(Q_{p,n}^{\mathrm{st,c}},\,\widetilde{\mathcal{C}}_1^{(1)}) &= \lim_{n\to\infty} n\cdot\int_0^{2\pi}|\widetilde{K}_1^c(x)|\,dx\\ &= \lim_{n\to\infty} 2\int_0^{2\pi}\left|B_1^*\left(\frac{x}{h}\right)\right|\cdot|\cos px|\,dx\\ &= 2\int_0^1|B_1(x)|\,dx\cdot\int_0^{2\pi}|\cos px|\,dx \;=\; 2.\end{aligned}$$

A more important result is the following bound.

THEOREM 8.5.2 (Ehlich 1966). Let $1\le p<n/2$. Then

$$\rho(Q_{p,n}^{\mathrm{st,c}},\,\widetilde{\mathcal{C}}_1^{(1)}) \le \frac{3.43}{n}.$$

With more work, the constant 3.43 can be replaced by 2.92 (Brass 1991), but it cannot be replaced by 2.90 as the example of $Q_{1000,2001}^{\mathrm{st,c}}$ above shows.

PROOF. We have to find a bound for

$$\rho(Q_{p,n}^{\mathrm{st,c}},\,\widetilde{\mathcal{C}}_1^{(1)}) = \int_0^{2\pi}|\widetilde{K}_1^c(x)|\,dx$$

where $\widetilde{K}_1^c$ is given by (8.39). The unwieldy integral becomes tractable upon applying the Cauchy–Schwarz inequality:

$$\begin{aligned}&\int_0^{2\pi}|\widetilde{K}_1^c(x)|\,dx\\ &\quad\le\sqrt{2\pi}\left(\int_0^{2\pi}\left[\widetilde{K}_1^c(x)\right]^2dx\right)^{1/2}dx\\ &\quad=\frac{1}{p}\sqrt{\frac{2}{\pi}}\left(\sum_{\nu=1}^{n}\int_{x_\nu}^{x_{\nu+1}}\left[\sin px-\frac{ph}{2}\cdot\sin\Big(\nu-\frac{1}{2}\Big)ph\cdot\left(\sin\frac{ph}{2}\right)^{-1}\right]^2dx\right)^{1/2}.\end{aligned}$$

After some obvious calculations and simplifications, this expression transforms into

$$\frac{\sqrt{2}\,\pi}{n}\left(\frac{ph}{2}\right)^{-1}\left[\left(\frac{ph}{2}\right)^2\sin^{-2}\left(\frac{ph}{2}\right)-1\right]^{1/2}.$$

Since

$$x^{-2}\left[x^2\sin^{-2}x-1\right]$$

is increasing for increasing x (think of the power series!), we have

$$\begin{aligned}\sqrt{2}\,\pi\left(\frac{ph}{2}\right)^{-1}&\left[\left(\frac{ph}{2}\right)^{2}\sin^{-2}\left(\frac{ph}{2}\right)-1\right]^{1/2}\\ &\le \frac{\sqrt{2}\,\pi}{\pi/2}\cdot\left[\frac{\pi^2}{4}-1\right]^{1/2}=\sqrt{2\pi^2-8}\,<\,3.43,\end{aligned}$$

and this proves the assertion. □

Up to now, the calculations have been done only for cosine coefficients. For sine coefficients, we obtain

$$\widetilde{K}_1^{\mathrm{s}}\big|_{]x_\nu,x_{\nu+1}[}(x)=\frac{1}{p\pi}\left[\cos px-\frac{ph}{2}\cdot\cos\Big(\nu-\frac{1}{2}\Big)ph\cdot\Big(\sin\frac{ph}{2}\Big)^{-1}\right],$$

where we have modified the Peano kernel by an additive constant, so that

$$\int_0^{2\pi}\widetilde{K}_1^{\mathrm{s}}(x)\,dx=0$$

holds. Unfortunately, the analogue of (8.40) does not hold for $\widetilde{K}_1^{\mathrm{s}}$, and we are only able to obtain

$$\rho(Q^{\mathrm{st,s}},\,\widetilde{\mathcal{C}}_1^{(1)})\le\int_0^{2\pi}|\widetilde{K}_1^{\mathrm{s}}(x)|\,dx$$

in place of (8.41). This is sufficient for imitating the proof of Theorem 8.5.2, but it is not enough for the arguments in Example 8.5.1.

Higher Peano kernels in the ordinary sense do not exist for these problems, since $R^{\mathrm{st}}[\mathbb{P}_s]\ne\{0\}$ for $s\ge 1$. Nonetheless, a representation

$$R^{\mathrm{st}}[f]=\int_0^{2\pi}f^{(s)}(x)\widetilde{K}_s(x)\,dx \tag{8.42}$$

is possible for any s. In order to derive it, we define $-\widetilde{K}_{s+1}$ to be the primitive of $\widetilde{K}_s$ that has mean value zero, and we do repeated integration by parts starting with $s=1$ in (8.42).

It is easy to determine the Fourier series of $\widetilde{K}_s^{\mathrm{c}}$. We have to use

$$\int_0^{2\pi}(\cos q\,\cdot)^{(s)}\widetilde{K}_s^{\mathrm{c}}(x)\,dx=R^{\mathrm{c}}[\cos q\,\cdot]$$

and (8.35). For $s\ge 1$, this yields

$$\begin{aligned}\widetilde{K}_{2s}^{\mathrm{c}}(x)&=\frac{(-1)^{s+1}}{\pi}\sum_{\lambda=1}^{\infty}\left(\frac{\cos(\lambda n-p)x}{(\lambda n-p)^{2s}}+\frac{\cos(\lambda n+p)x}{(\lambda n+p)^{2s}}\right),\\ \widetilde{K}_{2s+1}^{\mathrm{c}}(x)&=\frac{(-1)^{s}}{\pi}\sum_{\lambda=1}^{\infty}\left(\frac{\sin(\lambda n-p)x}{(\lambda n-p)^{2s+1}}+\frac{\sin(\lambda n+p)x}{(\lambda n+p)^{2s+1}}\right).\end{aligned}\tag{8.43}$$

Similar arguments apply to the case of sine coefficients.

From these expansions, we can deduce many further results concerning the error of the standard rules; see Brass (1991). The most important conclusion is the following (Quade 1940; Ehlich 1966):

$$\rho(Q^{\mathrm{st}},\,\widetilde{\mathcal{C}}_1^{(s)})\le\frac{4}{\pi}\sum_{\lambda=1}^{\infty}\left(\frac{1}{(\lambda n-p)^s}+\frac{1}{(\lambda n+p)^s}\right). \tag{8.44}$$

This relation follows immediately from (8.43) and its sine counterpart together with

$$\int_0^{2\pi} |\cos qx|\, dx = \int_0^{2\pi} |\sin qx|\, dx = 4 \quad \text{for } q \geq 1.$$

We can now prove our main result on error bounds.

THEOREM 8.5.3 (Brass 1991). Let $0 < p < n/2$. Then

$$\frac{1.27}{p^s + (n-p)^s} \leq \rho_n^{\text{opt}}(\widetilde{C}_1^{(s)}) \leq \rho(Q_n^{\text{st}}, \widetilde{C}_1^{(s)}) \leq \frac{3.75}{p^s + (n-p)^s}$$

for the functional I_p.

The consequence

$$\frac{\rho(Q_n^{\text{st}}, \widetilde{C}_1^{(s)})}{\rho_n^{\text{opt}}(\widetilde{C}_1^{(s)})} \leq 3$$

is worth emphasizing.

PROOF. (i) We start with the lower bound. Evidently,

$$\rho_n^{\text{opt}}(\widetilde{C}_1^{(s)}) \geq \left|I[f_0]\right|$$

for any $f_0 \in \widetilde{C}_1^{(s)}$ such that $f_0(x_\nu) = 0$ for all ν. For I_p^{c} we choose

$$f_0(x) := \frac{4}{\pi} \frac{1}{p^s + (n-p)^s} \sum_{\nu=0}^{\infty} (-1)^{\nu(s+1)} \frac{\cos{(2\nu+1)px} - \cos{(2\nu+1)(n-p)x}}{(2\nu+1)^{s+1}},$$

and for the corresponding sine functional we take

$$f_0(x) := \frac{4}{\pi} \frac{1}{p^s + (n-p)^s} \sum_{\nu=0}^{\infty} (-1)^{\nu s} \frac{\sin{(2\nu+1)px} + \sin{(2\nu+1)(n-p)x}}{(2\nu+1)^{s+1}}.$$

The necessary property of the zeros follows from the trigonometric identities

$$\cos{(2\nu+1)px} - \cos{(2\nu+1)(n-p)x} = 2\sin{(2\nu+1)\frac{n}{2}x} \cdot \sin{(2\nu+1)\frac{n-2p}{2}x}$$

and

$$\sin{(2\nu+1)px} + \sin{(2\nu+1)(n-p)x} = 2\sin{(2\nu+1)\frac{n}{2}x} \cdot \cos{(2\nu+1)\frac{n-2p}{2}x}.$$

To prove $f_0 \in \widetilde{C}_1^{(s)}$, we use the identities

$$\left|\sum_{\nu=0}^{\infty} \frac{\sin{(2\nu+1)x}}{2\nu+1}\right| = \frac{\pi}{4}$$

and

$$\left|\sum_{\nu=0}^{\infty} \frac{(-1)^\nu \cos{(2\nu+1)x}}{2\nu+1}\right| = \frac{\pi}{4},$$

which hold for all but finitely many $x \in [0, 2\pi[$ and whose proofs are well established in Fourier series theory. In any case, we obtain

$$\rho_n^{\text{opt}}(\widetilde{C}_1^{(s)}) \geq \frac{4}{\pi} \cdot \frac{1}{p^s + (n-p)^s}.$$

(ii) The upper bound in the special case where $s = 1$ follows from Theorem 8.5.2 and the remarks after its proof.

(iii) Let $s \geq 2$ and $4p \leq n$. We use Theorem 8.5.1 to obtain

$$\rho(Q_n^{\mathrm{st}}, \widetilde{\mathcal{C}}_1^{(s)}) \leq \frac{3.37}{(n-p)^s} = \frac{3.37}{p^s + (n-p)^s} \cdot \left(1 + \left(\frac{p}{n-p}\right)^s\right)$$
$$\leq \frac{3.37}{p^s + (n-p)^s} \cdot \left(1 + \frac{1}{3^s}\right) \leq \frac{3.75}{p^s + (n-p)^s}.$$

(iv) For the remaining cases, we use (8.44) and need to show that

$$\frac{4}{\pi}\left(p^s + (n-p)^s\right) \sum_{\lambda=1}^{\infty} \left(\frac{1}{(\lambda n - p)^s} + \frac{1}{(\lambda n + p)^s}\right) \leq 3.75$$

or, upon setting $x := p/n$,

$$\frac{4}{\pi}\left(x^s + (1-x)^s\right) \sum_{\lambda=1}^{\infty} \left(\frac{1}{(\lambda - x)^s} + \frac{1}{(\lambda + x)^s}\right) \leq 3.75 \quad \text{for } \frac{1}{4} \leq x \leq \frac{1}{2}.$$

The function on the left-hand side is decreasing as a function of s, so it suffices to show that

$$\frac{4}{\pi}\left(x^2 + (1-x)^2\right) \sum_{\lambda=1}^{\infty} \left(\frac{1}{(\lambda - x)^2} + \frac{1}{(\lambda + x)^2}\right) \leq 3.75.$$

Differentiating the well-known partial fraction expansion of cot (see Abramowitz and Stegun 1964: p. 75) gives

$$\frac{\pi^2}{\sin^2 \pi x} - \frac{1}{x^2} = \sum_{\lambda=1}^{\infty} \left(\frac{1}{(\lambda - x)^2} + \frac{1}{(\lambda + x)^2}\right) \qquad \text{for } 0 < x < 1.$$

Our assertion is therefore reduced to

$$\left(x^2 + (1-x)^2\right)\left(\frac{\pi^2}{\sin^2 \pi x} - \frac{1}{x^2}\right) \leq 3.75 \cdot \frac{\pi}{4} = 2.94\ldots \quad \text{for } \frac{1}{4} \leq x \leq \frac{1}{2}.$$

The left-hand side is of the form $F(x)G(x)$ where $F(x)$ is decreasing and $G(x)$ is increasing as x increases. In this situation the following method can be used: For $x \in [z_1, z_2]$ we have

$$F(x)G(x) \leq F(z_1)G(z_2),$$

so an appropriate partition of $[1/4, 1/2]$ (five parts are sufficient) gives the desired result. □

8.6. Other rules for Fourier coefficients

It is natural to seek optimal rules for the computation of Fourier coefficients. We begin with an example.

EXAMPLE 8.6.1. The functional I_1^{s} is especially suitable for illustrating the concept of optimality. For even n, we may apply the method used in Example 2.1.1. In $\widetilde{\mathcal{C}}_M^{(1)}$, we have

$$(8.45) \qquad Q^{\mathrm{so}}[f] = \frac{2M}{\pi} \sum_{\nu=1}^{n} \cos \frac{x_\nu + x_{\nu+1}}{2} \sin \frac{f(x_{\nu+1}) - f(x_\nu)}{2M},$$

where $x_\nu := (\nu - 1)h$ with $h = 2\pi/n$, and

$$\rho^{\mathrm{intr}} = \frac{2M}{\pi} \sum_{\nu=1}^{n} \left| \sin \frac{x_\nu + x_{\nu+1}}{2} \cos \frac{f(x_{\nu+1}) - f(x_\nu)}{2M} - \sin x_\nu \right|.$$

Now

$$\rho_n^{\text{opt}} = \frac{4M}{\pi} \tan \frac{\pi}{2n}$$

follows from Theorem 2.1.1 and the relation

$$Q_n^{\text{opt}}[f] = \frac{2}{\pi} \sin \frac{\pi}{n} \sum_{\nu=1}^{n} f(x_\nu) \sin x_\nu \tag{8.46}$$

from Theorem 2.1.5. The corresponding results for odd n can be obtained with a trick, which involves computing the strongly optimal rule Q^{so} for the set of nodes $\{0, \frac{2\pi}{n}, \frac{4\pi}{n}, \ldots, (n-1)\frac{2\pi}{n}\} \cup \{\pi\}$. One again obtains (8.45), but this time with the modified x_ν. From this, a formula for Q_n^{opt} with 0 as the coefficient of $f(\pi)$ follows, which is really a formula for the case of n equidistant nodes $x_\nu = (\nu - 1)h$, $\nu = 1, \ldots, n$, where $h := \frac{2\pi}{n}$. The formula is

$$Q_n^{\text{opt}}[f] = \frac{2}{\pi} \sin \frac{3\pi}{4n} \sin \frac{5\pi}{4n} \big[f(x_m) - f(x_{m+1})\big] + \frac{2}{\pi} \sin \frac{\pi}{n} \sum_{\substack{\nu=1 \\ \nu \notin \{m, m+1\}}}^{n} f(x_\nu) \sin x_\nu,$$

where $m = (n+1)/2$, with error bound

$$\rho_n^{\text{opt}} = \frac{4M}{\pi} \sin \frac{\pi}{2n}.$$

A comparison with Example 8.5.1 shows that Q^{st} is asymptotically optimal in this setting. This is also true for the functionals I_p, although it is more difficult to obtain the optimal rules (see Brass 1991).

The strongly optimal rule Q^{so} is nonlinear. A linear rule Q_n which is not much worse can be obtained as an consequence of Theorem 4.5.1. For this purpose, we have to construct the projection rule corresponding to the space of periodic broken line functions with knots $\nu \cdot 2\pi/n$, $\nu \in \mathbb{Z}$. A simple calculation gives

$$Q_n[f] = \Big(\frac{n}{\pi} \sin \frac{\pi}{n}\Big)^2 \cdot \frac{2}{n} \sum_{\nu=0}^{n-1} f\Big(\nu \frac{2\pi}{n}\Big) \sin \Big(\nu \frac{2\pi}{n}\Big). \tag{8.47}$$

By Theorem 2.2.4, we have

$$\sup_{f \in \widetilde{C}_M^{(1)}} \frac{\big|I_1^{\text{s}}[f] - Q_n[f]\big|}{\rho^{\text{intr}}} \leq 2.$$

For even n, there is no other quadrature rule for which the quotient is bounded.

The rules (8.46) and (8.47) have the simple form

$$Q = \tau Q^{\text{st}}. \tag{8.48}$$

In other words, they are derived from the standard rules after multiplying by an "attenuation factor" τ. For the functional I_0^{c}, we have shown in Theorem 8.1.1 that all optimal rules for any translation-invariant co-observation have this special form; but this does not hold for arbitrary I_p, as Example 8.6.1 for odd n shows. Nevertheless, there are always rules of the form (8.48) that are almost optimal. To make this statement more precise, we need another definition.

DEFINITION 8.6.1. A co-observation $\mathcal{C}$ is said to be *reflection invariant* if $f \in \mathcal{C}$ implies $\widehat{f} \in \mathcal{C}$ where $\widehat{f}(x) := f(-x)$.

THEOREM 8.6.1 (Brass 1988b). Suppose that $\mathcal{C}$ is translation and reflection invariant, and let Q be a rule with n equidistant nodes. Then

$$\rho(\tau_c Q_p^{\text{st,c}},\, \mathcal{C}) \le \frac{4}{3}\rho(Q,\, \mathcal{C}) \quad \text{with } \tau_c = Q[\cos p\,\cdot\,]$$

for the functional I_p^c and

$$\rho(\tau_s Q_p^{\text{st,s}},\, \mathcal{C}) \le \frac{4}{3}\rho(Q,\, \mathcal{C}) \quad \text{with } \tau_s = Q[\sin p\,\cdot\,]$$

for the functional I_p^s.

PROOF. Let $f \in \mathcal{C}$ and define the function $F = F(f)$ by

$$F(f)(x) := \sum_{\lambda=0}^{n-1} \big[f(x+\lambda h) + f(-x+\lambda h)\big] \cos p\lambda h.$$

From the symmetry, convexity and invariance properties of $\mathcal{C}$, we find that

$$\left(2\sum_{\lambda=0}^{n-1} |\cos p\lambda h|\right)^{-1} \cdot F \,\in\, \mathcal{C}.$$

Now, Lemma 8.5.1 yields

$$\big|R[F]\big| \le \rho(Q,\mathcal{C}) \cdot 2\sum_{\lambda=0}^{n-1} |\cos p\lambda h| \le \frac{4}{3}\, n\rho(Q,\mathcal{C}). \tag{8.49}$$

We have

$$\begin{aligned}
I_p^c[F] &= \frac{1}{\pi}\sum_{\lambda=0}^{n-1} \cos p\lambda h \cdot \left[\int_0^{2\pi} f(x+\lambda h)\cos px\, dx + \int_0^{2\pi} f(-x+\lambda h)\cos px\, dx\right] \\
&= \frac{1}{\pi}\sum_{\lambda=0}^{n-1} \cos p\lambda h \cdot \left[\int_0^{2\pi} f(y)\cos p(y-\lambda h)\, dy + \int_0^{2\pi} f(y)\cos p(-y+\lambda h)\, dy\right] \\
&= \frac{2}{\pi}\sum_{\lambda=0}^{n-1} \cos^2 p\lambda h \cdot \int_0^{2\pi} f(y)\cos py\, dy \\
&\quad + \frac{2}{\pi}\sum_{\lambda=0}^{n-1} \cos p\lambda h\, \sin p\lambda h \cdot \int_0^{2\pi} f(y)\sin py\, dy,
\end{aligned}$$

and this equation, together with

$$\sum_{\lambda=0}^{n-1} \cos^2 p\lambda h = \frac{n}{2} \qquad \text{and} \qquad \sum_{\lambda=0}^{n-1} \cos p\lambda h \sin p\lambda h = 0,$$

implies that

$$I_p^c[F] = nI_p^c[f]. \tag{8.50}$$

Furthermore,

$$\begin{aligned} F(\nu h) &= \sum_{\lambda=0}^{n-1} \Big[f\big((\lambda+\nu)h\big) + f\big((\lambda-\nu)h\big)\Big] \cos p\lambda h \\ &= \sum_{\mu=0}^{n-1} f(\mu h)\big[\cos p(\mu-\nu)h + \cos p(\mu+\nu)h\big] \\ &= 2\cos p\nu h \sum_{\mu=0}^{n-1} f(\mu h)\cos p\mu h \\ &= n\cos p\nu h \cdot Q_p^{\mathrm{st,c}}[f], \end{aligned}$$

which shows that

$$Q[F] = nQ[\cos p\,\cdot\,]Q_p^{\mathrm{st,c}}[f]. \tag{8.51}$$

So the cosine case of the theorem is proved upon combining (8.49), (8.50) and (8.51). The proof in the sine case is similar. □

Another reason for our special interest in formulas of the form (8.48) stems from the following theorem.

THEOREM 8.6.2 (Gautschi 1972). Suppose that the subspace $U \subset \widetilde{\mathcal{C}}$ has the following properties:

(i) $\dim U = n$;
(ii) $f \in U$ implies $f\Big(\cdot + \dfrac{2\pi}{n}\Big) \in U$;
(iii) there is an $l \in U$ such that

$$l(0) = 1 \quad \text{and} \quad l\left(\frac{2\pi}{n}\right) = l\left(2\cdot\frac{2\pi}{n}\right) = \cdots = l\left((n-1)\cdot\frac{2\pi}{n}\right) = 0.$$

Then, for $0 < p < n/2$, there is a projection rule Q for I_p corresponding to U. This rule has the form

$$Q = \tau Q^{\mathrm{st}} \quad \text{with } \tau = \frac{n}{2}\, I_p^{\mathrm{c}}[l].$$

A first example is the space U of all piecewise linear continuous periodic functions with knots $\nu \cdot 2\pi/n$. In this case, we obtain

$$Q[f] := \left(\frac{2}{ph}\sin\frac{ph}{2}\right)^2 \cdot \frac{2}{n}\sum_{\nu=0}^{n-1} f(\nu h)\cdot \begin{cases}\cos p\nu h \\ \sin p\nu h\end{cases} \qquad \text{where } h = \frac{2\pi}{n}.$$

We will prove a more general result later.

PROOF OF THEOREM 8.6.2. For each $f \in \widetilde{\mathcal{C}}$, there is a unique $u \in U$ such that

$$u(\nu h) = f(\nu h) \quad \text{for } \nu \in \mathbb{Z},$$

and this u is given by

$$u = \sum_{\nu=0}^{n-1} f(\nu h)\, l(\,\cdot - \nu h),$$

where uniqueness follows from the dimension. By Theorem 2.2.2 and Lemma 2.2.1, there exists a projection rule; it can be written as

$$I_p[u] = \sum_{\nu=0}^{n-1} f(\nu h) I_p[l(\,\cdot - \nu h)]. \tag{8.52}$$

We have

$$\begin{aligned} I_p^{\mathrm{c}}[l(\,\cdot - \nu h)] &= \frac{1}{\pi}\int_0^{2\pi} l(u-\nu h)\cos pu\,du \\ &= \frac{1}{\pi}\int_0^{2\pi} l(y)\cos p(y+\nu h)\,dy \\ &= \frac{\cos p\nu h}{\pi}\int_0^{2\pi} l(y)\cos py\,dy - \frac{\sin p\nu h}{\pi}\int_0^{2\pi} l(y)\sin py\,dy. \end{aligned}$$

The function l must be even, or else its reflection $\widetilde{l}(x) := l(-x)$ would also satisfy the interpolation conditions (iii) of the theorem, contradicting uniqueness. Therefore

$$\int_0^{2\pi} l(y)\sin p(y)\,dy = 0,$$

and so

$$I_p^{\mathrm{c}}[l(\,\cdot - \nu h)] = \cos p\nu h \cdot I_p^{\mathrm{c}}[l].$$

Combining this with (8.52) gives the claim for I_p^{c}. The I_p^{s} case is similar. □

Eagle (1928), Gautschi (1972) and Locher (1981) gave many examples illustrating Theorem 8.6.2. In the best-known one (see Quade and Collatz 1938), the subspace $U \subset \widetilde{\mathcal{C}}$ is chosen to be

$$\mathcal{S}_r^{\mathrm{per}} = \left\{ f \in \widetilde{\mathcal{C}} : f^{(r-1)} \in \widetilde{\mathcal{C}} \text{ and } f\big|_{[\nu\frac{2\pi}{n},\,(\nu+1)\frac{2\pi}{n}]} \in \mathbb{P}_r \text{ for } \nu \in \mathbb{Z} \right\},$$

i.e. the space of periodic spline functions of odd degree r with knots $\nu \cdot 2\pi/n$.

THEOREM 8.6.3 (Quade and Collatz 1938). For $0 < p < n/2$ and odd r, there is a projection rule for I_p corresponding to $\mathcal{S}_r^{\mathrm{per}}$. This rule has the form

$$\tau \cdot Q^{\mathrm{st}} \quad \text{with } \tau = \tau_r(p,n) = \left[\sum_{\lambda=-\infty}^{\infty} \left(\frac{p}{\lambda n + p} \right)^{r+1} \right]^{-1}. \tag{8.53}$$

Before starting the proof, we make two comments on the attenuation factors τ:

(i) We have $\tau < 1$. For moderately large p, τ is almost 1; for example, when $r \geq 5$ and $p < n/4$, we have $0.998 < \tau < 1$. For p near $n/2$, τ is significantly smaller. One can easily prove that

$$\lim_{r\to\infty} \tau_r\left(\frac{n}{2}, n\right) = \frac{1}{2}.$$

(ii) The formula (8.53) for the attenuation factors is due to Ehlich (1966). Using (8.53) along with the identity

$$\sum_{\lambda=-\infty}^{\infty} \left(\frac{x}{\lambda + x} \right)^2 = \left(\frac{\pi x}{\sin \pi x} \right)^2$$

and its derivatives, we may derive expressions not involving infinite series. In this way, one finds the following formulas, which were already given in Eagle (1928) and Quade and Collatz (1938):

$$\tau_1 = \left(\frac{\sin z}{z}\right)^2 \qquad \text{and} \qquad \tau_3 = \left(\frac{\sin z}{z}\right)^4 \frac{3}{2\cos^2 z + 1},$$

where $z = p\pi/n$.

Proof of Theorem 8.6.3. Let $\widetilde{K}_{2s}^{[p]}$ denote the modified $(2s)$th Peano kernel (8.43). With (8.39), it can be seen that

$$\widetilde{K}_1^{\mathrm{c}} + \frac{\sin(p\,\cdot)}{p\pi}$$

is piecewise constant with jumps at multiples of $2\pi/n$. Therefore

$$G_p(x) := \widetilde{K}_{2s}^{[p]}(x) - (-1)^s \frac{\cos px}{\pi p^{2s}}, \qquad p = 1, 2, \ldots, n-1,$$

is a periodic spline of degree $2s-1$. The Fourier series (8.43) of $\widetilde{K}_{2s}^{[p]}$ implies that

$$G_p(x) = \frac{(-1)^{s-1}}{\pi} \sum_{\lambda=-\infty}^{\infty} \frac{\cos(\lambda n + p)x}{(\lambda n + p)^{2s}},$$

and so

$$G_p(\nu h) = G_p(0) \cdot \cos p\nu h.$$

Now, we define

$$l(x) := \frac{1}{n}\left[1 + \sum_{q=1}^{n-1} \frac{G_q(x)}{G_q(0)}\right]$$

and find that

$$l(0) = 1 \qquad \text{and} \qquad l(\nu h) = \frac{1}{n}\left[1 + \sum_{q=1}^{n-1} \cos q\nu h\right] = 0 \quad \text{for } \nu = 1, 2, \ldots, n-1.$$

Thus, l is the function required in Theorem 8.6.2(iii). We finally obtain

$$\begin{aligned}\tau &= \frac{n}{2} I_p^{\mathrm{c}}[l] = \frac{n}{2\pi}\int_0^{2\pi} l(x)\cos px\,dx \\ &= \frac{n}{2\pi}\cdot\frac{2}{np^{2s}|G_p(0)|} = \left[p^{2s}\sum_{\lambda=-\infty}^{\infty}\frac{1}{(\lambda n + p)^{2s}}\right]^{-1},\end{aligned}$$

as asserted. □

The Quade–Collatz rules Q_n^{QC} are particularly interesting because they have a natural optimality property: If $r = 2s-1$, the rule Q_n^{QC} is optimal for the co-observation

$$\text{(8.54)} \qquad \left\{f \in \widetilde{C} : \int_0^{2\pi} \left[f^{(s)}\right]^2 dx \le M^2\right\}.$$

This is a consequence of Theorem 2.3.1, provided we can show that $\mathcal{S}_{2s-1}^{\mathrm{per}}$ is the spline space for the inner product

$$\varphi(f, g) := \int_0^{2\pi} f^{(s)}(x) g^{(s)}(x)\,dx.$$

But the above claim follows easily from repeated integration by parts.

It is now possible to derive from Theorem 2.3.1 some reasonably simple expressions for ρ^{intr}.

THEOREM 8.6.4 (Ehlich 1966; Golomb 1968). For the functional I_p and co-observation (8.54), we have

$$\rho^{\text{intr}}[f] = \left(1 - \frac{\|P[f]\|^2}{M^2}\right)^{1/2} \rho^{\text{opt}} \tag{8.55}$$

with

$$\rho^{\text{opt}} = \sqrt{\frac{1-\tau_{2s-1}}{\pi p^{2s}}} \tag{8.56}$$

and

$$\|P[f]\|^2 = \pi \sum_{\nu=1}^{\lfloor n/2 \rfloor}{}^{*} \nu^{2s} \tau_{2s-1} (\alpha_\nu^2 + \beta_\nu^2), \tag{8.57}$$

where

$$\alpha_\nu = \frac{2}{n} \sum_{\mu=0}^{n-1} f(\mu h) \cos \mu\nu h \quad \text{and} \quad \beta_\nu = \frac{2}{n} \sum_{\mu=0}^{n-1} f(\mu h) \sin \mu\nu h.$$

Here, $\sum^*$ means that the last summand has to be halved if n is even.

PROOF. Equation (8.55) is the same as (2.35). In order to prove (8.56), we need to determine the Peano kernel. This is not difficult, since

$$R^{\text{QC}} = I - Q^{\text{QC}} = I - \tau_{2s-1} Q^{\text{st}} = (1 - \tau_{2s-1}) I + \tau_{2s-1} R^{\text{st}}.$$

So we find that

$$\begin{aligned} \left(\rho^{\text{opt}}\right)^2 &= \int_0^{2\pi} \left[K_s^{\text{QC}}(x)\right]^2 dx \\ &= \pi \left[\frac{(1-\tau_{2s-1})^2}{\pi^2 p^{2s}} + \frac{\tau_{2s-1}^2}{\pi^2} \sum_{\substack{\lambda=-\infty \\ \lambda \neq 0}}^{\infty} \frac{1}{(\lambda n + p)^{2s}}\right] = \frac{1-\tau_{2s-1}}{\pi p^{2s}}. \end{aligned}$$

To show (8.57), we start with

$$\begin{aligned} \|P[f]\|^2 &= \varphi\big(P[f],\, P[f]\big) \\ &= \varphi\big(P[f],\, P[f] - \text{intpol}^{\text{tr}}[f]\big) + \varphi\big(P[f],\, \text{intpol}^{\text{tr}}[f]\big) \\ &= \varphi\big(P[f],\, \text{intpol}^{\text{tr}}[f]\big), \end{aligned}$$

where we have used (2.33) in the last step. Now, Theorem 8.6.3 implies that the Fourier coefficients a_ν and b_ν of $P[f]$ have the form

$$a_\nu = \tau_{2s-1} \alpha_\nu \qquad \text{and} \qquad b_\nu = \tau_{2s-1} \beta_\nu,$$

which holds even for $\nu = n/2$. Substituting these into the previous equation, we obtain

$$\varphi\big(P[f],\, \text{intpol}^{\text{tr}}[f]\big) = \pi \sum_{\nu=1}^{\lfloor n/2 \rfloor}{}^{*} \nu^{2s} \tau_{2s-1} (\alpha_\nu^2 + \beta_\nu^2),$$

which is (8.57). □

The same method by which we proved (8.56) can also be used to determine $\rho(Q^{\mathrm{st}})$ for the co-observation (8.54). We find that

$$\rho(Q^{\mathrm{st}}) = \frac{\rho^{\mathrm{opt}}}{\sqrt{\tau_{2s-1}}},$$

so the optimal formula Q^{QC} offers only a small advantage over Q^{st}.

CHAPTER 9

Variance and Chebyshev-type Rules

In many practical applications, the function values used in the determination of $Q[f]$ are obtained from experiments and contain measurement errors. We quantify the effect of these errors by the notion of variance from probability theory. The goal will be to construct rules with small variance and a high degree of polynomial exactness. Among all rules with $R[\mathbb{P}_0] = \{0\}$, the ones that involve equal coefficients (the so-called "Chebyshev-type rules") have the smallest variance. We thoroughly investigate this class of rules, but find disappointing results with regard to the degree condition.

9.1. Fundamentals

If there are observational errors in the function values, we may assume that these errors are independent identically distributed random variables with variance σ^2. Then, the random variable $Q_n[f]$ has variance

$$\sigma^2 \cdot \sum_{\nu=1}^{n} a_\nu^2. \tag{9.1}$$

We wish to bound the sensitivity of $Q_n[f]$ to accidental measurement errors, so it is of interest to see how we can make the variance of $Q_n[f]$ small. However, the only part of (9.1) that we have direct influence on is the second factor.

DEFINITION 9.1.1. We define the *variance of the quadrature rule* Q_n to be the number

$$\operatorname{Var} Q := \sum_{\nu=1}^{n} a_\nu^2.$$

We start with a simple result.

THEOREM 9.1.1. For any Q_n with $R[\mathbb{P}_0] = \{0\}$, we have

$$\operatorname{Var} Q_n \geq \frac{1}{n} \left(I[1]\right)^2,$$

and equality holds if and only if

$$a_\nu = \frac{I[1]}{n} \quad \text{for } \nu = 1, \ldots, n. \tag{9.2}$$

PROOF. We calculate

$$\sum_{\nu=1}^{n} a_\nu^2 = \sum_{\nu=1}^{n} \left(a_\nu - \frac{I[1]}{n}\right)^2 + 2\frac{I[1]}{n}\sum_{\nu=1}^{n} a_\nu - \frac{1}{n}\left(I[1]\right)^2$$
$$= \sum_{\nu=1}^{n} \left(a_\nu - \frac{I[1]}{n}\right)^2 + \frac{1}{n}\left(I[1]\right)^2,$$

which proves the assertion. □

To avoid trivial statements, from now on we shall assume that $I[1] \neq 0$.

DEFINITION 9.1.2. A quadrature rule satisfying (9.2) is called a *Chebyshev-type rule.*

Why would anyone be interested in Chebyshev-type rules? Two arguments have been put forward: first, the saving of many multiplication operations in the computation of $Q[f]$; and, second, the favourable behaviour in regard to stochastic error propagation. However, in this era of fast computers, the first argument has become antiquated; also, the second argument does not call for the smallest possible variance. So, in reality, the study of Chebyshev-type rules is mainly a problem in pure mathematics. Nevertheless, it has attracted interest from many mathematicians and has generated a body of deep results. For surveys of this area, see Gautschi (1976) and Förster (1993b).

Chebyshev-type rules include the best rules in $\mathcal{C}_M^{\mathrm{Var}}$ as discussed in Theorem 4.4.3. Specific examples are the midpoint rules Q_n^{Mi} and the rules

$$Q_n[f] = \frac{\pi}{n}\sum_{\nu=1}^{n} f\left(-\cos\left(\frac{2\nu-1}{2n}\pi\right)\right) \tag{9.3}$$

for

$$I[f] = \int_{-1}^{1} \frac{f(x)}{\sqrt{1-x^2}}\,dx \tag{9.4}$$

(see Example 5.1.1).

Among all Chebyshev-type rules, those with good approximation properties are naturally of greatest interest. Therefore, we require that

$$R_n[\mathbb{P}_s] = \{0\}$$

with maximal s. The next two theorems will show that $s = n$ is reasonable and that $s = 2n-1$, as in (9.3), is an exception. This suggests a definition:

DEFINITION 9.1.3. A Chebyshev-type rule Q_n with

$$R_n[\mathbb{P}_n] = \{0\} \tag{9.5}$$

is called a *Chebyshev rule.*

THEOREM 9.1.2. *For any combination of n and I, there exists at most one Chebyshev rule.*

This statement is not true if the requirement (9.5) is replaced by $R_n[\mathbb{P}_{n-1}] = \{0\}$, as can be demonstrated by simple examples.

The proof of Theorem 9.1.2 will provide a quite explicit method for constructing Chebyshev rules (when they exist). It requires a lemma from algebra.

LEMMA 9.1.1 (Newton's identities). Let

$$f(x) = \prod_{\nu=1}^{n} (x - x_\nu) = \sum_{\nu=0}^{n} c_\nu x^\nu$$

be any monic polynomial, and set

$$q_0 := n \qquad \text{and} \qquad q_\lambda := \sum_{\nu=1}^{n} x_\nu^\lambda \quad \text{for } \lambda = 1, 2, \ldots.$$

Then

$$\sum_{\nu=\mu+1}^{n} c_\nu q_{\nu-\mu} = (\mu - n) \cdot c_\mu \quad \text{for } \mu = 0, \ldots, n-1. \tag{9.6}$$

PROOF. The special case where $\mu = 0$ is none other than the trivial identity

$$\sum_{\nu=1}^{n} f(x_\nu) = 0.$$

For the other cases we argue as follows:

$$\begin{aligned} f'(x) &= \sum_{i=1}^{n} \frac{f(x)}{x - x_i} = \sum_{i=1}^{n} \frac{f(x) - f(x_i)}{x - x_i} \\ &= \sum_{i=1}^{n} \sum_{\nu=1}^{n} c_\nu \frac{x^\nu - x_i^\nu}{x - x_i} = \sum_{i=1}^{n} \sum_{\nu=1}^{n} c_\nu \sum_{\mu=1}^{\nu} x^{\mu-1} x_i^{\nu-\mu} \\ &= \sum_{\mu=1}^{n} x^{\mu-1} \sum_{\nu=\mu}^{n} c_\nu \sum_{i=1}^{n} x_i^{\nu-\mu} = \sum_{\mu=1}^{n} x^{\mu-1} \sum_{\nu=\mu}^{n} c_\nu q_{\nu-\mu}. \end{aligned}$$

By comparing the above with

$$f'(x) = \sum_{\mu=1}^{n} c_\mu \cdot \mu x^{\mu-1},$$

we obtain

$$\sum_{\nu=\mu}^{n} c_\nu q_{\nu-\mu} = \mu \cdot c_\mu \quad \text{for } \mu = 1, 2, \ldots, n,$$

so the lemma follows. □

PROOF OF THEOREM 9.1.2. Define

$$\widetilde{q}_\lambda := \frac{n}{I[1]} I[p_\lambda] \quad \text{for } \lambda = 1, \ldots, n, \quad \text{where } p_\lambda(x) = x^\lambda.$$

For any Chebyshev rule, we have

$$\widetilde{q}_\lambda = \frac{n}{I[1]} Q_n[p_\lambda] = \sum_{\nu=1}^{n} x_\nu^\lambda;$$

that is, the $\widetilde{q}_\lambda$ are the power sums q_λ corresponding to the nodes of the Chebyshev rule. So we need only show that the given q_λ uniquely determine the x_ν. First, the

q_λ uniquely determine $c_{n-1}, c_{n-2}, \ldots, c_0$ via the application of (9.6):

$$\begin{aligned} c_n &= 1, \\ -c_{n-1} &= c_n q_1, \\ -2c_{n-2} &= c_{n-1}q_1 + c_n q_2, \\ -3c_{n-3} &= c_{n-2}q_1 + c_{n-1}q_2 + c_n q_3, \end{aligned}$$

and so on. Now, the x_ν are the zeros of the nth degree polynomial with coefficients c_ν, and we are done. □

If we do not know beforehand that a Chebyshev rule exists, we compute the moments $\widetilde{q}_\lambda$, set $q_\lambda = \widetilde{q}_\lambda$, apply the above procedure to determine the c_ν, and calculate the zeros of the resulting polynomial. If these zeros are simple and in $[a, b]$, then we have a Chebyshev rule.

EXAMPLE 9.1.1. We apply the method described above, with $n = 3$, to

$$I[f] := \int_{-1}^{1} f(x)w(x)\,dx,$$

where w is such that $w(x) = w(-x)$. This assumption gives $I[p_1] = I[p_3] = 0$. So we have

$$q_1 = 0, \qquad q_2 = \frac{3}{I[1]}\,I[p_2], \qquad q_3 = 0.$$

Newton's identities then give

$$c_3 = 1, \qquad c_2 = 0, \qquad c_1 = -\frac{3}{2I[1]}\,I[p_2], \qquad c_0 = 0.$$

Therefore

$$f(x) = x^3 - \frac{3}{2I[1]}\,I[p_2] \cdot x.$$

This polynomial has three distinct zeros in $[-1, 1]$ if and only if

$$0 < \frac{3}{2I[1]}\,I[p_2] \leq 1,$$

that is,

$$0 < \frac{\displaystyle\int_{-1}^{1} x^2 w(x)\,dx}{\displaystyle\int_{-1}^{1} w(x)\,dx} \leq \frac{2}{3}.$$

The second inequality holds for $w = 1$ and leads to the Chebyshev rule

$$Q_3^{\mathrm{Ch}}[f] = \frac{2}{3}\left[f\left(-\frac{\sqrt{2}}{2}\right) + f(0) + f\left(\frac{\sqrt{2}}{2}\right)\right]$$

The inequality does not hold for $w(x) = x^{2r}$ when $r \geq 2$, so there is no Chebyshev rule with three nodes if we treat I as a functional on $C[-1, 1]$ (rather than one defined on $C\big[-\sqrt{1.5}, \sqrt{1.5}\,\big]$).

Theorem 9.1.2 does not preclude the existence of Chebyshev rules with a higher degree of polynomial exactness, but for methods (i.e. sequences $Q_1, Q_2, \ldots$) having that property there is a strong restriction:

THEOREM 9.1.3 (Peherstorfer 1988a). Let $(Q_n)_{n=2,3,\dots}$ be a sequence of Chebyshev rules with

$$R_n[\mathbb{P}_{n+1}] = \{0\} \quad \text{for all } n. \tag{9.7}$$

Then we have

$$I[f] = \alpha \cdot \int_\beta^\gamma f(x) \frac{dx}{\sqrt{(\gamma - x)(x - \beta)}} \tag{9.8}$$

for some constants α, β and γ.

PROOF. We write $p_\nu(x) = x^\nu$. First, observe that I is obtained from (9.4) via a simple transformation. The existence of a Chebyshev method satisfying (9.8) with, even, $R_n[\mathbb{P}_{2n-1}] = \{0\}$ follows for this special I.

The main idea of the proof is to show that if there is a sequence satisfying (9.7), then the underlying functional I^* must be determined by $I^*[p_i]$ for $i = 0, 1, 2$. If we can then choose α, β and γ so that

$$I^*[p_i] = I[p_i] \quad \text{for } i = 0, 1, 2, \tag{9.9}$$

then $I^* = I$ will be proved. From our knowledge of $I^*[p_0]$, $I^*[p_1]$ and $I^*[p_2]$, we can determine the rule Q_2 using Newton's identities, obtaining the value $I^*[p_3] = Q_2[p_3]$. Based on this information, we determine Q_3, which gives $I^*[p_4]$. Continuing in this manner, we obtain $I^*[p_\nu]$ for all ν and hence I^* by means of the Weierstrass approximation theorem.

Choosing

$$\begin{aligned}
\alpha &= \frac{I^*[p_0]}{\pi}, \\
\beta &= \frac{I^*[p_1] - \sqrt{2I^*[p_2]I^*[p_0] - 2\big(I^*[p_1]\big)^2}}{I^*[p_0]}, \\
\gamma &= \frac{I^*[p_1] + \sqrt{2I^*[p_2]I^*[p_0] - 2\big(I^*[p_1]\big)^2}}{I^*[p_0]},
\end{aligned}$$

the correctness of (9.9) is easily verified. Note that β and γ are real distinct numbers only if

$$I^*[p_2]I^*[p_0] - \big(I^*[p_1]\big)^2 > 0.$$

Owing to the fact that

$$I^*[p_i] = Q_2[p_i] = \frac{I[p_0]}{2}\,(x_1^i + x_2^i) \quad \text{for } i = 0, 1, 2,$$

the above inequality is equivalent to

$$2(x_1^2 + x_2^2) - (x_1 + x_2)^2 > 0.$$

The left-hand side equals $(x_2 - x_1)^2$. If this were zero, we would have constructed a rule of degree greater than or equal to two with just one node, which is not possible. We remark that this proof is due to Förster (1994). $\square$

9.2. Chebyshev methods

A Chebyshev quadrature method in the general sense is a sequence $Q_{n_1}, Q_{n_2}, \ldots$ of Chebyshev rules, where $n_1 < n_2 < \cdots$. A Chebyshev method in the narrower sense is a Chebyshev method with $n_\nu = \nu$. Functionals besides (9.8) that admit a method in the narrower sense are not easy to find. Here we give an example:

THEOREM 9.2.1 (Förster 1987b). *For the weight function*

$$w(x) = \left(|x| - x^2\right)^{-1/2} \quad \text{on } [-1,1],$$

there is a Chebyshev method in the narrower sense.

PROOF. (i) Let $n = 2m-1$. We start with the Radau rules

$$\int_{-1}^{1} \frac{f(x)}{\sqrt{1-x^2}}\,dx = \frac{2\pi}{2m-1}\left[\frac{1}{2}f(1) + \sum_{\nu=1}^{m-1} f\left(-\cos\left(\frac{2\nu-1}{2m-1}\pi\right)\right)\right]$$

and

$$\int_{-1}^{1} \frac{f(x)}{\sqrt{1-x^2}}\,dx = \frac{2\pi}{2m-1}\left[\frac{1}{2}f(-1) + \sum_{\nu=2}^{m} f\left(-\cos\left(\frac{2\nu-2}{2m-1}\pi\right)\right)\right],$$

valid for $f \in \mathbb{P}_{2m-2}$ (see Example 6.1.2). Transforming the first rule to $[-1,0]$ and of the second rule to $[0,1]$ gives

$$\int_{-1}^{0} \frac{f(y)}{\sqrt{-y-y^2}}\,dy = \frac{2\pi}{2m-1}\left[\frac{1}{2}f(0) + \sum_{\nu=1}^{m-1} f\left(-\sin^2\frac{\nu\pi}{2m-1}\right)\right]$$

and

$$\int_{0}^{1} \frac{f(y)}{\sqrt{y-y^2}}\,dy = \frac{2\pi}{2m-1}\left[\frac{1}{2}f(0) + \sum_{\nu=1}^{m-1} f\left(\sin^2\frac{\nu\pi}{2m-1}\right)\right],$$

which are valid also for $f \in \mathbb{P}_{2m-2}$. Upon adding these two equations, we obtain

$$\int_{-1}^{1} \frac{f(y)}{\sqrt{|y|-y^2}}\,dy = \frac{2\pi}{2m-1}\left[f(0) + \sum_{\nu=1}^{m-1}\left\{f\left(-\sin^2\frac{\nu\pi}{2m-1}\right) + f\left(\sin^2\frac{\nu\pi}{2m-1}\right)\right\}\right]$$

for $f \in \mathbb{P}_{2m-2}$. From the symmetry, it follows that we have equality for all odd functions, which implies exactness on $\mathbb{P}_{2m-1}$. Thus, we have constructed a Chebyshev rule.

(ii) Let $n = 2m$. We start with the Gaussian rule

$$\int_{-1}^{1} \frac{f(x)}{\sqrt{1-x^2}}\,dx = \frac{\pi}{m}\sum_{\nu=1}^{m} f\left(-\cos\left(\frac{2\nu-1}{2m}\pi\right)\right)$$

for $f \in \mathbb{P}_{2m-1}$ (see Example 6.1.2). As in case (i), we transform this rule to $[-1,0]$ and to $[0,1]$ and then add the results. This leads to

$$\int_{-1}^{1} \frac{f(y)}{\sqrt{|y|-y^2}}\,dy = \frac{\pi}{m}\sum_{\nu=1}^{m}\left\{f\left(-\cos^2\frac{2\nu-1}{4m}\pi\right) + f\left(\cos^2\frac{2\nu-1}{4m}\pi\right)\right\}. \tag{9.10}$$

This rule is exact for $f \in \mathbb{P}_{2m-1}$ but not for all $f \in \mathbb{P}_{2m}$, making the situtation more complicated than in the case of odd n.

The nodes in (9.10) are the zeros of

$$p^*(x) := T_m(2x+1)T_m(2x-1).$$

We set

$$p(x) := p^*(x) - \frac{1}{2}$$

and study the quadrature rule Q of interpolatory type which has the zeros of p as its nodes. We begin by showing the existence of $2m$ distinct zeros of p in $[-1, 1]$. Because

$$(-1)^\nu T_m(2z_\nu - 1) = 1 \qquad \text{and} \qquad T_m(2z_\nu + 1) \geq 1$$

for

$$z_\nu = \frac{1}{2}\left(1 + \cos\frac{\nu\pi}{m}\right), \qquad \nu = 0, 1, \dots, m,$$

p has m sign changes on $]0, 1]$ and the assertion follows.

The next step is to prove that Q is a Chebyshev-type rule. To this end, we use Newton's identities (Lemma 9.1.1). The numbers $c_n, c_{n-1}, \dots, c_1$ are the same for $f = p$ and for $f = p^*$, and so are the power sums $q_1, q_2, \dots, q_{n-1}$. Therefore, Q is of Chebyshev type since (9.10) is.

Finally, we prove that $Q[p] = I[p]$, so that Q is a Chebyshev rule:

$$\begin{aligned}
I[p] &- Q[p] \\
&= I[p] = \int_{-1}^{1} \frac{p(x)}{\sqrt{|x| - x^2}}\,dx \\
&= 2\int_0^1 \frac{p(x)}{\sqrt{x - x^2}}\,dx = 2\int_0^1 \frac{p(t^2)}{\sqrt{t^2 - t^4}} \cdot 2t\,dt = 2\int_{-1}^{1} \frac{p(t^2)}{\sqrt{1 - t^2}}\,dt \\
&= 2\int_{-1}^{1} \frac{T_m(2t^2 - 1)T_m(2t^2 + 1)}{\sqrt{1 - t^2}}\,dt - \pi = 2\int_{-1}^{1} \frac{T_{2m}(t)T_m(2t^2 + 1)}{\sqrt{1 - t^2}}\,dt - \pi \\
&= 2\int_{-1}^{1} \frac{T_{2m}(t) \cdot 2^{2m-1}t^{2m}}{\sqrt{1 - t^2}}\,dt - \pi = 2\int_{-1}^{1} \frac{T_{2m}^2(t)}{\sqrt{1 - t^2}}\,dt - \pi \;=\; 0,
\end{aligned}$$

where we have used

$$T_m(2t^2 - 1) = T_{2m}(t)$$

and the orthogonality of the Chebyshev polynomials. □

Other weight functions that admit Chebyshev quadrature methods in the narrower sense have been discovered by Ullman (1966), Byrd and Stalla (1984), Förster (1987b), Peherstorfer (1982, 1988a), Xu (1989) and Kuijlaars (1995c, 1995d). All of the explicitly given weight functions have a very special form, one of the simplest being

$$w(x) = \frac{1 + bx}{\sqrt{1 - x^2}} \quad \text{for } |b| < 0.27,$$

presented by Xu (1989); see also Kuijlaars (1995a).

If we allow Chebyshev methods in the broad sense, many further examples can be found; see Geronimus (1969) and Förster (1987b). We note that an equivalent of Theorem 9.1.3 does not stand in this case. There are many functionals admitting Chebyshev–Gauss methods; see Peherstorfer (1988b, 1990). The first example was given by Gautschi (1984):

THEOREM 9.2.2. For the weight function

$$w(x) = \begin{cases} |x|\left(x^2 - \xi^2\right)^{-1/2}\left(1 - x^2\right)^{-1/2} & \text{for } |x| \in]\xi, 1[, \\ 0 & \text{elsewhere,} \end{cases}$$

with $\xi \in]0,1[$, the Gaussian rules for even n are Chebyshev rules.

PROOF. Suppose that

$$\int_0^1 y^s w(y)\,dy = A\sum_{\nu=1}^{n} x_\nu^s \quad \text{for } s = 0,1,\ldots,m$$

with some $x_\nu \geq 0$. Then

$$(9.11)\quad \int_{-1}^{1} x^r w(x^2)\cdot|x|\,dx = \frac{A}{2}\sum_{\nu=1}^{n}\left[\left(\sqrt{x_\nu}\right)^r + \left(-\sqrt{x_\nu}\right)^r\right] \quad \text{for } r = 0,1,\ldots,2m+1$$

(for odd r this follows from the symmetry of $w(x^2)\cdot|x|$, and for even r it comes from substituting $y = x^2$).

The Chebyshev rule for the weight

$$w(x) = \begin{cases} (1-x)^{-1/2}(x-\xi^2)^{-1/2} & \text{for } x \in]\xi^2, 1[, \\ 0 & \text{elsewhere} \end{cases}$$

can be derived from (9.3) via an affine transformation. If we now use the construction above with $m = 2n-1$, we obtain the assertion. □

Surveying the known examples of Chebyshev methods leads to the impression that weight functions admitting Chebyshev methods are rare. This is supported by the following necessary condition.

THEOREM 9.2.3 (Kahaner 1969). Suppose that w has the following properties:

(i) w admits a Chebyshev method on $[-1,1]$;
(ii) w is continuous in $]-1,1[$;
(iii) there exists $c > 0$ such that $w(x) \geq c$ for $x \in]-1,1[$.

Then

$$(9.12)\qquad w(x) \geq \frac{1}{2\pi\sqrt{1-x^2}}\int_{-1}^{1} w(u)\,du \quad \text{for } x \in]-1,1[.$$

The proof makes use of the following lemma.

LEMMA 9.2.1. If a positive quadrature method converges for all $f \in C[a,b]$, then it converges for

$$(9.13)\qquad f_0(x) = \begin{cases} 1 & \text{for } x \in [c,d], \\ 0 & \text{elsewhere,} \end{cases}$$

where $a \leq c < d \leq b$.

PROOF. Let an arbitary $\varepsilon > 0$ be given. Then there exist continuous functions f_1 and f_2 such that

$$f_1(x) \leq f_0(x) \leq f_2(x) \qquad \text{and} \qquad I[f_2 - f_1] < \varepsilon.$$

The positivity of Q_n implies that

$$Q_n[f_1] \leq Q_n[f_0] \leq Q_n[f_2],$$

and the convergence gives

$$I[f_1] - \varepsilon \leq Q_n[f_0] \leq I[f_2] + \varepsilon$$

for sufficiently large n. Thus we obtain

$$-2\varepsilon \le I[f_1 - f_2] - \varepsilon \le I[f_1 - f_0] - \varepsilon \le Q_n[f_0] - I[f_0] \\ \le I[f_2 - f_0] + \varepsilon \le I[f_2 - f_1] + \varepsilon \le 2\varepsilon,$$

which proves the assertion. □

PROOF OF THEOREM 9.2.3. If Q_n is a Chebyshev rule, then we have

$$Q_n[f_0] = I[1]\,\frac{N_n(c,d)}{n},$$

where f_0 is the function defined in (9.13) and $N_n(c,d)$ is the number of nodes in $[c,d]$. Combining this relation with the above lemma gives

$$\lim_{n\to\infty} \frac{N_n(c,d)}{n} = \frac{1}{I[1]} \int_c^d w(u)\,du \tag{9.14}$$

since, by Theorem 3.1.2, a Chebyshev method is convergent. Theorem 5.6.1 yields

$$\liminf_{n\to\infty} \frac{N_n(c,d)}{n} \ge \frac{1}{2\pi} \int_c^d \frac{du}{\sqrt{1-u^2}}, \tag{9.15}$$

and a comparison of (9.14) and (9.15) gives

$$\int_c^d \left(\frac{w(u)}{I[1]} - \frac{1}{2\pi\sqrt{1-u^2}} \right) du \ge 0$$

for any interval $[c,d] \subset [a,b]$. Now, the continuity assumption leads to the conclusion (9.12). □

A first consequence of Theorem 9.2.3 is that no positive continuous weight function can admit a Chebyshev method. Another immediate conclusion is the non-existence of such a method for

$$w(x) = \left(1 - x^2\right)^\alpha \quad \text{if } \; -1/2 < \alpha < 0.$$

9.3. The special case of $w = 1$

For $w = 1$, it suffices to restrict our attention to the interval of integration $[-1,1]$. We start with a famous result.

THEOREM 9.3.1 (Bernstein 1937a). For $w = 1$, there exist Chebyshev rules only for $n \le 7$ and $n = 9$.

These rules are tabulated in Abramowitz and Stegun (1964: p. 920).

PROOF. According to Theorem 5.1.8, there is a node of the Chebyshev rule Q_n^{Ch} to the left of the first node of Q_s^{G}, where $s = \lfloor (n+1)/2 \rfloor$. An application of (6.29) gives $a_{1,n}^{\mathrm{Ch}} \le a_{1,s}^{\mathrm{G}}$, that is,

$$\frac{2}{n} \le a_{1,s}^{\mathrm{G}}. \tag{9.16}$$

This is a necessary condition for the existence of Q_n^{Ch}.

By combining Theorems 6.3.4 and 6.3.2, we get

$$a_{1,s}^{\mathrm{G}} \le \frac{\pi}{s+1/2} \sin \frac{\pi}{s+1/2} \le \frac{\pi^2}{(s+1/2)^2} \le \frac{4\pi^2}{(n+1)^2}.$$

Inserting this bound into (9.16) gives

$$\frac{2}{n} \leq \frac{4\pi^2}{(n+1)^2},$$

or $n \leq 17$.

The non-existence of Chebyshev rules for the remaining cases, $n = 8$ and $10 \leq n \leq 17$, can be proved by using (9.16) and the exact values of $a_{1,s}^{\mathrm{G}}$ as given in Stroud and Secrest (1966).

Finally, we have to prove the existence of Q_n^{Ch} for $n = 1, \ldots, 7, 9$. This can be done with the general method explained in the proof of Theorem 9.1.2; see, for example, Walther (1930) or Krylov (1962: pp. 187–191). □

The most interesting result on the error of Q_n^{Ch} concerns its definiteness; see Walther (1930). For the proof we use Theorem 4.2.9.

EXAMPLE 9.3.1. Let us examine Q_5^{Ch}. The method using Newton's identities (Theorem 9.1.2) gives

$$f(x) = x^5 - \frac{5}{6}x^3 + \frac{7}{12}x,$$

so that

$$-x_1 = x_5 = 0.83249\ldots, \qquad -x_2 = x_4 = 0.37454\ldots, \qquad x_3 = 0.$$

To prove definiteness by Theorem 4.2.9, we have to compute

$$F(x) := \int_x^1 f(u)\,du = -\frac{1}{6}\left(x^6 - \frac{5}{4}x^4 + \frac{7}{24}x^2 - \frac{1}{24}\right) = \frac{1-x^2}{6}\left(x^4 - \frac{1}{4}x^2 + \frac{1}{24}\right)$$

and show that $F(x) \geq 0$, which is easy. So we have

$$R_5^{\mathrm{Ch}}[g] = c \cdot \frac{g^{(6)}(\xi)}{6!} \quad \text{with } \xi \in [-1,1]$$

for any $g \in C^{(6)}[-1,1]$. We choose

$$g(x) = g_0(x) := xf(x)$$

and obtain

$$c = R_5^{\mathrm{Ch}}[g_0] = I[g_0] = \frac{13}{756}.$$

Our final result is therefore

$$R_5^{\mathrm{Ch}}[g] = \frac{13}{756} \cdot \frac{g^{(6)}(\xi)}{6!} = (2.388\ldots) \cdot 10^{-5} \cdot g^{(6)}(\xi).$$

The bound

$$\left|R_5^{\mathrm{Ch}}[g]\right| \leq 2.4 \cdot 10^{-5}\, \|g^{(6)}\|$$

follows immediately. The corresponding bounds for R_5^{Fi}, R_5^{Po} and R_5^{G} have smaller constants.

The disappointing result stated in Theorem 9.3.1 led to many suggested modifications of the Chebyshev rules. If equality of all coefficients is seen as the principal characteristic of Chebyshev rules, then the only possibility that remains is to weaken the requirement that $R_n[\mathbb{P}_n] = \{0\}$. Using the same method as in the preceding theorem, we easily obtain the following result.

THEOREM 9.3.2 (Bernstein 1936). If

$$R_n[\mathbb{P}_m] = \{0\}$$

holds for a Chebyshev-type rule, then

$$m \leq \pi\sqrt{2n}. \tag{9.17}$$

Bernstein (1937a) improved the constant $\pi\sqrt{2} = 4.44\ldots$ in (9.17) to 4. A further improvement to 3.8525 was made by Gatteschi and Vinardi (1978) and Korevaar (1992).

Theorem 9.3.2 implies that a Chebyshev-type rule with polynomial degree of exactness m must have number of nodes in the order of m^2 at least. Whether there really do exist rules that require only $O(m^2)$ nodes is a deeper question. Bernstein (1937c, 1937d) was able to prove that such rules exist provided some nodes are allowed to coincide; Kuijlaars (1993) has shown that multiple nodes can always be split up, resulting in Chebyshev rules in the ordinary sense.

For small n, more precise results can be obtained.

EXAMPLE 9.3.2. For $n = 8$ there is no Chebyshev rule. We can prove this special case of Theorem 9.3.1 with the method based on Newton's identities (Theorem 9.1.2). We obtain

$$f(x) = \sum_{\nu=0}^{n} c_\nu x^\nu = x^8 - \frac{4}{3}x^6 + \frac{22}{45}x^4 - \frac{148}{2835}x^2 - \frac{43}{42525}.$$

By Descartes' rule of signs, this polynomial can have at most three positive zeros, which proves the non-existence.

As a substitute, we seek a Chebyshev-type rule with eight nodes such that $R_8[\mathbb{P}_7] = \{0\}$. Proceeding as before, Newton's identities give the same values for $c_8, c_7, \ldots, c_1$, but c_0 remains undetermined. If there were to be a $c_0 \in \mathbb{R}$ such that

$$f_{c_0}(x) := x^8 - \frac{4}{3}x^6 + \frac{22}{45}x^4 - \frac{148}{2835}x^2 + c_0$$

has eight simple zeros in $[-1, 1]$, then the existence of the desired rule would be proved. Now, f_0 has six simple zeros and one double zero in $[-1, 1]$, since $g(x) := f_0(\sqrt{x})/x$ alternates sign at $x = 0,\ 1/5,\ 1/2,\ 1$. Therefore f_c will have eight simple zeros for sufficiently small $c > 0$. For a more precise discussion, see Gautschi and Monegato (1977). Hence, there is an infinite number of Chebyshev-type rules of interpolatory type with eight nodes. If we want to single out a "best" rule among these, we could use as our criterion the error for the first monomial that is not integrated exactly, namely $p_8(x) = x^8$. We have

$$R_8[p_8] = R_8[f_{c_0}] = I[f_{c_0}] = \frac{86}{42525} + 2c_0.$$

The infimum is not attained for $c_0 > 0$, so in this sense there is no best Chebyshev-type rule. Another way of stating the result is that the minimization leads to coinciding nodes.

For $n = 10$, a similar calculation leads to an analogous result. It can be proved that for $n > 10$, a rule being of Chebyshev type is incompatible with its being of interpolatory type.

Let us denote by $\mathbf{S}_n$ the set of Chebyshev-type rules with n nodes. For more information on the elements of $\mathbf{S}_n$ with highest possible degree, see Kahaner (1970), Anderson and Gautschi (1975), Gautschi and Monegato (1977), Förster (1982b) and Förster and Ostermeyer (1986). Optimality conditions frequently lead to multiple nodes; see Gautschi and Yanagiwara (1974) and Korevaar and Meyers (1993).

There are many other principles that could be used to single out a "best" rule in $\mathbf{S}_n$. Barnhill *et al.* (1969) proposed minimizing

$$\sum_{\nu=1}^{n} R^2[p_\nu], \quad \text{where } p_\nu(x) = x^\nu, \tag{9.18}$$

among all $Q \in \mathbf{S}_n$. Evidently, this leads to Chebyshev rules, if such rules exist.

THEOREM 9.3.3 (Gautschi and Yanagiwara 1974). *Any $Q \in \mathbf{S}_n$ that minimizes (9.18) is a Chebyshev rule.*

PROOF. We have to minimize

$$F(x_1, \ldots, x_n) := \sum_{\nu=1}^{n} R^2[p_\nu].$$

The partial derivatives

$$F_{x_i}(x_1, \ldots, x_n) = 2\sum_{\nu=1}^{n} R[p_\nu] \cdot \left(-\frac{2}{n}\,\nu x_i^{\nu-1}\right)$$

must vanish at any extremal point; that is,

$$\sum_{\nu=1}^{n} R[p_\nu]\nu \hat{x}_i^{\nu-1} = 0 \quad \text{for } i = 1, 2, \ldots, n$$

if the minimum is attained at the nodes $\hat{x}_1, \hat{x}_2, \ldots, \hat{x}_n$. This is a homogeneous linear system for the unknowns $\nu R[p_\nu]$, and the coefficient matrix is a Vandermonde matrix, which is regular. Therefore

$$R[p_\nu] = 0 \quad \text{for } \nu = 1, 2, \ldots, n,$$

and this is just the definition of a Chebyshev rule. □

If we are interested in finding a rule from $\mathbf{S}_n$ with good approximation properties, then it would seem natural to look for $Q^* \in \mathbf{S}_n$ such that

$$\rho(Q^*, \mathcal{C}) = \inf_{Q \in \mathbf{S}_n} \rho(Q, \mathcal{C}). \tag{9.19}$$

The midpoint rule Q_n^{Mi} is best for $\mathcal{C}_M^{(1)}$, $\mathcal{C}_M^{\mathrm{Var}}$, $\mathcal{C}_\omega^{\mathrm{mc}}$ and $\mathcal{C}_M^{\mathrm{Sa},1}$, so the solution of (9.19) is trivial since $Q_n^{\mathrm{Mi}} \in \mathbf{S}_n$. For $\mathcal{C}_M^{\mathrm{Sa},2}$, Franke (1971) exhibited numerically determined solutions for $n = 2, 3, \ldots, 11$. Rabinowitz and Richter (1970c) studied a similar problem for some Davis-type co-observations. The difference from our problem is that they use a generalized version of $\mathbf{S}_n$ where the common coefficient $2/n$ is replaced by an arbitrary number; this is quite natural if the constant functions are not in $\mathcal{C}$. Their results do not exclude the possibility of coinciding nodes, but for the numerically determined rules they do indeed obtain Chebyshev-type rules.

Finally, we could sacrifice the requirement of equal coefficients. Gatteschi and Vinardi (1978) discussed the possibility of rules of the form

$$Q_n[f] = b_n\big[f(-1) + f(1)\big] + c_n \sum_{\nu=2}^{n-1} f(x_\nu)$$

being exact for polynomials of degree $2\lfloor n/2 \rfloor + 1$. They proved existence for $n \le 13$ and non-existence otherwise. For the error of these rules, see Baratella (1977).

The non-existence for large n can be proved in a wider context: Erdős and Sharma (1965) were able to show that for any fixed k, a method with $n-k$ coefficients that are equal and k coefficients that can be chosen freely must have polynomial degree of at most $O(\sqrt{n})$.

9.4. Variance

For a quadrature rule to be practical, its variance should be small, though not necessarily minimal. For $w = 1$, the postulate of minimal variance enforces $\deg Q_n \le 4.44\sqrt{n}$, whereas for a slightly greater variance bound, namely $2\pi/n$ instead of $4/n$, even $\deg Q_n = 2n-1$ is possible; see Theorems 9.3.2 and 9.4.1. Förster (1988) discussed a more extreme example: For the weight function of Gautschi (see Theorem 9.2.2) he was able to show that

$$\lim_{n\to\infty} n \operatorname{Var} Q_n^{\mathrm{G}} = I^2[1],$$

which means asymptotically minimal variance, while we have

$$\deg Q_{2m+1} \le \mathrm{const} \cdot \ln m$$

for all Chebyshev-type methods (Förster 1986b).

In view of Theorem 9.1.1, we define "small variance" by

$$\operatorname{Var} Q_n = O(n^{-1}). \tag{9.20}$$

It should be mentioned that there are convergent methods with $\operatorname{Var} Q_n \ge c > 0$ for all n, for example $Q_n[f] := Q_n^{\mathrm{Tr}}[f] + \sqrt{c}\big(f(x_2^{\mathrm{Tr}}) - f(x_1^{\mathrm{Tr}})\big)$. But (9.20) holds for the methods that are usually applied as well as for a wide variety of weight functions. This is trivially true for the methods of Chapter 7: for the Romberg methods we use the fact that they are Riemann sums; for other methods, we have the following theorem.

THEOREM 9.4.1. For positive quadrature rules Q on $[-1,1]$, we have

$$\operatorname{Var} Q \le I[1] \cdot \frac{2\pi}{1+\deg Q} \sup_{x\in[-1,1]} w(x)\sqrt{1-x^2}.$$

PROOF. Start from

$$\sum_{\nu=1}^{n} a_\nu^2 \le \sup_\nu a_\nu \cdot \sum_{\nu=1}^{n} a_\nu = I[1] \sup_\nu a_\nu$$

and use Theorem 6.2.3. □

For $w = 1$, we obtain

$$\frac{4}{n} \le \operatorname{Var} Q_n^{\mathrm{G}} \le \frac{2\pi}{n}.$$

Also, there is a quite explicit expression for $\operatorname{Var} Q_n^{\mathrm{G}}$, namely

$$\operatorname{Var} Q_n^{\mathrm{G}} = \frac{6}{2n+1}\left(1 - \frac{1}{(2n-1)(2n+3)}\right)\sum_{\nu=1}^{n}\frac{1}{\nu^2} + \frac{12}{(2n-1)(2n+3)}$$

(Brass 1998b), which implies that

$$\lim_{n\to\infty} n \operatorname{Var} Q_n^{\mathrm{G}} = \frac{\pi^2}{2}.$$

The asymptotic formula (9.20) does not hold for Gaussian rules in the absence of any assumptions on w. In particular, the Gaussian rules for $w(x) = \left(1-x^2\right)^{\alpha}$ with $\alpha \le -3/4$ fail to satisfy (9.20); see Förster and Petras (1994).

Let $\mathbf{Q}^m(x_1,\ldots,x_n)$ denote the set of quadrature rules Q with nodes $x_1,\ldots,x_n$ and $\deg Q \ge m$. What is the rule in $\mathbf{Q}^m(x_1,\ldots,x_n)$ with minimal variance? A simple answer can be obtained, since there is a basis $u_0,\ldots,u_m$ for $\mathbb{P}_m$ with the property that

$$\sum_{\nu=1}^{n} u_i(x_\nu)u_j(x_\nu) = \begin{cases} 1 & \text{for } i=j, \\ 0 & \text{otherwise,} \end{cases} \tag{9.21}$$

if $m<n$. This basis can be constructed by orthogonalizing any basis with respect to the inner product

$$\varphi(f,g) := \sum_{\nu=1}^{n} f(x_\nu)g(x_\nu).$$

THEOREM 9.4.2 (Wilson 1970a; Chawla and Ramakrishnan 1976). *Let $m<n$. In $\mathbf{Q}^m(x_1,\ldots,x_n)$ there exists a uniquely determined rule Q^* with smallest variance. It is given by*

$$Q^*[f] = \sum_{\nu=1}^{n} a_\nu^* f(x_\nu) \tag{9.22}$$

with

$$a_\nu^* = \sum_{i=0}^{m} u_i(x_\nu) I[u_i], \tag{9.23}$$

and we have

$$\operatorname{Var} Q^* = \sum_{i=0}^{m} I^2[u_i]. \tag{9.24}$$

Here $u_0,\ldots,u_m$ denotes a basis for $\mathbb{P}_m$ satisfying (9.21).

PROOF. First, we show that $Q^* \in \mathbf{Q}^m(x_1,\ldots,x_n)$. We have

$$\begin{aligned} Q^*[u_\lambda] &= \sum_{\nu=1}^{n} a_\nu^* u_\lambda(x_\nu) = \sum_{\nu=1}^{n} u_\lambda(x_\nu)\sum_{i=0}^{m} u_i(x_\nu)I[u_i] \\ &= \sum_{i=0}^{m} I[u_i]\sum_{\nu=1}^{n} u_\lambda(x_\nu)u_i(x_\nu) = I[u_\lambda] \end{aligned}$$

for $\lambda \le m$. Now take $Q \in \mathbf{Q}^m(x_1,\ldots,x_n)$ with coefficients $a_1,\ldots,a_n$. Then

$$\sum_{\nu=1}^{n} a_\nu a_\nu^* = \sum_{\nu=1}^{n} a_\nu \sum_{i=0}^{m} u_i(x_\nu) I[u_i] = \sum_{i=0}^{m} I[u_i]\sum_{\nu=1}^{n} a_\nu u_i(x_\nu) = \sum_{i=0}^{m} I^2[u_i].$$

From this it follows that

$$\begin{aligned}\sum_{\nu=1}^{n} a_\nu^2 &= \sum_{\nu=1}^{n}(a_\nu - a_\nu^*)^2 + 2\sum_{\nu=1}^{n} a_\nu a_\nu^* - \sum_{\nu=1}^{n}(a_\nu^*)^2 \\ &= \sum_{\nu=1}^{n}(a_\nu - a_\nu^*)^2 + 2\sum_{i=0}^{m} I^2[u_i] - \sum_{i=0}^{m} I^2[u_i] \\ &= \sum_{\nu=1}^{n}(a_\nu - a_\nu^*)^2 + \sum_{i=0}^{m} I^2[u_i] \\ &= \sum_{\nu=1}^{n}(a_\nu - a_\nu^*)^2 + \operatorname{Var} Q^*,\end{aligned}$$

so uniqueness, minimality and (9.24) are verified. □

EXAMPLE 9.4.1. In the case where $[a, b] = [-1, 1]$, $x_\nu = -\cos\left(\frac{2\nu-1}{2n}\pi\right)$ and $m \le n-1$, we can choose

$$u_0 = \frac{1}{\sqrt{n}} T_0 \quad \text{and} \quad u_i = \sqrt{\frac{2}{n}}\, T_i \quad \text{for } i = 1, 2, \ldots.$$

Verification of (9.21) is done with the help of the Gauss–Chebyshev rule (see Example 5.1.1). We obtain

$$\operatorname{Var} Q = \frac{1}{n}\left(I^2[T_0] + 2\sum_{i=0}^{m} I^2[T_i]\right) = \frac{\pi}{n}\left(I^2\left[\frac{1}{\sqrt{\pi}} T_0\right] + 2\sum_{i=0}^{m} I^2\left[\sqrt{\frac{2}{\pi}}\, T_i\right]\right).$$

Bessel's inequality (Theorem A.8) for the expansion in Chebyshev polynomials gives

$$\operatorname{Var} Q \le \frac{\pi}{n}\int_{-1}^{1} w^2(x)\sqrt{1-x^2}\, dx, \tag{9.25}$$

provided the right-hand side exists.

For $m = n-1$, Q is of interpolatory type; for our special choice of nodes, it is the Polya rule (see Section 5.5). For $w = 1$, inequality (9.25) gives

$$\operatorname{Var} Q_n^{\text{Po}} \le \frac{\pi^2}{2n}.$$

We may characterize (9.22) in another way, which is apparently independent of the concept of variance.

THEOREM 9.4.3 (Wilson 1970a). *The rule (9.22) can be represented in the form*

$$Q^*[f] = I[u^*],$$

where $u^* \in \mathbb{P}_m$ is defined by

$$\sum_{\nu=1}^{n}\left[f(x_\nu) - u^*(x_\nu)\right]^2 = \inf\left\{\sum_{\nu=1}^{n}\left[f(x_\nu) - u(x_\nu)\right]^2 \,:\, u \in \mathbb{P}_m\right\}. \tag{9.26}$$

PROOF. Let $u_0, \ldots, u_m$ be the orthonormal system from the preceding proof. Then it is well known that

$$u^* = \sum_{i=0}^{m} \alpha_i u_i \quad \text{where } \alpha_i = \varphi(f, u_i).$$

Hence

$$I[u^*] = \sum_{i=0}^{m} \alpha_i I[u_i] = \sum_{i=0}^{m} I[u_i] \sum_{\nu=1}^{n} f(x_\nu) u_i(x_\nu) = \sum_{\nu=1}^{n} f(x_\nu) \sum_{i=0}^{m} I[u_i] u_i(x_\nu),$$

and (9.23) gives the result. □

EXAMPLE 9.4.2. In the most interesting case of Theorem 9.4.3, the nodes are equidistant. We have seen in Section 5.2 that, for this observation, integration of the interpolating polynomials leads to bad quadrature rules. It is therefore tempting to replace the interpolating polynomial by a polynomial based on the least squares approximation (9.26). The resulting rules have minimal variance, so we might expect that they will be useful, at least for small m.

For the special case where

$$I[f] = \int_0^1 f(x)\,dx, \qquad x_\nu = \frac{\nu - 1}{n-1}, \qquad m = 2,$$

after some calculations we find that

$$\alpha_i = \frac{1}{n} - \frac{5}{n(n-2)} + \frac{30i(n+1-i)}{(n-2)n(n+1)(n+2)}.$$

Upon specializing further to $n = 7$, we get

$$Q[f] = \frac{1}{12}[f(0) + f(1)] + \frac{1}{7}[f(1/6) + f(5/6)] + \frac{5}{28}[f(2/6) + f(4/6)] + \frac{4}{21} f(3/6),$$

which was first given in Sard (1949a). Duris (1971) presented some explicit rules for larger m.

CHAPTER 10

Problems

The field of quadrature theory is still very much alive, as we demonstrate in this chapter by compiling a dozen unsolved problems, all of which relate to the topics discussed earlier in the book.

Problem 1: Fast convergence

Let $(Q_n)_{n\in\mathbb{N}}$ be a convergent quadrature method of interpolatory type on $[-1,1]$ with respect to a weight function w that is positive almost everywhere. Setting

$$c_n := \sup\left\{\left|R_n[f]\right| \,:\, \|f^{(n)}\| \leq 1\right\},$$

we have the bound

$$\left|R_n[f]\right| \leq c_n\|f^{(n)}\|$$

where the constant c_n cannot be replaced by a smaller one for all functions having an nth derivative.

The number

$$u[(Q_n)] := \limsup_{n\in\mathbb{N}} \sqrt[n]{n!\, c_n}$$

is a measure of quality for the quadrature method. For positive methods of interpolatory type, it is easy to show that

$$u[(Q_n)] \leq \frac{1}{2}$$

(see Theorem 3.1.1 and Example 3.1.2).

PROBLEM. *Prove or disprove that*

$$u[(Q_n)] \geq \frac{1}{3\sqrt{3}}$$

for all quadrature methods.

We conjecture that the above inequality, whose right-hand side is attained by the Gaussian method as (6.62) shows, holds for all quadrature methods.

Some related references are Brass and Schmeisser (1981), Brass and Förster (1987) and Petras (1996b).

Problem 2: Universality of the Gaussian rule

We wish to extend Problem 1 to Peano kernels of more general order.

It was proved in Petras (1996b) that for the classes

$$\widetilde{\mathcal{C}}_M^{\mathrm{Sa},s} := \left\{ f : \int_{-1}^{1} \left[f^{(s)}(x)\right]^2 \left(1-x^2\right)^{s-1/2} dx \le M^2 \right\},$$

which are similar to the Sard classes, we have

$$\sup_{1\le s\le 2n} \frac{\rho(Q_n^{\mathrm{G}}, \widetilde{\mathcal{C}}_M^{\mathrm{Sa},s})}{\rho_n^{\mathrm{best}}(\widetilde{\mathcal{C}}_M^{\mathrm{Sa},s})} \le \frac{22}{3} n\sqrt{n+1}.$$

This is remarkable, given that the analogous quantity for rules such as the Clenshaw–Curtis rules, where the supremum is taken over $1 \le s \le n$ only, increases exponentially with n. Of course, these classes of functions are not of great practical relevance; they were chosen so as to make it possible to prove some kind of universality result (i.e. good behaviour in a wide range of classes). It would be much more valuable to establish an analogous result for the standard classes $\mathcal{C}_M^s$.

PROBLEM. *Prove that the quantities*

$$\sup_{1\le s\le 2n} \frac{\rho(Q_n^{\mathrm{G}}, \mathcal{C}_M^s)}{\rho_n^{\mathrm{best}}}$$

are bounded from above by a polynomial in n.

Problem 3: Overestimation for a wide range of functions

The best-known error bound for the Gaussian rule on $[-1,1]$ with $w=1$ is

$$\left|R_n^{\mathrm{G}}[f]\right| \le \frac{1}{2n+1} \cdot \frac{2^{2n+1}(n!)^4}{[(2n)!]^3} \cdot \|f^{(2n)}\| =: B_n[f].$$

But Example 6.6.1 showed that this could lead to drastic overestimation.

A conjecture, which if true would be a far-reaching generalization, is that this bound holds for any function that is analytic but not entire. More precisely, the question is stated as follows.

PROBLEM. *Prove that for every* f *which is analytic but not entire, there is a* $c<1$ *such that*

$$\left|R_n^{\mathrm{G}}[f]\right| \le c^n B_n[f]$$

for all n *(or at least for infinitely many* n*).*

The same conjecture can be stated for the Clenshaw–Curtis method and the error bound from Theorem 5.4.2. Additionally, one could ask whether there exist convergent interpolatory quadrature methods for which an analogous conjecture does not hold.

Problem 4:
A localized bound for the first Peano kernel

PROBLEM. *Determine whether or not the bound*

$$|K_1(x)| \leq \frac{\pi}{m+1}\sqrt{1-x^2} + \frac{8}{(m+1)(m+3)}$$

holds for the first Peano kernel K_1 of every positive quadrature rule with $\deg Q \geq m$ and weight function $w = 1$ on $[-1,1]$.

A slightly weaker result was stated in Theorem 5.7.2; see Brass (1993a), Petras (1992b) and DeVore and Scott (1984).

The factor $\pi/(m+1)$ is asymptotically the best possible, because the Peano kernel of Q^{G}_{2p-1} satisfies

$$K_1(0) = \frac{1}{2}\, a_p^{\mathrm{G}} < \frac{\pi}{4p-1}$$

by Corollary 6.3.2. The second summand in the conjectured bound is chosen in such a way that equality holds for $K_1(1-0)$, where K_1 is the Peano kernel of $Q^{\mathrm{Lo}}_{(m+3)/2}$ (cf. Theorem 3.4.2).

Problem 5:
A universal bound for Peano kernels

PROBLEM. *Prove or disprove the conjecture of Brass and Förster (1998) that if the functional R on $C[-1,1]$ has an sth Peano kernel, then the estimate*

$$|K_s(x)| \leq \frac{\left(1-x^2\right)^{s-1}}{2^s(s-1)!}\,\|R\|.$$

holds.

The form of the bound seems natural if one thinks of the $(s-1)$-fold zeros of K_s at the boundary. We have

$$\sup_x \frac{K_s(x)}{\left(1-x^2\right)^{s-1}} \geq \lim_{x\to 1-0} \frac{K_s(x)}{\left(1-x^2\right)^{s-1}} = \frac{1}{2^s(s-1)!}K_1(1-0),$$

where we have used the Bernoulli–l'Hospital rule. For any $\varepsilon > 0$, we can construct a functional R_ε (a suitable divided difference, for instance) which has a Peano kernel K_s and satisfies

$$K_1(1-0) \geq \left(\frac{1}{2} - \varepsilon\right)\|R_\varepsilon\|.$$

Then

$$\sup_x \frac{|K_s(x)|}{\left(1-x^2\right)^{s-1}} \geq \frac{1-2\varepsilon}{2^s(s-1)!}\,\|R\|,$$

which means that if the conjecture holds, it cannot be improved. It is not difficult to prove the conjecture for the cases of $s = 1, 2, 3$.

Problem 6:
$\|K_s\|$-extremality of the rule with smallest index

Markoff's theorem (Theorem 6.4.5) says that all $2n$th Peano kernels of rules in $\mathbf{Q}^+(I, \mathbb{P}_{2n-1})$ lie between those of Q_n^{G} and Q_{n+1}^{Lo}, and all $(2n-1)$st Peano kernels in $\mathbf{Q}^+(I, \mathbb{P}_{2n-2})$ lie between those of those of $Q_n^{\mathrm{Ra},a}$ and $Q_n^{\mathrm{Ra},b}$. In particular, the theorem implies that the supremum norms of Peano kernels are extremal for these Gauss, Lobatto and Radau rules. Furthermore, we know that for the norms of Peano kernels of lower order, the Krein rules are extremal among the rules in $\mathbf{Q}^+(I, \mathbb{P}_{2n-1})$ and $\mathbf{Q}^+(I, \mathbb{P}_{2n-2})$. It is natural to ask which of these Krein rules are extremal.

Problem. *For the weight function $w = 1$ on $[-1, 1]$, show that when n is odd,*

$$\sup_{Q \in \mathbf{Q}^+(I, \mathbb{P}_n)} \|K_s\|_\infty = \max\left\{\|K_s^{\mathrm{G}}\|_\infty, \|K_s^{\mathrm{Lo}}\|_\infty\right\},$$

where K_s^{G} is the sth Peano kernel of $Q_{(n+1)/2}^{\mathrm{G}}$ and K_s^{Lo} is the sth Peano kernel of $Q_{(n+3)/2}^{\mathrm{Lo}}$; and when n is even,

$$\sup_{Q \in \mathbf{Q}^+(I, \mathbb{P}_n)} \|K_s\|_\infty = \max\left\{\|K_s^{\mathrm{Ra},a}\|_\infty, \|K_s^{\mathrm{Ra},b}\|_\infty\right\},$$

where $K_s^{\mathrm{Ra},a}$ and $K_s^{\mathrm{Ra},b}$ are the sth Peano kernels of the respective Radau rules with $(n+2)/2$ nodes.

For illustration, define the functions

$$\underline{k}_{s,n}(x) := \inf_{Q \in \mathbf{Q}^+(I, \mathbb{P}_n)} K_s(x) \qquad \text{and} \qquad \overline{k}_{s,n}(x) := \sup_{Q \in \mathbf{Q}^+(I, \mathbb{P}_n)} K_s(x).$$

The following is a typical picture of the behaviour of these two functions and the Peano kernels of the Gaussian and Lobatto rules for odd n and $s > 2$.

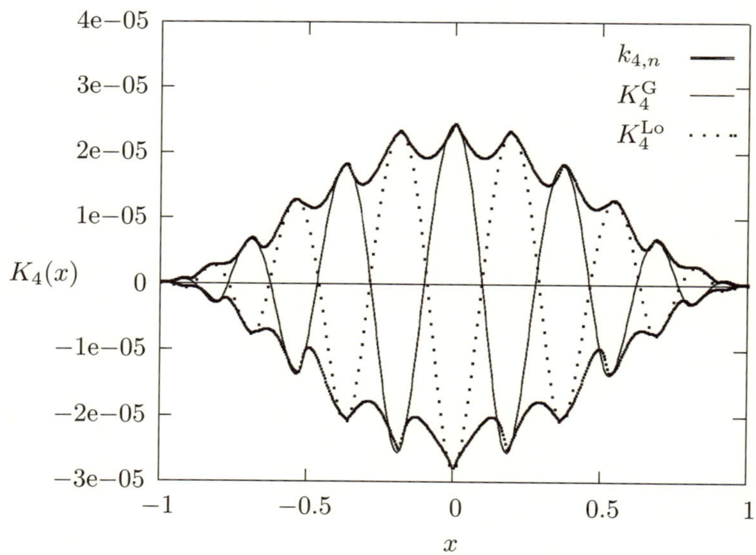

The case of $s = 4$ and $n = 8$.

For $\underline{k}_{1,n}(x)$, $\overline{k}_{1,n}(x)$ and $\overline{k}_{2,n}(x)$, we know that these values are attained by the Krein rule which has a node at x (for $\underline{k}_{1,n}(x)$ and $\overline{k}_{1,n}(x)$ as one-sided limits). Two typical situations are shown in the figures below.

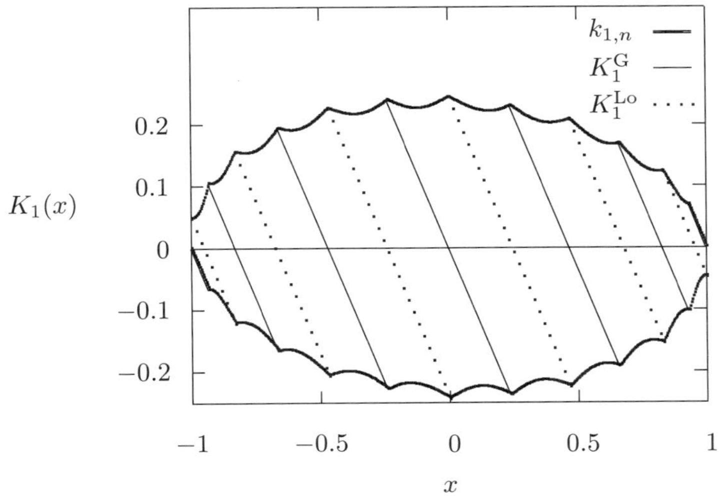

The case of $s = 1$ and $n = 6$.

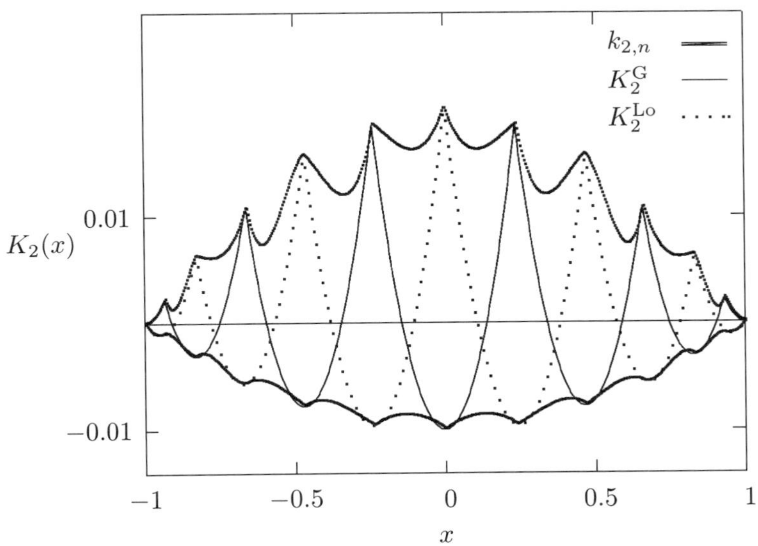

The case of $s = 2$ and $n = 6$.

It should be easier to prove the conjecture for these cases.

Problem 7: A counterpart of Theorem 6.5.3

In Theorem 6.5.3, the quantity

$$\sup\left\{\|K_s\|_\infty : Q \text{ is a positive quadrature rule of degree } s-1 \text{ or higher}\right\}$$

was estimated for even s.

PROBLEM. *Find similar bounds for odd s.*

It follows from Theorem 6.4.5 that the Radau rules are extremal for this problem. The missing symmetry, however, makes things more difficult for odd s than for even s.

Problem 8: Kronrod quadrature for special weight functions

PROBLEM. *For what values of α does the weight function $w_\alpha(x) = \left(1-x^2\right)^\alpha$ admit Kronrod quadrature? For which of these weight functions are Kronrod rules positive? Can these questions be answered at least for sufficiently large n?*

The functions $w_\alpha(x) = \left(1-x^2\right)^\alpha$ constitute the most intensively investigated class of weight functions in quadrature theory as well as in the related theory of orthogonal polynomials; see Szegö (1939).

The classical results in Kronrod quadrature theory are the existence of Kronrod quadrature rules for $\alpha \in [-1/2, 3/2]$ (Szegö 1935) and the positivity of Kronrod rules for $\alpha \in [-1/2, 1/2]$ (Monegato 1978a). It also follows from Szegö (1935) that for $n = 3$ and $\alpha < -1/2$, Kronrod quadrature is not possible. Moreover, it has been shown that for $\alpha = 5/2$ and sufficiently large n, Kronrod quadrature rules exist and are positive (Peherstorfer and Petras 2000; Petras 1999b). Finally, for $\alpha > 5/2$ and sufficiently large n, Kronrod quadrature does not exist (Peherstorfer and Petras 2000).

Thus, the "turning point" appears to be at $\alpha = 5/2$. If it could be shown that Kronrod quadrature behaves monotonically, in some sense, with respect to the parameter α, then one might conjecture that Kronrod quadrature exists and is positive for $\alpha \in [-1/2, 5/2]$, at least when n is sufficiently large.

Problem 9: The Gaussian error for Chebyshev polynomials

Let $w = 1$ and $[a,b] = [-1,1]$. The behaviour of the sequence $R_n^{\mathrm{G}}[T_k]$, $k = 1, 2, \ldots$, has attracted much interest over the years; see, for instance, Rabinowitz (1968), Nicholson *et al.* (1971), Curtis and Rabinowitz (1972), Brass and Förster (1987), Petras (1995), Hunter (1995) and Brass and Petras (1997). Hunter (1995) formulated the following remarkable conjecture:

$$\left|R_n^{\mathrm{G}}[T_k]\right| \le 2 \quad \text{for } n = 1, 2, \ldots \text{ and } k = 1, 2, \ldots.$$

However, for $n > 1$ there are counterexamples, albeit extremely rare (Brass and Petras 1997) . One counterexample for $n = 2$ is

$$R_2^{\mathrm{G}}[T_{1515982}] = -2.0000000000003388\ldots.$$

A modification of Hunter's conjecture leads to the following:

PROBLEM. *Show that there exists an* n_0 *such that*

$$\left|R_n^{\mathrm{G}}[T_k]\right| \leq 2 \quad \textit{for } k = 1, 2, \ldots$$

holds for $n > n_0$.

A possible candidate is $n_0 = 3$. For related results and further conjectures, see Hunter and Nikolov (1998).

Problem 10: Slow convergence

The sequence $\left(R_n^{\mathrm{Tr}}[f]\right)_{n\geq 2}$ tends to zero for each $f \in C[a, b]$. Can we say more; or, can every zero sequence $\left(\sigma_n\right)_{n\geq 2}$ be attained for an appropriate f? The latter is true at least if the sequence converges sufficiently fast: Brass (1979a) proved that if $\sigma_n = O(n^{-1-\varepsilon})$ for some $\varepsilon > 0$, then there exists an $f \in C[a, b]$ such that $R_n^{\mathrm{Tr}}[f] = \sigma_n$ for all n.

But error sequences can converge arbitrarily slowly—see Theorems 7.2.2 and 3.1.3 as well as Walsh and Sewell (1937), Filippi and Esser (1970), Lipow and Stenger (1972) and Schmetterer (1975)—although it is not known whether they always contain faster-converging partial sequences. We can formulate a concrete question as follows:

PROBLEM. *Is there an* $f \in C[a, b]$ *with*

$$R_n^{\mathrm{Tr}}[f] = n^{-1}, \qquad \textit{or at least } R_n^{\mathrm{Tr}}[f] \geq n^{-1},$$

for all $n \geq 2$?

Chui (1971) constructed such a function that is Riemann integrable but not continuous.

Problem 11: Chebyshev quadrature for special weight functions

PROBLEM. *For what values of* α *does the weight function* $w_\alpha(x) = \left(1 - x^2\right)^\alpha$ *admit Chebyshev methods in the narrower sense (as defined in Section 9.2)*

This seems quite a sensible question in view of the fact that the w_α form the most intensively studied class of weight functions in quadrature theory while the existence of Chebyshev methods in the narrower sense has been proved only for weight functions that do not (with the exception of $w_{-1/2}$) occur in other frameworks.

So far, non-existence has been proved for $\alpha > -1/2$ by Gatteschi (1963/64) and for $\alpha \in \,]-1, -0.8005\ldots]$ by Förster (1986a). Förster (1986a) also performed extensive numerical studies of the α-values in the gap, which led to the following conjecture:

There is an $\alpha_0 < -1/2$ *such that we have existence of Chebyshev quadrature for all* $\alpha \in [\alpha_0, -1/2]$.

A partial result was obtained by Kuijlaars (1995b).

Problem 12: A universal error bound for convex functions

Again take $w = 1$ and $[a, b] = [-1, 1]$. Theorem 4.10.2 implies that for the class $\mathcal{C}_M^{\mathrm{FP}}$, the error bound is given by M times the smallest positive coordinate h, so that the graph of the second Peano kernel lies in the quadrilateral with vertices $(-1, 0)$, $(0, -h)$, $(1, 0)$ and $(0, h)$. Drawing such quadrilaterals that enclose all graphs of second Peano kernels of quadrature rules from the class $\mathbf{Q}^+(I, \mathbb{P}_n)$ may lead us to conjecture that the descent of the edge from $(-1, 0)$ to $(0, -h)$ coincides with that of the second Peano kernel of the Lobatto rule for odd n and of the Radau rule $Q_{(n+2)/2}^{\mathrm{Ra},-1}$ for even n at -1. Here is one example for $n = 5$:

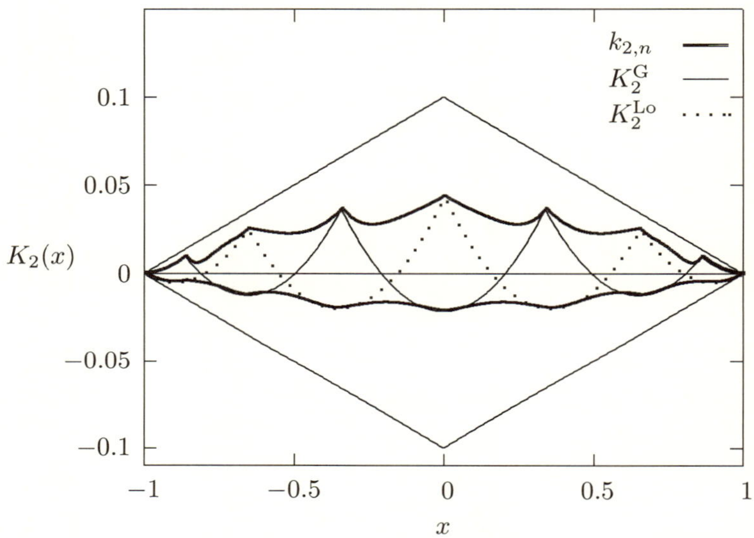

If this conjecture were to hold always, it would mean that the Lobatto and Radau rules are extremal (see Section 6.8) for the class $\mathcal{C}_M^{\mathrm{FP}}$ and that the largest error bound would be given by the coefficient of the Lobatto rule for the node -1. Since a_1 equals $\frac{8}{(n+3)(n+1)}$ for the Lobatto rule $Q_{(n+3)/2}^{\mathrm{Lo}}$ and $\frac{8}{(n+2)^2}$ for the Radau rule mentioned above, we can pose the following problem:

PROBLEM. *Prove that*

$$\sup_{Q \in \mathbf{Q}^+(I, \mathbb{P}_n)} \rho(Q, \mathcal{C}_M^{\mathrm{FP}}) = \rho(Q_{(n+3)/2}^{\mathrm{Lo}}, \mathcal{C}_M^{\mathrm{FP}}) = \frac{8M}{(n+3)(n+1)}$$

for odd n and

$$\sup_{Q \in \mathbf{Q}^+(I, \mathbb{P}_n)} \rho(Q, \mathcal{C}_M^{\mathrm{FP}}) = \rho(Q_{(n+2)/2}^{\mathrm{Ra},\pm 1}, \mathcal{C}_M^{\mathrm{FP}}) = \frac{8M}{(n+2)^2}$$

for even n.

APPENDIX A

Orthogonal Polynomials

There is an extensive theory of orthogonal polynomials; see, for example, Szegö (1939) or Freud (1969). In this appendix we review only those results that are needed in this book.

In what follows, W will be a non-negative weight function on $[a, b]$ with

$$\int_a^b W(x)\,dx > 0.$$

We use the abbreviation

$$I[g] := \int_a^b g(x)W(x)\,dx.$$

THEOREM A.1. There is a unique function q_n with the following properties:

(i) $q_n \in \mathbb{P}_n$;
(ii) $I[q_n p] = 0$ for all $p \in \mathbb{P}_{n-1}$ (if $n > 0$);
(iii) $I[q_n^2] = 1$;
(iv) the leading coefficient of q_n is positive.

PROOF. Let

$$q_n = \sum_{\nu=0}^{n} \alpha_\nu p_\nu \quad \text{where } p_\nu(x) := x^\nu.$$

Property (ii) has to be verified only for $p = p_i$, $i = 0, \ldots, n-1$. The condition then reads

$$\sum_{\nu=0}^{n} \alpha_\nu I[p_{\nu+i}] = 0 \quad \text{for } i = 0, \ldots, n-1.$$

These are n linear homogeneous equations for the $n+1$ unknowns $\alpha_0, \ldots, \alpha_n$. Hence there must be a non-trivial solution. A solution with $\alpha_n = 0$, i.e. $q_n \in \mathbb{P}_{n-1}$, would contradict (ii) upon setting $p = q_n$.

Now let q_n and $\hat{q}_n$ be polynomials having leading coefficients δ_n and $\hat{\delta}_n$ and which satisfy properties (i) and (ii). We set $g := \hat{\delta}_n q_n - \delta_n \hat{q}_n$ ($\in \mathbb{P}_{n-1}$). Then (ii) shows that $I[g^2] = 0$ and hence $\hat{\delta}_n q_n = \delta_n \hat{q}_n$. Therefore, (i) and (ii) determine q_n up to a constant factor and, obviously, this factor is given by (iii) and (iv). □

DEFINITION A.1. The polynomial q_n described in Theorem A.1 is called the nth orthogonal polynomial with respect to the weight function W. We will always denote its leading coefficient by δ_n.

The most important result for our purposes is the following:

THEOREM A.2. q_n has n distinct zeros in $]a, b[$.

PROOF. Denote by $x_1, \ldots, x_l$ the locations where q_n changes sign in $]a, b[$, and set

$$s(x) := \prod_{\nu=1}^{l} (x - x_\nu)$$

(with the convention that an empty product is set to equal 1). Suppose $l < n$. Then property (ii) in Theorem A.1 would imply $I[q_n s] = 0$, contradicting the fact that $q_n s$ has no sign changes. □

THEOREM A.3 (Three-term recurrence relation). There are constants A_n, B_n and C_n such that

$$q_n(x) = (A_n x + B_n) q_{n-1}(x) - C_n q_{n-2}(x).$$

In particular, we have

$$A_n = \frac{\delta_n}{\delta_{n-1}} \qquad \text{and} \qquad C_n = \frac{\delta_n \delta_{n-2}}{\delta_{n-1}^2}.$$

PROOF. Obviously, $q_0, q_1, \ldots, q_n$ form a basis for $\mathbb{P}_n$. Hence, there exists a representation of the form

$$x q_{n-1}(x) = \sum_{\nu=0}^{n} \alpha_\nu q_\nu(x).$$

Multiplying both sides by q_ν and applying I gives

$$\alpha_\nu = I[p_1 q_{n-1} q_\nu],$$

and hence $\alpha_\nu = 0$ for $\nu \leq n-3$. The formula $A_n = \delta_n / \delta_{n-1}$ follows directly from the definition of the leading coefficient, and for C_n we obtain

$$\begin{aligned} C_n &= C_n \cdot I[q_{n-2}^2] = I\big[(A_n p_1 q_{n-1} + B_n q_{n-1} - q_n) \cdot q_{n-2}\big] \\ &= A_n \cdot I[p_1 q_{n-1} q_{n-2}] = A_n \cdot \frac{\delta_{n-2}}{\delta_{n-1}} I[q_{n-1}^2] = A_n \cdot \frac{\delta_{n-2}}{\delta_{n-1}}. \end{aligned}$$

Therefore the theorem is proved. □

An important consequence is the following formula, which can be proved by induction.

THEOREM A.4 (Christoffel–Darboux formula).

$$\sum_{\nu=0}^{n} q_\nu(x) q_\nu(y) = \frac{\delta_n}{\delta_{n+1}} \frac{q_{n+1}(x) q_n(y) - q_n(x) q_{n+1}(y)}{x - y}.$$

Integration of the Christoffel–Darboux formula yields the formula below:

THEOREM A.5.

$$\int_a^b \frac{q_{n+1}(x) q_n(y) - q_n(x) q_{n+1}(y)}{x - y} W(x)\, dx = \frac{\delta_{n+1}}{\delta_n}.$$

A further consequence of the Christoffel–Darboux formula comes from considering the limit as $y \to x$:

THEOREM A.6.

$$\sum_{\nu=0}^{n}[q_\nu(x)]^2 = \frac{\delta_n}{\delta_{n+1}}\left(q'_{n+1}(x)q_n(x) - q'_n(x)q_{n+1}(x)\right).$$

One important application of the concept of orthogonality is, of course, the representation of functions as orthogonal series. If $f \in C[a,b]$ has a uniformly convergent series representation

$$f = \sum_{\nu=0}^{\infty} \alpha_\nu q_\nu, \tag{A.1}$$

then we necessarily have

$$\alpha_\nu = I[fq_\nu]. \tag{A.2}$$

This comes from multiplying (A.1) by q_ν and applying I. We say that

$$\sum_{\nu=0}^{\infty} I[fq_\nu]q_\nu$$

is the orthogonal polynomial expansion of f. However, checking uniform convergence can be a rather involved problem. If we do not insist on uniform convergence, the question of convergence becomes considerably simpler, as the next theorem shows.

THEOREM A.7. Suppose $I[f^2] < \infty$ and let the α_ν be as in (A.2). Then

$$\lim_{n\to\infty} I\left[\left(f - \sum_{\nu=0}^{n}\alpha_\nu q_\nu\right)^2\right] = 0.$$

PROOF. For arbitrarily chosen β_ν, we have

$$\begin{aligned}
I&\left[\left(f - \sum_{\nu=0}^{n}\beta_\nu q_\nu\right)^2\right]\\
&= I\left[\left(f - \sum_{\nu=0}^{n}\alpha_\nu q_\nu + \sum_{\nu=0}^{n}(\alpha_\nu - \beta_\nu)q_\nu\right)^2\right]\\
&= I\left[\left(f - \sum_{\nu=0}^{n}\alpha_\nu q_\nu\right)^2\right] + 2I\left[\left(f - \sum_{\nu=0}^{n}\alpha_\nu q_\nu\right)\sum_{\nu=0}^{n}(\alpha_\nu - \beta_\nu)q_\nu\right]\\
&\quad + I\left[\left(\sum_{\nu=0}^{n}(\alpha_\nu - \beta_\nu)q_\nu\right)^2\right].
\end{aligned}$$

For $\mu \le n$,

$$I\left[\left(f - \sum_{\nu=0}^{n}\alpha_\nu q_\nu\right)q_\mu\right] = I[fq_\mu] - \alpha_\mu = 0$$

and therefore

$$I\left[\left(f - \sum_{\nu=0}^{n}\beta_\nu q_\nu\right)^2\right] = I\left[\left(f - \sum_{\nu=0}^{n}\alpha_\nu q_\nu\right)^2\right] + \sum_{\nu=0}^{n}(\alpha_\nu - \beta_\nu)^2. \tag{A.3}$$

This means that

$$I\left[\left(f-\sum_{\nu=0}^{n}\alpha_\nu q_\nu\right)^2\right] \le I[(f-p)^2]$$

for every $p \in \mathbb{P}_n$. Using approximation theory, one then shows that the right-hand side can be arbitrarily close to zero for sufficiently large n and appropriately chosen p. □

Choosing $\beta_\nu = 0$ for all ν in (A.3) implies that

$$I\left[\left(f-\sum_{\nu=0}^{n}\alpha_\nu q_\nu\right)^2\right] = I[f^2] - \sum_{\nu=0}^{n}\alpha_\nu^2.$$

Combining this with Theorem A.7 gives the following identity.

THEOREM A.8 (Parseval's identity). If $I[f^2] < \infty$, then

$$I[f^2] = \sum_{\nu=0}^{\infty} I^2[fq_\nu].$$

The corollary

$$\sum_{\nu=0}^{n} I^2[fq_\nu] \le I[f^2]$$

is known as Bessel's inequality.

For our purposes, the most important systems of orthogonal polynomials are the ones with respect to the weight functions

$$\begin{aligned} W(x) &= \left(1-x^2\right)^{-1/2} && \text{on } [-1,1],\\ W(x) &= \left(1-x^2\right)^{1/2} && \text{on } [-1,1],\\ W(x) &= 1 && \text{on } [-1,1]. \end{aligned}$$

DEFINITION A.2. The function T_n defined on $[-1,1]$ by

$$T_n(x) := \cos(n \arccos x)$$

is called the nth Chebyshev polynomial (of the first kind).

We have $T_0(x) = 1$, $T_1(x) = x$ etc., and we will show that all the T_n are indeed polynomials. The definition can then be extended to the whole complex plane.

THEOREM A.9. Some properties of the Chebyshev polynomials are:

(i) T_n is a polynomial of degree n; for $n \ge 1$, its leading coefficient is 2^{n-1}.

(ii) $T_n(x) = 2xT_{n-1}(x) - T_{n-2}(x)$.

(iii) The zeros of T_n are $\xi_\nu = -\cos\left(\dfrac{2\nu-1}{2n}\pi\right)$, $\nu = 1, 2, \ldots, n$.

(iv) $\sup\limits_{x\in[-1,1]} |T_n(x)| = 1$.

(v) $\dfrac{1}{2}\left(\dfrac{T_{n+1}(x)}{n+1} - \dfrac{T_{n-1}(x)}{n-1}\right)' = T_n(x)$.

(vi) $T_n\left(\dfrac{1}{2}x + \dfrac{1}{2}x^{-1}\right) = \dfrac{1}{2}x^n + \dfrac{1}{2}x^{-n}$.

(vii) $T_n(x) = \dfrac{1}{2}\left[\left(x+\sqrt{x^2-1}\right)^n + \left(x-\sqrt{x^2-1}\right)^n\right]$ for $|x| \ge 1$.

PROOF. Properties (ii), (iii), (iv) and (v) follow immediately from the definition. Induction using (ii) gives (i) and (vi), and (vi) implies (vii). □

THEOREM A.10. The system of orthogonal polynomials with respect to $W(x) = (1-x^2)^{-1/2}$ on $[-1,1]$ consists of

$$q_0 = \frac{1}{\sqrt{\pi}} T_0, \qquad q_n = \sqrt{\frac{2}{\pi}}\, T_n \quad \text{for } n = 1, 2, \ldots$$

where

$$\delta_0 = \frac{1}{\sqrt{\pi}}, \qquad \delta_n = \sqrt{\frac{2}{\pi}}\, 2^{n-1} \quad \text{for } n = 1, 2, \ldots$$

PROOF. This can easily be shown by using the definition of T_n (and the substitution $x = \cos t$) together with Theorem A.9(i). □

Another important example of orthogonal polynomials that allow a simple explicit representation is related to the following special polynomials.

DEFINITION A.3. The polynomials

$$U_n := \frac{T'_{n+1}}{n+1}$$

are called Chebyshev polynomials of the second kind.

The theory of Chebyshev polynomials of the second kind can—like that for Chebyshev polynomials of the first kind—be based on trigonometric identities. To be specific, the definition of U_n gives

$$U_n(x) = \frac{\sin\big((n+1)\arccos x\big)}{\sqrt{1-x^2}} \quad \text{or, equivalently,} \quad U_n(\cos t) = \frac{\sin{(n+1)t}}{\sin t}.$$

Our interest in U_n arises from the following fact:

THEOREM A.11. The system of orthogonal polynomials with respect to the weight function $W(x) = (1-x^2)^{1/2}$ consists of

$$q_n = \sqrt{\frac{2}{\pi}}\, U_n \quad \text{with} \quad \delta_n = \sqrt{\frac{2}{\pi}}\, 2^n, \quad n = 0, 1, \ldots$$

PROOF. Orthogonality can be proved via the substitution $x = \cos t$, and the statement concerning the leading coefficients follows from the corresponding relation for T_{n+1}. □

We now turn to the most important weight function $W = 1$. As in the previous cases, we are mainly interested in certain constant multiples of the orthogonal polynomials.

DEFINITION A.4.

$$P_n(x) := \frac{1}{2^n n!} \left(\frac{d}{dx}\right)^n \left[\left(x^2-1\right)^n\right]$$

is called the nth Legendre polynomial.

THEOREM A.12. The system of orthogonal polynomials with respect to $W = 1$ on $[-1, 1]$ consists of

$$q_n = \sqrt{\frac{2n+1}{2}}\, P_n \quad \text{with} \quad \delta_n = \sqrt{\frac{2n+1}{2}}\, \frac{(2n)!}{2^n (n!)^2}, \qquad n = 0, 1, \ldots.$$

PROOF. The definition of P_n immediately tells us that P_n has degree n and leading coefficient $(2n)!/[2^n(n!)^2]$. Therefore, it remains to show that

$$\int_{-1}^{1} P_n(x) p(x)\, dx = 0 \quad \text{for all } p \in \mathbb{P}_{n-1} \tag{A.4}$$

and

$$\int_{-1}^{1} [P_n(x)]^2\, dx = \frac{2}{2n+1}. \tag{A.5}$$

The integral on the left-hand side of (A.4) can be transformed via repeated integration by parts to

$$\frac{(-1)^n}{2^n n!} \int_{-1}^{1} \left(x^2 - 1\right)^n p^{(n)}(x)\, dx. \tag{A.6}$$

For each integration, one uses the fact that

$$\left(\frac{d}{dx}\right)^{\nu} \left[\left(x^2 - 1\right)^n\right]$$

has a zero of order $n - \nu$ at ± 1. Assertion (A.4) then follows from $p^{(n)} = 0$.

To prove (A.5), note that taking $p = P_n$ in (A.6) yields

$$\begin{aligned}
\int_{-1}^{1} P_n(x) p(x)\, dx &= \frac{(-1)^n}{2^n n!} \int_{-1}^{1} \left(x^2 - 1\right)^n P_n^{(n)}(x)\, dx \\
&= \frac{(2n)!}{\left(2^n n!\right)^2} \int_{-1}^{1} \left(1 - x^2\right)^n dx \\
&= \frac{(2n)!}{\left(2^n n!\right)^2} \int_{0}^{\pi} \sin^{2n+1} t\, dt.
\end{aligned}$$

By using induction over n together with trigonometric identities, one can show that the integral on the right-hand side equals $2^{2n+1}(n!)^2/(2n+1)!$. This gives the normalization, and the asserted expression for δ_n is an immediate consequence. □

We can now specialize Theorem A.3 to the case of $W = 1$. For this purpose, note that P_n is odd for odd n and even for even n; hence $B_n = 0$ for all n. The coefficients A_n and C_n are found by applying Theorem A.12.

THEOREM A.13. The Legendre polynomials satisfy $P_0(x) = 1$, $P_1(x) = x$ and

$$n P_n(x) = (2n - 1) x P_{n-1}(x) - (n - 1) P_{n-2}(x) \quad \text{for } n \geq 2.$$

A first consequence is that

$$P_n(1) = 1 \quad \text{for all } n. \tag{A.7}$$

Many relations exist among the Legendre polynomials. The next theorem lists some of these.

THEOREM A.14.

(i) $(1-x^2)P_n''(x) - 2xP_n'(x) + n(n+1)P_n(x) = 0$.

(ii) $(1-x^2)P_n'(x) = n\big(P_{n-1}(x) - xP_n(x)\big)$.

(iii) $(1-x^2)P_n'(x) = (n+1)\big(xP_n(x) - P_{n+1}(x)\big)$.

(iv) $(1-x^2)P_n'(x) = \dfrac{n(n+1)}{2n+1}\big(P_{n-1}(x) - P_{n+1}(x)\big)$.

(v) $P_n'(x) - P_{n-2}'(x) = (2n-1)P_{n-1}(x)$.

(vi) $nP_n(x) = xP_n'(x) - P_{n-1}'(x)$.

(vii) $(1 \pm x)\big(P_n'(x) \mp P_{n-1}'(x)\big) = \pm n\big(P_n(x) \pm P_{n-1}(x)\big)$.

PROOF. We shall prove (i), (ii) and (vi). The other relations can be deduced from these three with the aid of the three-term recurrence relation (Theorem A.13).

(i) For any $p \in \mathbb{P}_{n-1}$, we have

$$\int_{-1}^{1} \big[(1-x^2)P_n'(x)\big]' p(x)\,dx = -\int_{-1}^{1} (1-x^2)P_n'(x)p'(x)\,dx = \int_{-1}^{1} P_n(x)\big[(1-x^2)p'(x)\big]'\,dx = 0.$$

Therefore

$$\big[(1-x^2)P_n'(x)\big]' = \alpha P_n(x) \quad \text{for some } \alpha \in \mathbb{R}.$$

Comparing the leading coefficients gives $\alpha = -n(n+1)$ and hence the claimed identity.

(ii) $(1-x^2)P_n'(x) + nxP_n(x)$ is a polynomial of degree at most $n-1$, because the coefficient of x^{n+1} vanishes (compare the leading coefficients) and, owing to symmetry, the monomial x^n is not contained in both summands. Furthermore, for any $p \in \mathbb{P}_{n-2}$, we have

$$\int_{-1}^{1} \big[(1-x^2)P_n'(x) + nxP_n(x)\big]p(x)\,dx = \int_{-1}^{1} P_n'(x)(1-x^2)p(x)\,dx = -\int_{-1}^{1} P_n(x)\big[(1-x^2)p(x)\big]'\,dx = 0.$$

Hence

$$(1-x^2)P_n'(x) + nxP_n(x) = \alpha P_{n-1}(x) \quad \text{for some } \alpha \in \mathbb{R}.$$

Setting $x = 1$ gives $\alpha = n$.

(vi) For any $p \in \mathbb{P}_{n-1}$, we have

$$\int_{-1}^{1} \big[P_n(x) + xP_n'(x) - P_{n-1}'(x)\big]p(x)\,dx = \int_{-1}^{1} \big[\big(xP_n(x)\big)' - P_{n-1}'(x)\big]p(x)\,dx = -\int_{-1}^{1} \big[xP_n(x) - P_{n-1}(x)\big]p'(x)\,dx = 0.$$

Therefore,

$$P_n(x) + xP_n'(x) - P_{n-1}'(x) = \alpha P_n(x) \quad \text{for some } \alpha \in \mathbb{R}.$$

Comparing the leading coefficients gives $\alpha = n+1$, which yields the statement. □

We need another formula of a different type. Writing

$$g_0 := 1, \qquad g_\nu := \frac{1 \cdot 3 \cdots (2\nu-1)}{2 \cdot 4 \cdots (2\nu)} \quad \text{for } \nu \geq 1,$$

we have the following expression for P_n.

THEOREM A.15.

$$P_n(x) = \sum_{\nu=0}^{\lfloor n/2 \rfloor}{}^{*}\, 2g_\nu g_{n-\nu} T_{n-2\nu}(x),$$

where $\sum^*$ means that the last summand has to be halved if n is even.

PROOF. The proof is by induction over n with help from the three-term recurrence relations for the Legendre polynomials and the Chebyshev polynomials (in the form $2xT_{n-2\nu}(x) = T_{n-2\nu+1}(x) + T_{n-2\nu-1}(x)$). The details are left to the reader. □

A simple consequence of this identity is that

$$\sup_{-1\le x\le 1} |P_n(x)| \le \sum_{\nu=0}^{\lfloor n/2 \rfloor}{}^{*}\, 2g_\nu g_{n-\nu} = P_n(1) = 1.$$

The next theorem is obtained by setting $x = (z^2+1)/(2z)$ in Theorem A.15 and using Theorem A.9(vi).

THEOREM A.16.

$$P_n\left(\frac{z^2+1}{2z}\right) = z^n \sum_{\nu=0}^{n} g_\nu g_{n-\nu} z^{-2\nu}.$$

The asymptotic behaviour of the Legendre polynomials is described in the following theorem.

THEOREM A.17. As $n \to \infty$, we have

$$P_n\left(\frac{x^2+1}{2x}\right) = \frac{1+o(1)}{\sqrt{n\pi}\, x^n \sqrt{1-x^2}} \quad \text{for } |x| < 1$$

and

$$P_n(c) = \frac{\left(c+\sqrt{c^2-1}\right)^{n+1/2}}{\sqrt{2n\pi}\,\sqrt[4]{c^2-1}} \cdot (1+o(1)) \quad \text{for } c > 1. \tag{A.8}$$

PROOF. By a standard argument, it can be shown that if $\lim_{n\to\infty} \alpha_\nu^{(n)} = \alpha_\nu$ and $\sup_{\nu,n} |\alpha_\nu^{(n)}| < \infty$, then

$$\lim_{n\to\infty} \sum_{\nu=0}^{\infty} \alpha_\nu^{(n)} x^\nu = \sum_{\nu=0}^{\infty} \alpha_\nu x^\nu.$$

We apply this result to Theorem A.16 in the form

$$x^n \sqrt{n\pi}\, P_n\left(\frac{x^2+1}{2x}\right) = \sum_{\nu=0}^{n} g_{n-\nu} \sqrt{n\pi}\, g_\nu x^{2\nu}.$$

The necessary assumptions follow from Wallis's inequality

$$\sqrt{\frac{2}{\pi(2\nu+1)}} \le g_\nu \le \sqrt{\frac{1}{\pi\nu}}.$$

Thus we obtain

$$\lim_{n\to\infty} x^n \sqrt{n\pi}\, P_n\left(\frac{x^2+1}{2x}\right) = \sum_{\nu=0}^{\infty} g_\nu x^{2\nu} = \sum_{\nu=0}^{\infty}(-1)^\nu \binom{-1/2}{\nu} x^{2\nu} = \frac{1}{\sqrt{1-x^2}},$$

which is the first assertion. The second assertion follows upon taking $x = c - \sqrt{c^2-1}$. □

APPENDIX B

Bernoulli Polynomials

DEFINITION B.1. The polynomials B_ν defined by

$$B_0(x) = 1, \tag{B.1}$$

$$B'_{\nu+1}(x) = B_\nu(x) \quad \text{for } \nu = 0, 1, \ldots, \tag{B.2}$$

$$\int_0^1 B_\nu(x)\,dx = 0 \quad \text{for } \nu = 1, 2, \ldots \tag{B.3}$$

are called the Bernoulli polynomials.

Some easy calculations give

$$B_1(x) = x - \frac{1}{2},$$
$$B_2(x) = \frac{1}{2}x^2 - \frac{1}{2}x + \frac{1}{12},$$
$$B_3(x) = \frac{1}{6}x^3 - \frac{1}{4}x^2 + \frac{1}{12}x.$$

We are mainly interested in the behaviour of B_ν on the interval $[0, 1]$. The first sequence of theorems will show that for $\nu > 0$, $B_{2\nu}$ restricted to $[0, 1]$ has roughly the shape of $(-1)^{\nu+1}\cos(2\pi\,\cdot)$ while $B_{2\nu+1}$ has the shape of $(-1)^{\nu+1}\sin(2\pi\,\cdot)$.

THEOREM B.1. $B_\nu(x) = (-1)^\nu B_\nu(1 - x)$.

PROOF. $\widetilde{B}_\nu(x) := (-1)^\nu B_\nu(1 - x)$ satisfies (B.1), (B.2) and (B.3). □

THEOREM B.2. $B_\nu(1) = B_\nu(0)$ if $\nu \geq 2$.

PROOF.

$$B_\nu(1) - B_\nu(0) = \int_0^1 B_{\nu-1}(x)\,dx = 0$$

by (B.3). □

THEOREM B.3. $B_{2\nu+1}(0) = B_{2\nu+1}(1/2) = B_{2\nu+1}(1) = 0$ if $\nu > 0$.

PROOF. Theorem B.1 says that

$$B_{2\nu+1}(x) = -B_{2\nu+1}(1 - x).$$

For $x = 1/2$, this gives $B_{2\nu+1}(1/2) = 0$. For $x = 0$, we have $B_{2\nu+1}(0) = -B_{2\nu+1}(1)$. Together with Theorem B.2, this yields $B_{2\nu+1}(0) = B_{2\nu+1}(1) = 0$. □

There is no simple expression for $B_{2\nu}(0)$, but we can at least determine the sign of $B_{2\nu}(0)$:

THEOREM B.4. $B_{2\nu}(0) = (-1)^{\nu-1}\displaystyle\int_0^1 \left[B_\nu(x)\right]^2 dx \quad$ for $\nu \geq 1$.

PROOF. Repeated integration by parts together with Theorem B.2 gives

$$\int_0^1 \left[B_\nu(x)\right]^2 dx = (-1)^{\nu-1}\int_0^1 B_{2\nu-1}(x)B_1(x)\,dx.$$

Another integration by parts, this time using $B_1(1) = -B_1(0) = 1/2$ and (B.3), completes the proof. □

THEOREM B.5. For positive ν, $B_{2\nu+1}$ has exactly one and $B_{2\nu}$ exactly two simple zeros in $]0,1[$.

PROOF. Suppose that $B_{2\nu+1}$ has two zeros in $]0,1[$, i.e. four zeros in $[0,1]$. Then, according to Rolle's theorem, $B_{2\nu}$ has three zeros in $]0,1[$ and therefore $B_{2\nu-1}$ has two zeros in $]0,1[$. On repeating this argument we get that B_2 has three zeros, which is impossible. In a similar way, one can see that $B_{2\nu}$ having more than two zeros in $]0,1[$ will also lead to a contradiction; on the other hand, Rolle's theorem applied to Theorem B.3 implies that there are at least two zeros of $B_{2\nu}$ in $]0,1[$. □

Many further results require a periodic continuation of Bernoulli polynomials.

DEFINITION B.2. Let $B_\nu^*(x) := B_\nu(x - \lfloor x \rfloor)$.

Evidently, B_ν^* is not a polynomial but rather a 1-periodic function. While B_1^* is not continuous, it can easily be shown that $B_\nu^* \in C^{(\nu-2)}$. The relations (B.1) and (B.3) remain valid with B_ν^* in place of B_ν; (B.2) holds for B_ν^* as well, except when $\nu \in \{0,1\}$ and $x \in \mathbb{Z}$.

The first situation where B_ν^* comes in handy is the Bernoulli polynomial expansion.

THEOREM B.6. If $g^{(s-1)}$ is absolutely continuous and $y \in [0,1]$, then we have

$$g(y) = \sum_{\mu=0}^{s} B_\mu(y)\int_0^1 g^{(\mu)}(u)\,du - (-1)^s \int_0^1 B_s^*(u-y)g^{(s)}(u)\,du.$$

PROOF. We use the definition of B_s^* and integration by parts to obtain

$$\begin{aligned}
\int_0^1 B_s^*(u-y)g^{(s)}(u)\,du &= \int_0^y B_s^*(1+u-y)g^{(s)}(u)\,du + \int_y^1 B_s^*(u-y)g^{(s)}(u)\,du \\
&= B_s(1+u-y)g^{(s-1)}(u)\Big|_0^y + B_s(u-y)g^{(s-1)}(u)\Big|_y^1 \\
&\quad - \int_0^1 B_{s-1}^*(u-y)g^{(s-1)}(u)\,du \\
&= \big[B_s(1) - B_s(0)\big]g^{(s-1)}(y) \\
&\quad + (-1)^s\big[g^{(s-1)}(1) - g^{(s-1)}(0)\big]B_s(y) \\
&\quad - \int_0^1 B_{s-1}^*(u-y)g^{(s-1)}(u)\,du.
\end{aligned} \tag{B.4}$$

For $s \neq 1$, the first summand on the right-hand side vanishes, so that

$$\begin{aligned}(-1)^s &\int_0^1 B_s^*(u-y) g^{(s)}(u)\,du \\ &= \big[g^{(s-1)}(1) - g^{(s-1)}(0)\big] B_s(y) - (-1)^s \int_0^1 B_{s-1}^*(u-y) g^{(s-1)}(u)\,du.\end{aligned}$$

By repeatedly applying this identity, we obtain

$$\begin{aligned}(-1)^s &\int_0^1 B_s^*(u-y) g^{(s)}(u)\,du \\ &= \sum_{\mu=2}^{s} B_\mu(y) \int_0^1 g^{(\mu)}(u)\,du - \int_0^1 B_1^*(u-y) g'(u)\,du\end{aligned}$$

Taking $s = 1$ in (B.4) gives

$$\int_0^1 B_1^*(u-y) g'(u)\,du = g(y) - [g(1) - g(0)] B_1(y) - \int_0^1 g(u)\,du,$$

and substituting this into the previous equation gives our assertion. □

Using in Theorem B.6 an affine transformation $g(y) = f\big(c + y(d-c)\big)$ of the fundamental interval $[0,1]$ onto $[c,d]$ gives

$$\text{(B.5)} \qquad \begin{aligned} f(z) = &\sum_{\mu=0}^{s} (d-c)^{\mu-1} B_\mu\left(\frac{z-c}{d-c}\right) \int_c^d f^{(\mu)}(u)\,du \\ &- (-1)^s (d-c)^{s-1} \int_c^d B_s^*\left(\frac{u-z}{d-c}\right) f^{(s)}(u)\,du \end{aligned}$$

on $[c,d]$ for absolutely continuous $f^{(s-1)}$ and $z \in [c,d]$.

Fourier series are used as an important tool for deeper investigation of B_ν^*. The calculation of Fourier coefficients using integration by parts is quite simple. All the B_ν^* are of bounded variation on one period. We therefore have uniform convergence of the Fourier series except at discontinuities, which exist only for B_1^*. Thus, we have the following result.

THEOREM B.7. For $s \geq 1$ we have

$$\begin{aligned} B_{2s+1}^*(x) &= 2(-1)^{s+1} \sum_{\nu=1}^{\infty} \frac{\sin 2\nu\pi x}{(2\nu\pi)^{2s+1}}, \\ B_{2s}^*(x) &= 2(-1)^{s+1} \sum_{\nu=1}^{\infty} \frac{\cos 2\nu\pi x}{(2\nu\pi)^{2s}}. \end{aligned}$$

The first representation also holds for $s = 0$ and $x \notin \mathbb{Z}$.

As a first application of this theorem, we derive a connection between the Favard constants

$$\mathcal{K}_s := \frac{4}{\pi} \sum_{\nu=0}^{\infty} \frac{(-1)^{\nu(s+1)}}{(2\nu+1)^{s+1}}$$

(used in Theorems 4.3.4 and 8.2.1) and the Bernoulli polynomials.

THEOREM B.8. For $s > 0$ we have

$$\begin{aligned}\mathcal{K}_{2s-1} &= (2\pi)^{2s-1}\int_0^1 |B_{2s-1}(u)|\,du,\\ \mathcal{K}_{2s} &= (2\pi)^{2s}\int_0^1 \left|B_{2s}(u) - B_{2s}\left(\frac{1}{4}\right)\right|\,du.\end{aligned}$$

PROOF. We prove only the first assertion:

$$\begin{aligned}\int_0^1 |B_{2s-1}(u)|\,du &= 2\left|\int_0^{1/2} B_{2s-1}(u)\,du\right| = 2\left|B_{2s}(0) - B_{2s}\left(\frac{1}{2}\right)\right|\\ &= 4\left|\sum_{\nu=1}^{\infty}\frac{1-\cos\nu\pi}{(2\nu\pi)^{2s}}\right| = \frac{1}{(2\pi)^{2s-1}}\cdot\frac{4}{\pi}\sum_{\nu=0}^{\infty}\frac{1}{(2\nu+1)^{2s}}.\end{aligned}$$

The proof of the assertion for $\mathcal{K}_{2s}$ is similar. □

APPENDIX C

Validation of Co-observations

Throughout this book, we often use error estimates that involve bounds for derivatives or function values in the complex plane. One might believe that such bounds are not readily available. However, if the integrand is made up of elementary functions, and sometimes also special functions, combined with arithmetic operations, then the task of finding bounds can be left to the computer. Since the computation of such bounds is fundamental to the application of many of the results in this text, here we give a brief overview of the three main techniques.

C.1. Automatic generation of Taylor coefficients

This technique is part of a field called "automatic differentiation"; see, for instance, Griewank (2000). The aim is to compute higher derivatives efficiently, and this is done by developing an arithmetic for partial sums of Taylor expansions,

$$\mathcal{T}_n[f;x_0](x) := \sum_{\nu=0}^{n} t_\nu \cdot (x-x_0)^\nu \quad \text{where } t_\nu := t_\nu[f;x_0] := \frac{f^{(\nu)}(x_0)}{\nu!}$$

(if they exist). Given an expression for $f(x)$, we interpret each occurrence of x as $\mathrm{id}(x)$ where id is the identity, with

$$\mathcal{T}_n[\mathrm{id};x_0](x) = x_0 + (x-x_0) \quad \text{for } n \geq 1,$$

i.e. $t_0 = x_0$, $t_1 = 1$ and $t_2 = t_3 = \cdots = 0$.

Addition and subtraction formulas are particularly simple:

$$t_\nu[f \pm g;x_0] = t_\nu[f;x_0] \pm t_\nu[g;x_0].$$

A formula for multiplication follows from Leibniz's rule:

$$\begin{aligned}
t_\nu[f\cdot g;x_0] &= \frac{1}{\nu!}(f\cdot g)^{(\nu)}(x_0) \\
&= \frac{1}{\nu!}\sum_{\mu=0}^{\nu}\binom{\nu}{\mu} f^{(\mu)}(x_0)\cdot g^{(\nu-\mu)}(x_0) \\
&= \sum_{\mu=0}^{\nu}\frac{f^{(\mu)}(x_0)}{\mu!}\cdot\frac{g^{(\nu-\mu)}(x_0)}{(\nu-\mu)!} \\
&= \sum_{\mu=0}^{\nu} t_\mu[f;x_0]\cdot t_{\nu-\mu}[g;x_0].
\end{aligned}$$

As we can see, here and in the cases below, writing a formula in terms of Taylor coefficients is usually simpler than expressing the same formula in terms of derivatives. Division is slightly more complicated. Let $h = f/g$, i.e. $f = g \cdot h$, and suppose that the Taylor coefficients of f and g are known and that we have already calculated the numbers $t_0[h;x_0],\ldots,t_{\nu-1}[h;x_0]$. The multiplication formula gives

$$t_\nu[f;x_0] = \sum_{\mu=0}^{\nu} t_\mu[g;x_0]\cdot t_{\nu-\mu}[h;x_0].$$

Solving for the first summand on the right-hand side and dividing by $t_0[g;x_0] = g(x_0)$ then gives the following result.

THEOREM C.1. For $h(x) = f(x)/g(x)$, the recursion formula that allows us to derive arbitrary-order Taylor coefficients of h from those of f and g is

$$t_\nu[h;x_0] = \frac{1}{t_0[g;x_0]}\Big(t_\nu[f;x_0] - \sum_{\mu=1}^{\nu} t_\mu[g;x_0]\cdot t_{\nu-\mu}[h;x_0]\Big).$$

For elementary functions, as well as some special functions, calculations using Taylor coefficients are quite easy to do. For example, let $f(x) = e^{g(x)}$ and suppose that the Taylor coefficients of g are known. From the chain rule we have $f'(x) = g'(x)e^{g(x)} = g'(x)f(x)$; therefore

$$t_0[f;x_0] = e^{t_0[g;x_0]} \qquad\text{and}\qquad t_1[f;x_0] = t_1[g;x_0]\cdot t_0[f;x_0].$$

Furthermore, for $\nu > 1$, Leibniz's rule shows that

$$\frac{f^{(\nu)}}{\nu!} = \frac{(g'f)^{(\nu-1)}}{\nu!} = \frac{1}{\nu!}\sum_{\mu=0}^{\nu-1}\binom{\nu-1}{\mu} g^{(\nu-\mu)}f^{(\mu)} = \frac{1}{\nu}\sum_{\mu=0}^{\nu-1}(\nu-\mu)\frac{g^{(\nu-\mu)}}{(\nu-\mu)!}\cdot\frac{f^{(\mu)}}{\mu!}.$$

After shifting the index of summation, we obtain the following formula.

THEOREM C.2. For $f(x) = e^{g(x)}$, we have

$$t_\nu[f;x_0] = \frac{1}{\nu}\sum_{\mu=1}^{\nu}\mu\cdot t_\mu[g;x_0]\cdot t_{\nu-\mu}[f;x_0].$$

Recall that in Example 6.5.1 the integrand contained a logarithm. We now demonstrate how easy it is to obtain Taylor coefficients of such inverse functions. For $f(x) = \ln g(x)$, we have $g(x) = e^{f(x)}$ and can therefore apply Theorem C.2 to get

$$t_\nu[g;x_0] = \frac{1}{\nu}\sum_{\mu=1}^{\nu}\mu\cdot t_\mu[f;x_0]\cdot t_{\nu-\mu}[g;x_0].$$

For $\nu \geq 1$, we can then solve for $t_\nu[f;x_0]$.

THEOREM C.3. For $f(x) = \ln\big(g(x)\big)$, we have $t_0[f;x_0] = \ln t_0[g;x_0]$ and

$$t_\nu[f;x_0] = \frac{1}{t_0[g;x_0]}\Big(t_\nu[g;x_0] - \frac{1}{\nu}\sum_{\mu=1}^{\nu-1}\mu\cdot t_\mu[f;x_0]\cdot t_{\nu-\mu}[g;x_0]\Big)$$

for $\nu \geq 1$.

Other elementary functions can be treated similarly, since they can all be thought of as solutions of simple differential equations. A table of the resulting automatic differentiation formulas is given in Krämer *et al.* (2009). For arithmetic operations and elementary functions, the computation of all Taylor coefficients up to order d consists of zero, one or two evaluations of elementary functions plus at most $O(d^2)$ arithmetic operations. This $O(d^2)$ complexity holds for all functions that are composed of elementary functions and arithmetic operations.

Many higher transcendental functions satisfy differential equations of higher order. These can also be used to obtain recursion formulas for the Taylor coefficients, based on knowing Taylor coefficients of order lower than that of the differential equation.

C.2. Real interval arithmetic

In this section, we show how a computer can be used to obtain bounds for a real function on intervals. In the context of this book, we wish to not merely find bounds for the function itself but also combine our method with the techniques of the previous section to obtain bounds for the function's derivatives. Typical co-observations are defined by such bounds, so we want a method for checking automatically by computer whether any explicitly given function belongs to a particular co-observation.

With this purpose in mind, we first define an arithmetic on intervals. If $x \in [a, b]$ and $y \in [c, d]$, then

$$x + y \in [a + c,\ b + d], \qquad x - y \in [a - d,\ b - c],$$

$$x \cdot y \in \big[\min\{ac, ad, bc, bd\},\ \max\{ac, ad, bc, bd\}\big],$$

and

$$\frac{1}{x} \begin{cases} \in \left[\min\left\{\frac{1}{a}, \frac{1}{b}\right\}, \max\left\{\frac{1}{a}, \frac{1}{b}\right\}\right] & \text{if } 0 \notin [a, b], \\ \text{is not always defined} & \text{if } 0 \in [a, b]. \end{cases}$$

All values in the right-hand side intervals may occur as x and y run through all values in $[a, b]$ and $[c, d]$, respectively. We therefore define

$$[a, b] + [c, d] := [a + c,\ b + d], \quad [a, b] - [c, d] := [a - d,\ b - c],$$

$$[a, b] \cdot [c, d] := \big[\min\{ac, ad, bc, bd\},\ \max\{ac, ad, bc, bd\}\big],$$

and

$$\frac{1}{[a, b]} := \begin{cases} \left[\min\left\{\frac{1}{a}, \frac{1}{b}\right\}, \max\left\{\frac{1}{a}, \frac{1}{b}\right\}\right] & \text{if } 0 \notin [a, b], \\ \text{not an interval} & \text{if } 0 \in [a, b]. \end{cases}$$

If "not an interval" occurs as an intermediate result in an interval computation, we define the final result as "not an interval". Furthermore, we replace any number a by the interval $[a, a]$ and set

$$\frac{[a, b]}{[c, d]} := [a, b] \cdot \frac{1}{[c, d]}.$$

This completes the definition of arithmetic operations on intervals.

Monotone functions such as exponential functions or logarithms are easily applied to intervals:

$$e^{[a,b]} := [e^a,\ e^b] \qquad \text{and} \qquad \ln\,[a, b] := [\ln a,\ \ln b],$$

where the latter is defined if and only if $a > 0$. For non-monotonic functions such as sin, cos or cosh, as well as for functions with singularities such as tan or cot, we have to do case distinctions. For example, to calculate $\cos[a, b]$, we check:

(1) whether a multiple of 2π is in $[a, b]$. If so, then $\cos[a, b]$ is of the form $[c, 1]$; if not, then the result is $\big[c, \max\{\cos a, \cos b\}\big]$ for some c.
(2) whether an odd multiple of π is in $[a, b]$. If so, then $\cos[a, b]$ is of the form $[-1, c]$; if not, then the result is $\big[\min\{\cos a, \cos b\}, c\big]$ for some c.

By combining (1) and (2) we get a complete description of the cosine of an interval. Similarly, $\tan[a, b]$ is the bounded interval $[\tan a,\ \tan b]$ if and only if no multiple of π is in $[a, b]$, otherwise it is "not an interval".

Note that, in general, a computer will produce rounding errors if it computes the bounds of intervals with a floating point arithmetic. We have to ensure that the result of an interval operation or function on a computer really does contain the true resulting interval. This is achieved by setting the so-called rounding mode of the computer in a certain way; then we will always be on the safe side and can compute guaranteed bounds.

It is also possible to deal with unbounded intervals. This can still yield finite bounds in the end, as we can see from the example

$$\frac{1}{1 + e^{[1,\infty]}} = [0,\ 1/(1 + e)].$$

However, we will keep things simple here and not delve further into the details.

Finally, we demonstrate how to combine real interval arithmetic with the Taylor coefficients of the previous section. We simply replace all real arguments by intervals and thus get enclosures for Taylor coefficients.

EXAMPLE C.1. We shall compute bounds for $f(x) = e^{x/(1+x^2)}$ and its first derivative on $[1, 2]$. Applying the rules of the present and preceding sections, we obtain

$$
\begin{aligned}
g(x) &= x && \Longrightarrow\ t_0[g; [1,2]] = [1,2]\\
& && \ t_1[g; [1,2]] = [1,1]\\
h(x) &= x^2 = g(x)\cdot g(x) && \Longrightarrow\ t_0[h; [1,2]] = [1,2]\cdot[1,2] = [1,4]\\
& && \ t_1[h; [1,2]] = [1,1]\cdot[1,2] + [1,2]\cdot[1,1] = [2,4]\\
i(x) &= 1 + x^2 = 1 + h(x) && \Longrightarrow\ t_0[i; [1,2]] = [2,5]\\
& && \ t_1[i; [1,2]] = t_1[h; [1,2]] = [2,4]\\
j(x) &= x/(1+x^2) && \Longrightarrow\ t_0[j; [1,2]]\\
&= g(x)/i(x) && \ = [1,2]/[2,5] = [1,2]\cdot[1/5, 1/2] = [1/5, 1]\\
& && \ t_1[j; [1,2]]\\
& && \ = (t_1[g; [1,2]] - t_1[i; [1,2]]\cdot t_0[j; [1,2]])/t_0[i; [1,2]]\\
& && \ = ([1,1] - [2,4]\cdot[1/5,1])/[2,5] = [-3/2, 3/10]\\
f(x) &= e^{j(x)} && \Longrightarrow\ t_0[f; [1,2]] = [e^{1/5}, e] = [1.22\ldots, 2.71\ldots]\\
& && \ t_1[f; [1,2]] = t_1[j; [1,2]]\cdot t_0[f; [1,2]]\\
& && \ = [-3/2, 3/10]\cdot[e^{1/5}, e]\\
& && \ = [-3e/2, 3e/10] = [-4.07\ldots, 0.81\ldots]
\end{aligned}
$$

These bounds are not optimal, the main reason being that the arithmetic does not take into account the fact that the x's in the numerator and denominator of the exponent in the function are identical. In other words, it computes $F([1,2],[1,2])$ where $F(x_1, x_2) = e^{x_1/(1+x_2^2)}$, and then sets $f(x) = F(x,x)$. Strictly speaking, each equality sign on the right of a Taylor coefficient should be replaced by "$\subseteq$". Tighter bounds can be obtained if we apply the arithmetic to the two halves of the interval $[1,2]$ separately; that is, $t_0[f;[1,1.5]] \cup t_0[f;[1.5,2]] \subset [1.34, 2.12]$ and $t_1[f;[1,1.5]] \cup t_1[f;[1.5,2]] \subset [-1.33, 0.41]$.

Interval arithmetic usually gives only rough bounds. On the other hand, the calculation of these bounds is very fast. If the computed bounds are not sufficient, one could always subdivide the interval and apply interval arithmetic on all the subintervals, as mentioned in the example above; this usually reduces the overestimation. A typical application of this technique is "integration with automatic result verification" or "self-validating numerical quadrature" (see Corliss and Rall 1987), where the computer receives a function form for f, an interval $[a,b]$ and an error tolerance $\varepsilon > 0$ and then returns V such that

$$\left| \int_a^b f(x)\,dx - V \right| < \varepsilon$$

is guaranteed.

C.3. Complex interval arithmetic

In complex interval arithmetic, the "simple intervals" from the previous section are replaced by "standard regions" in the complex plane. We want to be able to calculate with such regions efficiently; in particular, we would like it to be easy to give an enclosure for the image of a function on a standard region provided that the enclosing region is again required to be of standard type.

Two types of standard regions have been studied in the literature: circular discs and rectangles whose edges are parallel to the coordinate axes. Since in this book we deal with real functions for which we want to show analyticity and compute bounds over regions in the complex plane, we may restrict our attention to intervals that are symmetric with respect to the real axis.

Let us present some examples of operations on the following special type of rectangle:

$$([a,b], M) := \{x + iy \in \mathbb{C} \,:\, x \in [a,b],\ |y| \le M\}.$$

For addition and subtraction, $u \in r_1 := ([a,b], M)$ and $v \in r_2 := ([c,d], N)$ imply that

$$u \pm v \in \big([a,b] \pm [c,d],\ M + N\big),$$

For multiplication,

$$\begin{aligned} u \cdot v &\in \operatorname{mul}(a,b,M;\, c,d,N) \\ &:= \big([a,b]\cdot[c,d] + [-MN, MN],\ N\max\{|a|,|b|\} + M\max\{|c|,|d|\}\big), \end{aligned}$$

and for reciprocals or division,

$$\frac{1}{u} \in \operatorname{rec}(a,b,M) := \begin{cases} \left(\left[\frac{b}{b^2+M^2}, \frac{1}{a}\right], \frac{M}{a^2+M^2}\right) & \text{if } a > M > 0, \\ \left(\left[\min\left\{\frac{a}{a^2+M^2}, \frac{b}{b^2+M^2}\right\}, \frac{1}{a}\right], \frac{1}{2a}\right) & \text{if } M > a > 0, \\ \operatorname{rec}(-b,-a,M) & \text{if } b < 0 \end{cases}$$

(but not that $1/u$ is in a fixed complex interval otherwise). Furthermore, we have

$$e^v \in \exp(a,b,M) := \begin{cases} \left(\left[e^a \cos M, e^b\right], e^b \sin M\right) & \text{if } M \le \pi/2, \\ \left(\left[e^b \cos M, e^b\right], e^b\right) & \text{if } \pi > M > \pi/2, \\ \left(\left[-e^b, e^b\right], e^b\right) & \text{if } M \ge \pi \end{cases}$$

and, if $a > 0$,

$$\ln v \in \ln(a,b,M) := \left(\left[\ln a, \frac{1}{2}\ln(b^2+M^2)\right], \arctan\frac{M}{a}\right).$$

For all of these operations, the sets on the right-hand side cannot be replaced by smaller rectangles. For the operations of multiplication and taking the reciprocal, as well as for all elementary functions, we see that the image of a rectangle is usually not a rectangle. If we wish to continue working with rectangles after such an operation, we would need to calculate the smallest rectangle containing the image. Hence, we define

$$\begin{aligned} \rho(r_1 \pm r_2) &:= \big([a,b] \pm [c,d],\ M+N\big), \\ \rho(r_1 \cdot r_2) &:= \operatorname{mul}(a,b,M;\ c,d,N), \\ \rho(1/r_1) &:= \operatorname{rec}(a,b,M), \\ e_1^r &:= \exp(a,b,M), \\ \ln r_1 &:= \ln(a,b,M). \end{aligned}$$

Then, for a given function f which is composed of these operations and functions, we can define $\rho(f(r_1))$ recursively.

EXAMPLE C.2. As in the previous section, we consider $f(x) = e^{x/(1+x^2)}$, but now on the rectangle $r := ([7/8,\ 17/8],\ 3/8)$. This is the smallest standard rectangle that contains the ellipse with foci 1 and 2 and axes whose lengths sum to 2. Intermediate results are

$$\begin{aligned} \rho(1+r^2) &= \big([104/64,\ 362/64],\ 102/64\big), \\ \rho\big(1/(1+r^2)\big) &= \big([2896/17681,\ 8/13],\ 1632/5305\big), \\ \rho\big(r/(1+r^2)\big) &\subset \big([-0.225,\ 3.972],\ 2.721\big), \end{aligned}$$

giving

$$\rho(f(r)) \subset \big([-48.5,\ 53.1],\ 53.1\big).$$

Hence, on the ellipse mentioned above, we have derived automatically that the absolute value of f is bounded by $53.1 \cdot \sqrt{2} < 75.1$, and we can apply the corresponding error bounds in numerical integration. Of course, a subdivision strategy as in Example C.1 can also be helpful.

Bibliography

Books on Quadrature

BRASS, H. (1977) Quadraturverfahren. Vandenhoeck & Ruprecht, Göttingen.

DAVIS, P.J.; RABINOWITZ, P. (1984) Methods of Numerical Integration, second edition. Academic Press, Orlando.

ENGELS, H. (1980) Numerical Quadrature and Cubature. Academic Press, London.

EVANS, G. (1993) Practical Numerical Integration. Wiley, Chichester.

GHIZZETTI, A.; OSSICINI, A. (1970) Quadrature Formulae. Birkhäuser, Basel.

KOWALEWSKI, G. (1932) Interpolation und genäherte Quadratur. Teubner, Leipzig.

KROMMER, A.R.; UEBERHUBER, C.W. (1994) Numerical Integration on Advanced Computer Systems. Springer, Berlin.

KROMMER, A.R.; UEBERHUBER, C.W. (1998) Computational Integration. SIAM, Philadelphia.

KRONROD, A.S. (1964) Nodes and Weights of Quadrature Formulas (in Russian). Izdat Nauka, Moscow. English translation: Consultants Bureau, New York (1965).

KRYLOV, V.J. (1962) Approximate Calculation of Integrals. (Translated from the Russian by A.H. Stroud.) Macmillan, New York.

KYTHE, P.K.; SCHÄFERKOTTER, M.R. (2005) Handbook of Computational Methods for Integration. Chapman & Hall/CRC Press, Boca Raton, FL.

LEVIN, M.; GIRSHOVICH, J. (1979) Optimal Quadrature Formulas. Teubner, Leipzig.

NIKOL'SKII, S.M. (1964) Quadrature Formulae. Hindustan Publishing Corporation, Delhi.

PIESSENS, R.; DE DONCKER-KAPENGA, E.; ÜBERHUBER, C.W.; KAHANER, D.K. (1983) QUADPACK: A Subroutine Package for Automatic Integration. Springer, Berlin.

SMITH, H.V. (1993) Numerical Methods of Integration. Chartwell-Bratt, Bromley.

STROUD, A.H.; SECREST, D. (1966) Gaussian Quadrature Formulas. Prentice-Hall, Englewood Cliffs, NJ.

ZIELAS, W. (1977) Verallgemeinerte Interpolation und Quadraturen. Akademie-Verlag, Berlin.

References

ABRAMOWITZ, M.; STEGUN, I.A. (1964) Handbook of Mathematical Functions. National Bureau of Standards, Washington, DC.

ACHIESER, N.J. (1956) Theory of Approximation. Frederick Ungar Publishing Co., New York.

ACHIESER, N.J.; KREIN, M.G. (1938) Some remarks on the coefficients in quadrature formulae of the type of Gauss (in Ukrainian). *Trudi Odes. Derz. Univ. Mat.* **2**, 29–38.

AKRIVIS, G. (1984) Fehlerabschätzungen für Gauss-Quadraturformeln. *Numer. Math.* **44**, 261–278.

AKRIVIS, G. (1985) The error norm of certain Gaussian quadrature formulae. *Math. Comp.* **45**, 513–519.

AKRIVIS, G.; FÖRSTER, K.-J. (1984) On the definiteness of quadrature formulae of Clenshaw-Curtis type. *Computing* **33**, 363–366.

ALBRECHT, J. (1972) Intervallschachtelungen beim Romberg-Verfahren. *Z. Angew. Math. Mech.* **52**, 433–435.

ANDERSON, L.A.; GAUTSCHI, W. (1975) Optimal weighted Chebyshev-type quadrature formulas. *Calcolo* **12**, 211–248.

ARNOLD, R.W. (1977) Untersuchung optimaler Quadraturformeln für Funktionen von beschränkter Schwankung. Dissertation, Technische Universität Clausthal.

ASKEY, R. (1972) Positivity of the Cotes numbers for some Jacobi abscissas. *Numer. Math.* **19**, 46–48.

ASKEY, R. (1979) Positivity of the Cotes numbers for some Jacobi abscissas II. *J. Inst. Math. Appl.* **24**, 95–98.

ASKEY, R.; FITCH, J. (1968) Positivity of the Cotes numbers for some ultraspherical abscissas. *SIAM J. Numer. Anal.* **5**, 199–201.

BABENKO, A.G. (1984) An extremal problem for polynomials. *Math. Notes* **35**, 181–186.

BABUSKA, J. (1966) Über optimale Quadraturformeln im Raume periodischer Funktionen. *Apl. Mat.* **11**, 259–263.

BABUSKA, J. (1968) Über universal optimale Quadraturformeln. *Apl. Mat.* **13**, 304–338 and 388–403.

BAKER, C.T.H. (1968) On the nature of certain quadrature formulas and their errors. *SIAM J. Numer. Anal.* **5**, 783–804.

BAKHVALOV, N.S. (1967) On the optimal speed of integrating analytic functions (in Russian). *Zh. Vychisl. Mat. Mat. Fiz.* **7**, 1011–1020. English translation: *USSR Comput. Math. Math. Phys.* **7**, 63–75.

BAKHVALOV, N.S. (1971) On the optimality of linear methods for operator approximation in convex classes of functions. *USSR Comput. Math. Math. Phys.* **11**, 244–249.

BARATELLA, P. (1977) Valutazione dell' errore per una formula di quadratura alla Tchebycheff di tipo chiuso. *Calcolo* **14**, 237–242.

BARNHILL, R.E. (1967) Optimal quadratures in $L^2[E_\rho]$. I. *SIAM J. Numer. Anal.* **4**, 390–397.

BARNHILL, R.E. (1968) Asymptotic properties of minimum norm and optimal quadratures. *Numer. Math.* **12**, 384–393.

BARNHILL, R.E.; DENNIS JR., J.E.; NIELSON, G.M. (1969) A new type of Chebyshev quadrature. *Math. Comp.* **23**, 437–441.

BARRAR, R.B.; LOEB, H.L.; WERNER, H. (1974) On the existence of optimal integration formulas for analytic functions. *Numer. Math.* **23**, 105–117.

BARRET, W. (1952) On the remainder term in numerical integration formulae. *J. Lond. Math. Soc.* **27**, 465–470.

BARRET, W. (1960) Convergence properties of Gaussian quadrature formulae. *Comput. J.* **3**, 272–277.

BARRET, W. (1964) On the convergence of Cotes quadrature formulae. *J. Lond. Math. Soc.* **39**, 296–302.

BARY, N.K. (1964) A Treatise on Trigonometric Series. Pergamon Press, Oxford.

BASU, N.K. (1970) Error estimates for a Chebyshev quadrature method. *Math. Comput.* **24**, 863–867.

BAUER, F.L. (1961) La méthode d'intégration numérique de Romberg. In: Centre Belge Rech. Math., Colloque Analyse Numerique, Mons, pp. 119–129.

BAUER, F.L.; RUTISHAUSER, H.; STIEFEL, E. (1963) New aspects in numerical quadrature. In: Proceedings of Symposia in Applied Mathematics, vol. XV, pp. 199–218. American Mathematical Society, Providence, RI.

BELWARD, J.A. (1986) An exponential version of Filon's rule. *J. Comp. Appl. Math.* **14**, 461–466.

BERNSTEIN, S. (1936) Sur la formule de quadrature approchée de Tschebycheff. *C. R. Acad. Sci. Paris* **203**, 1305–1306.

BERNSTEIN, S. (1937a) Sur les formules de quadrature de Cotes et Tschebycheff. *C. R. (Dokl.) Acad. Sci. USSR* **14**, 323–326.

BERNSTEIN, S. (1937b) Sur les formules de quadrature à coefficients non négatifs et abscisses équidistantes. *C. R. Acad. Sci. Paris* **204**, 1294–1296.

BERNSTEIN, S. (1937c) On quadrature formulas with positive coefficients (in Russian). *Izv. Acad. Nauk SSSR, Ser. Mat.* **4**, 479–503.

BERNSTEIN, S. (1937d) Modification de la formule de quadrature de Tchebycheff. *C. R. Acad. Sci. Paris* **204**, 1526–1529.

BLAKELEY, G.R.; BOROSH, J.; CHUI, C.K. (1973) A two-dimensional mean problem. *J. Approx. Theory* **22**, 11–26.

BLOOM, T.; LUBINSKY, D.S.; STAHL, H. (1993a) Distribution of points in convergent sequences of interpolatory integration rules: the rates. In: H. Brass and G. Hämmerlin (eds.) Numerical Integration IV, pp. 5–37. Birkhäuser, Basel.

BLOOM, T.; LUBINSKY, D.S.; STAHL, H.(1993b) What distributions of points are possible for convergent sequences of interpolatory integration rules? *Constr. Approx.* **9**, 41–58.

BOJANOV, B.D. (1974a) Best quadrature formula for a certain class of analytic functions. *Zastosowania Matematyki* **14**, 441–447.

BOJANOV, B.D. (1974b) On an optimal quadrature formula. *C. R. Acad. Bulg. Sci.* **27**, 619–621.

BOJANOV, B.D. (1981) Uniqueness of the optimal nodes of quadrature formulae. *Math. Comp.* **36**, 525–546.

BOJANOV, B.D.; PETROV, P.P. (2001) Gaussian interval quadrature formula. *Numer. Math.* **87**, 625–643.

BOJANOV, B.D.; PETROV, P.P. (2005) Gaussian interval quadrature formulae for Tchebycheff systems. *SIAM J. Numer. Anal.* **43**, 787–795.

BRAKHAGE, H.; BROMBEER, R. (1969) Konvergenzfragen für äquidistante Diskretisierungen hoher Ordnung. *Z. Angew. Math. Mech.* **49**, 101–103.

BRASS H. (1968) Divergenz beim Newton-Cotesschen Quadraturverfahren. *Abh. Math. Sem. Univ. Hamb.* **32**, 7–19.

BRASS, H. (1973) Eine Fehlerabschätzung zum Quadraturverfahren von Clenshaw und Curtis. *Numer. Math.* **21**, 397–403.

BRASS, H. (1974a) Eine Restdarstellung bei Gregoryschen Quadraturverfahren ungerader Ordnung. *Z. Angew. Math. Mech.* **54**, 123–125.

BRASS, H. (1974b) Asymptotisch optimale Quadraturverfahren. In: J. Albrecht and L. Collatz (eds.) Numerische Methoden bei Differentialgleichungen und mit funktionalanalytischen Hilfsmitteln, ISNM 19, pp. 47–65. Birkhäuser, Basel.

BRASS, H. (1975) Zur Theorie der definiten Funktionale. *Z. Angew. Math. Mech.* **55**, T230–T231.

BRASS, H. (1976) Quadraturverfahren vom Gregoryschen Typ. In: J. Albrecht and L. Collatz (eds.) Moderne Methoden der Numerischen Mathematik, ISNM 32, pp. 49–66. Birkhäuser, Basel.

BRASS, H. (1977a) Quadraturverfahren. Vandenhoeck & Ruprecht, Göttingen.

BRASS, H. (1977b) Ein Gegenbeispiel zum Newton-Cotes-Verfahren, *Z. Angew. Math. Mech.* **57**, 609.

BRASS, H. (1978a) Umkehrsätze beim Trapezverfahren. *Aequationes Math.* **18**, 338–344.

BRASS, H. (1978b) Monotonie bei den Quadraturverfahren von Gauß und Newton-Cotes. *Numer. Math.* **30**, 349–354.

BRASS, H. (1979a) Der Wertebereich des Trapezverfahrens. In: G. Hämmerlin (ed.) Numerische Integration, ISNM 45, pp. 98–108. Birkhäuser, Basel.

BRASS, H. (1979b) Zur Approximation einer Klasse holomorpher Funktionen im Reellen. In: G. Meinardus (ed.) Appoximation in Theorie und Praxis, pp. 103–121. Bibliographisches Institut, Mannheim.

BRASS, H. (1982) Zur Quadraturtheorie konvexer Funktionen. In: G. Hämmerlin (ed.) Numerical Integration, ISNM 57, pp. 34–47. Birkhäuser, Basel.

BRASS, H. (1984) Restabschätzungen zur Polynomapproximation. In: L. Collatz, G. Meinardus and H. Werner (eds.) Numerical Methods of Approximation Theory, vol. 7, pp. 19–31. Birkhäuser, Basel.

BRASS, H. (1985) Eine Fehlerabschätzung für positive Quadraturformeln. *Numer. Math.* **47**, 395–399.

BRASS, H. (1988a) Universal quadrature rules in the space of periodic functions. In: H. Brass and G. Hämmerlin (eds.) Numerical Integration III, ISNM 85, pp. 16–24. Birkhäuser, Basel.

BRASS, H. (1988b) Fast-optimale Formeln zur harmonischen Analyse *Z. Angew. Math. Mech.* **68**, T484–T485.

BRASS, H. (1990) Optimal estimation rules for functions of high smoothness. *IMA J. Numer. Anal.* **10**, 129–136.

BRASS, H. (1991) Practical Fourier analysis, error bounds and complexity. *Z. Angew. Math. Mech.* **71**, 3–20.

BRASS, H. (1992a) Error bounds based on approximation theory. In: T.O. Espelid and A. Genz (eds.) Numerical Integration: Recent Developments, Software and Applications, pp. 147–163. Kluwer Academic Publishers, Dordrecht.

BRASS, H. (1992b) Error estimation for the Clenshaw-Curtis quadrature method. *Abh. Braunschw. Wiss. Ges.* **43**, 45–53.

BRASS, H. (1993a) Bounds for Peano kernels. In: H. Brass and G. Hämmerlin (eds.) Numerical Integration IV, ISNM 112, pp. 39–55. Birkhäuser, Basel.

BRASS, H. (1993b) Asymptotically optimal error bounds for quadrature rules of given degree. *Math. Comp.* **61**, 785–798.

BRASS, H. (1995a) On the principle of avoiding the singularity in quadrature. *Z. Angew. Math. Mech.* **75**, 617–618.

BRASS, H. (1995b) Zur Konstruktion fast-optimaler Algorithmen in der Numerik. *Abh. Braunschw. Wiss. Ges.* **46**, 71–78.

BRASS, H. (1996) On the quality of algorithms based on spline interpolation. *Numer. Algorithms* **13**, 159–177.

BRASS, H. (1998a) Linear algorithms with finite deviation. *J. Complexity* **14**, 85–101.

BRASS, H. (1998b) On the variance of the Gaussian quadrature rule. *Calcolo* **35**, 125–129.

BRASS, H.; FISCHER, J.-W. (1999a) Error bounds for Romberg quadrature. *Numer. Math.* **82**, 389–408.

BRASS, H.; FISCHER, J.-W. (1999b) Fehlerschranken für Quadraturformeln gegebenen Grades. *Abh. Braunschw. Wiss. Ges.* **49**, 251–261.

BRASS, H.; FISCHER, J.-W.; PETRAS, K. (1996) The Gaussian quadrature method. *Abh. Braunschw. Wiss. Ges.* **47**, 115–150.

BRASS, H.; FÖRSTER, K.-J. (1987) On the estimation of linear functionals. *Analysis* **7**, 237–258.

BRASS, H.; FÖRSTER, K.-J. (1989) Error bounds for quadrature formulas near Gaussian quadrature. *J. Comput. Appl. Math.* **28**, 145–154.

BRASS, H.; FÖRSTER, K.-J. (1998) On the application of the Peano representation of linear functionals in numerical analysis. In: G.V. Milovanović (ed.) Recent Progress in Inequalities, pp. 175–202. Kluwer Academic Publishers, Dordrecht.

BRASS, H.; PETRAS, K. (1997) On a conjecture of D.B. Hunter. *BIT* **37**, 227–231.

BRASS, H.; PETRAS, K. (2003) Bounds for norms of Peano Kernels. *East J. Approx.* **9**, 195–207.

BRASS, H.; SCHMEISSER, G. (1979) The definiteness of Filippi's quadrature formulae and related problems. In: G. Hämmerlin (ed.) Numerische Integration, ISNM 45, pp. 109–119. Birkhäuser, Basel.

BRASS, H.; SCHMEISSER, G. (1981) Error estimates for interpolatory quadrature formulae. *Numer. Math.* **37**, 371–386.

BROMBEER, R. (1975) Zur Koppelung zwischen Verfahrensfehler und numerischer Instabilität bei Diskretisierungsformeln. *Z. Angew. Math. Mech.* **55**, T231–T234.

BROMBEER, R. (1978) Über optimale Quadraturformeln mit freien Knoten zu Hilberträumen periodischer Funktionen. *Numer. Math.* **30**, 149–164.

BULIRSCH, R. (1964) Bemerkungen zur Romberg-Integration. *Numer. Math.* **6**, 6–16.

BYRD, P.F.; STALLA, L. (1984) Chebyshev quadrature rules for a new class of weight functions. *Math. Comp.* **42**, 173–181.

CHAWLA, M.M. (1969) On Davis' method for the estimation of errors of Gauss-Chebyshev quadratures. *SIAM J. Numer. Anal.* **6**, 108–117.

CHAWLA, M.M. (1970) Hilbert spaces for estimating errors of quadratures for analytic functions. *BIT* **10**, 145–155.

CHAWLA, M.M.; JAIN, M.K. (1968a) Error estimates for Gauss quadrature formulas for analytic functions. *Math. Comp.* **22**, 82–90.

CHAWLA, M.M.; JAIN, M.K. (1968b) Asymptotic error estimates for the Gaussian quadrature formula. *Math. Comp.* **22**, 91–97.

CHAWLA, M.M.; RAINA, B.L. (1972) Optimal quadratures for analytic functions. *BIT* **12**, 489–502.

CHAWLA, M.M.; RAMAKRISHNAN, T.R. (1976) Minimum variance approximate formulas. *SIAM J. Numer. Anal.* **13**, 113-128

CHENEY, E.W. (1966) Introduction to Approximation Theory. McGraw-Hill, New York.

CHUI, C.K. (1971) Concerning rates of convergence of Riemann sums. *J. Approx. Theory* **4**, 279–287.

CHUYAN, O.R.; SUKHAREV, A.G.: On adaptive and nonadaptive stochastic and deterministic algorithms. *J. Complexity* **6**, 119–127 (1990).

CLENSHAW, C.W.; CURTIS, A.R. (1960) A method for integration on an automatic computer. *Numer. Math.* **2**, 197–205.

CORLISS, G.F.; RALL, L.B. (1987) Adaptive, self-validating numerical quadrature. *J. Sci. Stat. Comput.* **8**, 831–847.

CORLISS, G.F.; RALL, L.B. (1991) Computing the Range of Derivatives. In: E.W. Kaucher, S.M. Markov and G. Mayer (eds.) Computer Arithmetic, Scientific Computation and Mathematical Modelling, pp. 195–212. Baltzer, Basel.

CURTIS, A.R.; RABINOWITZ, P. (1972) On the Gaussian integration of Chebyshev polynomials. *Math. Comp.* **26**, 207–211.

DAVENPORT, H. (1937) On some infinite series involving arithmetical functions. *Quart. J. Math. Oxford Ser.* **8**, 8–13.

DAVIS, P. (1953) Errors of numerical approximation for analytic functions. *J. Ration. Mech. Anal.* **2**, 303-313

DAVIS, P.J. (1959) On the numerical integration of periodic analytic functions. In: E. Langer (ed.) On Numerical Approximation, pp. 45–59. University of Wisconsin Press, Madison.

DAVIS, P.J. (1962) Errors of numerical approximation for analytic functions. In: J. Todd (ed.) Survey of Numerical Analysis, pp. 468–484. McGraw-Hill, New York.

DAVIS, P.J. (1963): Interpolation and Approximation. Blaisdell Publishing Company, New York.

DAVIS, P.J.; RABINOWITZ, P. (1954) On the estimation of quadrature errors for analytic functions. *Math. Tables Aids Comput.* **8**, 193–203.

DEVORE, R.A.; LORENTZ, G.G. (1993) Constructive Approximation. Springer, Berlin.

DEVORE, R.A.; SCOTT, L.R. (1984) Error bounds for Gaussian quadrature and weighted-L_1 polynomial approximation. *SIAM J. Numer. Anal.* **21**, 400–412.

DUBUC, S.; TODOR, F. (1983) La règle du trapèze appliquée à quelques fonctions sans dérivées. *Canad. Math. Bull.* **24**, 1–5

DUREN, P.L. (1970) Theory of H_p-spaces. Academic Press, London.

DURIS, C.S. (1971) Optimal quadrature formulas using generalized inverses. Part I: General theory and minimum variance formulas. *Math. Comp.* **25**, 495–504.

EAGLE, A. (1928) On the relations between the Fourier constants of a periodic function and the coefficients determined by harmonic analysis. *Philos. Magazine*, Ser. 7, **5**, 113–132.

EHLICH, H. (1966) Untersuchungen zur numerischen Fourieranalyse. *Math. Z.* **91**, 380–420.

EHLICH, H.; ZELLER, K. (1964) Schwankung von Polynomen zwischen Gitterpunkten. *Math. Z.* **86**, 41–44.

EHRENMARK, U.T. (1987) On the error term of the Filon quadrature formulae. *BIT* **27**, 85–97.

EHRICH, S. (1993): Einige neue Ergebnisse zu den Fehlerkonstanten der Gauss-Kronrod-Quadraturformel. *Z. Angew. Math. Mech.* **73**, T882–T886.

EHRICH, S. (1994a) Error bounds for Gauss-Kronrod quadrature formulae. *Math. Comp.* **62**, 295–304.

EHRICH, S. (1994b) Gauss-Kronrod quadrature error estimates for analytic functions. *Z. Angew. Math. Mech.* **74**, T691–T693.

EHRICH, S. (1995) Asymptotic properties of Stieltjes polynomials and Gauss-Kronrod quadrature formulae. *J. Approx. Theory* **82**, 287–303.

EHRICH, S. (1996a) High order error constants of Gauss-Kronrod quadrature formulas. *Analysis* **16**, 335–345.

EHRICH, S. (1996b) A note on Peano constants of Gauss-Kronrod quadrature schemes. *J. Comput. Appl. Math.* **66**, 177–183.

EHRICH, S. (1998) On product integration with Gauss-Kronrod nodes. *SIAM J. Numer. Anal.* **35**, 78–92.

EHRICH, S. (1999) Stieltjes polynomials and the error of Gauss-Kronrod quadrature formulas. In: W. Gautschi, G.H. Golub and G. Opfer (eds.) Application and Computation of Orthogonal Polynomials, pp. 57–77. Birkhäuser, Basel.

ELLIOT, D. (1964) The evaluation and estimation of the coefficients in the Chebyshev series expansion of a function. *Math. Comp.* **18**, 274–284.

ENGELS, H. (1977) Eine Familie interpolatorischer Quadraturformeln mit ableitungsfreien Fehlerschranken. *Numer. Math.* **28**, 49–58.

ENGELS, H.; ECKHARDT, U. (1977) Algorithm 33 Wilf-Quadrature. *Computing* **18**, 271–279.

ENGELS, H.; ECKHARDT, U. (1979) The determination of nodes and weights in Wilf quadrature formulas. *Abh. Math. Sem. Univ. Hamburg* **48**, 34–41.

ENGELS, H.; MANGOLD, R. (1971) Über eine Klasse Wilfscher Quadraturformeln. *Berichte der Kernforschungsanlage Jülich*, **789**.

ERDELYI, A. (ED.) (1953) Higher transcendental functions I. McGraw-Hill, New York.

ERDELYI, A. (ED.) (1955) Higher transcendental functions II, III. McGraw-Hill, New York.

ERDŐS, P.; TURÁN, P. (1937) On interpolation. I. *Ann. Math.* **38**, 142–155.

ERDŐS, P.; TURÁN, P. (1938) On interpolation. II. *Ann. Math.* **39**, 702–724.

ERDŐS, P.; TURÁN, P. (1940a) On uniformly dense distribution of certain sequences of points. *Ann. Math.* **42**, 162–173.

ERDŐS, P.; TURÁN, P. (1940b) On interpolation. III. *Ann. Math.* **41**, 510–553.

ERDŐS, P.; SHARMA, A. (1965) On Tschebycheff quadrature. *Can. J. Math.* **17**, 652–658.

FÉJER, L. (1933) Mechanische Quadraturen mit positiven Cotesschen Zahlen. *Math. Z.* **37**, 287–309.

FIEDLER, H. (1987) Das asymptotische Verhalten der Peanokerne einiger interpolatorischer Quadraturverfahren. *Numer. Math.* **51**, 571–581.

FILIPPI, S. (1964) Angenäherte Tschebyscheff-Approximation einer Stammfunktion – eine Modifikation des Verfahrens von Clenshaw and Curtis. *Numer. Math.* **6**, 320–328.

FILIPPI, S. (1966) Neue Gauß-Typ-Quadraturverfahren. *Elektron. Datenverarb.* **8**, 174–180.

FILIPPI, S.; ESSER, H. (1970) Darstellungs- und Konvergenzsätze für Quadraturverfahren auf C and C^m. Forschungsberichte des Landes Nordrhein-Westfalen, 2137. Westdeutscher Verlag, Köln.

FILON, L.N.G. (1928/29) On a quadrature formula for trigonometric integrals. *Proc. Roy. Soc. Edinburgh* **49**, 38–47.

FISCHER, J.-W. (1997) Zur Approximation linearer Funktionale bei periodischen Funktionen. *Abh. Braunschw. Wiss. Ges.* **48**, 67–74.

FISCHER, J.-W. (2002) Romberg quadrature using the Bulirsch sequence. *Numer. Math.* **90**, 509–519.

FÖRSTER, K.-J. (1982a) Fehlerschranken bei der Romberg-Quadratur. *Z. Angew. Math. Mech.* **62**, 133–135.

FÖRSTER, K.-J. (1982b) Bemerkungen zur optimalen Tschebyscheff-Typ Quadratur. *Numer. Math.* **38**, 421–425.

FÖRSTER,K.-J. (1982c) A comparison theorem for linear functionals and its application in quadrature. In: G. Hämmerlin (ed.) Numerical Integration, ISNM 57, pp. 66–76. Birkhäuser, Basel.

FÖRSTER, K.-J. (1986a) On Chebyshev quadrature for ultraspherical weight functions. *Calcolo* **23**, 355–381.

FÖRSTER, K.-J. (1986b) On Chebyshev quadrature for a special class of weight functions. *BIT* **26**, 327–333.

FÖRSTER, K.-J. (1987a) Schranken für die Varianz und die Gewichte von Quadraturformeln. Habilitationsschrift, Technische Universität Braunschweig.

FÖRSTER, K.-J. (1987b) On weight functions admitting Chebyshev quadrature. *Math. Comp.* **49**, 251–258.

FÖRSTER, K.-J. (1987c) Über Monotonie und Fehlerkontrolle bei den Gregoryschen Quadraturverfahren. *Z. Angew. Math. Mech.* **67**, 257–266.

FÖRSTER, K.-J. (1988) On Chebyshev quadrature and variance of quadrature formulas. *BIT* **28**, 360–363.

FÖRSTER, K.-J. (1992) On quadrature formulae near Gaussian quadrature. In: T.O. Espelid and A. Genz (eds.) Numerical Integration: Recent Developments, Software and Applications, pp. 45–54. Kluwer Academic Publishers, Dordrecht.

FÖRSTER, K.-J. (1993a) Inequalities for ultraspherical polynomials and application to quadrature. *J. Comp. Appl. Math.* **49**, 59–70.

FÖRSTER, K.-J. (1993b) Variance in quadrature – a survey. In: H. Brass and G. Hämmerlin (eds.) Numerical Integration IV, pp. 91–110. Birkhäuser, Basel.

FÖRSTER, K.-J. (1994) On a theorem of C. Posse concerning Gaussian quadrature of Chebyshev type. *Math. Comp.* **62**, 719–724.

FÖRSTER, K.-J. (1995) On the weights of positive quadrature formulas for ultraspherical weight functions. *Ann. Numer. Math.* **2**, 35–78.

FÖRSTER, K.-J.; OSTERMEYER, G.-P. (1986) On weighted Chebyshev-type quadrature formulas. *Math. Comp.* **46**, 591–599.

FÖRSTER, K.-J.; PETRAS, K. (1990a) On estimates for the weights in Gaussian quadrature in the ultraspherical case. *Math. Comp.* **55**, 243–264.

FÖRSTER, K.-J.; PETRAS, K. (1990b) On a problem proposed by H. Brass concerning the remainder term in quadrature for convex functions. *Numer. Math.* **57**, 763–777.

FÖRSTER, K.-J.; PETRAS, K. (1991) Error estimates in Gaussian quadrature for functions of bounded variation. *SIAM J. Numer. Anal.* **28**, 880–889.

FÖRSTER, K.-J.; PETRAS, K. (1994) On the variance of Gaussian quadrature formulae in the ultraspherical case. *Calcolo* **31**, 1–33.

FORST, W. (1976) Zur Quadratur analytischer periodischer Funktionen. *J. Approx. Theory* **18**, 350–356.

FRANKE, R. (1971) Best Chebyshev quadratures. *Rocky Mountain J. Math.* **1**, 499–509.

FREUD, G. (1969) Orthogonale Polynome. Birkhäuser, Basel.

FREUD, G. (1973) Error estimates for Gauss-Jacobi quadrature formulae. In: J.J.H. Miller (ed.) Topics in Numerical Analysis, pp. 113–121. Academic Press, London.

GAIER, D. (1970/71) Saturation bei Spline-Approximation und Quadratur. *Numer. Math.* **16**, 129–140.

GAL, S.; MICCHELLI, C.A. (1980) Optimal sequential and non-sequential procedures for evaluating a functional. *Appl. Anal.* **10**, 105–120.

GANELIUS, T. (1956) On one-sided approximation by trigonometric polynomials. *Math. Scand.* **4**, 247–258.

GATTESCHI, L. (1963/64) Su di un problema connesso alle formule di quadratura di Tschebyscheff. *Univ. e Politec. Torino, Rend. Sem. Mat.* **23**, 75–87.

GATTESCHI, L. (1988) Uniform approximation of Christoffel numbers for Jacobi weight. In: H. Brass and G. Hämmerlin (eds.) Numerical Integration III, ISNM 85, pp. 49–59. Birkhäuser, Basel.

GATTESCHI, L.; VINARDI, G. (1978) Sul grado di precisione di formule di quadratura del tipo di Tchebycheff. *Calcolo* **15**, 59–85.

GAUSS, C.F. (1816) Methodus nova integralium valores per approximationem inveniendi. Comment. soc. regiae sci. Gottingensis rec. III, Göttingen, or Werke 3, pp. 163–196.

GAUTSCHI, W. (1968) Construction of Gauss-Christoffel quadrature formulas. *Math. Comp.* **22**, 251–270.

GAUTSCHI, W. (1972) Attenuation factors in practical Fourier Analysis. *Numer. Math.* **18**, 373–400.

GAUTSCHI, W. (1976) Advances in Chebyshev quadrature. In: G.A. Watson (ed.) Numerical Analysis, Lecture Notes in Mathematics 506, pp. 100–121. Springer, Berlin.

GAUTSCHI, W. (1979) On generating Gaussian quadrature rules. In: G. Hämmerlin (ed.) Numerische Integration, ISNM 45, pp. 147–154. Birkhäuser, Basel.

GAUTSCHI, W. (1981) A survey of Gauss-Christoffel quadrature formulae. In: P.L. Butzer and F. Fehér (eds.) E.B. Christoffel: The Influence of His Work in Mathematics and the Physical Sciences, pp. 72–147. Birkhäuser, Basel.

GAUTSCHI, W. (1983) How to check and how not to check Gaussian quadrature formulae. *BIT* **23**, 209–216.

GAUTSCHI, W. (1984) On some orthogonal polynomials of interest in theoretical chemistry. *BIT* **24**, 473–483.

GAUTSCHI, W. (1988) Gauss-Kronrod quadrature – a survey. In: G. Milovanović (ed.) Numerical Methods and Approximation Theory III, pp. 39–66. University of Niš, Niš.

GAUTSCHI, W. (1992) Remainder estimates for analytic functions. In: T.O. Espelid and A. Genz (eds.) Numerical Integration: Recent Developments, Software and Applications, pp. 133–145. Kluwer Academic Publishers, Dordrecht.

GAUTSCHI, W.; MONEGATO, G. (1977) On optimal Chebyshev-type quadratures. *Numer. Math.* **28**, 59–67.

GAUTSCHI, W.; NOTARIS, S.E. (1989) Gauss-Kronrod quadrature formulae for weight functions of Bernstein-Szegö type. *J. Comput. Appl. Math.* **25**, 199–224. Corrigendum: *J. Comput. Appl. Math.* **27**, 429 (1989).

GAUTSCHI, W.; RIVLIN, TH.J. (1988) A family of Gauss-Kronrod quadrature formulae. *Math. Comp.* **51**, 749–754.

GAUTSCHI, W.; VARGA, R.S. (1983) Error bounds for Gaussian quadrature of analytic functions. *SIAM J. Numer. Anal.* **20**, 1170–1186.

GAUTSCHI, W.; YANAGIWARA, H. (1974) On Chebyshev-type quadratures. *Math. Comput.* **28**, 125–134.

GERONIMUS, J.L. (1969) On the Čebyšev quadrature formula. *Math. USSR-Ivzestija* **3**, 1115–1138

GLINKIN, I.A. (1984) Best quadrature formula in the class of convex functions. *Math. Notes* **36**, 368–374.

GOLOMB, M. (1968) Approximation by periodic spline interpolants on uniform meshes. *J. Approx. Theory* **1**, 26–65.

GOLOMB, M.; WEINBERGER, H.F. (1959) Optimal approximation and error bounds. In: R.E. Langer (ed.) On Numerical Approximation, pp. 117–190. University of Wisconsin Press, Madison.

GOLUB, G.H.; WELSCH, J.H. (1969) Calculation of Gauss quadrature rules. *Math. Comp.* **23**, 221–230.

GRIEWANK, A. (2000) Evaluating Derivatives: Principles and Techniques of Algorithmic Differentiation. SIAM, Philadelphia.

HAHN, H. (1918) Über das Interpolationsproblem. *Math. Z.* **1**, 115–142.

HÄMMERLIN, G. (1965) Über ableitungsfreie Schranken für Quadraturfehler II. *Numer. Math.* **7**, 232–237.

HÄMMERLIN, G. (1966) Zur Abschätzung von Quadraturfehlern für analytische Funktionen. *Numer. Math.* **8**, 334–344.

HAMMING, R.W.; PINKHAM, R.S. (1966) A class of integration formulas. *J. Assoc. Comput. Mach.* **13**, 430–438.

HARDY, G.H.; WRIGHT, E.M. (1960) An Introduction to the Theory of Numbers. Clarendon Press, Oxford.

HEINDL, G. (1982) Optimal quadrature of convex functions. In: G. Hämmerlin (ed.) Numerical integration, ISNM 57, pp. 138–147. Birkhäuser, Basel.

HOLLADAY, J.C. (1957) A smoothest curve approximation. *Math. Tables Aids Comput.* **11**, 233–243.

HUERTA, I.P. (1986) Adaption helps for some nonconvex classes. *J. Complexity* **2**, 333–352.

HUNTER, D.B. (1995) Some error expansions for Gaussian quadrature. *BIT* **35**, 62–82.

HUNTER, D.B.; NIKOLOV, G. (1998) Gaussian quadrature of Chebyshev polynomials. *J. Comput. Appl. Math.* **94**, 123–131.

IMHOF, J.P. (1963) On the method for numerical integration of Clenshaw and Curtis. *Numer. Math.* **5**, 138–141.

JETTER, K. (1976) Optimale Quadraturformeln mit semidefiniten Peanokernen. *Numer. Math.* **25**, 239–249.

JETTER, K. (1982) Ein elementarer Beweis des Satzes von Krein. In: G. Hämmerlin (ed.) Numerical Integration, ISNM 57, pp. 34–47. Birkhäuser, Basel.

KAHANER, D.K. (1969) On equal and almost equal weight quadrature formulas. *SIAM J. Numer. Anal.* **6**, 551–556.

KAHANER, D.K. (1970): Chebyshev type quadrature formulas. *Math. Comp.* **24**, 571–574.

KAMBO, N.S. (1970) Error of the Newton-Cotes and Gauss-Legendre quadrature formulas. *Math. Comp.* **24**, 261–269.

KARLIN, S.; STUDDEN, W.J. (1966) Tchebycheff Systems: With Applications in Analysis and Statistics. Interscience Publishers, New York.

KERSHAW, D. (1972) Sard's best quadrature formulas of order two. *J. Approx. Theory* **6**, 466–474

KIEFER, J. (1957) Optimum sequential search and approximation methods under minimum regularity assumptions. *J. Soc. Indust. Appl. Math.* **5**, 105–136.

KNAUFF, W. (1977a) Gewichtsoptimale Quadraturformeln bei analytischen Funktionen. *Computing* **18**, 59–66.

KNAUFF, W. (1977b) Fehlernormen zur Quadratur analytischer Funktionen. *Computing* **17**, 309–322.

KNAUFF, W.; KRESS, R. (1974) Optimale Approximation linearer Funktionale auf periodischen Funktionen. *Numer. Math.* **22**, 187–205.

KNAUFF, W.; PAULIK, A. (1978) A note on Davis type error bounds. *BIT* **18**, 175–183.

KOBER, H. (1957) Dictionary of Conformal Representations. Dover Publications, Mineola, NY.

KÖHLER, P. (1988a) Optimale Quadraturformeln für Funktionen mit beschränkter zweiter Ableitung bei äquidistanten Stützstellen. In: H. Brass and G. Hämmerlin (eds.) Numerical Integration III, ISNM 85, pp. 79–90. Birkhäuser, Basel.

KÖHLER, P. (1988b) On the weights of Sard's quadrature formulas. *Calcolo* **25**, 169–186.

KÖHLER, P. (1989a) On Sard's quadrature formulas of order two. *J. Approx. Theory* **59**, 247–257.

KÖHLER, P. (1989b) Estimates for Sard's best formulas for linear functionals on $C^s[a,b]$. *J. Comput. Appl. Math.* **28**, 259–267.

KÖHLER, P.(1993) On the error of Filon's quadrature formula. *Z. Angew. Math. Mech.* **73**, T886–T889.

KÖHLER, P. (1994) Estimates for linear remainder functionals by the modulus of continuity. In: B. Bojanov (ed.) Open problems in Approximation Theory (Voneshta Voda, Bulgaria, 1993), pp. 109–124. Science Culture Technology Publishing, Singapore.

KÖHLER, P. (1995) Construction of asymptotically best quadrature formulas. *Numer. Math.* **72**, 93–116.

KÖHLER, P. (1996) A reduction method for linear functionals and its applications. *Analysis* **16**, 181–193.

KÖHLER, P.; NIKOLOV, G. (1995) Error bounds for optimal definite quadrature formulae. *J. Approx. Theory* **81**, 397–405.

KOREVAAR, J. (1992) Behavior of Cotes numbers and other constants, with an application to Chebyshev-type quadrature. *Indag. Math.* **3**, 391–402.

KOREVAAR, J.; MEYERS, J.L.H. (1993) Massive coalescence of nodes in optimal Chebyshev-type quadrature on $[-1,1]$. *Indag. Math.*, **4**, 327–338.

KORNEICHUK, N. (1991) Exact Constants in Approximation Theory. Cambridge University Press, Cambridge.

KORNEICHUK, N.; LUSPAI, N.E. (1969) Best quadrature formulas for classes of differentiable functions and piecewise polynomial approximation. *Math. USSR Izv.* **3**, 1335–1355.

KOWALEWSKI, G. (1896) Über eine Art von simultaner Darstellung bestimmter Integrale. *J. Reine Angew. Math.* **117**, 267-272

KOWALEWSKI, G. (1932) Interpolation und genäherte Quadratur. Teubner, Leipzig.

KRÄMER, W.; KULISCH, U.; LOHNER, R. (2009) Numerical Toolbox for Verified Computing II: Advanced Numerical Problems. Springer, Berlin.

KREIN, M.G. (1951) The ideas of P.L. Čebišev and A.A. Markoff in the theory of limiting values of integrals and their further developments (in Russian). *Uspeki Mat. Nauk (N.S.)* **6**, 3–120. English translation: *AMS Transl.*, ser. 2, **12**, 1–121 (1959).

KRESS, R. (1971) Ein ableitungsfreies Restglied für die trigonometrische Interpolation periodischer analytischer Funktionen. *Numer. Math.* **16**, 389–396.

KRESS, R. (1972) Zur numerischen Integration periodischer Funktionen nach der Rechteckregel. *Numer. Math.* **20**, 87–92.

KRESS, R. (1978) On error norms of the trapezoidal rule. *SIAM J. Numer. Anal.* **15**, 433–443.

KRONROD, A.S. (1964) Nodes and Weights of Quadrature Formulas (in Russian). Izdat Nauka, Moscow. English translation: Consultants Bureau, New York (1965).

KRYLOV, V.J. (1962) Approximate Calculation of Integrals (translated from the Russian by A.H. Stroud). Macmillan, New York.

KRYLOV, V.J.; SUL'GINA, L.T. (1964) On the remainder in the numerical integration of periodic functions of one variable (in Russian). *Dokl. Akad. Nauk BSSR* **8**, 617–620.

KUIJLAARS, A. (1993) The minimal number of nodes in Chebyshev-type quadrature formulas. *Indag. Math.* (N.S.) **4**, 339–362.

KUIJLAARS, A. (1995a) Chebyshev-type quadrature and partial sums of the exponential series. *Math. Comp.* **64**, 251–263.

KUIJLAARS, A. (1995b) Chebyshev-type quadrature and zeros of Faber polynomials. *J. Comp. Appl. Math.* **62**, 155–179.

KUIJLAARS, A. (1995c) Chebyshev-type quadrature for Jacobi weight functions. *J. Comp. Appl. Math.* **57**, 171–180.

KUIJLAARS, A. (1995d) Chebyshev quadrature for measures with a strong singularity. *J. Comp. Appl. Math.* **65**, 207–214.

KUMAR, R. (1974) A class of quadrature formulas. *Math. Comp.* **28**, 769–778.

KUNTZMANN, J. (1949) Formules de quadrature approchée pour les fonctions continues à derivée première continue et a derivée seconde bornée. *C. R. Acad. Sci. Paris* **228**, 38–40.

KÜTZ, M. (1982) On the positivity of certain Cotes numbers. *Aequationes Math.* **24**, 110–118.

KÜTZ, M. (1984a) Asymptotic error bounds for a class of interpolatory quadratures. *SIAM J. Numer. Anal.* **21**, 167–175.

KÜTZ, M. (1984b) A note on error estimates for the Clenshaw-Curtis and other interpolatory quadratures. *Analysis* **4**, 45–51.

KÜTZ, M. (1986) On the asymptotic behaviour of some Cotes numbers. *Z. Angew. Math. Mech.* **66**, 373–375.

LANGE, G. (1977) Beste und optimale definite Quadraturformeln. Dissertation, Technische Universität Clausthal.

LANGE, G. (1979) Optimale definite Quadraturformeln. In: G. Hämmerlin (ed.) Numerische Integration, ISNM 45, pp. 187–197. Birkhäuser, Basel.

LAURIE, D.P. (1999) Questions related to Gaussian quadrature formulas and two-term recursions. In: W. Gautschi, G.H. Golub and G. Opfer (eds.) Applications and computation of orthogonal polynomials, pp. 133–144. Birkhäuser, Basel.

LEBED, G.K. (1968) Quadrature formulas with smallest remainder estimate on certain function classes (in Russian). *Mat. Zametki* **3**, 577–586. English translation: *Math. Notes* **3**, 368–373 (1968).

LETHER, F.G. (1978) On the construction of Gauss-Legendre quadrature rules. *J. Comp. Appl. Math.* **4**, 47–52.

LETHER, F.G.; WENSTON, P.R. (1995) Minimax approximations to the zeros of $P_n(x)$ and Gauss-Legendre quadrature. *J. Comp. Appl. Math.* **59**, 245–252.

LETHER, F.G.; WILHELMSEN, D.; FRAZIER, R. (1979) On the positivity of Cotes numbers for interpolatory quadratures with Jacobi abscissas. *Computing* **21**, 171–175.

LIGUN, A.A. (1976) Inequalities for upper bounds of functionals. *Analysis Mathematica* **2**, 11–40.

LIGUN, A.A. (1978) Best quadrature formulas for certain classes of periodic functions. *Math. Notes* **24**, 853–857.

LIPOW, P.R.; STENGER, F. (1972) How slowly can quadrature formulas converge? *Math. Comp.* **26**, 917–922.

LO, Y.T.; LEE, S.W.; SUN, B. (1965) On Davis' method of estimating quadrature errors. *Math. Comp.* **19**, 133–138.

LOCHER, F. (1969) Fehlerabschätzung für das Quadraturverfahren von Clenshaw und Curtis. *Computing* **4**, 304–315.

LOCHER, F. (1973a) Zur Struktur von Quadraturformeln. *Numer. Math.* **20**, 317–326.

LOCHER, F. (1973b) Norm bounds of quadrature processes. *SIAM J. Numer. Anal.* **10**, 553–558.

LOCHER, F. (1981) Interpolation on uniform meshes by the translates of one function and related attenuation factors. *Math. Comp.* **37**, 403–416.

LOCHER, F.; ZELLER, K. (1968) Approximationsgüte und numerische Integration. *Math. Z.* **104**, 249–251.

LOEB H.L. (1974) A note on optimal integration in H_∞. *C. R. Acad. Bulg. Sci.* **27**, 615–618.

LÖTZBEYER, W.A. (1971) Über asymptotische Eigenschaften linearer und nichtlinearer Quadraturformeln. Dissertation, Universität Karlsruhe.

LÖTZBEYER, W.A. (1972) Über asymptotische Eigenschaften linearer und nichtlinearer Quadraturformeln. *Z. Angew. Math. Mech.* **52**, T211–T214.

LOWAN, A.N.; SALZER, H. (1943) Table of coefficients in numerical integration formulae. *J. Math. Phys.* **22**, 49–50.

LUKE, Y.L. (1954) On the computation of oscillatory integrals. *Proc. Cambridge Philos. Soc.* **50**, 269–277.

MALOZEMOV, V.N. (1967a) Estimation of the accuracy of a quadrature formula for periodic functions (in Russian). *Vestnik Leningradskogo Universiteta, Ser. Matematika, Mechanika, Astronomija* **22**, 52–59.

MALOZEMOV, V.N. (1967b) Accuracy of a rectangle quadrature formula for periodic functions. *Math. Notes* **2**, 703–705.

MARKOFF, A.A. (1884) Démonstration de certaines inégalités de M. Tchébychef. *Math. Ann.* **24**, 172–180.

MARKOFF, A.A. (1885) Sur la méthode de Gauss pour le calcul approché des intégrales. *Math. Ann.* **25**, 427–432.

MARSDEN, M.J.; TAYLOR, G.D. (1972) Numerical evaluation of Fourier integrals. In: L. Collatz and G. Meinardus (eds.) Numerische Methoden der Approximationstheorie I, ISNM 16, pp. 61–76. Birkhäuser, Basel.

MARTENSEN, E. (1964) Optimale Fehlerschranken für die Quadraturformel von Gregory. *Z. Angew. Math. Mech.* **44**, 159–168.

MARTENSEN, E. (1970) Zur Restglieddarstellung der Gregoryschen Quadraturformel ungerader Ordnung. *Numer. Math.* **15**, 229–233.

MARTENSEN, E. (1973) Darstellung und Entwicklung des Restgliedes der Gregoryschen Quadraturformel mit Hilfe von Spline-Funktionen. *Numer. Math.* **21**, 70–80.

MARTENSEN, E. (2005) On the generalized Euler-Maclaurin formula. *Z. Angew. Math. Mech.* **85**, 858–863.

MASON, J.C.; VENTURINO, E. (1997) Integration methods of Clenshaw-Curtis type, based on four kinds of Chebyshev polynomials. In: J.W. Schmidt and G. Walz (eds.) Multivariate Approximation and Splines, pp. 153–165. Birkhäuser, Basel.

MEINARDUS, G. (1967) Approximation of Functions. Springer, New York.

MESCHKOWSKI, H. (1962) Hilbertsche Räume mit Kernfunktion. Springer, Berlin.

MEYERS, L.F.; SARD, A. (1950) Best approximate integration formulas. *J. Math. Phys.* **29**, 118–123.

MICCHELLI, C.A. (1980) Some positive Cotes numbers for the Chebyshev weight function. *Aequationes Math.* **21**, 105–109.

MICCHELLI, C.A.; RIVLIN, T.J. (1973) Numerical integration rules near Gaussian quadrature. *Israel J. Math.* **16**, 287–299.

MILLS, T.M.; SMITH, S.J. (1991) A note on the Newton-Cotes integration formula. *J. Approx. Theory* **66**, 98–105.

VON MISES, R. (1933) Zur mechanischen Quadratur. *Z. Angew. Math. Mech.* **13**, 53–56.

VON MISES, R. (1936) Über allgemeine Quadraturformeln. *J. Reine Angew. Math.* **174**, 56–67.

MONEGATO, G. (1976) A note on extended Gaussian quadrature rules. *Math. Comp.* **30**, 812–817.

MONEGATO, G. (1978a) Positivity of the weights of extended Gauss-Legendre quadrature rules. *Math. Comp.* **32**, 243–245.

MONEGATO, G. (1978b) Some remarks on the construction of extended Gaussian quadrature rules. *Math. Comp.* **32**, 247–252.

MONEGATO, G. (2001) An overview of the computational aspects of Kronrod quadrature rules. *Numer. Algorithms* **26**, 173–196.

MOORE, R.E. (1966) Interval Analysis. Prentice Hall, Englewood Cliffs, NJ.

MOTORNYI, V.P. (1974) On the best quadrature formula of the form $\sum_{k=1}^{n} p_k f(x_k)$ for some classes of differentiable periodic functions. *Math. USSR Izv.* **8**, 591–620.

MOTORNYI, V.P. (1998) On the best interval quadrature formula in the class of functions with bounded rth derivative. *East J. Approx.* **4**, 459–478.

NAVOT, J. (1961) An extension of the Euler-Maclaurin summation formula to functions with a branch singularity. *J. Math. Phys.* **40**, 271–276.

NEWMAN, D.J. (1974) Monotonicity of quadrature approximations. *Proc. Amer. Math. Soc.* **42**, 251–257.

NICHOLSON, D.; RABINOWITZ, P.; RICHTER, N.; ZEILBERGER, D. (1971) On the error in the numerical integration of Chebyshev polynomials. *Math. Comp.* **25**, 79–86.

NIKOLOV, G. (1993) Gaussian quadrature formulae for splines. In: H. Brass and G. Hämmerlin (eds.) Numerical Integration IV, pp. 267–281. Birkhäuser, Basel.

NIKOLOV, G. (1996) On certain definite quadrature formulae. *J. Comput. Appl. Math.* **75**, 329–343.

NIKOLOV, G. (2001) On the weights of nearly Gaussian quadrature formulae. *East J. Approx.* **7**, 115–120.

NIESSNER, H. (1985) Zur optimalen Quadratur konvexer Funktion. Dissertation, Universität Wuppertal.

NOTARIS, S.E. (1990a) The error norm of Gaussian quadrature formulae for weight functions of Bernstein-Szegö type. *Numer. Math.* **57**, 271–283.

NOTARIS, S.E. (1990b) Gauss-Kronrod quadrature formulae for weight functions of Bernstein-Szegö type, II. *J. Comput. Appl. Math.* **29**, 161–169.

NOTARIS, S.E. (1993) Error bounds for Gauss-Kronrod quadrature formulae of analytic functions. *Numer. Math.* **64**, 371–380.

NOTARIS, S.E. (1997) Interpolatory quadrature formulae with Chebyshev abscissae of the third or fourth kind. *J. Comput. Appl. Math.* **81**, 83–99.

NOTARIS, S.E. (2001) Interpolatory quadrature formulae with Chebyshev abscissae. *J. Comput. Appl. Math.* **133**, 507–517.

NOTARIS, S.E. (2002) Positivity of the weights of interpolatory quadrature formulae with Jacobi abscissae. *BIT* **42**, 440–446.

NOTARIS, S.E. (2003) New interpolatory quadrature formulae with Gegenbauer abscissae. *J. Comput. Appl. Math.* **161**, 295–317.

NOVAK, E. (1988) Deterministic and Stochastic Error Bounds in Numerical Analysis, Lecture Notes in Mathematics 1349. Springer, Berlin.

NOVAK, E. (1992) Quadrature formulas for monotone functions. *Proc. Amer. Math. Soc.* **115**, 59–68.

NÜRNBERGER, G. (1985) Unicity in one-sided L_1-approximation and quadrature formulae. *J. Approx. Theory* **45**, 271–279.

OMLADIČ, M. (1992) Average quadrature formulas of Gauss type. *IMA J. Numer. Anal.* **12**, 189–199.

OMLADIČ, M.; PAHOR, S.; SUHADOLC, A. (1976) On a new type of quadrature formulas. *Numer. Math.* **25**, 421–426.

OSIPENKO, K.YU. (1988) On best and optimal quadrature formulas on classes of bounded analytic functions (in Russian). *Izv. Akad. Nauk SSSR Ser. Mat.* **52**, 79–99. English translation: *Math. USSR Izv.* **31**, 77–97 (1989).

OSIPENKO, K.YU. (1995) Exact values of n-widths and optimal quadratures on classes of bounded analytic and harmonic functions. *J. Approx. Theory* **82**, 156–175.

OSTROWSKI, A.M. (1969) On the remainder term of the Euler-Maclaurin formula. *J. Reine Angew. Math.* **240**, 268–286.

OUSPENSKY, J. (1925) Sur les valeurs asymptotiques des coefficients de Cotes. *Bull. Amer. Math. Soc.* **31**, 145–156.

PATTERSON, T.N.L. (1968) The optimum addition of points to quadrature formulae. *Math. Comp.* **22**, 847–856. Errata: *Math. Comp.* **23**, 892 (1969).

PAULIK, A. (1977) Zur Existenz optimaler Quadraturformeln mit freien Knoten bei Integration analytischer Funktionen. *Numer. Math.* **27**, 395–405.

PEANO, G. (1913) Resto nelle formule di quadrature, espresso con un integrale definito. *Rom. Acc. L. Rend.* (5) **22**, 562–569.

PEHERSTORFER, F. (1981) Characterization of positive quadrature formulas. *SIAM J. Math. Anal.* **12**, 935–942.

PEHERSTORFER, F. (1982) Weight functions which admit Tchebycheff quadrature. *Bull. Austral. Math. Soc.* **26**, 29–37

PEHERSTORFER, F. (1984) Characterization of quadrature formulas II. *SIAM J. Math. Anal.* **15**, 1021–1030.

PEHERSTORFER, F. (1988a) On Tchebycheff quadrature formulas. In: H. Brass and G. Hämmerlin (eds.) Numerical Integration III, pp. 172–185. Birkhäuser, Basel.

PEHERSTORFER, F. (1988b) On Gauss quadrature formulas with equal weights. *Numer. Math.* **52**, 317–327.

PEHERSTORFER, F. (1990) Gauss-Tchebycheff quadrature formulas. *Numer. Math.* **58**, 273–286.

PEHERSTORFER, F. (1992) On the asymptotic behaviour of functions of the second kind and Stieltjes polynomial and on the Gauss-Kronrod quadrature formulas. *J. Approx. Theory* **70**, 156–190.

PEHERSTORFER, F. (1993a) On positive quadrature formulas. In: H. Brass and G. Hämmerlin (eds.) Numerical Integration IV, pp. 297–313. Birkhäuser, Basel.

PEHERSTORFER, F. (1993b) On the remainder of Gaussian quadrature formulas for Bernstein-Szegö weight functions. *Math. Comp.* **60**, 317–325.

PEHERSTORFER, F. (1995) Stieltjes polynomials and functions of the second kind. *J. Comput. Appl. Math.* **65**, 319–338.

PEHERSTORFER, F. (2008) Positive quadrature formulas III: asymptotics of weights. *Math. Comp.* **77**, 2241–2259.

PEHERSTORFER, F.; PETRAS, K. (2000) Ultraspherical Gauss-Kronrod quadrature is not possible for $\lambda > 3$. *SIAM J. Numer. Anal.* **37**, 927–948.

PETRAS, K. (1988) Asymptotic behaviour of Peano kernels of fixed order. In: H. Brass and G. Hämmerlin (eds.) Numerical Integration III, ISNM 85, pp. 186–198. Birkhäuser, Basel.

PETRAS, K. (1989) Normabschätzungen für die ersten Peanokerne der Gauß-Formeln. *Z. Angew. Math. Mech.* **69**, T81–T82

PETRAS, K. (1990) Error estimates for Filon quadrature formulae. *BIT* **30**, 529–541.

PETRAS, K. (1992a) One-sided L1-approximation and bounds for Peano kernels. In: T.O. Espelid and A. Genz (eds.) Numerical Integration, NATO ASI Series C vol. 357, pp. 165–174. Kluwer Academic Publishers, Dordrecht.

PETRAS, K. (1992b) Error bounds for Gaussian and related quadrature with applications to r-convex functions. *SIAM J. Numer. Anal.* **29**, 578–585.

PETRAS, K. (1993a) Asymptotics for the remainder of a class of positive quadratures for integrands with an interior singularity. *Numer. Math.* **65**, 121–133.

PETRAS, K. (1993b) Quadrature theory of convex functions. In: H. Brass and G. Hämmerlin (eds.) Numerical Integration IV, pp. 315–329. Birkhäuser, Basel.

PETRAS, K. (1993c) Gaussian quadrature formulae – second Peano kernels, nodes, weights and Bessel functions. *Calcolo* **30**, 1–28.

PETRAS, K. (1994a) High order error constants of Clenshaw-Curtis product rules. *Analysis* **14**, 1–18.

PETRAS, K. (1994b) Quadrature errors for functions with derivatives of bounded variation. *J. Comput. Inform. Technol.* **4**, 123–143.

PETRAS, K. (1995) Gaussian integration of Chebyshev polynomials and analytic functions. *Numer. Algorithms* **10**, 187–202.

PETRAS, K. (1996a) An asymptotic expansion for the weights of Gaussian quadrature formula. *Acta Math. Hung.* **70**, 89–100.

PETRAS, K. (1996b) On the universality of the Gaussian quadrature formula. *East J. Approx.* **2**, 427–438.

PETRAS, K. (1998a) Gaussian versus optimal integration of analytic functions. *Constructive Approximation* **14**, 231–245.

PETRAS, K. (1998b) Quadrature error estimates involving moduli of continuity of derivatives. *Suppl. Rend. Circ. Mat. Palermo, Ser. II* **52**, 706–722.

PETRAS, K. (1999a) On the computation of the Gauss-Legendre quadrature formula with a given precision. *J. Comput. Appl. Math.* **112**, 253–267.

PETRAS, K. (1999b) Positivity of Gauss-Kronrod formulae for a certain ultraspherical weight function. In: W. Gautschi, G.H. Golub and G. Opfer (eds.) Applications and computation of orthogonal polynomials, ISNM 131, pp. 195–202. Birkhäuser, Basel.

PETRAS, K. (1999c) Bounds for the third Peano constants of Gaussian quadrature formulae. *Acta Math. Hung.* **83**, 241–250.

PETRAS, K. (2005) On Proinov's bound for the quadrature error of functions having a given modulus of continuity. *Anal. Math.* **31**, 43–61.

PIESSENS, R.; DE DONCKER-KAPENGA, E.; ÜBERHUBER, C.W.; KAHANER, D.K. (1983) QUADPACK: A Subroutine Package for Automatic Integration. Springer, Berlin.

PITTNAUER, F.; REIMER, M. (1979) Intervallfunktionale vom Gauß-Legendre-Typ. *Math. Nachr.* **87**, 239–248.

POLYA, G. (1933) Über die Konvergenz von Quadraturverfahren. *Math. Z.* **37**, 264–286.

PONOMARENKO, A.K. (1983) Estimation of the error functional of quadrature formulas with Chebyshev weight (in Russian). *Metody Vychisl.* **13**, 116–121.

PRÁGER, M. (1979) Universally optimal approximation of functionals. *Apl. Mat.* **24**, 406–420.

PROINOV, P.D. (1988) Integration of smooth functions and ϕ-discrepancy. *J. Approx. Theory* **52**, 284–292.

PTÁK, V. (1958) A remark on approximation of continuous functions. *Czech. Math. J.* **8**, 251–256.

QUADE, W. (1940) Abschätzungen zur trigonometrischen Interpolation. *Deutsche Math.* **5**, 482–512.

QUADE, W.; COLLATZ, L. (1938) Zur Interpolationstheorie der reellen periodisschen Funktionen. *Sitzungsber. Preuss. Akad. Wiss. Phys.-Math. Klasse* **30**, 383–429.

RABINOWITZ, P. (1968) Error bounds in Gaussian integration of functions of low-order continuity. *Math. Comp.* **22**, 431–434.

RABINOWITZ, P. (1969) Rough and ready error estimates in Gaussian integration of analytic functions. *Commun. ACM* **12**, 268–270.

RABINOWITZ, P. (1979) On avoiding the singularity in the numerical integration of proper integrals. *BIT* **19**, 104–110.

RABINOWITZ, P. (1980) The exact degree of precision of generalized Gauss-Kronrod integration rules. *Math. Comp.* **35**, 1275–1283. Corrigendum: *Math. Comp.* **46**, 226 (1986).

RABINOWITZ, P. (1986a) The convergence of interpolatory product integration rules. *BIT* **26**, 131–1344.

RABINOWITZ, P. (1986b) On the definiteness of Gauss-Kronrod integration rules. *Math. Comp.* **46**, 225–227.

RABINOWITZ, P. (1987) On the convergence of closed interpolatory integration rules based on the zeros of Gegenbauer polynomials. *J. Comput. Appl. Math.* **17**, 43–46.

RABINOWITZ, P.; RICHTER, N. (1970a) New error coefficients for estimating quadrature errors for analytic functions. *Math. Comp.* **24**, 561–570.

RABINOWITZ, P.; RICHTER, N. (1970b) Asymptotic properties of minimal integration rules. *Math. Comp.* **24**, 593–609.

RABINOWITZ, P.; RICHTER, N. (1970c) Chebyshev-type integration rules of minimum norm. *Math. Comp.* **24**, 831–845.

RAHMAN, Q.I.; SCHMEISSER, G. (1990) Characterization of the speed of convergence of the trapezoidal rule. *Numer. Math.* **57**, 123–138.

RAHMAN, Q.I.; SCHMEISSER, G. (1991) Characterization of functions in terms of rate of convergence of a quadrature process. *Proc. Amer. Math. Soc.* **111**, 85–94.

RALSTON, A. (1959) A family of quadrature formulas, which achieve high accuracy in composite rules. *J. Assoc. Comput. Mach.* **6**, 384–394.

RIESS, R.D.; JOHNSON, L.W. (1972) Error estimates for Clenshaw-Curtis Quadrature. *Numer. Math.* **18**, 345–353.

RIESZ, F.; SZ.-NAGY, B. (1956) Functional Analysis. Blackie & Son, London.

ROMBERG, W. (1955) Vereinfachte numerische Integration. *Det Kong. Norske Vid. Selsk. Forhdl.* **28**, 30–36.

ROWLAND, J.H.; VAROL, Y.L. (1972) Exit criteria for Simpson's compound rule. *Math. Comp.* **26**, 699–703.

SARD, A. (1949a) Smoothest Approximation Formulas. *Ann. Math. Stat.* **20**, 612–615

SARD, A. (1949b) Best approximate integration formulas; best approximation formulas. *Amer. J. Math.* **71**, 80–91.

SARD, A. (1963) Linear Approximation. American Mathematical Society, Providence, RI.

SCHERER, R.; SCHIRA, T. (2000) Estimating quadrature errors for analytic functions using kernel representations and biorthogonal systems. *Numer. Math.* **84**, 497–518.

SCHERER, R.; ZELLER, K. (1979) Lobatto-Quadratur und Fixpunkte. In: G. Hämmerlin (ed.) Numerische Integration, ISNM 45, pp. 245–251. Birkhäuser, Basel.

SCHIRA, T. (1997) The remainder term for analytic functions of symmetric Gaussian quadratures. *Math. Comp.* **66**, 297–310.

SCHMEISSER, G. (1972) Optimale Quadraturformeln mit semidefiniten Kernen. *Numer. Math.* **20** , 32–53.

SCHMEISSER, G. (1980) Two contributions to numerical quadrature. In: L. Collatz, G. Meinardus and H. Werner (eds.) Numerical Methods of Approximation Theory, vol. 5, pp. 289–318. Birkhäuser, Basel.

SCHMETTERER, L. (1975) Zur numerischen Quadratur. *Z. Angew. Math. Mech.* **55**, 59–61.

SCHMID, H.J. (1989) A note on positive quadrature rules. *Rocky Mountain J. Math.* **19**, 395–404.

SCHMIDT, R.(1935) Die allgemeine Newtonsche Quadraturformel und Quadraturformeln für Stieltjesintegrale. *J. Reine Angew. Math.* **173**, 52–59.

SCHMIDT, R.(1949) Mechanische Quadratur nach Gauß für periodische Funktionen. *Sitzungsber. Math.-Naturw. Kl. Bayer. Akad. Wiss. München* **1947**, 155–173.

SCHNEIDER, C. (1988) An analysis of Ralston's quadrature. *Numer. Math.* **52**, 597–603.

SCHOENBERG, I.J. (1958) Spline functions, convex curves and mechanical quadrature. *Bull. Amer. Math. Soc.* **64**, 352–357.

SCHOENBERG, I.J.(1964) Spline interpolation and best quadrature formulae. *Bull. Amer. Math. Soc.* **70**, 143–148.

SCHOENBERG, I.J. (1972) Notes on Spline functions I. *Indag. Math.* **34**, 412–422.

SCHOENBERG, I.J. (1974) Cardinal interpolation and spline functions VI: semicardinal interpolation and quadrature formulae. *J. Anal. Math.* **27**, 159–204.

SCHÖNHAGE, A. (1971) Approximationstheorie. De Gruyter, Berlin.

SCHÖNHAGE, A. (1975) Zur Quadratur holomorpher periodischer Funktionen. *J. Approx. Theory* **13**, 341–347.

SECREST, D. (1964) Numerical integration of arbitrarily spaced data and estimation of errors. *SIAM J. Numer. Anal.* **2**, 52–68.

SINWEL, H.F. (1981) Uniform approximation of differentiable functions by algebraic polynomials. *J. Approx. Theory* **32**, 1–8.

SLOAN, I.H. (1978) On the numerical evaluation of singular integrals. *BIT* **18**, 91–102.

SLOAN, I.H.; SMITH, W.E. (1978) Product integration with the Clenshaw-Curtis and related points. *Numer. Math.* **30**, 415–428.

SLOAN, I.H.; SMITH, W.E. (1982) Properties of interpolatory product integration rules. *SIAM J. Numer. Anal.* **19**, 427–442.

SMITH, W.E.; SLOAN, I.H. (1980) Product integration based on the zeros of Jacobi polynomials. *SIAM J. Numer. Anal.* **17**, 1–13.

SMOLYAK, S.A. (1965) On optimal restoration of functions and functionals of them (in Russian). Candidate dissertation, Moscow State University.

SOTTAS, G. (1982) On the positivity of quadrature formulas with Jacobi abscissas. *Computing* **29**, 83–88.

SOTTAS, G. (1989) On the positivity of ultraspherical type quadratures with Jacobi abscissas. *Aequationes Math.* **38**, 10–27.

SOTTAS, G.; WANNER, G. (1982) The number of positive weights of a quadrature formula. *BIT* **22**, 339–352.

STEFFENSEN, J.F. (1921) On the remainder form of certain formulas of mechanical quadrature. *Skand. Aktuarietidskrift* **4**, 201–209.

STEFFENSEN, J.F.(1924) On Laplace's and Gauss' summation formulas. *Skand. Aktuarietidskrift* **7**, 1–15.

STEFFENSEN, J.F. (1925) Interpolationslaere (in Danish). G.E.C. Gad. English translation: Interpolation, second edition. Chelsea Publishing Company, New York (1950).

STEFFENSEN, J.F. (1933) Das Restglied der Cotesschen Formel zur numerischen Integration. *Jahresber. Deutsche Math. Ver.* **42**, 141–143.

STEKLOV, V. (1918) Remarques sur les quadratures. *Petrograd. Bull. Acad. Sci.* (6) **12**, 99–118

STENGER, F. (1966) Bounds on the error of Gauss-type quadratures. *Numer. Math.* **8**, 150–160.

STENGER, F. (1973) Integration formulae based on the trapezoidal formula. *J. Inst. Math. Appl.* **12**, 103–114.

STENGER, F. (1978) Optimal convergence of minimum norm approximation in H^p. *Numer. Math.* **29**, 345–362.

STERN, M.D. (1967) Optimal quadrature formulae. *Comput. J.* **9**, 396–403.

STIELTJES, T. (1884) Quelques recherches sur la théorie des quadratures dites mécaniques. *Ann. Sci. Ecole Norm. Sup.* (3) **1**, 409–426.

STRAUSS, H. (1984) Monotonicity of quadrature formulae of Gauss type and comparison theorems for monosplines. *Numer. Math.* **44**, 337–347.

STRÖM, T. (1967) Strict error bounds in Romberg quadrature. *BIT* **7**, 314–321.

STRÖM, T. (1972) Monotonicity in Romberg quadrature. *Math. Comp.* **26**, 461–465.

STROUD, A.H.; SECREST, D. (1966) Gaussian Quadrature Formulas. Prentice-Hall, Englewood Cliffs, NJ.

SUKHAREV, A.G. (1979) A sequentially optimal algorithm for numerical integration. *J. Optim. Theory Appl.* **28**, 363–373.

SUKHAREV, A.G.(1985) The equality of errors in classes of passive and sequential algorithms. *USSR Comput. Math. Math. Phys.* **25**, 193–195.

SUKHAREV, A.G. (1986) On the existence of optimal affine methods for approximating linear functionals. *J. Complexity* **2**, 317–322.

SUKHAREV, A.G. (1987) The concept of sequential optimality for problems in numerical analysis. *J. Complexity* **3**, 347–357.

SWARZTRAUBER, P.N. (2002) On computing the points and weights for Gauss-Legendre quadrature. *SIAM J. Sci. Comp.* **24**, 945–954.

VON SYDOW, B. (1977) Error estimates for Gaussian quadrature formulae. *Numer. Math.* **29**, 59–64.

SZÁSZ, P. (1960) On a theorem of L. Fejér concerning trigonometric interpolation. *Acta Sci. Math. Szeged* **21**, 164–165.

SZEGÖ, G. (1933/34) Asymptotische Entwicklungen der Jacobischen Polynome. *Schriften der Königsberger Gelehrten Gesellschaft, naturwissenschaftliche Klasse* **10**, 35–112.

SZEGÖ, G. (1935) Über gewisse orthogonale Polynome, die zu einer oszillierenden Belegungsfunktion gehören. *Math. Ann.* **110**, 501–513.

SZEGÖ, G. (1936) Inequalities for the zeros of Legendre polynomials and related functions. *Trans. Amer. Math. Soc.* **39**, 1–17.

SZEGÖ, G. (1939) Orthogonal Polynomials. American Mathematical Society, Providence, RI.

TCHAKALOFF, V. (1957) Formules de cubatures mécaniques à coefficients non négatifs. *Bull. Sci. Math.* (2) **81**, 123–134.

TIMAN, A.F. (1963) Theory of Approximation of Functions of a Real Variable. Pergamon Press, Oxford.

TRAUB, J.F.; LEE, D. (1985) Optimal integration for functions of bounded variation. *Math. Comp.* **45**, 505–512.

TRAUB, J.F.; WASILKOWSKI, G.W.; WOŹNIAKOWSKI, H. (1988) Information-Based Complexity. Academic Press, New York.

TREFETHEN, L.N. (2008) Is Gaussian quadrature better than Clenshaw-Curtis? *SIAM Review* **50**, 67–87.

TUCK, E.O. (1967) A simple "Filon-trapezoidal" rule. *Math. Comp.* **21**, 239–241.

ULLMANN, J.L. (1966) A class of weight functions that admit Tchebycheff quadrature. *Michigan Math. J.* **13**, 417–423.

USPENSKY, J.V. (1935) On the expansion of the remainder in the Newton-Cotes Formula. *Trans. Amer. Math. Soc.* **37**, 381–396.

VALENTIN, R.A. (1968) The use of the hypercircle inequality in deriving a class of numerical approximation rules for analytic functions. *Math. Comp.* **22**, 110–117.

VELLEMAN, D.J. (2005) The generalized Simpson's rule. *Amer. Math. Monthly* **112**, 342–350.

DE VILLIERS, J.M. (1993) A nodal spline interpolant for the Gregory rule of even order. *Numer. Math.* **66**, 123–137.

VAN DE VOOREN, A.J.; VAN LINDE, H.J. (1966) Numerical calculation of integrals with strongly oscillating integrand. *Math. Comp.* **20**, 232–245.

WALTHER, A. (1930) Bemerkungen über das Tschebyscheffsche Verfahren zur numerischen Integration. *Skand. Aktuarietidskrift* **13**, 168–192.

WALSH, J.L.; SEWELL, W.E. (1937) Note on degree of approximation to an integral by Riemann Sums. *Amer. Math. Monthly* **44**, 155–160.

WEBSTER, R. (1994) Convexity. Oxford University Press, Oxford.

WEIDEMAN, J.A.C. (2002) Numerical integration of periodic functions: A few examples. *Amer. Math. Monthly* **109**, 21–36.

WEIDEMAN, J.A.C.; TREFETHEN, L.N. (2007) The kink phenomenon in Fejér and Clenshaw-Curtis quadrature. *Numer. Math.* **107**, 707–727.

WEINBERGER, H.F. (1978) Some remarks on good, simple and optimal quadrature formulas. In: C. de Boor and G.H. Golub (eds.) Recent Advances in Numerical Analysis, pp. 207–229. Academic Press, New York.

WILF, H.S. (1964) Exactness conditions in numerical quadrature. *Numer. Math.* **6**, 315–319.

WILF, H.S. (1967) Advances in numerical quadrature. In: A. Ralston and H.S. Wilf (eds.) Mathematical Methods for Digital Computers, vol. 2, pp. 133–144. Wiley, New York.

WILSON, M.W. (1969) A general algorithm for nonnegative quadrature formulas. *Math. Comp.* **23**, 253–258.

WILSON, M.W. (1970a) Discrete least squares and quadrature formulas. *Math. Comp.* **24**, 271–282.

WILSON, M.W. (1970b) Necessary and sufficient conditions for equidistant quadrature formula. *SIAM J. Numer. Anal.* **7**, 134–141.

WINSTON, C. (1934) On mechanical quadratures formulae involving the classical orthogonal polynomials. *Ann. Math.* **35**, 658–677.

WOLFE, J.M. (1959) An adjusted trapezoidal rule using function values within the range of integration. *Amer. Math. Monthly* **66**, 125–127.

WOŹNIAKOWSKI, H. (1985) A survey of information-based complexity. *J. Complexity* **1**, 11–44.

XU, Y. (1989) Weight functions for Chebyshev quadrature. *Math. Comp.* **53**, 297–302.

XU, Y. (1994) A characterization of positive quadrature formulas. *Math. Comp.* **62**, 703–718.

Z̃ENSYKBAEV, A.A. (1977a) Best quadrature formula for the class $W^r L_2$. *Analysis Mathematica* **3**, 83–93.

Z̃ENSYKBAEV, A.A. (1977b) Best quadrature formula for some classes of periodic differentiable functions. *Math. USSR Izv.* **11**, 1055–1071.

Z̃ENSYKBAEV, A.A. (1981) Optimal recovery methods for the integral on classes of differentiable functions. *Analysis Mathematica* **7**, 303–318.

ZIELKE, R. (1979) Discontinuous Čebyšev Systems, Lecture Notes in Mathematics 707. Springer, New York.

ZUBRZYCKI, S. (1963/64) Some approximate integration formulas of statistical interest. *Coll. Math.* **XI**, 123–135.

ZWICK, D. (1988) Optimal quadrature for convex functions and generalizations. In: H. Brass and G. Hämmerlin (eds.) Numerical Integration III, ISNM 85, pp. 310–315. Birkhäuser, Basel.

ZYGMUND, A. (1977) Trigonometric Series. Cambridge University Press, Cambridge.

Symbols

$\lfloor x \rfloor$	largest integer less than or equal to x	
$\lceil x \rceil$	smallest integer greater than or equal to x	
$f\big\|_X$	restriction of the function f to the set X	
$U^{\perp}$	orthogonal complement of a subspace U	
$(u)_+^r$	truncated rth power	50
$\\|f\\|$	supremum norm of a function f	29
$\\|L\\|$	norm of a linear functional L	29
$\partial\mathcal{G}$	boundary of a domain $\mathcal{G}$ in the complex plane	
$A(\eta, Q)$	the coefficient of $f(\eta)$ in Q	40
B_ν	Bernoulli polynomial	325
B_ν^*	Bernoulli function	326
$\mathbb{C}$	the set of complex numbers	
$C[X]$	the space of continuous functions	29
$\mathrm{conv}(M)$	convex hull of $M \subset \mathbb{R}^m$	34
$\mathcal{C}$	co-observation	7
$\mathcal{C}_M^{(r)}$	co-observation defined by the rth derivative being bounded	47
$\mathcal{C}_M^{\mathrm{Var}}$	co-observation comprising functions of bounded variation	48
$\mathcal{C}_M^{\mathrm{Var},r}$	co-observation comprising functions whose rth derivatives are of bounded variation	49
$\mathcal{C}_\omega^{\mathrm{mc}}$	co-observation defined by a modulus of continuity	48
$\widetilde{\mathcal{C}}_M^{(r)}$	co-observation comprising periodic functions from $\mathcal{C}_M^{(r)}$	49
$\mathcal{C}_M^{\mathrm{Sa},r}$	Sard's co-observation	73
$\mathcal{C}_M^{\mathrm{Da}}$	Davis's co-observation	78

$\mathcal{C}_M(\mathcal{G})$	co-observation comprising functions holomorphic in $\mathcal{G}$	82
$\mathcal{C}_r^{\mathrm{Kr}}$	Kress's co-observation	266
$\mathcal{C}_M^{\mathrm{FP}}$	co-observation comprising convex functions	93
$\deg Q$	degree of exactness of the quadrature rule Q	101
$\dim U$	dimension of a linear space U	
$\mathrm{dist}(f, U)$	best approximation error	31
dvd	divided difference	121
Δ	difference operator	106
E_n	Stieltjes polynomial	201
$\mathcal{E}_r$	ellipse in the complex plane	77
I	functional to be estimated	1, 47
$\mathrm{ind}(Q)$	index of a quadrature rule Q	42
$\mathrm{Info}(f)$	available information about f	8
intpol	interpolation polynomial	100
K_s	sth Peano kernel	53
$\mathcal{K}_r$	Favard constant	63
$\mathrm{Ker}\, O$	kernel of a mapping O	
L_λ	Laplace coefficient	105
$\mathrm{Lip}_M\, \alpha$	the set of Lipschitz continuous functions	220
$\mathbb{N}$	the set of positive integers	
O	observation	7
$\omega(f; \delta)$	modulus of continuity	48
P_n	Legendre polynomial	319
$\mathbb{P}_n$	space of algebraic polynomials of degree at most n	
q_n	normalized orthogonal polynomial	315
Q	quadrature rule, estimation rule	1, 13
$\check{Q}$, $\hat{Q}$	best lower bound and best upper bound for I	8
$\mathbf{Q}(O)$	set of all estimation rules based on the observation O	26
$\mathbf{Q}^{\mathrm{gen}}$	set of all estimation rules	13
$\mathbf{Q}_n$	set of all quadrature rules with n nodes	4

$\mathbf{Q}^+(I, U)$	set of positive estimation rules for I that are exact for all $u \in U$	40
Q_n^{best}	best quadrature rule	26
Q^{opt}	optimal estimation rule	3, 13
Q^{so}	strongly optimal estimation rule	8
Q^η	Krein rule with node η	43
Q_n^{G}	Gaussian rule	100
$Q_n^{\text{Gr},r}$	Gregory rule	237
Q_n^{Kro}	Kronrod rule	202
Q_n^{Lo}	Lobatto rule	153
Q_n^{Mi}	midpoint rule	10
$Q_n^{\text{Ra},a}$, $Q_n^{\text{Ra},b}$	Radau rule	151, 152
$Q_{2^{m+\sigma}+1}^{\text{Ro}(\sigma)}$	Romberg rule	246
Q_{2m+1}^{Si}	Simpson's rule	230
Q_n^{Tr}	trapezoidal rule	10
$\mathbb{R}$	the set of real numbers	
R	remainder functional	1, 16
ρ_n^{best}	infimum of worst-case errors over n-dimensional observations	26
ρ^{intr}	intrinsic error	8
ρ^{opt}	error of a standard estimation framework	8
$\rho(Q)$	(worst-case) error of Q	13
$\rho(Q, \mathcal{C})$	(worst-case) error of Q in the class $\mathcal{C}$	1
$\Re z$, $\Im z$	real and imaginary parts of a complex number z	
$\mathcal{S}_r(\xi_1, \ldots, \xi_m)$	set of spline functions	39
T_n	Chebyshev polynomials of the first kind	318
$\mathbb{T}_m$	space of trigonometric polynomials of degree at most m	137
U_n	Chebyshev polynomials of the second kind	319
$\operatorname{Var} f$	total variation of a function f	
$\operatorname{Var} Q$	variance of a quadrature rule Q	291
$\mathbb{Z}$	the set of integers	

Index

Titles in This Series

178 **Helmut Brass and Knut Petras,** Quadrature theory: The theory of numerical integration on a compact interval, 2011

177 **Alexei Myasnikov, Vladimir Shpilrain, and Alexander Ushakov,** Non-commutative cryptography and complexity of group-theoretic problems, 2011

176 **Peter E. Kloeden and Martin Rasmussen,** Nonautonomous Dynamical Systems, 2011

175 **Warwick de Launey and Dane Flannery,** Algebraic design theory, 2011

174 **Lawrence S. Levy and J. Chris Robson,** Hereditary Noetherian prime rings and idealizers, 2011

173 **Sariel Har-Peled,** Geometric approximation algorithms, 2011

172 **Michael Aschbacher, Richard Lyons, Stephen D. Smith, and Ronald Solomon,** The classification of finite simple groups: Groups of characteristic 2 type, 2011

171 **Leonid Pastur and Mariya Shcherbina,** Eigenvalue distribution of large random matrices, 2011

170 **Kevin Costello,** Renormalization and effective field theory, 2011

169 **Robert R. Bruner and J. P. C. Greenlees,** Connective real K-theory of finite groups, 2010

168 **Michiel Hazewinkel, Nadiya Gubareni, and V. V. Kirichenko,** Algebras, rings and modules: Lie algebras and Hopf algebras, 2010

167 **Michael Gekhtman, Michael Shapiro, and Alek Vainshtein,** Cluster algebra and Poisson geometry, 2010

166 **Kyung Bai Lee and Frank Raymond,** Seifert fiberings, 2010

165 **Fuensanta Andreu-Vaillo, José M. Mazón, Julio D. Rossi, and J. Julián Toledo-Melero,** Nonlocal diffusion problems, 2010

164 **Vladimir I. Bogachev,** Differentiable measures and the Malliavin calculus, 2010

163 **Bennett Chow, Sun-Chin Chu, David Glickenstein, Christine Guenther, James Isenberg, Tom Ivey, Dan Knopf, Peng Lu, Feng Luo, and Lei Ni,** The Ricci flow: Techniques and applications, Part III: Geometric-analytic aspects, 2010

162 **Vladimir Maz′ya and Jürgen Rossmann,** Elliptic equations in polyhedral domains, 2010

161 **Kanishka Perera, Ravi P. Agarwal, and Donal O'Regan,** Morse theoretic aspects of p-Laplacian type operators, 2010

160 **Alexander S. Kechris,** Global aspects of ergodic group actions, 2010

159 **Matthew Baker and Robert Rumely,** Potential theory and dynamics on the Berkovich projective line, 2010

158 **D. R. Yafaev,** Mathematical scattering theory: Analytic theory, 2010

157 **Xia Chen,** Random walk intersections: Large deviations and related topics, 2010

156 **Jaime Angulo Pava,** Nonlinear dispersive equations: Existence and stability of solitary and periodic travelling wave solutions, 2009

155 **Yiannis N. Moschovakis,** Descriptive set theory, 2009

154 **Andreas Čap and Jan Slovák,** Parabolic geometries I: Background and general theory, 2009

153 **Habib Ammari, Hyeonbae Kang, and Hyundae Lee,** Layer potential techniques in spectral analysis, 2009

152 **János Pach and Micha Sharir,** Combinatorial geometry and its algorithmic applications: The Alcála lectures, 2009

151 **Ernst Binz and Sonja Pods,** The geometry of Heisenberg groups: With applications in signal theory, optics, quantization, and field quantization, 2008

150 **Bangming Deng, Jie Du, Brian Parshall, and Jianpan Wang,** Finite dimensional algebras and quantum groups, 2008

149 **Gerald B. Folland,** Quantum field theory: A tourist guide for mathematicians, 2008

148 **Patrick Dehornoy with Ivan Dynnikov, Dale Rolfsen, and Bert Wiest,** Ordering braids, 2008

TITLES IN THIS SERIES

147 **David J. Benson and Stephen D. Smith,** Classifying spaces of sporadic groups, 2008

146 **Murray Marshall,** Positive polynomials and sums of squares, 2008

145 **Tuna Altinel, Alexandre V. Borovik, and Gregory Cherlin,** Simple groups of finite Morley rank, 2008

144 **Bennett Chow, Sun-Chin Chu, David Glickenstein, Christine Guenther, James Isenberg, Tom Ivey, Dan Knopf, Peng Lu, Feng Luo, and Lei Ni,** The Ricci flow: Techniques and applications, Part II: Analytic aspects, 2008

143 **Alexander Molev,** Yangians and classical Lie algebras, 2007

142 **Joseph A. Wolf,** Harmonic analysis on commutative spaces, 2007

141 **Vladimir Maz′ya and Gunther Schmidt,** Approximate approximations, 2007

140 **Elisabetta Barletta, Sorin Dragomir, and Krishan L. Duggal,** Foliations in Cauchy-Riemann geometry, 2007

139 **Michael Tsfasman, Serge Vlăduţ, and Dmitry Nogin,** Algebraic geometric codes: Basic notions, 2007

138 **Kehe Zhu,** Operator theory in function spaces, 2007

137 **Mikhail G. Katz,** Systolic geometry and topology, 2007

136 **Jean-Michel Coron,** Control and nonlinearity, 2007

135 **Bennett Chow, Sun-Chin Chu, David Glickenstein, Christine Guenther, James Isenberg, Tom Ivey, Dan Knopf, Peng Lu, Feng Luo, and Lei Ni,** The Ricci flow: Techniques and applications, Part I: Geometric aspects, 2007

134 **Dana P. Williams,** Crossed products of C^*-algebras, 2007

133 **Andrew Knightly and Charles Li,** Traces of Hecke operators, 2006

132 **J. P. May and J. Sigurdsson,** Parametrized homotopy theory, 2006

131 **Jin Feng and Thomas G. Kurtz,** Large deviations for stochastic processes, 2006

130 **Qing Han and Jia-Xing Hong,** Isometric embedding of Riemannian manifolds in Euclidean spaces, 2006

129 **William M. Singer,** Steenrod squares in spectral sequences, 2006

128 **Athanassios S. Fokas, Alexander R. Its, Andrei A. Kapaev, and Victor Yu. Novokshenov,** Painlevé transcendents, 2006

127 **Nikolai Chernov and Roberto Markarian,** Chaotic billiards, 2006

126 **Sen-Zhong Huang,** Gradient inequalities, 2006

125 **Joseph A. Cima, Alec L. Matheson, and William T. Ross,** The Cauchy Transform, 2006

124 **Ido Efrat, Editor,** Valuations, orderings, and Milnor K-Theory, 2006

123 **Barbara Fantechi, Lothar Göttsche, Luc Illusie, Steven L. Kleiman, Nitin Nitsure, and Angelo Vistoli,** Fundamental algebraic geometry: Grothendieck's FGA explained, 2005

122 **Antonio Giambruno and Mikhail Zaicev, Editors,** Polynomial identities and asymptotic methods, 2005

121 **Anton Zettl,** Sturm-Liouville theory, 2005

120 **Barry Simon,** Trace ideals and their applications, 2005

119 **Tian Ma and Shouhong Wang,** Geometric theory of incompressible flows with applications to fluid dynamics, 2005

118 **Alexandru Buium,** Arithmetic differential equations, 2005

117 **Volodymyr Nekrashevych,** Self-similar groups, 2005

116 **Alexander Koldobsky,** Fourier analysis in convex geometry, 2005

For a complete list of titles in this series, visit the AMS Bookstore at **www.ams.org/bookstore/**.